Microbiome Engineering

This reference book compiles the latest techniques and applications of microbiome engineering. Microbial communities interact dynamically with their hosts, creating a considerable impact on the host and their ecosystem. This book introduces readers to microbiomes and microbiome engineering. It covers topics like omics tools in microbial research, strategies to engineer human microbiomes, the application of synthetic biology to build smart microbes, and the future of microbiome engineering. It includes the application of microbiome engineering in improving human health, livestock, and agricultural productivity. The book is intended for researchers and students in the fields of microbiology and biotechnology.

Dr. Nimmy Srivastava is an assistant professor in the Department of Biotechnology, at Amity University Jharkhand, India. She completed her Ph.D. in the field of genetics.

Salam A. Ibrahim is a professor of food science at North Carolina A & T State University. His areas of expertise are food safety, microbiology, dairy food fermentation, and biotechnology.

Dr. Mohamed Hussein Arbab Nasr is an associate professor of medical microbiology at Omdurman Ahlia University Sudan, Khartoum. His research is focused on medical microbiology.

Microbiome Engineering

The New Dimension of Biotechnology

Edited by

Nimmy Srivastava, Salam A. Ibrahim
and Mohamed Hussein Arbab Nasr

CRC Press
Taylor & Francis Group
Boca Raton London New York

CRC Press is an imprint of the
Taylor & Francis Group, an **informa** business

Designed cover image: Shutterstock

First edition published 2025
by CRC Press
2385 NW Executive Center Drive, Suite 320, Boca Raton FL 33431

and by CRC Press
4 Park Square, Milton Park, Abingdon, Oxon, OX14 4RN
CRC Press is an imprint of Taylor & Francis Group, LLC

ISBN: 9781032479606 (hbk)
ISBN: 9781032496221 (pbk)
ISBN: 9781003394662 (ebk)

DOI: 10.1201/9781003394662

Typeset in Times
by Deanta Global Publishing Services, Chennai, India

Contents

Contributors

Vijaya Chitra A
Department of Microbiology
Sri Ramakrishna College of Arts and Science for Women
Coimbatore, Tamil Nadu, India

Safoora Ahmed
Department of Biotechnology and Bioinformatics
JSS Academy of Higher Education and Research
Mysore, Karnataka, India

Anshu
All India Institute of Medical Sciences
New Delhi, India

Swastika Banerjee
Department of Botany
Sanjivani Arts, Commerce and Science College
Kopargaon, Maharashtra, India

Sudipta Sankar Bora
DBT-North East Centre for Agricultural Biotechnology (DBT-NECAB)
Assam Agricultural University
Jorhat, Assam, India

Ankan Chakraborty
Department of Biotechnology and Bioinformatics
JSS Academy of Higher Education and Research
Mysore, Karnataka, India

Machindra J. Chavan
Amrutvahini College of Pharmacy
Sangamner, Maharashtra, India

Abhra Abir Chowdhury
Department of Botany
Darjeeling Government College
Darjeeling, West Bengal, India

Ankita Chowdhury
Heritage Institute of Technology
Kolkata, West Bengal, India

Rajanyaa Chowdhury
Heritage Institute of Technology
Kolkata, West Bengal, India

Nivethini D
Department of Microbiology
Sri Ramakrishna College of Arts and Science for Women
Coimbatore, Tamil Nadu, India

Jugabrata Das
College of Horticulture and Farming System Research
Assam Agricultural University, Nalbari, Assam, India

Pratyusha De
Department of Botany
University of Burdwan
Burdwan, West Bengal, India

Anuvab Dey
Department of Biosciences and Bioengineering
IIT Guwahati
North Guwahati, Assam, India

Walaa A. Eid
Food Science and Technology Department
Agriculture Faculty
New Valley University
New Valley Governorate, Egypt

Sumitha Elayaperumal
Department of Biotechnology and Bioinformatics
JSS Academy of Higher Education and Research
Mysore, Karnataka, India

Md Faiyazuddin
School of Pharmacy
Al-Karim University
Katihar, Bihar, India

Sharmistha Ganguly
Department of Dravyaguna
Institute of Post Graduate Ayurvedic Education and Research
Kolkata, West Bengal, India

Amol D. Gholap
Department of Pharmaceutics
St. John Institute of Pharmacy and Research
Palghar, Maharashtra, India

Subhrojyoti Ghosh
Department of Biotechnology
IIT Madras
Chennai, India

Rahul Gogoi
Department of Agricultural Biotechnology
Assam Agricultural University
Jorhat, Assam, India

Navnath T. Hatvate
Institute of Chemical Technology Mumbai
Marathwada Campus, Jalna, Maharashtra, India

Rajashree HS
Department of Biotechnology
Karunya Institute of Technology and Sciences
Coimbatore, Tamil Nadu, India

Anu Jacob
Department of Biotechnology
Karunya Institute of Technology and Sciences
Coimbatore, Tamil Nadu, India

Karthikeyan K
Gujarat Institute of Desert Ecology
Bhuj, Gujarat, India

Nanthakumar K
The Energy and Resources Institute (TERI)
New Delhi, India

Pankaj R. Khuspe
Department of Pharmaceutics
Shriram Shikshan Sanstha's, College of Pharmacy
Solapur, Maharashtra, India

Afrah P Kounin
Department of Biotechnology and Bioinformatics
JSS Academy of Higher Education and Research
Mysore, Karnataka, India

Veena Kumari
Department of Botany
Buddha P.G. College
Kushinagar, Uttar Pradesh, India

Hany G. EL-Masry
Nutrition and Food Science Department
Faculty of Home Economics
Helwan University
Cairo, Egypt

Jissin Mathew
Department of Biotechnology
Karunya Institute of Technology and Sciences
Coimbatore, Tamil Nadu, India

Arun Kumar Maurya
Department of Botany
Multanimal Modi College
Modinagar, Ghaziabad, India

Hemashenpagam N
Department of Microbiology
Hindusthan College of Arts & Science
Coimbatore, Tamil Nadu, India

Romen Singh Naorem
Multi Disciplinary Research Unit
Jorhat Medical College and Hospital
Jorhat, Assam, India

Chandan Naskar
Department of Botany
Darjeeling Government College
Darjeeling, West Bengal, India

Shaimaa H. Negm
Home Economic Department
Specific Education Faculty
Port Said University
Port Fouad City, Egypt

Ayyasamy P M
Periyar University
Salem, Tamil Nadu, India

Sagar R. Pardeshi
Department of Pharmaceutics
St. John Institute of Pharmacy and Research
Palghar, Maharashtra, India

Emi Grace Mary Gowshika Rajendran
Women's Christian College
Chennai, Tamil Nadu, India

Mayuri Rastogi
Department of Nutrition and Dietetic, SSAHS
Sharda University
Gr. Noida, Uttar Pradesh, India

Selvajeyanthi S
Department of Microbiology
Theeran Chinnamalai College of Arts and Science for Women
Tirupur, Tamil Nadu, India

Sampada D. Sawant
Department of Pharmaceutics
St. John Institute of Pharmacy and Research
Palghar, Maharashtra, India

Sadikali F. Sayyad
Amrutvahini College of Pharmacy
Sangamner, Maharashtra, India

Sanchita Seal
Department of Botany
Polba Mahavidyalaya
Polba, India

Prakriti Seth
Department of Biotechnology and Medical Engineering
National Institute of Technology
Rourkela, Odisha, India

Anmol Kumar Sharma
Department of Biosciences
Jamia Millia Islamia
New Delhi, India

Madhurjya Ranjan Sharma
Department of Agricultural Biotechnology
Assam Agricultural University
Jorhat, Assam, India

Silva Shihab
Department of Biotechnology
Karunya Institute of Technology and Sciences
Coimbatore, Tamil Nadu, India

Eskalin Shiny
Department of Biotechnology
Karunya Institute of Technology and Sciences
Coimbatore, Tamil Nadu, India

Vandana Singh
Department of Microbiology, SSAHS
Sharda University
Gr. Noida, Uttar Pradesh, India

Dwaipayan Sinha
Department of Botany
Government General Degree College
Mohanpur, West Bengal, India

Yuvaraj Sivamani
Crescent School of Pharmacy
B.S. Abdur Rahman Crescent Institute of Science & Technology
Chennai, Tamil Nadu, India

Subhashree Sunderay
Department of Biotechnology and Bioinformatics
JSS Academy of Higher Education and Research
Mysore, Karnataka, India

Md Jasim Uddin
Department of Pharmaceutical Technology
Faculty of Pharmacy
Universiti Malaya
Kuala Lumpur, Malaysia

1 An Introduction to the Microbiome

Shaimaa H. Negm
Home Economic Dept., Specific Education Faculty,
Port Said University
Egypt

Walaa A. Eid
Food Science and Technology Dept., Agriculture
Faculty, New Valley University
Egypt

1.1 INTRODUCTION

The word "microbiome" is not totally new: it combines the word "microbe" with the suffix "-ome", a Latinized form of the "Greek suffix", which means "total", "aggregate", or "collective action". As a result, the microbiome is the total of all microorganisms that share a common point, generally a common place/habitat throughout a given time frame (Aleman and Valenzano, 2019). This is a unique microbial community that lives in a rather well-defined environment and has specialized physio-chemical characteristics. As a result, the word not only relates to the microbes engaged but also to the environment in which they operate (Barko et al., 2018). The term "microbiome" has been used to characterize the combined genome of our endogenous microorganisms (microflora) with the notion that a full genetic perspective of *Homo sapiens* as an entity must encompass the genes in our microbiome (Berg et al., 2020; Bilen, 2020).

The ecological population of mutualistic, commensal, and pathogenic microorganisms occupying a specific niche in the body is referred to as the microbiota (Peterson et al., 2009). It is mainly composed of bacteria but also harbors fungi, archaea, viruses, and some single-celled eukaryotes (Ho and Ross, 2017). The complete assemblage of the genes and genomes found within the microbiota makes up the microbiome. Joshua Lederberg first used the word "microbiome" (Lederberg and McCray, 2001) to emphasize the relevance of the microbiota inhabiting the human body in normal, healthy, and diseased conditions.

How does our body acquire its microbiota? It was long thought that as a neonate slides through the mother's vaginal canal, the process of inhabitation of the microbiota in the neonate's gut was initiated (Ho and Ross, 2017). However, it was found that gut microbial colonization may commence even earlier, before birth, as evident from a recent study in premature babies, where both their meconium and placenta possessed their own distinct microbiomes (Slattery et al., 2016). The nature of the microbiota population and functions of the microbiome in the gut are influenced by multiple elements such as heredity, age, delivery mode—either by vaginal (natural) or by Caesarean section—diet (breast-feeding or cup/bottle feeding for new-borns) (Sharon et al., 2013), medication, antibiotics, and illness (Ho and Ross, 2017).

The microbiota contributes to trophic functions, metabolism, barrier function, immune stimulation, and signalization in virtually all organs of the body. The intestinal innate and adaptive immunity

DOI: 10.1201/9781003394662-1

coordinate and interact with the symbionts, contributing to intestinal homeostasis through the establishment of a mutually beneficial relationship by tolerating symbiotic microbiota and retaining the ability to exert a pro-inflammatory response towards invasive pathogens. Maintenance of symbiosis is essential to health and well-being. Disruption of gut microbiota increases susceptibility to infection and sepsis through several mechanisms, including (a) allowing for the expansion of pathogenic intestinal bacteria, (b) priming the immune system for a robust pro-inflammatory response, and (c) decreasing production of beneficial microbial products such as short-chain fatty acids (SCFAs) (Adelman et al., 2020). If there is disruption at the level of barrier immunity, there will be a loss of immune homeostasis with an associated risk of immune or inflammatory disorders. Therefore, it is essential to understand symbiosis and develop the tools to monitor and fine-tune it to prevent or mitigate the risks, or even for use in microbiome-targeted therapeutics (Malard et al., 2021).

1.2 MICROBIOME: THEN AND NOW

The study of the microbiome is as old as the field of microbiology itself. Studies on the microbiome began as early as the 1680s, when Antoine van Leeuwenhoek compared his oral and fecal microbiota. He noted differences in the nature of the microbes found in these two sites. Also, when he analyzed the oral and fecal samples from healthy and diseased individuals, he observed distinguishing features (Chattopadhyay and Mythili, 2018). Initially, isolation and direct culturing of microbes on simple media were employed to study different microorganisms (Morgan and Huttenhower, 2012). However, there were certain demerits in the qualitative and quantitative studies of the uncultured microorganisms. To overcome these problems, culture-independent procedures were developed in the late 1980s. In such methods, extraction of DNA occurs directly from a sample rather than from individually cultivated microbes. This has enabled researchers to develop a broad understanding and to investigate the various features of the microbial population (Morgan and Huttenhower, 2012). One of the significant aspects of these procedures includes next generation sequencing (NGS) technologies, which have received special attention over the past ten years. These technologies are comparatively faster, more precise, and more reproducible than direct cultivation methods (Gevers et al., 2012). Using both culture-independent and conventional procedures for the whole genomic characterization of specific microorganisms would provide a deep understanding of the role of intestinal microbiota. This, in turn, would help in the identification of novel gene targets for the diagnosis and treatment of many incurable diseases (D'Argenio and Salvatore, 2015).

1.2.1 Metabolic Components of the Microbiota and Diet

Infant nutrition, particularly breastfeeding, has been found to enhance colonization of *Lactobacilli* and *Bifidobacteria*, which is another important variable determining gut microbiome biodiversity. Maternal milk includes oligosaccharides and a variety of fatty acids that influence the gut microbiota and its ability to generate metabolites that safeguard against asthma and allergies by promoting the formation of Treg cells. This impact is also caused by raw milk consumption in the first year of birth, which is most likely due to increased amounts of proteins in the serum component as well as unsaturated omega-3 fatty acids. Polyphenols and fish oils, among other nutritional ingredients, are crucial for microbiome diversity (Ferrer et al., 2017; Finlay et al., 2019). By fermenting different nutritional polysaccharides, bacteria such as *Ruminococcaceae* and *Lachnospiraceae* can impact the gut microbiota by generating SCFAs such as acetate, butyrate, and propionate. Notably, these acids have anti-inflammatory effects on the immune system that can attenuate fibrosing alveolitis (FA) and lung diseases by enhancing epithelial barrier function and triggering dendritic cell (DC) analogues, Treg cells, and synthesis of IL-10, in addition to acting as a vital source of energy for bowel colonocytes (FitzGerald and Spek, 2020; Garud and Pollard, 2020).

1.2.2 The Significance of the Microbiome in Childhood

Early childhood exposure is critical for the microbiota, and there is mounting evidence that gut microbiota dysbiosis has a significant impact on immune system functioning. Perinatal food intake of the mother or infant, antibiotic usage, and interaction with older siblings are all possible influences on how the immune system functions. The highest inter-individual microbial diversity is seen in the first three years of life, according to evidence from various populations. The presence of bacteria in tiny amounts in the meconium and placenta implies that interactions with the microbiome may start before birth. Bacterial exposure in the first months of life can trigger the innate immune system in a number of ways, affecting FA. Early vaccination with class IV and XIV *Clostridium* spores and other bacteria will reduce the IgE levels circulating in adulthood (Giles and Couper, 2020; Gould et al., 2018). In contrast, three-week-old babies with a greater *Clostridium difficile* fecal load and a higher *Bifidobacterium* to *C. difficile* ratio had a higher number of skin-test-positive responses to aeroallergens and food. Similarly, high levels of fecal *E. coli* in babies' feces are linked to IgE-mediated eczema in their first month. Surprisingly, the same colonization pattern might have various outcomes depending on age. Colonization of *M. catarrhalis*, *H. influenzae*, or *S. pneumoniae* in the first month of birth, for example, raises the chance of asthma by causing a rise in atopic indicators such as serum IgE and eosinophils, but not at 12 months (Greenhill, 2020; Gulati and Plosky, 2020).

1.2.3 The Significance of Antibiotic Exposure

The use of antibiotics in the 1950s has been linked to an increase in allergy cases. This is considered to be caused by antibiotics inducing dysbiosis, which has been linked to the onset of asthma and Alzheimer's disease. Because maternal antibiotic use during gestation enhances the incidence of allergies in children, and antibiotic usage in the first month after birth has been linked to cow's milk allergies, the age of early introduction may be crucial. Antibiotics used intrapartum have been found to alter the microbiota of children aged 3–12 months. Antibiotics have been shown to have an impact on the microbiota of elderly people, according to research. In newborns, antibiotic treatment is linked to chronic allergic airway inflammation, but not in adults (Haque and Mande, 2019). Antibiotics, even at modest dosages, can alter microbiome composition; nevertheless, the links between antibiotic use and allergy disorders grow stronger as the number of antibiotics administered increases, and various drug families have distinct impacts. According to several studies, beta-lactam antibiotics are the most prevalent cause of FA in children under the age of two, but macrolides are linked to FA in children beyond the age of two. In the case of asthma, additional research is needed to determine whether the infection or the medicines themselves increase susceptibility (Hintze, 2019; Hitchings and Kelly, 2019).

1.3 GUT MICROBIOME

The most abundant and varied microbiota population exists in the gut within the human body. Firmicutes and Bacteroidetes are the two predominant phyla that dominate the gut microbiota. They represent more than 98% of the 16S rRNA sequences observed in the intestinal microbiota (Hinzey et al., 2016). As far as our health is concerned, these intestinal minions direct many normal host physiologic activities such as digestion, synthesis of essential vitamins, detoxification, maintenance of epithelial integrity, regulation of the immune and endocrine systems, influencing behavioral responses, and defending against possible pathogens, thereby enabling an understanding of the consequences of metabolic imbalances (D'Argenio and Salvatore, 2015). The information for all these physiologic processes is contained in the gut microbial genes, which resemble that of an organ but are lacking within the genome of humans. This has led to the gut microbiome being referred to as the "neglected" organ (O'Hara and Shanahan, 2006). Humans are known as "superorganisms"

(Walsh et al., 2014) because they possess two types of genomes: one inherited from their parents and the other acquired through the microbiome. The genome inherited from the parents is stable throughout one's lifetime, while the microbiome, the second genome, is dynamic and influenced by alterations in environmental parameters. Gut microbiome dysbiosis has been linked to a wide range of health issues, such as asthma, functional bowel disorders, allergies, obesity, diabetes, anxiety, depression, etc. (Slattery et al., 2016).

1.4 DISEASE AND THE HUMAN MICROBIOME

1.4.1 Infectious Diseases

Microbes that can execute de-conjugation and transformation processes transform entangled bile salts into unconjugated primary and secondary bile acids when they are released by the liver into the gastrointestinal tract. Some of these molecular species help *C. difficile* spores germinate, while others stop the organism's vegetative form from growing. The discovery of innovative therapies has resulted from a better understanding of molecular mechanisms (Jagodzinski et al., 2019; Johnson et al., 2020). Microbiota replacement treatment, particularly fecal microbiota transplant, is an important field of research because the pathophysiology of *C. difficile* infection, particularly persistent infection, is linked to a lack of natural microbial communities and functionality. Numerous additional infections and inflammatory disorders can be influenced by the gut microbiota. The microbiota condition of patients receiving allogeneic stem cell is linked to the likelihood of acquiring bacteremia. The presence of gastrointestinal microorganisms in the airways of individuals with sepsis and acute respiratory distress syndrome (ARDS) appears to promote chronic lung inflammation. Ultimately, the gut microbiota's makeup may have a significant impact on the repair of surgical intestinal anastomoses. These findings might affect therapy as well as prognosis and outcome (King et al., 2020; Klassen, 2018).

1.4.2 Cardiovascular (CVS) Diseases

The gut microbiota produces compounds that not only impact the gut but also have systemic effects. Some gastrointestinal (GI) bacteria may produce trimethylamine N-oxide (TMAO) metabolites, which may be linked to heart disease. Liver flavins comprising monooxygenase help the gut microbiota convert trimethylamine from phosphatidylcholine, choline, and l-carnitine-rich diets to TMAO. In experimental animals, TMAO disrupts lipid transport and causes the secretion of progenitors that induce foam cell development and artery stiffness. It has been revealed that intestinal dysbiosis is linked to CVS disease. A clinical study was conducted on two groups of people: those who had a low risk of CVS disease and those who had a high risk of CVS disease. A disturbed gut flora was linked to a greater risk of CVS disease, according to their results. CVS illnesses have been linked to the predominance of particular species. In germ-free mice animal studies, feces transplantation from hypertension patients with overexpressed *Klebsiella* and *Prevotella* raised blood pressure. Additionally, the *Firmicutes* to *Bacteroidetes* ratio in the feces microbiota of hypertensive mice increased significantly (Koontz et al., 2019; Kumar et al., 2019; Marijnissen et al., 2020).

1.4.3 Metabolic Disease and Obesity

Decades ago, ground-breaking research found a link between obesity and the gut microbiota in both human and animal illness models. The use of leptin-defective mice in a study looking at the function of the microbiota and obesity showed the strong relationship between host and microbial variables in the complicated pathophysiology of diseases such as obesity. Moreover, according to meta-analysis of numerous recent studies, the direct link between the microbiome and obesity might be less than formerly thought. Whatever the magnitude of the impact, it is apparent that the microbiota may alter the digestive tract's nutrient processing (Maruvada et al., 2017; Mills et al., 2019). The expression of key metabolic regulating peptides such as glucagon-like peptide YY and peptide 1 can be influenced by microbially generated products including bile acids and short-chain fatty acids. Some of the ways through which the microbiota might impact human energy metabolism have been elicited in recent research. Other studies have found that altering the host diet has an impact on the gut microbiota, establishing a complicated system in which extrinsic and intrinsic microbiome connections can affect host metabolism. How accidental modification of the microbiota—for example, by antimicrobial therapy—disturbs the natural equilibrium and inclines towards the genesis of metabolic syndrome and obesity is an intriguing area of study that has attracted a lot of interest. The impacts of microbial metabolism on different organ systems have been studied recently (Mitchell, 2019; Mohajeri et al., 2018).

1.4.4 Cancer

The gut microbiota has a big influence on its host's health. These organisms engage in biochemical processes that influence tumorigenesis, neoplasm growth, and immune treatment reaction. Persistent intra-abdominal infections, antibiotic medications, or both may raise colorectal cancer risk, based on a well-studied model on variables that could promote dysbiosis in the gut. In addition, the final products produced by the gut microbiota have an effect on intestinal cell coverage, either promoting or inhibiting tumorigenesis. Besides colorectal cancer, the microbiome of the intestine has been demonstrated to have a contribution to extra-intestinal cancers such as hepatic carcinoma by allowing organisms to spread to other regions of the body. *H. pylori* also increases the risk of stomach cancer in humans. *Clostridium* and *Fusobacterium* are highly represented in people with stomach cancer, according to new research on the human microbiota and cancer (Myers et al., 2019; Neugent et al., 2020). In the context of breast cancer, ecological and host variables have a direct impact on the disease's development. Bacterial populations, on the other hand, have the potential to cause breast cancer. When compared to healthy people, those with breast cancer have more *Staphylococcus*, *Enterobacteriaceae,* and *Bacillus* bacteria in their breast tissue. In addition, bacteria such as *Staphylococcus* and *Escherichia coli epidermidis* isolated from cancer patients caused a double-stranded break in HeLa cells' DNA. *Lactobacillus spp.*, which has a variety of health advantages, was not identified in the breast tissue of breast cancer patients. A greater abundance of *Bacteroides massiliensis* has been linked to prostate cancer. The intricate connections between cancer and the human microbiota have been attributed to a change in the human microbiome (Neuman and Koren, 2017; Paoli et al., 2019).

1.4.5 Lung Disease

Despite the fact that the lungs were once believed to be a sterile environment, the use of culture-independent techniques shows that the lungs are home to a small biomass of microorganisms that are rather varied. Initial research questioned the significance of this tiny community of bacteria in a healthy lung; however, most current research suggests that the makeup of the lung microbiota might influence baseline inflammatory tone in healthy persons (Paxton, 2020; Peirce and Alviña, 2019). In addition, it is known that microbial populations are prevalent and physiologically relevant in particular pulmonary illness conditions. Although it has generally been established that several cystic fibrosis patients are persistently colonized by harmful organisms, the airways of these patients have recently been discovered to contain a far more varied ecosystem than initially assumed. Although the significance of this discovery for the etiology of lung illness and cystic fibrosis has yet to be determined, it is fair to assume that microbe–microbe interaction in this milieu is just as significant as those in the GI tract. The significance of microbial populations in the development of lung illnesses including chronic obstructive pulmonary disease (COPD) and asthma is a hot topic of research right now. Several of the previous research studies indicate correlation instead of causation;

however, most current research is looking at how the lung microbiome could promote the inflammatory processes that are so important in COPD pathogenesis. Further research is anticipated to shed more light on the function of changed microbial populations in the development of various lung illnesses (Perez-Muñoz et al., 2017; Peterson et al., 2020). Polymicrobial relationships in chronic and acute rhinosinusitis have been studied in the upper airway. Pathogens and other microorganisms have been studied with respect to their capacity to alter host physiology, much as they have been in the lower airways. Some microorganisms are known to be prevalent in those with sinusitis. According to previous research, *Corynebacterium pseudotuberculosis* was shown to be present in individuals with sinusitis. The possible pathogenicity of this bacterium has been demonstrated in an animal model of sinusitis. Other members of the native sinus population modulate resistance to colonization by this organism, according to a study of the native microbiome of the upper airway in individuals either with or without sinus illness (Picardo et al., 2019; Rabb et al., 2018).

1.4.6 Allergic Diseases

It has been discovered that the human microbiota may have a role in allergy disorders. However, very little is known about the impact of the pulmonary microbiome on pulmonary tract immune modulation. A healthy microbiota, on the other hand, influences the mucosa of the lungs and shapes the pulmonary tract. By microaspiration, a dysbiotic flora immediately impacts the microbiota of the lungs, increasing the development of pulmonary illnesses in people. This was demonstrated in germ-free mice by researchers. The absence of an immunological regulation system in experimental mice resulted in pulmonary and allergy disorders. Caesarean (CS) delivery of neonates has also been identified as a risk factor for allergic diseases (Rajagopala et al., 2017; Robinson et al., 2019). Children are predisposed to such illnesses because of the absence of normal mother flora during CS. Children who are born by CS have reduced numbers of good flora (*Bacteroidetes*) in their stomach, according to molecular research. CS delivery can decrease *Bacteroidetes*' anti-inflammatory activity and promote local tissue inflammation (allergic rhinitis and asthma) caused by environmental and genetic factors. Recent empirical studies found a link between the generation of allergy antigens and dysbiotic gut flora in children, leading to airway illness (Chellappan et al., 2019; Singhvi et al., 2018).

At the age of four, children with reduced *Faecalibacterium*, *Akkermansia*, and *Bifidobacterium* microbiota diversity were more vulnerable to various allergen pulmonary hypersensitivities, which may be connected to asthma (Salazar et al., 2020; Salvucci, 2019). Mice that were not exposed to germs were more vulnerable to allergic airway illness. The susceptibility was restored after microbial colonization, and there was a reduction in allergen sensitivity. Clinical investigations of allergy prevalence in Europe revealed that individuals inhabiting farming environments with varied microbial ecosystems had a reduced occurrence of airway allergies. The stimulation of the innate immune response in pulmonary epithelial cells has been related to the cause of this condition. Contact with agricultural dust, including that containing *Lactococcus lactis* G121 and *Acinetobacter lwoffii* F78 microbial diversity, has been shown to decrease pulmonary inflammation in mice (Schwartz et al., 2020; Selway et al., 2020).

1.5 MICROBIOME-BASED INNOVATIONS

At present, there is great interest in the innovative health outcomes of the research and discoveries of the gut microbiome. Dore et al. (2017) identified four main areas of application: (a) stratification and monitoring; (b) target for modulation; (c) source of novel biologics and targets; and (d) development of microbiota-derived drugs. Several studies of different disease conditions suggest a predictive value of microbiome profiling and richness. The authors concluded that the diversity of the intestinal microbiota at engraftment is an independent predictor of mortality at three years in allo-HSCT recipients, indicating that the intestinal microbiota may be an important factor in the success or failure of allo-HSCT (Taur et al., 2014).

This finding was confirmed in a large multicenter study comprising 1362 patients from the United States, Germany, and Japan, where higher diversity of intestinal microbiota at the time of neutrophil engraftment was associated with lower mortality (Peled et al., 2020). Furthermore, it was shown that early use of broad-spectrum antibiotics (i.e., before allo-HSCT) was associated with a decrease in fecal abundance of commensal *Clostridiales* and an increased risk of transplant-related mortality (Weber et al., 2017). Similarly, it has been shown that the administration of antibiotics to cancer patients significantly shortens overall survival (OS). In a study of patients with non-small cell lung cancer (NSCLC) (n = 140), renal cell cancer (n = 67), and urothelial carcinoma (n = 42), median OS was 20.6 months versus 11.5 months ($P < 0.001$) for no antibiotic versus antibiotic given in the two months that precede or the month following initiation of immunotherapy, respectively (Routy et al., 2018).

Other work has previously shown that Gram-positive symbionts act as adjuvants in anticancer therapy. Working with mouse models led to the hypothesis that gut permeability induced by chemotherapy treatment would result in dysbiosis and allow the translocation of bacteria into secondary lymphoid organs, thereby priming a Th1/Th17 response that would favor the efficacy of the treatment in tumor beds. Antibiotic use actually knocked down the adjuvant microbes able to promote the efficacy of anticancer therapy (Viaud et al., 2013). Disruption of the microbiota was also shown to impair the response of subcutaneous tumors to CpG-oligonucleotide immunotherapy and platinum chemotherapy (Iida et al., 2013). Furthermore, intestinal microbiota richness was found to be associated with a better response to anti PD-1 immunotherapies, namely in NSCLC, renal cell carcinoma, urothelial carcinoma, and melanoma. In the former, the metagenomic species (MGS) count was higher in patients with long (>6 months) progression-free survival (PFS) than in those with short (<6 months) PFS. This was similar for gene count (Routy et al., 2018). The above-mentioned study in 112 melanoma patients documenting the modulation of the anticancer response to anti-PD-1 by the intestinal microbiota found a significantly longer PFS in patients with a high abundance of *Faecalibacterium* species, after 600 days, compared with those with a low abundance ($P = 0.03$). Conversely, there was a shorter PFS in patients with a high abundance of bacteria of the order Bacteroidales, compared with those patients with a low abundance ($P = 0.05$) (Gopalakrishnan et al., 2018). Similarly, stool sample metagenomics analysis of patients with non-small cell lung cancer, renal cell carcinoma, and urothelial carcinoma at diagnosis revealed correlations between clinical responses to immune checkpoint inhibitors and the relative abundance of *Akkermansia muciniphila* (Routy et al., 2018). Oral supplementation with *A. muciniphila* after fecal microbiota transplant (FMT) with nonresponder feces restored the efficacy of PD-1 blockade in an interleukin-12-dependent manner by increasing the recruitment of CCR9+ CXCR3+ CD4+ T lymphocytes into mouse tumor beds. These mouse model studies and the recent finding that primary resistance to immune checkpoint inhibitors can be attributed to abnormal gut microbiome composition (Routy et al., 2018) suggest that the gut microbiota helps shape the anticancer immune response (Routy et al., 2018).

1.6 THE MICROBIOME AS A THERAPEUTIC TARGET

The microbiota can contribute to a number of illnesses, such as when a microbiome lacks a useful function or when harmful microbial activity is present. As a result, it is hopeful to believe that restoring a favorable microbial function and structure may be a unique therapy for some illnesses. To achieve this, a number of different techniques have been presented. While the effectiveness of this innovative technique for illness protection and management has been limited to a few diseases and treatments thus far, the potential of what the future may hold justifies a consideration of further research in this area (Silbergeld, 2017; Skowron et al., 2018).

1.6.1 Antibiotics

Antibiotic-mediated alterations in microbial flora modify the disease-related microbiota and can assist in recovering health, but ancillary disruption to indigenous gut flora produced by therapeutic antibiotics plays a significant part in the genesis of *C. difficile* infection. Antibiotic therapy studies, for example, were used to treat pouchitis, irritable bowel syndrome, and liver encephalopathy in individuals who had a colon resection for ulcerative colitis. It was thought that there were no hidden typical microbial pathogens involved in such initial treatment trials. This method has the apparent drawback of being empirical in nature. We still do not know how a specific antibiotic regimen would affect a particular microbial population (Suraya et al., 2020; Thahir et al., 2020). For the avoidance of repeated *C. difficile* infection, a variation of this strategy has been recommended. Some of the most modern antibiotics produced for the therapy of *C. difficile* infection are intended to be more tightly limited to the pathogen with the goal of reducing collateral harm to the endogenous microbiome, which is linked to recurring illness. Fidaxomicin, which has a lesser propensity for altering the microbiome, has similarly been linked to a decreased risk of recurring illness while retaining strong pathogen effectiveness. This technique is confined to treating *C. difficile* infection; however, when managing a known bacterial disease, the use of broad-range antibiotics must be restricted to preserve the microbiome. As a result, proper antibiotic management can assist in restricting the growth or selection of antibiotic-resistant organisms while also protecting the native microbiome (Ticinesi et al., 2019; Unger et al., 2020). Using bacteriophage treatments to treat infections is another option that is considered to have little impact on the microbiome. Bacteriophage treatments have been designed to treat particular bacterial infections, and they are unlikely to have off-target impacts on other microbiome members due to their unique characteristics. While bacteriophage is recognized to select for resistant bacterial variations, these resistant bacteria frequently have changed surface features that, in addition to phage resistance, reduce pathogenicity in the host. Although much more research is required until bacteriophage can be turned into therapeutic agents, there is a surge of interest in developing new treatments to reduce microbiome disruption (van der Goot et al., 2020).

1.6.2 Microbial Biotherapies and Probiotics

Since most microbiota-related diseases are considered to be caused by a lack of helpful organisms, replacing "missing" microbiome components is an approach that precedes contemporary interest in the microbiota. The WHO defines probiotics as "alive microbes that, when supplied in appropriate numbers, impart a health advantage on the host". Additionally, regulatory authorities such as the US FDA have permitted many probiotics to be classified as nutritional supplements as long as they are not "designed to diagnostic, heal, ameliorate, cure, or protect a human illness" (van Tilburg et al., 2020; Vorholt et al., 2017). However, if the official WHO definition of probiotics is followed, the new definition may not be needed because it requires formal controls and verification of new drugs. However, many studies have used probiotics in the treatment of various diseases. In fact, Ilya Metchnikoff suggested that microorganisms could have beneficial and detrimental effects on the host. He suggested that the intake of fermented milk products could be beneficial to health, which led to members of *Lactobacillus* and *Bifidobacterium* being developed as possible probiotics. These organisms are generally present in therapeutic foods, such as kefir and yogurt, which are fermented milk products. Children's acute gastroenteritis can be prevented and treated with these medicines, according to studies. Results include minimizing the development of antibiotic-related diarrhea and preventing *C. difficile* infection.

A major double-blind study found that a probiotic combination of *Lactobacilli* and *Bifidobacteria* sought to stop antibiotic-associated diarrhea or *C. difficile* infection (Walter et al., 2020). Furthermore, probiotics have shown potential in aiding various medical conditions, including but not limited to constipation, diarrhea, polycystic ovary syndrome, ulcerative colitis, stress and anxiety, inflammatory bowel disease, breast cancer, and diabetes. Many of the classic probiotic microbes were obtained from fermented food items, as previously stated. They were picked for criteria other than theory-based or empirically proven modes of action as a result. Emerging microbiome research, particularly that which looks at microbial activity, has resulted in the design and animal evaluation of organisms that might be utilized clinically for certain conditions. Turning to *C. difficile* infection, the relevance of the metabolism of bile acid in pathophysiology has led to clinical studies taking place in connection to bile acids, their analogues, and organisms that could change the metabolism of bile acid inside the GI tract. Even though this therapy is currently in the early stages of research, the concept of generating live biotherapeutics based on intelligently determined modes of action will ideally become a key method for probiotic development in the future (Wheeler and Liss, 2019; Williams et al., 2020).

1.7 MICROBIAL RESTORATION

A logical implication of the probiotic method is the substitution or repair of an unhealthy population. There are a few distinctions to be made, however. Microbiota transplant, which involves transferring an entire microbiome from a healthy person to a person suffering from a microbiota-related illness, has piqued curiosity. Such therapies, notably the transplant of entire feces or material generated from feces, have a long history. New attention in FMT as a therapy for recurring *C. difficile* infection has prompted a number of investigations into this type of microbiota treatment. In 1958, FMT was initially used to treat antibiotic-associated pseudomembranous colitis, which was thought to be caused by *C. difficile* infection. Human studies of FMT for recurring *C. difficile* infection have subsequently employed several feces formulations and evaluated various administration methods. The patient's own feces were given to the placebo group in one placebo-controlled experiment (Garud and Pollard, 2020; Marijnissen et al., 2020). The positive development of all types of FMT in treating recurring *C. difficile* infection has sparked speculation that microbiota substitution may be utilized to treat other illnesses. Unfortunately, this result has yet to be reproduced in other diseases such as obesity and inflammatory bowel disease (IBD). Small experiments have had contradictory results. This therapy may not be immediately applicable to other disorders, according to some. The spore-forming portion of the gut microbiota appears to be the source of FMT's therapeutic impact in recurring *C. difficile* infection. FMT as produced for *C. difficile*, which favors the injection of spore-forming organisms, may not always be effective in other situations (Ferrer et al., 2017; Peirce and Alviña, 2019).

1.8 CONCLUSION

The investigation of the human microbiome is crucial because it would provide an in-depth understanding of the interactions between humans and their microbiota. This information will be useful in future research investigations aimed at improving these organisms' ability to resist life-threatening illnesses. Microbiota modulation should be considered as a target for personalized nutrition and therapy. Specific nutrients and live microbes can be considered strategic bioactives for the maintenance, preservation, or restoration of host–microbe symbiosis. A major way of leveraging prevention in chronic diseases and iatrogenic conditions will hence be related to the management and monitoring of intestinal ecology. It is worth noting that long-term usage of broad-spectrum antibiotics has the potential to alter the human microbiome. As a result, the indigenous microbial population becomes unbalanced, allowing invading diseases to thrive. Treatments that include pre- and probiotics, on the other hand, should be recommended. Humans each have a core intestinal microbiome, yet each microbiome differs according to genes, species, enterotypes (ecology), and gene count (microbial diversity).

REFERENCES

Adelman MW, Woodworth MH, Langelier C, Busch LM, Kempker JA, Kraft CS and Martin GS (2020). The gut microbiome's role in the development, maintenance, and outcomes of sepsis. *Critical Care* 24(1):1–10.

Aleman FDD and Valenzano DR (2019). Microbiome evolution during host aging. *PLoS Pathog* 15(7):e1007727.

Barko PC, McMichael MA, Swanson KS and Williams DA (2018). The gastrointestinal microbiome: a review. *J Vet Intern Med* 32(1):9–25.

Berg G, Rybakova D, Fischer D, Cernava T, Vergès MC, Charles T et al (2020). Microbiome definition re-visited: old concepts and new challenges. *Microbiome* 8(1):103.

Bilen M (2020). Strategies and advancements in human microbiome description and the importance of cul-turomics. *Microb Pathog* 149:104460.

Chattopadhyay A and Mythili S. (2018). The journey of gut microbiome – An introduction and its influence on metabolic disorders. *Front Biol* 13(5):327–341 https://doi.org/10.1007/s11515-018-1490-6.

Chellappan DK, Ning QLS, Min SKS, Bin SY, Chern PJ, Shi TP et al (2019). Interactions between microbiome and lungs: paving new paths for microbiome based bio-engineered drug delivery systems in chronic respiratory diseases. *Chem Biol Interact* 310:108732.

D'Argenio V and Salvatore F (2015). The role of the gut microbiome in the healthy adult status. *Clin Chim Acta* 451(Pt A):97–102.

Dore J, Multon MC and Behier JM. (2017). The human gut microbiome as source of innovation for health: which physiological and therapeutic outcomes could we expect? *Therapie* 72:21–38.

Ferrer M, Méndez-García C, Rojo D, Barbas C and Moya A (2017). Antibiotic use and microbiome function. *Biochem Pharmacol* 134:114–126.

Finlay BB, Pettersson S, Melby MK and Bosch TCG (2019). The microbiome mediates environmental effects on aging. *BioEssays* 41(10):e1800257.

FitzGerald MJ and Spek EJ (2020). Microbiome therapeutics and patent protection. *Nat Biotechnol* 38(7):806–810.

Garud NR and Pollard KS (2020). Population genetics in the human microbiome. *Trends Genet* 36(1):53–67.

Gevers D, Knight R, Petrosino JF, Huang K, McGuire AL, Birren BW, Nelson KE, White O, Methé BA and Huttenhower C (2012). The human microbiome project: a community resource for the healthy human microbiome. *PLoS Biol* 10(8):e1001377.

Giles EM and Couper J (2020). Microbiome in health and disease. *J Paediatr Child Health* 56(11):1735–1738.

Gopalakrishnan V et al. (2018). Gut microbiome modulates response to anti-PD-1 immunotherapy in mela-noma patients. *Science* 359:97–103.

Gould AL, Zhang V, Lamberti L, Jones EW, Obadia B and Korasidis N et al (2018). Microbiome interactions shape host fitness. *Proc Natl Acad Sci USA* 115(51):E11951–E11e60.

Greenhill C (2020). Gut microbiome influences exercise response. *Nat Rev Endocrinol* 16(2):68–69.

Gulati M and Plosky B (2020). As the microbiome moves on toward mechanism. *Mol Cell* 78(4):567.

Haque MM and Mande SS (2019). Decoding the microbiome for the development of translational applications: overview, challenges and pitfalls. *J Biosci* 44(5):118.

Hintze G (2019). Microbiome – growing importance in medicine. *Deutsch Med Wochenschr* 144(14):929.

Hinzey EM, R D, L DN (2016). Firmicutes and Bacteroidetes. *Nutrition* 411:8–11. Retrieved from: http://www.nutrition411.com/articles/fir- micutes-and-bacteroidetes.

Hitchings R and Kelly L (2019). Predicting and understanding the human microbiome's impact on pharmacol-ogy. *Trends Pharmacol Sci* 40(7):495–505.

Ho P and Ross DA (2017). More than a gut feeling: the implications of the gut microbiota in psychiatry. *Biol Psychiatry* 81(5):e35–e37.

Iida N et al. (2013). Commensal bacteria control cancer response to therapy by modulating the tumor micro environment. *Science* 342:967–970.

Jagodzinski A, Zielinska E, Laczmanski L and Hirnle L (2019). The early years of life. Are they influenced by our microbiome? *Ginekol Pol* 90(4):228–232.

Johnson D, Letchumanan V, Thurairajasingam S and Lee LH (2020). A revolutionizing approach to autism spectrum disorder using the microbiome. *Nutrients* 12(7):1983.

King KC, Stevens E and Drew GC (2020). Microbiome: evolution in a world of interaction. *Curr Biol* 30(6):R265–R2r7.

Klassen JL (2018). Defining microbiome function. *Nat Microbiol* 3(8):864–869.

Koontz JM, Dancy BCR, Horton CL, Stallings JD, DiVito VT and Lewis JA (2019). The role of the human microbiome in chemical toxicity. *Int J Toxicol* 38(4):251–264.

Kumar M, Ji B, Zengler K and Nielsen J (2019). Modelling approaches for studying the microbiome. *Nat Microbiol* 4(8):1253–1267.

Lederberg J and McCray AT (2001). 'Ome sweet' omics- a genealogical treasury of words genealogical treasury of words. *Scientist* 15(7):8. https://lhncbc.nlm.nih.gov/publication/lhncbc-2001-047 URL.

Malard F, Dore J, Gaugler B et al. (2021). Introduction to host microbiome symbiosis in health and disease. *Mucosal Immunol* 14:547–554 https://doi.org/10.1038/s41385-020-00365-4.

Marijnissen GM, Zwittink RD, Kuijper EJ and van Furth EF (2020). Microbiome and psychiatry: autism as an example. *Tijdschr Psychiatr* 62(2):131–140.

Maruvada P, Leone V, Kaplan LM and Chang EB (2017). The human microbiome and obesity: moving beyond associations. *Cell Host Microbe* 22(5):589–599.

Mills S, Stanton C, Lane JA, Smith GJ and Ross RP (2019). Precision nutrition and the microbiome, part I: current state of the science. *Nutrients* 11(4):923.

Mitchell AB (2019). The lung microbiome and transplantation. *Curr Opin Organ Transplant* 24(3):305–310.

Mohajeri MH, Brummer RJM, Rastall RA, Weersma RK, Harmsen HJM and Faas M et al (2018). The role of the microbiome for human health: from basic science to clinical applications. *Eur J Nutr* 57(Suppl 1):1–14.

Morgan XC, Tickle TL, Sokol H, Gevers D, Devaney KL, Ward DV, Reyes JA, Shah SA, LeLeiko N, Snapper SB, Bousvaros A, Korzenik J, Sands BE, Xavier RJ and Huttenhower C (2012). Dysfunction of the intestinal microbiome in inflammatory bowel disease and treatment. *Genome Biol* 13(9):R79.

Myers B, Brownstone N, Reddy V, Chan S, Thibodeaux Q, Truong A et al (2019). The gut microbiome in psoriasis and psoriatic arthritis. *Best Pract Res Clin Rheumatol* 33(6):101494.

Neugent ML, Hulyalkar NV, Nguyen VH, Zimmern PE and De Nisco NJ (2020). Advances in understanding the human urinary microbiome and its potential role in urinary tract infection. *mBio.* 11(2):e00218–e00220.

Neuman H and Koren O (2017). The pregnancy microbiome. *Nestle Nutr Inst Workshop Ser* 88:1–9.

O'Hara AM and Shanahan F (2006). The gut flora as a forgotten organ. *EMBO Rep* 7(7): 688–693.

Paoli A, Mancin L, Bianco A, Thomas E, Mota JF and Piccini F (2019). Ketogenic diet and microbiota: friends or enemies? *Genes* 10(7):534.

Paxton RJ (2020). A microbiome silver bullet for honey bees. *Science* 367(6477):504–506.

Peirce JM and Alviña K (2019). The role of inflammation and the gut microbiome in depression and anxiety. *J Neurosci Res* 97(10):1223–1241.

Peled JU et al. (2020). Microbiota as predictor of mortality in allogeneic hematopoietic-cell transplantation. *N Engl J Med* 382:822–834.

Perez-Muñoz ME, Arrieta MC, Ramer-Tait AE and Walter J (2017). A critical assessment of the "sterile womb" and "in utero colonization" hypotheses: implications for research on the pioneer infant microbiome. *Microbiome* 5(1):48.

Peterson J, Garges S, Giovanni M, McInnes P, Wang L, Schloss JA and NIH HMP Working Group. (2009). The NIH human microbiome project. *Genome Res* 19(12):2317–2323.

Peterson SN, Bradley LM and Ronai ZA (2020). The gut microbiome: an unexpected player in cancer immunity. *Curr Opin Neurobiol* 62:48–52.

Picardo SL, Coburn B and Hansen AR (2019). The microbiome and cancer for clinicians. *Crit Rev Oncol Hematol* 141:1–12.

Rabb H, Pluznick J and Noel S (2018). The microbiome and acute kidney injury. *Nephron* 140(2):120–123.

Rajagopala SV, Vashee S, Oldfield LM, Suzuki Y, Venter JC, Telenti A et al (2017). The human microbiome and cancer. *Cancer Prev Res* 10(4):226–234.

Robinson CD, Bohannan BJ and Britton RA (2019). Scales of persistence: transmission and the microbiome. *Curr Opin Microbiol* 50:42–49.

Routy B et al. (2018). Gut microbiome influences efficacy of PD-1-based immu- notherapy against epithelial tumors. *Science* 359:91–97.

Salazar N, González S, Nogacka AM, Rios-Covián D, Arboleya S, Gueimonde M et al (2020). Microbiome: effects of ageing and diet. *Curr Issues Mol Biol* 36:33–62.

Salvucci E (2019). The human-microbiome superorganism and its modulation to restore health. *Int J Food Sci Nutr* 70(7):781–795.

Schwartz DJ, Langdon AE and Dantas G (2020). Understanding the impact of antibiotic perturbation on the human microbiome. *Genome Med* 12(1):82.

Selway CA, Eisenhofer R and Weyrich LS (2020). Microbiome applications for pathology: challenges of low microbial biomass samples during diagnostic testing. *J Pathol Clin Res* 6(2):97–106.

Sharon I, Morowitz MJ, Thomas BC, Costello EK, Relman DA and Banfield JF (2013). Time series community genomics analysis reveals rapid shifts in bacterial species, strains, and phage during infant gut colonization. *Genome Res* 23(1):111–120.

Silbergeld EK (2017). The microbiome. *Toxicol Pathol* 45(1):190–194.

Singhvi G, Girdhar V, Patil S, Gupta G, Hansbro PM and Dua K (2018). Microbiome as therapeutics in vesicular delivery. *Biomed Pharmacother* 104:738–741.

Skowron KB, Shogan BD, Rubin DT and Hyman NH (2018). The new frontier: the intestinal microbiome and surgery. *J Gastrointest Surg* 22(7):1277–1285.

Slattery J, MacFabe DF and Frye RE (2016). The significance of the enteric microbiome on the development of childhood disease: a review of prebiotic and probiotic therapies in disorders of childhood. *Clin Med Insights Pediatr* 10: CMPed.S38338-107.

Suraya R, Nagano T, Kobayashi K and Nishimura Y (2020). Microbiome as a target for cancer therapy. *Integr Cancer Ther* 19:1534735420920721.

Taur Y et al (2014). The effects of intestinal tract bacterial diversity on mortality following allogeneic hematopoietic stem cell transplantation. *Blood* 124:1174–1182.

Thahir AIA, Gordon A and Salam A (2020). Does gut microbiome associate with the growth of infants? A review of the literature. *Enferm Clin* 30(Suppl 4):66–70.

Ticinesi A, Lauretani F, Tana C, Nouvenne A, Ridolo E and Meschi T (2019). Exercise and immune system as modulators of intestinal microbiome: implications for the gut-muscle axis hypothesis. *Exerc Immunol Rev* 25:84–95.

Unger MM, Becker A, Keller A, Schäfer KH, Schwiertz A and Oertel WH (2020). The role of the gut microbiome in idiopathic Parkinson's disease. *Nervenarzt* 91(12):1085–1095.

van der Goot E, van Spronsen FJ, Falcão Salles J and van der Zee EA (2020). A microbial community ecology perspective on the gut-microbiome-brain axis. *Front Endocrinol* 11:611.

van Tilburg BE, Gutierrez MW and Arrieta MC (2020). The fungal microbiome and asthma. *Front Cell Infect Microbiol* 10:583418.

Viaud S et al. (2013). The intestinal microbiota modulates the anticancer immune effects of cyclophosphamide. *Science* 342:971–976.

Vorholt JA, Vogel C, Carlström CI and Müller DB (2017). Establishing causality: opportunities of synthetic communities for plant microbiome research. *Cell Host Microbe* 22(2):142–155.

Walsh CJ, Guinane CM, O'Toole PW and Cotter PD (2014). Beneficial modulation of the gut microbiota. *FEBS Lett* 588(22):4120–4130.

Walter J, Armet AM, Finlay BB and Shanahan F (2020). Establishing or exaggerating causality for the gut microbiome: lessons from human microbiota-associated rodents. *Cell* 180(2):221–232.

Weber D et al. (2017). Microbiota disruption induced by early use of broad-spectrum antibiotics is an independent risk factor of outcome after allogeneic stem cell transplantation. *Biol Blood Marrow Transplant* 23:845–852.

Wheeler KM and Liss MA (2019). The microbiome and prostate cancer risk. *Curr Urol Rep* 20(10):66.

Williams CL, Garcia-Reyero N, Martyniuk CJ, Tubbs CW and Bisesi JH Jr (2020). Regulation of endocrine systems by the microbiome: perspectives from comparative animal models. *Gen Comp Endocrinol* 292:113437.

2 Microbiome Engineering and its Applications

Yuvaraj Sivamani
Crescent School of Pharmacy,
B.S. Abdur Rahman Crescent Institute of Science & Technology
Chennai, Tamil Nadu, India

Ankan Chakraborty, Afrah Kounin P and Sumitha Elayaperumal
Department of Biotechnology and Bioinformatics, JSS Academy of Higher Education and Research
Mysore, Karnataka, India

2.1 INTRODUCTION

The term "microbiome" refers to the diverse community of microbes that live in a given habitat, such as the human gut, soil, oceans, or plants ("A Review of 10 Years of Human Microbiome Research Activities at the US National Institutes of Health, Fiscal Years 2007–2016" 2019). This community may include bacteria, viruses, fungi, archaea, and other microbes. These communities of microbes play essential roles in various ecosystems, affecting nutrient cycling, disease resistance, and overall ecosystem health. Microbiome engineering involves deliberate interventions to modify and control the composition, structure, and function of microbial communities. This can be done through various techniques, including genetic modification, microbial consortia design, and ecological management strategies. Complex interactions between these microbes and their host or environment are also present (Berg et al. 2020a). In a wide range of habitats and organisms, microbiomes are essential. Understanding microbiomes is essential for understanding the complex communities of microbes that live in many places on Earth, including the human body and terrestrial and aquatic ecosystems. Human health, environmental systems, and ecosystem functioning are all critically dependent on microbiomes. Numerous microbiomes can be found inside the human body, with the gut microbiome attracting the most attention (Berg et al. 2020a). The gastrointestinal tract, skin, oral cavity, and respiratory system are just a few of the body parts where these communities of microbes can be found. The human microbiome is crucial for the development of the immune system, digestion, and general health (Shreiner, Kao, and Young 2015). For instance, the gut microbiota in people has a role in immune system control, nutrient absorption, digestion, and even mental health. Agriculture, environmental biogeochemical cycles, and biotechnological applications are all impacted by microbiomes. Microbiome engineering entails purposefully changing the makeup, function, or activity of microbial communities within a given microbiome to produce desired results (Grice and Segre 2012). Improved production, better health, or environmental restoration are some examples of these results. Microbiome engineering is an exciting and rapidly evolving field at the intersection of microbiology, genetics, biotechnology, and environmental science (Cullen et al. 2020). It focuses on manipulating and optimizing the communities of microorganisms that inhabit various environments, including the human body, soil, water, and industrial processes, to achieve

DOI: 10.1201/9781003394662-2

specific desired outcomes. This emerging discipline holds tremendous promise for a wide range of applications, from improving human health and agriculture to environmental remediation and industrial processes (Cullen et al. 2020). In the context of the human microbiome, researchers are exploring ways to engineer microbial communities in the gut, skin, and other body sites to improve health. This includes developing probiotics, prebiotics, and fecal microbiota transplants (FMT) to treat diseases such as irritable bowel syndrome (IBS), obesity, and autoimmune disorders (Fan and Pedersen 2021; Flint, Scott, Louis et al. 2012; Amabebe et al. 2020; Hou et al. 2022). Microbiome engineering has significant implications for agriculture. By optimizing soil microbiomes, scientists can enhance crop productivity, reduce the need for chemical fertilizers and pesticides, and promote sustainable farming practices (Suman et al. 2022; Cullen et al. 2020). Microbiome engineering can help address environmental challenges by harnessing the power of microorganisms to clean up polluted sites, degrade toxic chemicals, and restore ecosystems. Microbes play critical roles in various industrial processes, such as biofuel production, wastewater treatment, and bioremediation (Visibelli et al. 2023). Engineering microbial communities can improve the efficiency and sustainability of these processes. Researchers in microbiome engineering use a variety of tools and techniques, including DNA sequencing, synthetic biology, CRISPR-Cas9 gene editing, and computational modeling, to understand, manipulate, and predict microbial interactions and behaviors (Visibelli et al. 2023). As with any emerging field, microbiome engineering faces challenges and ethical considerations, including potential unintended consequences, ecological impacts, and questions about biosecurity and bioterrorism.

2.2 CHEMISTRY OF THE MICROBIOME

To improve health and treat a range of disorders, the field of microbiome engineering in humans entails changing the makeup and operation of the microbial communities that live in and on the human body (Grice and Segre 2012; Amabebe et al. 2020; Fan and Pedersen 2021; Flint, Scott, Louis et al. 2012; Hou et al. 2022; Ley et al. 2006; Hou et al. 2022). Principles from the fields of medicine, genetics, synthetic biology, and microbiology are all combined in the present area of study. Microbiomes contain organisms that have high metabolic activity (Veseli et al. 2023). The carbon, nitrogen, sulfur, and phosphorus cycles are just a few of the nutrient cycles in which they participate. For instance, several bacteria can break down organic material into simpler chemicals through reactions such as fermentation or respiration, producing carbon dioxide and other byproducts in the process. Anaerobic microbes, or those that can survive in settings with little to no oxygen, make up a large portion of the microbiomes (Veseli et al. 2023). These organisms frequently engage in fermentation reactions, which result in the metabolic byproducts of diverse organic acids, alcohols, and gases. In ecosystems like the human gastrointestinal tract, these metabolic pathways are crucial. Chemical signaling molecules are used by microorganisms in a microbiome to communicate with one another and with the organisms that serve as their hosts (Fischbach and Segre 2016). These signaling molecules have an impact on host-microbe interactions and microbial community dynamics. For instance, certain bacteria employ quorum sensing, a form of communication, to coordinate gene expression and behavior in response to population density. Antibiotics, toxins, and bioactive substances are only a few of the diverse secondary metabolites that microorganisms are known to make (Danquah et al. 2022). These compounds have the potential to significantly affect both the host and the microbial ecosystem. To compete with other germs for resources, certain bacteria create antibiotics. Numerous diseases in humans and other species have been connected to the makeup and operation of microbiomes. Inconsistencies in the gut microbiome, for instance, have been linked to diseases including inflammatory bowel disease (IBD), obesity, and even mental health issues (Figure 2.1) (Bonder et al. 2016; Brahe et al. 2015; Durack and Lynch 2019). For the development of treatments and interventions that advance health, it is essential to comprehend the chemical interactions that occur inside microbiomes. Microbiomes also play a role in environmental processes such as bioremediation, where particular bacteria can help in the breakdown of toxins

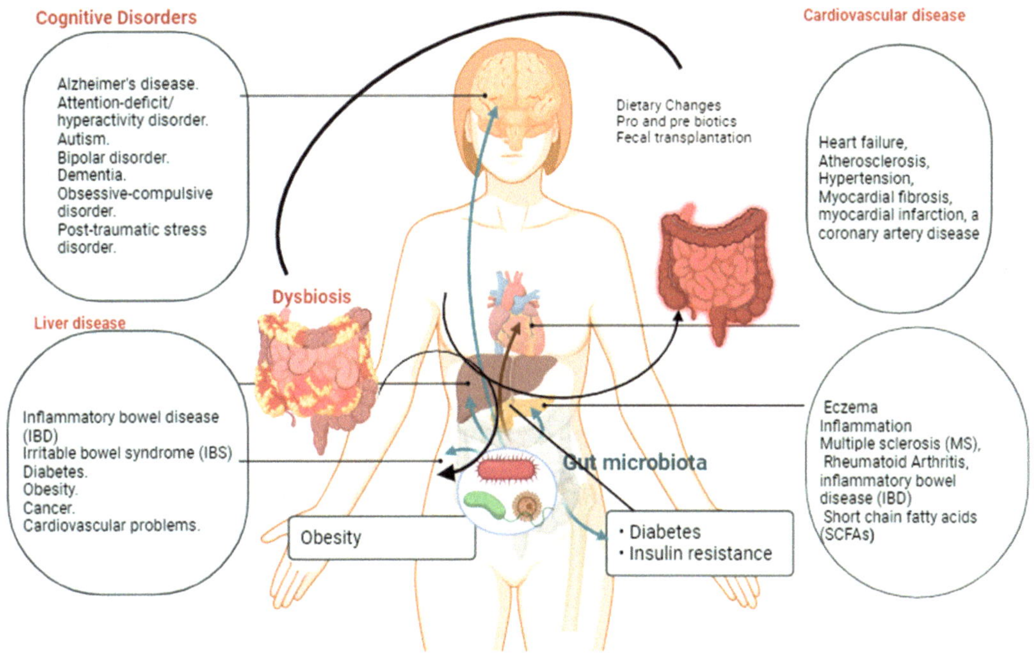

FIGURE 2.1 Chemistry of microbiome in health and disease.

and the cleanup of contaminated areas (Bonder et al. 2016). The availability of nutrients, the standard of the water, and the overall health of the ecosystems are all impacted by the chemistry of the microbiomes in soil and aquatic habitats. Microbiomes have the potential to be used in a variety of biotechnological applications, according to researchers (Foo et al. 2017). This includes modifying microorganisms to make biofuels, chemicals, and medications as well as to increase agricultural productivity through interactions between plants and microbes. Foreseeing and regulating microbiomes' stability and resilience in the face of environmental changes, such as climate change or perturbations brought on by human activity, requires a thorough understanding of their chemistry (Lozupone et al. 2012).

There are connections between the, as shown in Figure 2.1. For disorders linked to gut dysbiosis, probiotics, prebiotics, dietary changes, and fecal transplantation are the possible existing therapeutic options.

2.3 UNDERSTANDING THE HUMAN MICROBIOME

The gut microbiome stands out as one of the most varied and significant microbiomes in the human body. It is made up of trillions of microorganisms and is found predominantly in the colon in the gastrointestinal tract (Chen, Zhou, and Wang 2021; Kho and Lal 2018). Individual differences in the gut microbiome can be attributed to dietary habits, genetics, and environmental influences. A single habitat is home to thousands of distinct microbial species, which gives microbiomes a high level of variety. As a result of multiple environmental circumstances, including dietary restrictions, medication, and lifestyle modifications, these communities are dynamic and prone to change (Visibelli et al. 2023). The microbes that make up the human microbiome interact intricately with the host body. These interactions include commensalism, in which microorganisms coexist without negatively impacting the host, and mutualism, in which germs perform crucial tasks, including food processing (Visibelli et al. 2023). The host's immune system is also modulated by the microbiome, which affects how receptive it is to pathogens and how it develops. The metabolism of nutrients and digestion both heavily depend on the gut microbiota. It aids in the metabolism of food components that the body is unable to digest on its own, the breakdown of complex carbohydrates, and vitamin synthesis (Visibelli et al. 2023). Effective nutrient absorption depends on a healthy gut microbiota. The immune system of the host is greatly influenced by the microbiome during its growth and operation. A healthy microbiome helps to reduce the likelihood of autoimmune responses and allergies by teaching the immune system to distinguish between helpful and harmful microbes (Visibelli et al. 2023). The gut-brain axis, which is the subject of a recent study, proposes a link between the microbiota of the gut and the brain. The microbiome may have an impact on one's mood, behavior, and mental health issues such as anxiety and depression. This complex interaction highlights the microbiome's wider-reaching effects beyond digestion (Geng et al. 2022). Weight control and metabolic health have been linked to the gut flora. Obesity and type 2 diabetes have both been linked to imbalances in the microbiome composition. Understanding and treating metabolic problems may be made possible by research in this field (Harsch and Konturek 2018). Despite being less studied than the gut microbiome, the skin's microbiome contributes to preserving skin health and warding off viruses. Acne, eczema, and skin infections can all be caused by imbalances in the skin's microbiota. An imbalance in the microbiome known as dysbiosis is indicated by changes in the makeup and function of the bacteria. Dysbiosis can be caused by several factors, such as using antibiotics or changing your diet (Harsch and Konturek 2018).

2.3.1 Microbiome Engineering—Gut Health

Microbiome engineering plays a crucial role in gut health and digestion by manipulating the composition and function of the gut microbiome, the complex community of microorganisms residing in the gastrointestinal tract (Berg et al. 2020a; Chen, Zhou, and Wang 2021; Kho and Lal 2018). A balanced and diverse gut microbiome is essential for optimal digestion and overall well-being. Microbiome engineering can be used to restore a healthy balance of microorganisms in the gut, especially after disruptions caused by factors such as antibiotics, infections, or dietary changes (Markowiak and Ślizewska 2017). This balance is important for efficient digestion. The gut microbiome helps break down complex carbohydrates, fibers, and other food components that the human digestive system alone cannot fully process. These microbes produce enzymes that aid in the digestion of these compounds, facilitating nutrient absorption. The gut microbiome plays a role in regulating metabolism and energy balance. Microbiome engineering may offer strategies to support healthy weight management and metabolic function (Foo et al. 2017; Inda et al. 2019; Albright et al. 2021). Gut bacteria ferment dietary fibers and produce short-chain fatty acids (SCFAs) as byproducts. SCFAs not only provide an energy source for colon cells but also help maintain a healthy gut environment and influence immune function (Lawson et al. 2019; Albright et al. 2021; Inda et al. 2019). Some gut bacteria produce digestive enzymes that assist in the breakdown of complex molecules, such as proteins and fats, into smaller, more easily absorbable components. Microbiome engineering may help prevent or manage gastrointestinal disorders such as IBS and IBD by promoting a balanced microbial community and reducing inflammation (Turnbaugh et al. 2007; Lawson et al. 2019; D. J. Sanders et al. 2021). The gut microbiome interacts with the immune system, influencing its development and function. By engineering the microbiome, researchers can modulate immune responses, potentially reducing the risk of autoimmune conditions related to gut health. Some gut bacteria produce beneficial compounds, such as vitamins (e.g., B vitamins and vitamin K) and bioactive molecules that support gut health (D. J. Sanders et al. 2021)

Microbiome engineering involves the use of prebiotics (compounds that nourish beneficial bacteria) and probiotics (beneficial live microorganisms) to support a healthy gut microbiome and improve digestion (D. J. Sanders et al. 2021). Microbiome profiling allows for personalized approaches to gut health. By analyzing an individual's microbiome composition, healthcare practitioners can recommend specific interventions and dietary modifications tailored to that person's unique microbial community. Understanding an individual's gut microbiome can inform dietary recommendations.

Microbiome engineering can involve dietary modifications to promote the growth of beneficial microbes (D. J. Sanders et al. 2021). In clinical settings, microbiome engineering techniques may be applied to restore gut health and alleviate digestive symptoms in patients with gastrointestinal conditions.

2.3.2 Microbiome Engineering—Immune System Regulation

The gut microbiome helps balance the immune system's responses. It teaches the immune system to distinguish between harmless substances (e.g., food and commensal bacteria) and harmful invaders (e.g., pathogens) (Macfarlane and Macfarlane 1997; Singh et al. 2014; Belkaid and Hand 2014). Microbiome engineering aims to promote this balanced immune response. Dysbiosis, an imbalance in the gut microbiome, has been linked to autoimmune diseases where the immune system mistakenly attacks the body's tissues (Visibelli et al. 2023). Microbiome engineering strategies can help modulate the gut microbiome to reduce the risk of autoimmune conditions. Certain gut bacteria promote immune tolerance, which prevents the immune system from overreacting to harmless substances. By engineering the microbiome, it's possible to encourage the development of immune tolerance (Belkaid and Hand 2014; Bhutia and Ganapathy 2015; Flint, Scott, Duncan et al. 2012). Some gut microbes produce anti-inflammatory compounds that help control excessive inflammation in the gut and systemically. Microbiome engineering may promote the growth of these beneficial bacteria (Yu et al. 2021). The gut microbiome "trains" the immune system by exposing it to a wide range of antigens. This exposure helps educate immune cells and contributes to a more responsive and adaptable immune system. Beneficial microbes in the gut can help protect against pathogenic invaders (Yu et al. 2021). Microbiome engineering can bolster the presence of these protective species, helping prevent infections. Certain microbiota can enhance the effectiveness of vaccinations by priming the immune system for a stronger and more targeted response. Microbiome engineering could potentially improve vaccine outcomes. Microbiome engineering is being explored as a potential treatment for immune disorders, such as IBD and allergies, where immune responses are dysregulated. Microbiome profiling allows for personalized approaches to immune system regulation. By analyzing an individual's microbiome, healthcare practitioners can develop tailored strategies to enhance immune function or reduce excessive immune activity. Microbiome engineering may enhance the efficacy of cancer immunotherapy by optimizing the gut microbiome to promote a more robust anti-tumor immune response. Early-life microbiome engineering interventions, such as breastfeeding and probiotics, can shape the development of a healthy immune system in infants (Ursell et al. 2014). Engineered live biotherapeutic products (eLBPs) containing specific microbial strains are being developed to directly modulate the immune system for therapeutic purposes.

2.3.3 Role of Gut Microbiome—Disorders

The gut microbiome, which consists of trillions of microorganisms residing in the gastrointestinal tract, plays a critical role in maintaining human health. Research has shown that alterations in the gut microbiome composition and function are associated with a wide range of disorders (Visibelli et al. 2023). Consequently, understanding and manipulating the gut microbiome has the potential to be a powerful tool for treating various disorders. Here are some examples of disorders where the gut microbiome plays a role in treatment. Gastrointestinal disorders include IBD, where alterations in the gut microbiome are implicated, including Crohn's disease and ulcerative colitis (Visibelli et al. 2023). Treatment strategies involve modulating the gut microbiome to reduce inflammation and restore microbial balance. Dysbiosis, an imbalance in the gut microbiome, is associated with IBS. Probiotics, prebiotics, and dietary modifications are used to manage symptoms and restore gut health. Concerning metabolic disorders in terms of obesity, the gut microbiome composition can differ between lean and obese individuals (Cerdó et al. 2019). Microbiome-based interventions aim to promote the growth of beneficial bacteria associated with weight management. Emerging research on type 2 diabetes suggests that gut microbiome manipulation may help regulate blood sugar levels and improve insulin sensitivity in individuals with diabetes (Xi and Xu 2021; Scheithauer et al. 2020; Patel, Patel, and Moochhala 2023). Studies on autoimmune disorders, including autoimmune diseases, indicate that dysbiosis in the gut microbiome may contribute to rheumatoid arthritis (RA). Hence, therapies aim to restore microbial balance to alleviate symptoms and reduce inflammation. Also, in multiple sclerosis (MS), researchers are investigating how gut microbiome alterations may influence the development and progression of MS. Microbiome-based interventions may hold promise for managing this autoimmune disorder. Concerning neurological disorders, the gut-brain axis, which involves bidirectional communication between the gut and the brain, is influenced by the gut microbiome (De Vadder et al. 2014; Rutsch, Kantsjö, and Ronchi 2020; Liang, Wu, and Jin 2018). Probiotics and dietary changes are explored for their potential in managing mood disorders. Some studies have linked gut microbiome dysbiosis to autism spectrum disorder (ASD). Microbiome interventions are being investigated as adjunct therapies for ASD. Food allergies, including early exposure to diverse gut microbes, may reduce the risk of developing food allergies. Microbiome engineering in infants could help prevent food allergies. The gut-lung axis, a relationship between the gut microbiome and respiratory health, is being explored for its potential in managing asthma symptoms. Infectious diseases, including Clostridium difficile infections (C. difficile), have shown that fecal microbiota transplants (FMT) are a well-established treatment for recurrent C. difficile infections, where healthy donor microbes are introduced to restore microbial balance (Tan and Johnson 2019). Studies have shown that the gut microbiome's composition can influence the response to cancer immunotherapy. Modulating the microbiome may improve the efficacy of these treatments. Emerging research also suggests that the manipulation of the gut microbiome may have an impact on stress and anxiety levels (Tan and Johnson 2019).

2.3.4 Role of Gut Microbiome in Metabolism

The gut microbiome plays a significant role in metabolism and weight management, and emerging research suggests that alterations in the composition and function of the gut microbiome can influence an individual's propensity to gain or lose weight (Aoun, Darwish, and Hamod 2020). The gut microbiome aids in the digestion of complex carbohydrates and fibers that the human digestive system cannot fully break down. Microbes in the colon ferment these dietary components, producing SCFAs and other metabolites. Some SCFAs can be absorbed and used as an energy source, contributing to calorie intake. The gut microbiome can influence appetite and food cravings through its interactions with the gut-brain axis, a bidirectional communication system between the gut and the brain (Carabotti et al. 2015). Certain gut microbes can produce compounds that affect appetite-regulating hormones, potentially influencing food intake. Microbes in the gut can affect nutrient absorption, including the absorption of fats. Changes in the gut microbiome can alter the efficiency of nutrient uptake and metabolism. Dysbiosis, an imbalance in the gut microbiome, has been associated with low-grade inflammation, which is linked to obesity and metabolic syndrome (Breton, Galmiche, and Déchelotte 2022). Chronic inflammation can disrupt insulin sensitivity and metabolism. Some gut bacteria produce bioactive molecules that can affect metabolism and weight. For example, certain bacteria can produce bile acid metabolites that influence lipid metabolism. The gut microbiome can affect fat storage by influencing the expression of genes involved in lipid metabolism and adipogenesis (the process of fat cell formation). Studies have shown differences in the gut microbiome composition between lean individuals and those with obesity. Modifying the gut microbiome may impact adiposity (body fat content) and body composition (Breton, Galmiche, and Déchelotte 2022). The gut microbiome can metabolize various dietary components, such as polyphenols and phytochemicals, into bioactive compounds that may have metabolic benefits. Bariatric surgery, such as gastric bypass, can lead to significant changes in the gut microbiome. These changes are believed to contribute to the metabolic improvements seen after surgery. Researchers are exploring interventions aimed at modifying the gut microbiome to promote weight loss and improve metabolic health.

These interventions may include probiotics, prebiotics, dietary modifications, and fecal microbiota transplantation (Nigam, Panwar, and Singh 2022). It's important to note that the impact of the gut microbiome on metabolism and weight management can vary from person to person. Factors such as genetics, diet, lifestyle, and baseline microbiome composition all contribute to this variability. Microbiome profiling may lead to personalized approaches to weight management. By analyzing an individual's gut microbiome, healthcare practitioners may develop tailored dietary and lifestyle recommendations to support weight loss or weight maintenance.

2.3.4.1 Engineered Live Biotherapeutic Products

Engineered live biotherapeutic products (eLBPs) are a class of therapeutic interventions that use living microorganisms that have been genetically modified, primarily bacteria, for medical or therapeutic purposes. These microbes are made to perform certain tasks and interact with other species or biological systems. eLBPs are anchored in the study of synthetic biology, which permits precise modification of microbial genetics, and they represent a cutting-edge approach to healthcare. The idea of employing live microbes for health advantages has a long history, and probiotic-rich diets and yogurt intake are just two examples. Using live microorganisms intentionally in engineering and medicine is a more recent discovery. Advances in genetic engineering, microbial genomics, and our growing understanding of the human microbiome are all responsible for the emergence of eLBPs (Cullen et al. 2020). The importance of eLBPs is found in their capacity to modulate the microbiome and the population of microbes residing inside the human body, and thereby to treat a wide range of health issues. An imbalance in microbial communities, known as microbiome dysbiosis, has been related to several ailments, including gastrointestinal problems, autoimmune diseases, and even mental health problems (Vyas and Ranganathan 2012; Valdes et al. 2018). eLBPs provide a focused strategy to improve health outcomes and reestablish microbial balance.

2.4 THE MICROBIOME AND MICROBIAL THERAPIES

Microbiome and microbial therapeutics highlight the complex interactions between the human body and microbes, and they offer an exciting area of research and medical intervention. These treatments take advantage of how microbial populations can improve human health (Aggarwal et al. 2023). The National Institutes of Health (NIH) introduced the ground-breaking Human Microbiome Project (HMP) in 2008. Its main objective was to characterize the composition, variety, and function of the human microbiome at distinct bodily regions, as well as to comprehend its involvement in health and disease. The goal of the study was to improve our knowledge of the microbial populations that live inside humans and how they affect our health. The HMP aimed to provide a thorough map of the microbial communities (microbiota) that live in various anatomical locations throughout the human body. The urogenital tract, oral cavity, nasal passages, skin, and stomach were among these locations. The human microbiome contains a variety of microbes, and researchers sought to sequence their genomes. To do this, the genes present in these microbial communities had to be identified and described (Athanasopoulou, Adamopoulos, and Scorilas 2023). The initiative acknowledged that there might be substantial individual variation in microbiome composition and attempted to explore the variables influencing this variation, including genetics, environment, diet, and lifestyle. The HMP developed sophisticated bioinformatic tools and computational methodologies to analyze and interpret microbial genomic data to study the enormous amounts of microbiome data generated. The establishment of a reference database of the microorganisms that live inside the human body was one of the HMP's most important accomplishments. The genetic data of thousands of bacterial species were contained in this database. The experiment gave precise information about how microbial communities are distributed among various body areas. For instance, it showed how diverse and rich the human gut's microbiome is. Researchers found that the microbiome makeup varies greatly between people and even within the same person over time (Valles-Colomer et al. 2023). The significance of individualized approaches in microbiome research and healthcare was highlighted by this variability. The HMP made tremendous progress in figuring out how changes in the microbiome can affect disease development. For instance, it found associations between altered gut microbiota and diseases including obesity and IBD. The project's results paved the way for the creation of microbiome-based treatments and interventions, including fecal microbiota transplantation (FMT) for some digestive disorders (Peterson et al. 2009)

2.5 DYSBIOSIS AND DISEASE

Dysbiosis is the term for an imbalance or disruption in the microbiome's makeup and function, which can result in several health problems (Kho and Lal 2018; Chen, Zhou, and Wang 2021). A variety of diseases and ailments have been linked to changes in the balance of the human microbiome, which is especially important in the gut. One distinguishing characteristic of these chronic inflammatory gastrointestinal illnesses is dysbiosis in IBD. IBD includes diseases such as Crohn's disease and ulcerative colitis, both of which are marked by recurring gastrointestinal tract inflammation. The genesis and course of IBD are significantly influenced by dysbiosis in the gut microbiota. The broad spectrum of the gut microbiome, or the number of various microbial species present, is frequently reduced in people with IBD. The less resilient microbiome that results from this lower variety is thought to be less able to respond to changes or stressors. During IBD, variations in the relative abundance of particular bacterial taxa are frequently used to describe dysbiosis (Santana et al. 2022). As an illustration, there might be a drop in helpful bacteria, such as some species of *Bacteroides* and Firmicutes, and an increase in potentially hazardous bacteria, such as Enterobacteriaceae. Dysbiotic IBD microbiomes contain several bacteria that have been connected to immune system activation and inflammation. This could be a factor in the ongoing inflammation found in IBD. Dysbiosis can make IBD symptoms worse and cause flare-ups of the condition. Gut microbiota imbalances can start or exacerbate inflammation, leading to symptoms such as diarrhea, abdominal pain, and bleeding. Dysbiosis may affect how people with IBD react to medication. For instance, the gut microbiome can be further disrupted by several drugs used to treat IBD, such as antibiotics (Santana et al. 2022). A patient's microbiome profile can be used to personalize treatment plans for better results. IBD issues include the development of strictures (narrowing of the intestines), fistulas (abnormal connections between the intestines and other organs), and an elevated risk of colorectal cancer, all of which are all linked to dysbiosis (Cai et al. 2023).

Live bacteria known as probiotics can improve health when given in sufficient doses. Probiotics with beneficial bacteria, such as certain strains of *Lactobacillus* and *Bifidobacterium*, can be given to patients with dysbiosis to aid in reestablishing microbial balance (Fijan 2014). Prebiotics are indigestible substances that are present in some foods and act as a food supply for the good bacteria in the stomach. People can encourage the development and activity of advantageous microorganisms in the gut by consuming meals high in prebiotics. FMT entails transferring feces from a donor who is healthy to a recipient who has dysbiosis (Ser et al. 2021).

This treatment, which is frequently used to treat severe cases of *Clostridium difficile* infection (CDI) or recurring CDI, tries to restore a diverse and balanced gut microbiome. The live microorganisms used to treat particular dysbiosis-related illnesses are known as eLBPs (Kachrimanidou and Tsintarakis 2020). To treat dysbiosis in diseases like IBD, researchers are developing eLBPs. These artificially created bacteria can influence immune responses, deliver medicinal compounds, or encourage a healthy microbiome (Shah et al. 2023). A more in-depth study of a person's gut microbiome is now possible thanks to developments in microbiome profiling and sequencing technologies. Through the identification of certain microbial imbalances and the development of specialized treatments, this knowledge can direct individualized interventions (Zheng, Liwinski, and Elinav 2020). Several diets are intended to reduce the symptoms of dysbiosis-related illnesses like IBS, such as the low FODMAP diet and specific carbohydrate diets. Certain carbohydrates that can ferment in the gut and cause pain are restricted by these diets. The significance of the microbiome in gastrointestinal illnesses is still being studied by researchers, who are also experimenting with novel microbiome engineering techniques (Boix-Amorós et al. 2022). Studying the effects of microbial metabolites, the function of fungi and viruses, and the creation of precision medicine techniques are all included in this.

2.5.1 MICROBIOME ENGINEERING IN ONCOLOGY

The manipulation of the gut microbiome to prevent, treat, or improve the efficacy of cancer therapies is the subject of the young and exciting field of study known as "microbiome engineering in cancer" (Matthewman et al. 2023). The billions of microorganisms in the gastrointestinal tract that make up the gut microbiome can affect the metabolism of cancer medications, immunological responses, and inflammation, among other aspects of cancer. Recent studies indicate that dysbiosis may be linked to different cancers, such as liver, pancreatic, breast, and lung cancer (Lee 2021). The gut microbiome contains certain bacteria that can improve the body's immune response to cancer cells. To enhance the antitumor immune response, microbiome engineering tries to encourage the proliferation of these advantageous microorganisms. The gut microbiota can affect how well immune checkpoint inhibitors and other cancer immunotherapies work (Humphries and Daud 2018). To improve the effectiveness of these medicines, researchers are investigating ways to modify the microbiome. Certain cancer medications may be altered in efficacy and toxicity by gut bacteria. Microbiome engineering aims to increase therapeutic advantages while reducing negative effects by optimizing drug metabolism in the stomach. Chronic inflammation and dysbiosis are linked to an elevated risk of several malignancies (Khan, Hauptman, and Kelly 2021). Probiotics and other dietary changes are only two examples of microbiome engineering techniques that attempt to lessen inflammation and perhaps even the risk of cancer (Figure 2.2). Modern microbiome profiling techniques enable the detection of microbial signatures linked to specific malignancies (Rahman et al. 2022). Microbiome-based therapies can be tailored using this information. The gut microbiome can be dramatically impacted by dietary choices. Promoting diets high in fiber, prebiotics, and specific fermented foods to support a healthy microbiome may be part of microbiome engineering in cancer. It is being investigated whether FMT, which includes giving feces from a healthy donor to a recipient, can help cancer patients regain a healthy gut flora. To transport therapeutic chemicals, control the immune system, or specifically target cancer cells, researchers are developing and manipulating unique microorganisms (Rahman et al. 2022). These eLBPs have the potential to be employed in

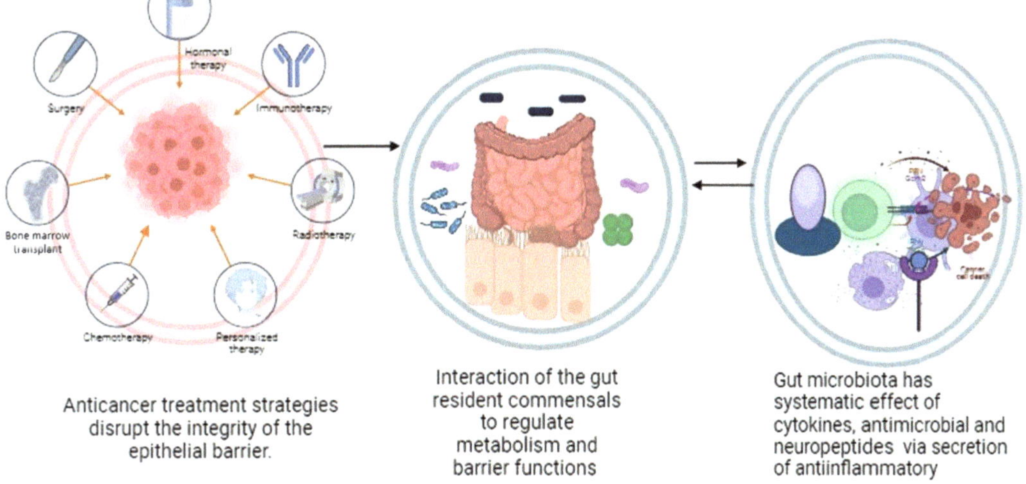

FIGURE 2.2 Schematic representation of microbiome at the interface of disease and physiology in cancer.

the treatment of cancer. Figure 2.2 is a representation of microbiome at the interface of disease and physiology in cancer.

The delicate balancing act between inflammation, infection, and tolerance of food and commensal antigens is fine-tuned by the gut microbiota, which is critical in the lifelong programming of innate and acquired immune responses. Several therapeutic techniques could be used to stop the spread of cancer and treat it while reestablishing the gut's and the metaorganism's equilibrium.

2.5.1.1 Colorectal Cancer

Colorectal cancer (CRC) has the most pronounced association between dysbiosis and cancer. A rise in potentially hazardous bacteria and a fall in helpful bacteria are characteristics of dysbiosis in CRC (Rahman et al. 2022). The development of toxic compounds in the gut and specific microbial imbalances have both been linked to an increased risk of CRC. Chronic inflammation brought on by the metabolic activities of the gut microbiome may aid in the growth of malignant tumors. As a preventative strategy, the microbiome balance should be restored and inflammation decreased (Gagliardi et al. 2018). The gut microbiome is significantly impacted by dietary decisions. Dietary changes to support a healthy microbiome may be used in microbiome engineering strategies for CRC prevention, such as adopting a diet high in fiber, prebiotics, and specific fermented foods (Haldar et al. 2023). To sustain a healthy gut microbiota and lower the risk of CRC, probiotic supplements containing advantageous bacteria and prebiotic substances can be used. Immune checkpoint inhibitors and other cancer immunotherapies may function more or less effectively depending on the gut flora. To improve the response to these treatments in CRC patients, microbiome engineering initiatives aim to optimize the gut microbiota (Belkaid and Hand 2014). To reestablish a balanced gut microbiome, FMT entails transferring fecal material from a healthy donor to a recipient. The precise microbial fingerprints connected to CRC are currently being determined by researchers. These characteristics could be used as prognostic or diagnostic biomarkers to aid in early diagnosis and treatment formulation. For patients with CRC, researchers are creating designed live biotherapeutic items. These "designer microbes" have been created to target and modify the gut microbiome in a way that aids in the treatment or prevention of CRC. The detection of unique microbial imbalances is now possible thanks to advancements in microbiome profiling. This allows for the customization of microbiome-based therapies to each patient's specific microbiome profile.

2.5.1.2 Breast Cancer

In an emerging field of study called "microbiome engineering in breast cancer", the gut microbiome is experimented with to better understand its potential role in the onset, progression, and treatment of breast cancer. The metabolism of estrogen is regulated by some gut microorganisms, and estrogen intake increases the risk of hormone receptor-positive breast cancer (Belkaid and Hand 2014). Researchers are looking into how the gut flora may affect estrogen levels and the risk of breast cancer linked to hormones. The immune system is significantly shaped by the gut flora. Research is ongoing to determine how microbiome engineering may affect the immune system's reaction to breast cancer cells, possibly improving the body's capacity to identify and destroy them (Viswanathan et al. 2023). An elevated risk of various malignancies, including breast cancer, is linked to chronic inflammation. Researchers are studying if microbiome engineering techniques can lessen inflammation and lower the risk of breast cancer because dysbiosis in the gut microbiome can cause it. The relationship between the gut microbiome and the liver is being studied by researchers as part of the gut-liver axis (Tripathi et al. 2018). Dysbiosis in the gut can alter liver performance, which may have an impact on the development and progression of breast cancer. To lessen the negative effects of chemotherapy on breast cancer patients, microbiome engineering is being investigated. Probiotics or dietary changes may be used as tactics to support a healthy gut microbiome while receiving treatment. Personalized microbiome profiling can aid in finding microbial signatures connected to specific breast cancer cases, much like it can with other cancer types (Bhatt, Redinbo, and Bultman 2017). The gut microbiome can be dramatically impacted by dietary choices. To support a healthy microbiome in breast cancer patients, researchers are looking into dietary interventions, such as those that encourage the consumption of fiber, prebiotics, and fermented foods. The use of microbiome-based diagnostics to find biomarkers linked to breast cancer risk or treatment response is being investigated by certain researchers (Jarman, Ribeiro-Milograna, and Kalle 2020).

2.5.1.3 Liver Cancer

Research in the domain of microbiome engineering in liver cancer, particularly hepatocellular carcinoma (HCC), is still in its early stages (Ganesan, Yoon, and Suk 2023). The possible impact of the gut microbiota on liver health, inflammation, and the emergence of liver disorders, including liver cancer, is coming to light more frequently. HCC and dysbiosis, or an imbalance in the gut microbiome, have been linked to liver conditions. Specific microbial imbalances that may contribute to the emergence or spread of liver cancer are the subject of research (Wan and El-Nezami 2018). Chronic inflammation is a risk factor for liver cancer and can be influenced by the gut flora. The ability of the body to identify and target malignant cells may be improved by manipulating the gut microbiome to boost the immune response against liver cancer cells. The health of the liver is influenced by the metabolism of bile acids, which is regulated by some gut flora. Dysbiosis may result in the creation of toxic byproducts or changes in bile acid metabolism that support the development of liver cancer. Toxins such as endotoxins, which may have systemic effects, including liver inflammation, can be produced as a result of dysbiosis (Wan and El-Nezami 2018). The gut-liver axis connects the two organs, and the gut microbiota can affect how well the liver functions. To better understand how microbiome engineering may affect liver cancer, researchers are examining this axis. To increase the efficacy of chemotherapy and immunotherapy in the treatment of liver cancer, microbiome engineering is being studied (Sieow et al. 2021). Modifying the gut microbiome may be one of the strategies used to improve therapeutic responsiveness and minimize negative effects. The gut microbiome is significantly impacted by diet. For patients with liver cancer, dietary therapies that support a healthy gut microbiome include those that advocate fiber-rich meals, prebiotics, and fermented goods. FMT, which involves transferring feces from a healthy donor to a recipient, is being researched as a potential method of reestablishing a healthy gut microbiome in people with liver cancer (Zhao et al. 2023).

2.5.1.4 Endometrial and Ovarian Cancer

Hormone-related diseases, such as endometrial cancer, are relevant to the gut microbiota because they can affect how estrogen is metabolized and other hormones (Boutriq et al. 2021). In shaping the immune system, the gut microbiome is of utmost importance. Patients with ovarian and endometrial cancer are being studied to determine how microbiome engineering can improve immune responses against cancer cells. Endometrial cancer is one of many malignancies that is a risk due to chronic inflammation (Huang et al. 2023). Researchers are looking at whether microbiome engineering can lessen inflammation and lower the risk of cancer. Dysbiosis in the gut microbiome can contribute to inflammation. Obesity raises the risk of developing ovarian and endometrial cancer (Belkaid and Hand 2014). The gut microbiome can affect metabolic functions and may contribute to the increased risk of cancer associated with obesity. Interventions to address obesity as a cancer risk factor may be included in microbiome engineering strategies (Elmore et al. 2021). The identification of microbial signatures connected to specific cases of ovarian and endometrial cancer is being done via personalized microbiome profiling. Microbiome-based therapies may be customized using this knowledge. The gut microbiome is significantly influenced by dietary decisions. To sustain healthy gut microbiota in people with ovarian and endometrial cancer, researchers are looking into dietary interventions, such as those that encourage the consumption of fiber, prebiotics, and specific fermented foods (Dahiya and Nigam 2022). The use of microbiome-based diagnostics to find biomarkers linked to cancer risk or treatment response in ovarian and endometrial cancer is being studied by certain researchers. The efficiency and safety of microbiome-based therapies in treating gynecological cancers are now being investigated in clinical trials and research investigations.

2.5.2 Microbiome Engineering in Celiac Disease

Microbiome engineering in celiac disease (CD) has the potential to shed light on the pathophysiology of the condition and create brand-new therapeutic strategies (Induri et al. 2022). The ingestion of gluten, a protein present in wheat, barley, and rye, causes CD, an autoimmune condition. Researchers are looking into various ways to modify the microbiome for improved disease management because dysbiosis in the gut microbiome has been seen in people with CD. Alterations in the gut microbiome's structure and functionality are a part of dysbiosis in CD. Researchers have found variations in the microbial profiles of CD patients compared to healthy individuals. Dysbiosis may contribute to the development of CD and the immune system's reaction to gluten.

CD treatment options are being researched using probiotics, which are living, helpful microorganisms. These probiotics might improve gut health, reestablish microbial balance, and control the immune system's gluten response (Induri et al. 2022). Particular strains, such as those of the *Lactobacillus* and *Bifidobacterium* species, are of interest. Prebiotics are indigestible food ingredients that help the formation of good bacteria in the stomach. People with CD may be able to encourage the development of advantageous microorganisms and maybe lessen dysbiosis by ingesting prebiotic-rich meals or supplements (Judkins et al. 2020). The use of microbial enzymes that can break down gluten is being investigated by certain researchers. To lessen gluten exposure and the ensuing immunological reaction in people with CD, these enzymes may be given together with gluten-containing meals. Researchers are looking into modifying gut microbes to express enzymes that break down gluten. These genetically modified bacteria may live in the stomach and aid in digesting gluten before it raises an immune response (Kõiv and Tenson 2021). A stringent gluten-free diet that excludes any products containing gluten is the main treatment for CD. This dietary intervention aids in preventing immune system activation brought on by gluten even though it is not a microbiome engineering strategy per se. Recent developments in microbiome profiling might make it easier to spot particular microbial imbalances in CD patients. Then, customized solutions can be created to deal with these imbalances. Researchers are investigating the possibility of FMT, also known as microbial community transplantation, to modify the gut microbiome and immune response in CD, even though it is not yet a common treatment for the condition (Kõiv and Tenson 2021).

2.5.3 Microbiome Engineering in Gastroesophageal Reflux Disease (GERD)

Microbiome engineering in gastroesophageal reflux disease (GERD) investigates the potential function of the gut microbiome in the occurrence and treatment of GERD (Shin et al. 2019). While lower esophageal sphincter (LES) failure and increased stomach acid production are the main causes of GERD, dysbiosis (an imbalance in the gut microbiome) may exacerbate the condition and worsen its symptoms. It is being researched whether probiotics with beneficial bacteria, including specific strains of *Lactobacillus* and *Bifidobacterium*, can alter the gut microbiota and lessen GERD symptoms (Cheng and Ouwehand 2020). These probiotics might assist in reestablishing microbial equilibrium and enhancing gut health. Prebiotics are indigestible food ingredients that support the development of good gut bacteria. Ingesting prebiotic-rich meals or supplements may help people with GERD encourage the development of healthy bacteria and perhaps reduce dysbiosis. A soluble fiber-rich diet can help maintain a balanced gut microbiota. By encouraging regular bowel movements and preserving gut health, high-fiber diets may help lessen the symptoms of GERD (D'Souza et al. 2021). Researchers are investigating the possibility of FMT to modify the gut microbiome, despite it not yet being a common treatment for GERD. To restore a healthy microbiome, FMT includes the transfer of feces from a donor to a recipient (Wang et al. 2019). GERD symptoms can worsen when certain meals and drinks are consumed. In this context, microbiome engineering may entail dietary changes to avoid items that harm the gut microbiome and cause acid reflux. The gut flora can change if proton pump inhibitors (PPIs) are used long-term to treat GERD symptoms

(Takagi et al. 2018). If necessary, PPI usage can be decreased to lessen its negative effects on the microbiome. Modern microbiome profiling and sequencing techniques may make it possible to pinpoint specific microbial imbalances in GERD patients. Then, customized solutions can be created to deal with these imbalances.

2.6 ENVIRONMENTAL MICROBIOMES

The variety of microorganisms that make up environmental microbiomes is astounding, and many of them have evolved to live and thrive in particular environmental circumstances. Microbes can be found in almost every ecosystem on Earth, including soil, water, air, extreme settings (such as hot springs and deep-sea hydrothermal vents), and even inside other creatures (as symbionts or pathogens) (Cavicchioli et al. 2019). Numerous crucial tasks carried out by the microbes in environmental microbiomes support the health and stability of ecosystems. These processes involve the cycling of nutrients, the breakdown of organic matter, and the synthesis of secondary metabolites, for example, nitrogen fixation and carbon degradation (Cavicchioli et al. 2019). In biogeochemical processes, such as the carbon and nitrogen cycles, microbes are also involved. Global climate control is significantly influenced by environmental microbiomes. For instance, microbial activity in soils helps reduce carbon emissions and slow down climate change. Additionally, some marine bacteria generate and absorb greenhouse gases, which affect the climate of Earth.

Environmental microbiomes' microbes can be used in bioremediation projects. The capacity of some microbes to break down toxins and pollutants makes them useful tools for cleaning up contaminated settings (Ayilara and Babalola 2023). On human health, environmental microbiomes may indirectly have an impact. For instance, indoor air quality and the spread of infectious diseases may be impacted by the microbiome of the built environment (such as homes and hospitals). The management of these microbiomes can have effects on public health. Microbiome ecology is the study of the composition and operation of environmental microbiomes (Berg et al. 2020b). Characterizing microbial communities, identifying important species, and analyzing how they interact with one another and their environment are all necessary steps in this process. Environmental microbiomes' function in ecosystem resilience, prospective uses in agriculture and biotechnology, and responses to environmental disturbances such as pollution and climate change are all being better understood as a result of ongoing studies (Suman et al. 2022). Our capacity to investigate environmental microbiomes has been completely transformed by developments in DNA sequencing and metagenomic analysis. Researchers can survey microbial communities in complicated environments using high-throughput sequencing techniques to find novel species and activities. Conservation initiatives must recognize the significance of environmental microbiomes (Combrink et al. 2023). Natural microbiota preservation and ecosystem restoration can offer significant ecological and financial advantages.

2.6.1 Microbiomes in Environmental Sustainability

Microbiomes have an important effect on the health of our world and play a big part in environmental sustainability in different habitats. For nutrient cycling, the breakdown of organic matter, and soil structure, soil microbiomes are essential. By supplying vital nutrients to plants and boosting their resistance to diseases, they promote plant growth. Reducing tillage and practicing organic farming are examples of sustainable agriculture techniques that can improve soil fertility and lessen the demand for chemical fertilizers (Tuğrul and Tuğrul 2019). Pollutants in soil and water can be broken down and detoxified by microbes. As an environmentally friendly substitute for expensive and harmful remediation approaches, bioremediation techniques use microbial communities to clean up contaminated places. The procedures used to remediate wastewater rely heavily on microbial ecosystems. They eliminate toxins and break down organic materials to create wastewater that is safe to discharge back into the environment. Optimizing microbial processes is key to sustainable wastewater treatment since it uses less energy and has a smaller environmental impact. Anaerobic digestion, which produces biogas, depends on microbes (Meegoda et al. 2018). Without oxygen, anaerobic bacteria decompose organic material, creating biogas that can be used as a sustainable energy source. This procedure lessens reliance on fossil fuels and greenhouse gas emissions. Rhizosphere microbiomes, or plant root microbiomes, have been shown to improve plant growth, nutrient uptake, and disease resistance. Ecosystem stability can be increased and chemical inputs can be minimized through the use of sustainable forestry and farming methods (Mendes, Garbeva, and Raaijmakers 2013). Emissions of greenhouse gases are influenced by microbial activity. Methane is a powerful greenhouse gas that is produced by methanogenic archaea, while other bacteria absorb it and store carbon. By lowering emissions and boosting carbon sequestration, understanding and regulating microbial communities in ecosystems can help slow the effects of climate change.

Agriculture can benefit from the use of microbial-based products such as biostimulants and fertilizers to stimulate plant development. These environmentally friendly substitutes for synthetic fertilizers and pesticides increase nutrient uptake, reduce chemical input, and boost crop resilience. Reintroducing important microbial populations helps to restart ecological processes when restoring damaged ecosystems (Lemke and DeSalle 2023). By utilizing this strategy, deteriorated habitats can recover more quickly, enhancing long-term sustainability. For an ecosystem to be resilient and adaptable, microbial diversity must be preserved. Microbial diversity and the health of the ecosystem as a whole are maintained by conservation initiatives that save natural habitats and lessen habitat damage.

2.6.2 Microbiome Engineering of Soil Microbiomes

Soil microbiome engineering is the deliberate modification of microbial communities in soil ecosystems to increase their functionality for certain ecological, environmental, or agricultural goals (Afridi et al. 2022; Leister 2012). This new field has enormous potential for enhancing plant growth, soil health, nutrient cycling, and environmental sustainability. Plant health and growth can be improved by locating and isolating helpful microorganisms (such as nitrogen-fixing bacteria, mycorrhizal fungi, and bacteria that promote plant growth) and then introducing these microbes to the soil. Microbiome engineering tries to return native microbial communities to their normal composition and function in degraded or disturbed soils. Important microbial species may need to be reintroduced to do this, or their recovery may be aided by habitat restoration. Creating artificial microbial communities involves putting together particular mixtures of microorganisms with complementary roles (Visibelli et al. 2023). For instance, creating a community that improves nutrient cycling or reduces plant infections. Developing microbial-based biostimulants that, when applied to soil, increase the development and activity of beneficial microbes is another approach. These biostimulants can increase plant resiliency and nutrient availability using naturally occurring communities of microbes that cooperate to carry out particular tasks, such as the suppression of disease or the breakdown of organic materials. Additionally, modifying the genomes of particular microorganisms through genetic engineering techniques can improve their capacity for pesticide breakdown, nutrient cycling, or other desirable tasks.

Altering soil microorganism metabolic processes enhances their capacity to create useful compounds or break down contaminants in the soil. Optimal environments for can be created for beneficial microorganisms by combining microbiome engineering with the addition of soil additives such as biochar or compost. Cutting-edge molecular methods and bioinformatics to monitor changes in the soil microbiome following engineering interventions can make sure the desired effects are realized. Implementing precise soil management techniques, such as targeted nutrient administration and irrigation, can maximize agricultural output while reducing adverse effects on the environment. Soil health and resilience can be improved by combining microbial engineering with regenerative agricultural techniques, including crop rotation, decreased tillage, and cover crops (Khangura et al. 2023; Gao et al. 2021). By encouraging carbon sequestration and lowering soil emissions of greenhouse gases, improving soil microbiomes can also aid in the mitigation of global warming. To stop soil erosion, enhance water retention, and boost overall soil fertility, microbiome engineering should be included in sustainable land management practices. Utilizing microbiome engineering to support ecosystem services, including pollination, soil water filtering, and disease suppression, would benefit both agriculture and the environment.

2.7 CHALLENGES AND CONSIDERATIONS IN MICROBIAL-BASED THERAPIES

Probiotics, prebiotics, FMT, and eLBPs, among with other microbial-based medicines, show significant potential for treating a variety of medical diseases by modifying the microbiome. Safety is the top priority in treatments based on microorganisms (M. E. Sanders et al. 2010). Live microorganisms can be dangerous for humans to consume, especially if they are immunocompromised or have underlying medical issues. In the case of FMT, stringent donor screening and quality control procedures for eLBPs are necessary to guarantee the security of these therapies. The long-term implications of microbial-based therapies are not well understood. While there might be immediate advantages, it's important to keep an eye out for any potential long-term negative effects or changes in the microbiota. Standardized methods for treatments based on microorganisms are nonexistent (M. E. Sanders et al. 2010). Because treatment outcomes may vary as a result, it may be difficult to generate reliable, repeatable results. Individual reactions to treatments based on microbes can differ greatly. Due to individual variances in their baseline microbiota, genetics, nutrition, and health state, what works for one person may not work for another. To maximize therapy success, personalized strategies are required. Therapies based on microorganisms are regulated to ensure their efficacy and safety. It can be difficult to strike a balance between the need for rapid access to cutting-edge medicines and regulatory restrictions. Informed consent and equitable access to these medicines are two more ethical difficulties that come up. The processes by which these medicines function need to be better understood. Standardization of microbial products is crucial for microbial-based therapeutics like eLBPs (Lozupone et al. 2012). This entails specifying the product's specific composition, assuring its stability, and putting in place quality control procedures. Treatments may be resistant to or modified by microorganisms, which eventually reduces their efficacy. This is an issue when it comes to chronic illnesses or repeated infections. One of the main objectives of many medicines based on microbes is to maintain or restore microbial diversity. A diversified and stable microbiome might be difficult to attain and maintain (Lozupone et al. 2012). Medication interactions may affect the efficacy or safety of microbial-based therapy. Potential drug-microbiome interactions should be understood by healthcare professionals. There may be issues with how the general public views microbial-based medicines, notably FMT. To encourage comprehension and acceptance, education and awareness efforts are required. Microbial-based treatments can be expensive, and for some people, access may be restricted. For greater acceptance, it is crucial to make efforts to lower costs and improve accessibility. Collaboration between researchers, healthcare professionals, regulators, and patients is necessary to address these issues. To fully utilize the potential of microbial-based therapies while assuring their safety and effectiveness in a variety of patient populations, ongoing research, the creation of guidelines, and the adoption of best practices are crucial.

2.8 FUTURE DIRECTIONS AND INNOVATIONS IN MICROBIAL ENGINEERING

Precision microbiome engineering is now possible thanks to developments in bioinformatics and microbiome profiling. Based on each person's particular microbiome composition, researchers are creating tailored medicines that will enable more focused interventions. Microbes with specialized therapeutic capabilities are known as eLBPs. These designer microorganisms can be directed to make medicinal compounds, alter immune responses, or focus on particular diseases. They show promise for a variety of uses, such as the treatment of cancer and the control of autoimmune diseases. The microbiome is being studied using synthetic biology approaches, which enable precise engineering and creation of microbial communities. To accomplish the desired purposes, this involves the capability of adding or deleting particular genes from microorganisms.

New bioactive chemicals are abundant in the microbiome. To discover new antibiotics, anti-inflammatory drugs, and metabolic regulators, researchers are studying the microbiome. New studies point to a significant relationship between gut microbiome health and neurological well-being. Future discoveries could result in microbiome-based treatments for ailments such as anxiety, depression, and neurodegenerative illnesses. A growing field of study looks into how the microbiome affects aging and longevity. Researchers are investigating methods for microbiome engineering to support healthy aging and lengthen longevity. The applications of microbial engineering go beyond human health. Engineered microbes are being used by scientists to tackle environmental issues such as cleaning up oil spills, decontaminating polluted locations, and enhancing soil health for sustainable agriculture. Research on the microbiome incorporates machine learning and artificial intelligence (AI). These tools can aid in the analysis of enormous volumes of microbiome data, the prediction of microbial interactions, and the creation of efficient therapies. Regulatory frameworks are changing as microbiome-based medicines develop to guarantee their effectiveness and safety. Regulatory science advancements will be crucial in bringing these treatments to the clinic. For space travel, where astronaut health is impacted by their gut microbiota, it is crucial to comprehend and engineer the microbiome. Long-duration space missions may require a focus on microbiome engineering. Agriculture is using microbial engineering to improve the health and yield of crops. Innovative methods for managing the soil microbiota can support sustainable agricultural practices. The number of patients actively participating in data gathering and sharing their experiences in microbiome research is rising. This patient-driven methodology could result in cutting-edge discoveries and therapeutic advances.

ABBREVIATIONS

HMP	Human Microbiome Project
FMT	Fecal microbiota transplants
DNA	Deoxyribonucleic acid
CRISPR	Clustered regularly interspaced short palindromic repeats
SCFAS	Short-chain fatty acids
IBS	Irritable bowel syndrome
IBD	Inflammatory bowel disease
ELBPS	Engineered live biotherapeutic products
MS	Multiple sclerosis
ASD	Autism spectrum disorder
NIH	National Institutes of Health
CDI	Clostridium difficile infection
FODMAP	Fermentable oligosaccharides, disaccharides, monosaccharides, and polyols
CD	Celiac disease
GERD	Microbiome engineering In gastroesophageal reflux disease
LES	Lower esophageal sphincter
CRC	Colorectal cancer
HCC	Hepatocellular carcinoma
AI	Artificial intelligence

REFERENCES

"A Review of 10 Years of Human Microbiome Research Activities at the US National Institutes of Health, Fiscal Years 2007-2016." 2019. *Microbiome* 7 (1). NLM (Medline): 31. doi:10.1186/S40168-019-0620-Y/FIGURES/8.

Afridi, Muhammad Siddique, Sher Ali, Abdul Salam, Willian César Terra, Aqsa Hafeez, Sumaira, Baber Ali, et al. 2022. “Plant Microbiome Engineering: Hopes or Hypes.” *Biology* 11 (12). Multidisciplinary Digital Publishing Institute (MDPI): 1782. doi:10.3390/BIOLOGY11121782.

Aggarwal, Nikhil, Shohei Kitano, Ginette Ru Ying Puah, Sandra Kittelmann, In Young Hwang, and Matthew Wook Chang. 2023. “Microbiome and Human Health: Current Understanding, Engineering, and Enabling Technologies.” *Chemical Reviews* 123 (1). American Chemical Society: 31–72. doi:10.1021/ACS.CHEMREV.2C00431/ASSET/IMAGES/LARGE/CR2C00431_0007.JPEG.

Albright, Michaeline B.N., Stilianos Louca, Daniel E. Winkler, Kelli L. Feeser, Sarah Jane Haig, Katrine L. Whiteson, Joanne B. Emerson, and John Dunbar. 2021. “Solutions in Microbiome Engineering: Prioritizing Barriers to Organism Establishment.” *The ISME Journal* 16 (2). Nature Publishing Group: 331–38. doi:10.1038/s41396-021-01088-5.

Amabebe, Emmanuel, Faith O. Robert, Tarimoboere Agbalalah, and Ebiowei S.F. Orubu. 2020. “Microbial Dysbiosis-Induced Obesity: Role of Gut Microbiota in Homoeostasis of Energy Metabolism.” *British Journal of Nutrition* 123 (10). Cambridge University Press: 1127–37. doi:10.1017/s0007114520000380.

Aoun, Antoine, Fatima Darwish, and Natacha Hamod. 2020. “The Influence of the Gut Microbiome on Obesity in Adults and the Role of Probiotics, Prebiotics, and Synbiotics for Weight Loss.” *Preventive Nutrition and Food Science* 25 (2). Korean Society of Food Science and Nutrition (KFN): 113. doi:10.3746/PNF.2020.25.2.113.

Athanasopoulou, Konstantina, Panagiotis G. Adamopoulos, and Andreas Scorilas. 2023. “Unveiling the Human Gastrointestinal Tract Microbiome: The Past, Present, and Future of Metagenomics.” *Biomedicines* 11 (3). Multidisciplinary Digital Publishing Institute: 827. doi:10.3390/BIOMEDICINES11030827.

Ayilara, Modupe S., and Olubukola O. Babalola. 2023. “Bioremediation of Environmental Wastes: The Role of Microorganisms.” *Frontiers in Agronomy* 5 (May). Frontiers Media S.A.: 1183691. doi:10.3389/FAGRO.2023.1183691/BIBTEX.

Belkaid, Yasmine, and Timothy W. Hand. 2014. “Role of the Microbiota in Immunity and Inflammation.” *Cell* 157 (1). NIH Public Access: 121. doi:10.1016/J.CELL.2014.03.011.

Berg, Gabriele, Daria Rybakova, Doreen Fischer, Tomislav Cernava, Marie Christine Champomier Vergès, Trevor Charles, Xiaoyulong Chen, et al. 2020a. “Microbiome Definition Re-Visited: Old Concepts and New Challenges.” *Microbiome* 8 (1). BMC: 103. doi:10.1186/S40168-020-00875-0.

Berg, Gabriele, Daria Rybakova, Doreen Fischer, Tomislav Cernava, Marie Christine Champomier Vergès, Trevor Charles, Xiaoyulong Chen, et al. 2020b. “Microbiome Definition Re-Visited: Old Concepts and New Challenges.” *Microbiome* 8 (1). BioMed Central Ltd: 1–22. doi:10.1186/S40168-020-00875-0/FIGURES/7.

Bhatt, Aadra P., Matthew R. Redinbo, and Scott J. Bultman. 2017. “The Role of the Microbiome in Cancer Development and Therapy.” *CA: A Cancer Journal for Clinicians* 67 (4). NIH Public Access: 326. doi:10.3322/CAAC.21398.

Bhutia, Yangzom D., and Vadivel Ganapathy. 2015. “Short, but Smart: SCFAs Train T Cells in the Gut to Fight Autoimmunity in the Brain.” *Immunity* 43 (4). Cell Press: 629–31. doi:10.1016/J.IMMUNI.2015.09.014.

Boix-Amorós, Alba, Hilary Monaco, Elisa Sambataro, and Jose C. Clemente. 2022. “Novel Technologies to Characterize and Engineer the Microbiome in Inflammatory Bowel Disease.” *Gut Microbes* 14 (1). Taylor & Francis. doi:10.1080/19490976.2022.2107866.

Bonder, Marc Jan, Alexander Kurilshikov, Ettje F. Tigchelaar, Zlatan Mujagic, Floris Imhann, Arnau Vich Vila, Patrick Deelen, et al. 2016. “The Effect of Host Genetics on the Gut Microbiome.” *Nature Genetics* 48 (11). Nature Publishing Group: 1407–12. doi:10.1038/NG.3663.

Boutriq, Soukaina, Alicia González-González, Isaac Plaza-Andrades, Aurora Laborda-Illanes, Lidia Sánchez-Alcoholado, Jesús Peralta-Linero, María Emilia Domínguez-Recio, et al. 2021. “Gut and Endometrial Microbiome Dysbiosis: A New Emergent Risk Factor for Endometrial Cancer.” *Journal of Personalized Medicine* 11 (7). Multidisciplinary Digital Publishing Institute (MDPI). doi:10.3390/JPM11070659.

Brahe, Lena K., Emmanuelle Le Chatelier, Edi Prifti, Nicolas Pons, Sean Kennedy, Trine Blædel, Janet Håkansson, et al. 2015. “Dietary Modulation of the Gut Microbiota - A Randomised Controlled Trial in Obese Postmenopausal Women.” *British Journal of Nutrition* 114 (3). Cambridge University Press: 406–17. doi:10.1017/S0007114515001786.

Breton, Jonathan, Marie Galmiche, and Pierre Déchelotte. 2022. “Dysbiotic Gut Bacteria in Obesity: An Overview of the Metabolic Mechanisms and Therapeutic Perspectives of Next-Generation Probiotics.” *Microorganisms* 10 (2). MDPI. doi:10.3390/MICROORGANISMS10020452/S1.

Cai, Ping, Hao Rong, Qiaoqiao Zhu, Xiaoyu Dai, and Jianpei Zhao. 2023. “The Potential Roles of Gut Microbiome in Anal Fistula.” *AMB Express* 13 (1). Springer. doi:10.1186/S13568-023-01560-9.

Carabotti, Marilia, Annunziata Scirocco, Maria Antonietta Maselli, and Carola Severi. 2015. “The Gut-Brain Axis: Interactions between Enteric Microbiota, Central and Enteric Nervous Systems.” *Annals of Gastroenterology : Quarterly Publication of the Hellenic Society of Gastroenterology* 28 (2). The Hellenic Society of Gastroenterology: 203. /pmc/articles/PMC4367209/.

Cavicchioli, Ricardo, William J. Ripple, Kenneth N. Timmis, Farooq Azam, Lars R. Bakken, Matthew Baylis, Michael J. Behrenfeld, et al. 2019. “Scientists’ Warning to Humanity: Microorganisms and Climate Change.” *Nature Reviews Microbiology* 17 (9). Nature Publishing Group: 569–86. doi:10.1038/s41579-019-0222-5.

Cerdó, Tomás, José Antonio García-Santos, Mercedes G. Bermúdez, and Cristina Campoy. 2019. “The Role of Probiotics and Prebiotics in the Prevention and Treatment of Obesity.” *Nutrients* 11 (3). Multidisciplinary Digital Publishing Institute (MDPI). doi:10.3390/NU11030635.

Chen, Yinwei, Jinghua Zhou, and Li Wang. 2021. “Role and Mechanism of Gut Microbiota in Human Disease.” *Frontiers in Cellular and Infection Microbiology* 11 (March). Frontiers Media S.A. doi:10.3389/FCIMB.2021.625913/FULL.

Cheng, Jing, and Arthur C. Ouwehand. 2020. “Gastroesophageal Reflux Disease and Probiotics: A Systematic Review.” *Nutrients* 12 (1). Multidisciplinary Digital Publishing Institute (MDPI). doi:10.3390/NU12010132.

Combrink, Leigh, Ian R. Humphreys, Quinn Washburn, Holly K. Arnold, Keaton Stagaman, Kristin D. Kasschau, Anna E. Jolles, Brianna R. Beechler, and Thomas J. Sharpton. 2023. “Best Practice for Wildlife Gut Microbiome Research: A Comprehensive Review of Methodology for 16S RRNA Gene Investigations.” *Frontiers in Microbiology* 14. Frontiers Media S.A. doi:10.3389/FMICB.2023.1092216/FULL.

Cullen, Chad M., Kawalpreet K. Aneja, Sinem Beyhan, Clara E. Cho, Stephen Woloszynek, Matteo Convertino, Sophie J. McCoy, et al. 2020. “Emerging Priorities for Microbiome Research.” *Frontiers in Microbiology* 11 (February). Frontiers Media S.A.: 491374. doi:10.3389/FMICB.2020.00136/BIBTEX.

D’Souza, Steve M., Kevin Houston, Lauren Keenan, Byung Soo Yoo, Parth J. Parekh, and David A. Johnson. 2021. “Role of Microbial Dysbiosis in the Pathogenesis of Esophageal Mucosal Disease: A Paradigm Shift from Acid to Bacteria?” *World Journal of Gastroenterology* 27 (18). Baishideng Publishing Group Inc: 2054. doi:10.3748/WJG.V27.I18.2054.

Dahiya, Divakar, and Poonam Singh Nigam. 2022. “The Gut Microbiota Influenced by the Intake of Probiotics and Functional Foods with Prebiotics Can Sustain Wellness and Alleviate Certain Ailments like Gut-Inflammation and Colon-Cancer.” *Microorganisms* 10 (3). Microorganisms. doi:10.3390/MICROORGANISMS10030665.

Danquah, Cynthia Amaning, Prince Amankwah Baffour Minkah, Isaiah Osei Duah Junior, Kofi Bonsu Amankwah, and Samuel Owusu Somuah. 2022. “Antimicrobial Compounds from Microorganisms.” *Antibiotics* 11 (3). Multidisciplinary Digital Publishing Institute (MDPI). doi:10.3390/ANTIBIOTICS11030285.

Durack, Juliana, and Susan V. Lynch. 2019. “The Gut Microbiome: Relationships with Disease and Opportunities for Therapy.” *The Journal of Experimental Medicine* 216 (1). The Rockefeller University Press: 20. doi:10.1084/JEM.20180448.

Elmore, Lynne W., Susanna F. Greer, Elvan C. Daniels, Charles C. Saxe, Michael H. Melner, Ginger M. Krawiec, William G. Cance, and William C. Phelps. 2021. “Blueprint for Cancer Research: Critical Gaps and Opportunities.” *CA: A Cancer Journal for Clinicians* 71 (2). American Cancer Society: 107–39. doi:10.3322/CAAC.21652.

Fan, Yong, and Oluf Pedersen. 2021. “Gut Microbiota in Human Metabolic Health and Disease.” *Nature Reviews Microbiology* 19 (1). Nature Research: 55–71. doi:10.1038/s41579-020-0433-9.

Fijan, Sabina. 2014. “Microorganisms with Claimed Probiotic Properties: An Overview of Recent Literature.” *International Journal of Environmental Research and Public Health* 11 (5). Multidisciplinary Digital Publishing Institute (MDPI): 4745. doi:10.3390/IJERPH110504745.

Fischbach, Michael A., and Julia A. Segre. 2016. “Signaling in Host-Associated Microbial Communities.” *Cell* 164 (6). NIH Public Access: 1288. doi:10.1016/J.CELL.2016.02.037.

Flint, Harry J., Karen P. Scott, Sylvia H. Duncan, Petra Louis, and Evelyne Forano. 2012. “Microbial Degradation of Complex Carbohydrates in the Gut.” *Gut Microbes* 3 (4): 289–306. doi:10.4161/GMIC.19897.

Flint, Harry J., Karen P. Scott, Petra Louis, and Sylvia H. Duncan. 2012. “The Role of the Gut Microbiota in Nutrition and Health.” *Nature Reviews Gastroenterology & Hepatology* 9 (10): 577–89. doi:10.1038/nrgastro.2012.156.

Foo, Jee Loon, Hua Ling, Yung Seng Lee, and Matthew Wook Chang. 2017. "Microbiome Engineering: Current Applications and Its Future." *Biotechnology Journal* 12 (3). John Wiley & Sons, Ltd: 1600099. doi:10.1002/BIOT.201600099.

Gagliardi, Antonella, Valentina Totino, Fatima Cacciotti, Valerio Iebba, Bruna Neroni, Giulia Bonfiglio, Maria Trancassini, Claudio Passariello, Fabrizio Pantanella, and Serena Schippa. 2018. "Rebuilding the Gut Microbiota Ecosystem." *International Journal of Environmental Research and Public Health* 15 (8). Multidisciplinary Digital Publishing Institute (MDPI). doi:10.3390/IJERPH15081679.

Ganesan, Raja, Sang Jun Yoon, and Ki Tae Suk. 2023. "Microbiome and Metabolomics in Liver Cancer: Scientific Technology." *International Journal of Molecular Sciences* 24 (1). Multidisciplinary Digital Publishing Institute (MDPI). doi:10.3390/IJMS24010537.

Gao, Jie, Miao Liu, Sixue Shi, Ying Liu, Yu Duan, Xianguo Lv, Tsing Bohu, et al. 2021. "Disentangling Responses of the Subsurface Microbiome to Wetland Status and Implications for Indicating Ecosystem Functions." *Microorganisms* 9 (2). Multidisciplinary Digital Publishing Institute: 211. doi:10.3390/MICROORGANISMS9020211.

Geng, Jiafeng, Qingqiang Ni, Wei Sun, Liangge Li, and Xiujing Feng. 2022. "The Links between Gut Microbiota and Obesity and Obesity Related Diseases." *Biomedicine & Pharmacotherapy* 147 (March). Elsevier Masson: 112678. doi:10.1016/J.BIOPHA.2022.112678.

Grice, Elizabeth A., and Julia A. Segre. 2012. "The Human Microbiome: Our Second Genome." *Annual Review of Genomics and Human Genetics* 13 (September): 151–70. doi:10.1146/annurev-genom-090711-163814.

Haldar, Swati, Snehal R Jadhav, Vandana Gulati, David J Beale, Acharya Balkrishna, Anurag Varshney, Enzo A Palombo, Avinash V Karpe, and Rohan M Shah. 2023. "Unravelling the Gut-Lung Axis: Insights into Microbiome Interactions and Traditional Indian Medicine's Perspective on Optimal Health." *FEMS Microbiology Ecology* 99 (10). Oxford Academic. doi:10.1093/FEMSEC/FIAD103.

Harsch, Igor Alexander, and Peter Christopher Konturek. 2018. "The Role of Gut Microbiota in Obesity and Type 2 and Type 1 Diabetes Mellitus: New Insights into 'Old' Diseases." *Medical Sciences* 6 (2). Multidisciplinary Digital Publishing Institute (MDPI). doi:10.3390/MEDSCI6020032.

Hou, Kaijian, Zhuo Xun Wu, Xuan Yu Chen, Jing Quan Wang, Dongya Zhang, Chuanxing Xiao, Dan Zhu, et al. 2022. "Microbiota in Health and Diseases." *Signal Transduction and Targeted Therapy* 7 (1). Nature Publishing Group: 1–28. doi:10.1038/s41392-022-00974-4.

Huang, Jing Yang, Kevin Sheng Kai Ma, Li Tzu Wang, Cho Han Chiang, Shun Fa Yang, Chun Hao Wang, and Po Hui Wang. 2023. "The Risk of Endometrial Cancer and Uterine Sarcoma Following Endometriosis or Pelvic Inflammatory Disease." *Cancers* 15 (3). MDPI: 833. doi:10.3390/CANCERS15030833/S1.

Humphries, Audrey, and Adil Daud. 2018. "The Gut Microbiota and Immune Checkpoint Inhibitors." *Human Vaccines & Immunotherapeutics* 14 (9). Taylor & Francis: 2178. doi:10.1080/21645515.2018.1442970.

Inda, María Eugenia, Esther Broset, Timothy K. Lu, and Cesar de la Fuente-Nunez. 2019. "Emerging Frontiers in Microbiome Engineering." *Trends Immunol* 40 (10). Elsevier Ltd: 952–73. doi:10.1016/j.it.2019.08.007.

Induri, Sri Nitya Reddy, Payalben Kansara, Scott C. Thomas, Fangxi Xu, Deepak Saxena, and Xin Li. 2022. "The Gut Microbiome, Metformin, and Aging." *Https://Doi.Org/10.1146/Annurev-Pharmtox-051920-093829* 62 (January). Annual Reviews . doi:10.1146/ANNUREV-PHARMTOX-051920-093829.

Jarman, Robyn, Sarah Ribeiro-Milograna, and Wouter Kalle. 2020. "Potential of the Microbiome as a Biomarker for Early Diagnosis and Prognosis of Breast Cancer." *Journal of Breast Cancer* 23 (6). Korean Breast Cancer Society: 579. doi:10.4048/JBC.2020.23.E60.

Judkins, Taylor C., Douglas L. Archer, Dean C. Kramer, and Rebecca J. Solch. 2020. "Probiotics, Nutrition, and the Small Intestine." *Current Gastroenterology Reports* 22 (1). Springer: 1–8. doi:10.1007/S11894-019-0740-3/TABLES/1.

Kachrimanidou, Melina, and Eleftherios Tsintarakis. 2020. "Insights into the Role of Human Gut Microbiota in Clostridioides Difficile Infection." *Microorganisms* 8 (2). Multidisciplinary Digital Publishing Institute: 200. doi:10.3390/MICROORGANISMS8020200.

Khan, Saad, Ruth Hauptman, and Libusha Kelly. 2021. "Engineering the Microbiome to Prevent Adverse Events: Challenges and Opportunities." *Annual Review of Pharmacology and Toxicology* 61 (January). NIH Public Access: 159. doi:10.1146/ANNUREV-PHARMTOX-031620-031509.

Khangura, Ravjit, David Ferris, Cameron Wagg, and Jamie Bowyer. 2023. "Regenerative Agriculture—A Literature Review on the Practices and Mechanisms Used to Improve Soil Health." *Sustainability* 15 (3). Multidisciplinary Digital Publishing Institute: 2338. doi:10.3390/SU15032338.

Kho, Zhi Y., and Sunil K. Lal. 2018. "The Human Gut Microbiome - A Potential Controller of Wellness and Disease." *Frontiers in Microbiology* 9 (August). Frontiers Media S.A.: 356589. doi:10.3389/FMICB.2018.01835/BIBTEX.

Kõiv, Viia, and Tanel Tenson. 2021. "Gluten-Degrading Bacteria: Availability and Applications." *Applied Microbiology and Biotechnology* 105 (8). Springer: 3045. doi:10.1007/S00253-021-11263-5.

Lawson, Christopher E., William R. Harcombe, Roland Hatzenpichler, Stephen R. Lindemann, Frank E. Löffler, Michelle A. O'Malley, Héctor García Martín, et al. 2019. "Common Principles and Best Practices for Engineering Microbiomes." *Nat Rev Microbiol* 17 (12). Nature Publishing Group: 725–41. doi:10.1038/s41579-019-0255-9.

Lee, Mong Hong. 2021. "Harness the Functions of Gut Microbiome in Tumorigenesis for Cancer Treatment." *Cancer Communications* 41 (10). John Wiley & Sons, Ltd: 937–67. doi:10.1002/CAC2.12200.

Leister, Dario. 2012. "Retrograde Signaling in Plants: From Simple to Complex Scenarios." *Frontiers in Plant Science* 3 (June). Frontiers Research Foundation. doi:10.3389/FPLS.2012.00135.

Lemke, Michael, and Rob DeSalle. 2023. "The Role of Microbial Ecology in Restoration Ecology in the Age of Genomics: A Summary of the Microbial Ecology Special Issue." *Microbial Ecology* 85 (3). Springer: 1136–41. doi:10.1007/S00248-023-02206-W/TABLES/1.

Ley, Ruth E., Peter J. Turnbaugh, Samuel Klein, and Jeffrey I. Gordon. 2006. "Human Gut Microbes Associated with Obesity." *Nature* 444 (7122): 1022–23. doi:10.1038/4441022a.

Liang, Shan, Xiaoli Wu, and Feng Jin. 2018. "Gut-Brain Psychology: Rethinking Psychology from the Microbiota–Gut–Brain Axis." *Frontiers in Integrative Neuroscience* 12 (September). Frontiers Media S.A. doi:10.3389/FNINT.2018.00033/FULL.

Lozupone, Catherine A., Jesse I. Stombaugh, Jeffrey I. Gordon, Janet K. Jansson, and Rob Knight. 2012. "Diversity, Stability and Resilience of the Human Gut Microbiota." *Nature* 489 (7415). NIH Public Access: 220. doi:10.1038/NATURE11550.

Macfarlane, G.T., and S. Macfarlane. 1997. "Human Colonic Microbiota: Ecology, Physiology and Metabolic Potential of Intestinal Bacteria." *Scandinavian Journal of Gastroenterology* 32 (Sup 222): 3–9. doi:10.1080/00365521.1997.11720708.

Markowiak, Paulina, and Katarzyna Śliżewska. 2017. "Effects of Probiotics, Prebiotics, and Synbiotics on Human Health." *Nutrients* 9 (9). MDPI AG: 1021. doi:10.3390/nu9091021.

Matthewman, Cristina, Alexandra Narin, Hannah Huston, and Christopher Edward Hopkins. 2023. "Systems to Model the Personalized Aspects of Microbiome Health and Gut Dysbiosis." *Molecular Aspects of Medicine* 91 (June). Pergamon: 101115. doi:10.1016/J.MAM.2022.101115.

Meegoda, Jay N., Brian Li, Kush Patel, and Lily B. Wang. 2018. "A Review of the Processes, Parameters, and Optimization of Anaerobic Digestion." *International Journal of Environmental Research and Public Health* 15 (10). Multidisciplinary Digital Publishing Institute (MDPI). doi:10.3390/IJERPH15102224.

Mendes, Rodrigo, Paolina Garbeva, and Jos M. Raaijmakers. 2013. "The Rhizosphere Microbiome: Significance of Plant Beneficial, Plant Pathogenic, and Human Pathogenic Microorganisms." *FEMS Microbiology Reviews* 37 (5). Oxford Academic: 634–63. doi:10.1111/1574-6976.12028.

Nigam, Manisha, Abhaya Shikhar Panwar, and Rahul Kunwar Singh. 2022. "Orchestrating the Fecal Microbiota Transplantation: Current Technological Advancements and Potential Biomedical Application." *Frontiers in Medical Technology* 4 (September). Frontiers Media S.A.: 961569. doi:10.3389/FMEDT.2022.961569/BIBTEX.

Patel, Bharati Kadamb, Kadamb Haribhai Patel, and Shabbir M. Moochhala. 2023. "Gut Microbiota Intervention Strategies Using Active Components from Medicinal Herbs to Evaluate Clinical Efficacy of Type 2 Diabetes – A Review." *Clinical and Translational Discovery* 3 (1). John Wiley & Sons, Ltd: e170. doi:10.1002/CTD2.170.

Peterson, Jane, Susan Garges, Maria Giovanni, Pamela McInnes, Lu Wang, Jeffery A. Schloss, Vivien Bonazzi, et al. 2009. "The NIH Human Microbiome Project." *Genome Research* 19 (12). Cold Spring Harbor Laboratory Press: 2317. doi:10.1101/GR.096651.109.

Rahman, Md Mominur, Md Rezaul Islam, Sheikh Shohag, Md Tanjimul Ahasan, Nadia Sarkar, Hosneara Khan, Alexandru Madalin Hasan, Simona Cavalu, and Abdur Rauf. 2022. "Microbiome in Cancer: Role in Carcinogenesis and Impact in Therapeutic Strategies." *Biomedicine & Pharmacotherapy* 149 (May). Elsevier Masson: 112898. doi:10.1016/J.BIOPHA.2022.112898.

Rutsch, Andrina, Johan B. Kantsjö, and Francesca Ronchi. 2020. "The Gut-Brain Axis: How Microbiota and Host Inflammasome Influence Brain Physiology and Pathology." *Frontiers in Immunology* 11 (December). Frontiers Media S.A.: 604179. doi:10.3389/FIMMU.2020.604179/BIBTEX.

Sanders, David J., Saskia Inniss, Gregory Sebepos-Rogers, Farooq Z. Rahman, and Andrew M. Smith. 2021. "The Role of the Microbiome in Gastrointestinal Inflammation." *Bioscience Reports* 41 (6). Portland Press Ltd: 20203850. doi:10.1042/BSR20203850.

Sanders, Mary Ellen, Louis M.A. Akkermans, Dirk Haller, Cathy Hammerman, James Heimbach, Gabriele Hörmannsperger, Geert Huys, et al. 2010. "Safety Assessment of Probiotics for Human Use." *Gut Microbes* 1 (3). Taylor & Francis: 164. doi:10.4161/GMIC.1.3.12127.

Santana, Patricia Teixeira, Siane Lopes Bittencourt Rosas, Beatriz Elias Ribeiro, Ygor Marinho, and Heitor S.P. de Souza. 2022. "Dysbiosis in Inflammatory Bowel Disease: Pathogenic Role and Potential Therapeutic Targets." *International Journal of Molecular Sciences* 23 (7). Multidisciplinary Digital Publishing Institute (MDPI). doi:10.3390/IJMS23073464.

Scheithauer, Torsten P.M., Elena Rampanelli, Max Nieuwdorp, Bruce A. Vallance, C. Bruce Verchere, Daniël H. van Raalte, and Hilde Herrema. 2020. "Gut Microbiota as a Trigger for Metabolic Inflammation in Obesity and Type 2 Diabetes." *Frontiers in Immunology* 11 (October). Frontiers Media S.A. doi:10.3389/FIMMU.2020.571731.

Ser, Hooi Leng, Vengadesh Letchumanan, Bey Hing Goh, Sunny Hei Wong, and Learn Han Lee. 2021. "The Use of Fecal Microbiome Transplant in Treating Human Diseases: Too Early for Poop?" *Frontiers in Microbiology* 12 (May). Frontiers Media S.A.: 519836. doi:10.3389/FMICB.2021.519836/BIBTEX.

Shah, Aabid Manzoor, Najeebul Tarfeen, Hassan Mohamed, and Yuanda Song. 2023. "Fermented Foods: Their Health-Promoting Components and Potential Effects on Gut Microbiota." *Fermentation* 9 (2). Multidisciplinary Digital Publishing Institute: 118. doi:10.3390/FERMENTATION9020118.

Shin, Andrea, Geoffrey A. Preidis, Robert Shulman, and Purna C. Kashyap. 2019. "The Gut Microbiome in Adult and Pediatric Functional Gastrointestinal Disorders." *Clinical Gastroenterology and Hepatology* 17 (2). Elsevier: 256–74. doi:10.1016/J.CGH.2018.08.054.

Shreiner, Andrew B., John Y. Kao, and Vincent B. Young. 2015. "The Gut Microbiome in Health and in Disease." *Curr. Opin. Gastroenterol.* 31 (1). Lippincott Williams and Wilkins: 69–75. doi:10.1097/mog.0000000000000139.

Sieow, Brendan Fu Long, Kwok Soon Wun, Wei Peng Yong, In Young Hwang, and Matthew Wook Chang. 2021. "Tweak to Treat: Reprograming Bacteria for Cancer Treatment." *Trends in Cancer* 7 (5). Elsevier: 447–64. doi:10.1016/J.TRECAN.2020.11.004.

Singh, Nagendra, Ashish Gurav, Sathish Sivaprakasam, Evan Brady, Ravi Padia, Huidong Shi, Muthusamy Thangaraju, et al. 2014. "Activation of Gpr109a, Receptor for Niacin and the Commensal Metabolite Butyrate, Suppresses Colonic Inflammation and Carcinogenesis." *Immunity* 40 (1): 128–39. doi:10.1016/J.IMMUNI.2013.12.007.

Suman, Jarupula, Amitava Rakshit, Siva Devika Ogireddy, Sonam Singh, Chinmay Gupta, and J. Chandrakala. 2022. "Microbiome as a Key Player in Sustainable Agriculture and Human Health." *Frontiers in Soil Science* 2 (April). Frontiers Media SA: 821589. doi:10.3389/FSOIL.2022.821589/BIBTEX.

Takagi, Tomohisa, Yuji Naito, Ryo Inoue, Saori Kashiwagi, Kazuhiko Uchiyama, Katsura Mizushima, Saeko Tsuchiya, et al. 2018. "The Influence of Long-Term Use of Proton Pump Inhibitors on the Gut Microbiota: An Age-Sex-Matched Case-Control Study." *Journal of Clinical Biochemistry and Nutrition* 62 (1). The Society for Free Radical Research Japan: 100. doi:10.3164/JCBN.17-78.

Tan, Xing, and Stuart Johnson. 2019. "Fecal Microbiota Transplantation (FMT) for C. Difficile Infection, Just Say 'No.'" *Anaerobe* 60 (December). Academic Press: 102092. doi:10.1016/J.ANAEROBE.2019.102092.

Tripathi, Anupriya, Justine Debelius, David A. Brenner, Michael Karin, Rohit Loomba, Bernd Schnabl, and Rob Knight. 2018. "The Gut-Liver Axis and the Intersection with the Microbiome." *Nature Reviews. Gastroenterology & Hepatology* 15 (7). Nat Rev Gastroenterol Hepatol: 397–411. doi:10.1038/S41575-018-0011-Z.

Tuğrul, Koç Mehmet, and Koç Mehmet Tuğrul. 2019. "Soil Management in Sustainable Agriculture." *Sustainable Crop Production* (September). IntechOpen. doi:10.5772/INTECHOPEN.88319.

Turnbaugh, Peter J., Ruth E. Ley, Micah Hamady, Claire M. Fraser-Liggett, Rob Knight, and Jeffrey I. Gordon. 2007. "The Human Microbiome Project." *Nature* 449 (7164). Nature Publishing Group: 804–10. doi:10.1038/NATURE06244.

Ursell, Luke K., Henry J. Haiser, Will Van Treuren, Neha Garg, Lavanya Reddivari, Jairam Vanamala, Pieter C. Dorrestein, Peter J. Turnbaugh, and Rob Knight. 2014. "The Intestinal Metabolome: An Intersection between Microbiota and Host." *Gastroenterology* 146 (6). W.B. Saunders: 1470–76. doi:10.1053/j.gastro.2014.03.001.

Vadder, Filipe De, Petia Kovatcheva-Datchary, Daisy Goncalves, Jennifer Vinera, Carine Zitoun, Adeline Duchampt, Fredrik Bäckhed, and Gilles Mithieux. 2014. "Microbiota-Generated Metabolites Promote Metabolic Benefits via Gut-Brain Neural Circuits." *Cell* 156 (1–2). Elsevier B.V.: 84–96. doi:10.1016/J.CELL.2013.12.016.

Valdes, Ana M., Jens Walter, Eran Segal, and Tim D. Spector. 2018. "Role of the Gut Microbiota in Nutrition and Health." *BMJ* 361 (June). British Medical Journal Publishing Group: 36–44. doi:10.1136/BMJ.K2179.

Valles-Colomer, Mireia, Aitor Blanco-Míguez, Paolo Manghi, Francesco Asnicar, Leonard Dubois, Davide Golzato, Federica Armanini, et al. 2023. "The Person-to-Person Transmission Landscape of the Gut and Oral Microbiomes." *Nature* 614 (7946). Nature Publishing Group: 125–35. doi:10.1038/s41586-022-05620-1.

Veseli, Iva, Yiqun T. Chen, Matthew S. Schechter, Chiara Vanni, Emily C. Fogarty, Andrea R. Watson, Bana Jabri, et al. 2023. "Microbes with Higher Metabolic Independence Are Enriched in Human Gut Microbiomes under Stress." *BioRxiv* (May). Cold Spring Harbor Laboratory Preprints. doi:10.1101/2023.05.10.540289.

Visibelli, Anna, Bianca Roncaglia, Ottavia Spiga, and Annalisa Santucci. 2023. "The Impact of Artificial Intelligence in the Odyssey of Rare Diseases." *Biomedicines* 11 (3). Multidisciplinary Digital Publishing Institute: 887. doi:10.3390/BIOMEDICINES11030887.

Viswanathan, Sathiyapriya, Sheetal Parida, Bhuvana Teja Lingipilli, Ramalingam Krishnan, Devendra Rao Podipireddy, and Nethaji Muniraj. 2023. "Role of Gut Microbiota in Breast Cancer and Drug Resistance." *Pathogens* 12 (3). Multidisciplinary Digital Publishing Institute (MDPI). doi:10.3390/PATHOGENS12030468.

Vyas, Usha, and Natarajan Ranganathan. 2012. "Probiotics, Prebiotics, and Synbiotics: Gut and Beyond." *Gastroenterol Research and Practice* 2012: 872716–872716. doi:10.1155/2012/872716.

Wan, Murphy L.Y., and Hani El-Nezami. 2018. "Targeting Gut Microbiota in Hepatocellular Carcinoma: Probiotics as a Novel Therapy." *Hepatobiliary Surgery and Nutrition* 7 (1). AME Publishing Company: 110–20. doi:10.21037/HBSN.2017.12.07.

Wang, Jiunn Wei, Chao Hung Kuo, Fu Chen Kuo, Yao Kuang Wang, Wen Hung Hsu, Fang Jung Yu, Huang Ming Hu, Ping I. Hsu, Jaw Yuan Wang, and Deng Chyang Wu. 2019. "Fecal Microbiota Transplantation: Review and Update." *Journal of the Formosan Medical Association* 118 (March). Elsevier: S23–31. doi:10.1016/J.JFMA.2018.08.011.

Xi, Yue, and Peng-Fei Xu. 2021. "Diabetes and Gut Microbiota." *World Journal of Diabetes* 12 (10). Baishideng Publishing Group Inc.: 1693–1703. doi:10.4239/WJD.V12.I10.1693.

Yu, Dahai, Xin Meng, Willem M. De Vos, Hao Wu, Xuexun Fang, and Amit K. Maiti. 2021. "Implications of Gut Microbiota in Complex Human Diseases." *International Journal of Molecular Sciences* 22 (23). Multidisciplinary Digital Publishing Institute: 12661. doi:10.3390/IJMS222312661.

Zhao, Yuanyuan, Chen Gong, Jing Xu, Dong Chen, Bo Yang, Zhishui Chen, and Lai Wei. 2023. "Research Progress of Fecal Microbiota Transplantation in Liver Diseases." *Journal of Clinical Medicine* 12 (4). Multidisciplinary Digital Publishing Institute: 1683. doi:10.3390/JCM12041683.

Zheng, Danping, Timur Liwinski, and Eran Elinav. 2020. "Interaction between Microbiota and Immunity in Health and Disease." *Cell Research* 30 (6). Nature Publishing Group: 492–506. doi:10.1038/s41422-020-0332-7.

3 Omics Tools for Determining Microbial Composition

A Systematic Review

Amol D. Gholap, Sampada D. Sawant and Sagar R. Pardeshi*
Department of Pharmaceutics, St. John
Institute of Pharmacy and Research
Palghar, Maharashtra, India

Pankaj R. Khuspe
Department of Pharmaceutics, Shriram Shikshan
Sanstha's, College of Pharmacy
Solapur, Maharashtra, India

Sadikali F. Sayyad and Machindra Chavan
Amrutvahini College of Pharmacy
Sangamner, Maharashtra, India

Navnath T. Hatvate
Institute of Chemical Technology Mumbai,
Marathwada Campus
Maharashtra, India

Md Faiyazuddin
School of Pharmacy, Al - Karim University
Katihar, Bihar, India

Md Jasim Uddin
Department of Pharmaceutical Technology,
Faculty of Pharmacy, Universiti Malaya
Kuala Lumpur, Malaysia

3.1 INTRODUCTION

Because of the connections between the microbiome, the quality of the soil and its inhabitants, and the sustainability of the environment, soil microbial features are being evaluated gradually and accurately as sensitive and complicated markers of soil quality. Scientists must have a better understanding of microbial traits like productivity, biomass, interaction, and variety to further develop knowledge as well as data that improves the general health of the soil (Bertola, Ferrarini, and Visioli 2021). Producers, researchers, and development personnel may find it useful to plan

* Corresponding author. Email: amolgholap16@gmail.com

DOI: 10.1201/9781003394662-3

and develop practical soil quality measures after analyzing these results (Gholap, Said et al. 2023; Gholap, Bhowmik et al. 2023; Gholap, Gupta et al. 2023; Gholap 2012). Studies on the characteristics of soil microbes have frequently been carried out at the process level. Biomass, oxygen consumption, signal transduction, and enzymatic activity have been examined in relation to organism or society reactions to modifications to soil parameters, as presented in Figure 3.1 (Giovannetti et al. 2023; Mishra, Singh, and Singh 2023). Even so, these investigations at the process level give us crucial knowledge about microbial activities, including their potential impact on soil quality. Investigations may provide information on qualitative society alterations since various taxa may participate in a single microbial activity. Furthermore, the capacity of these process-level metrics to identify a specific microbial community is constrained (Walder et al. 2022; De Corato 2020; Hartman et al. 2018).

Discrete species rely on the presence, function, interaction, and engagement with multiple other species in the complex and diverse phenomenon known as society-level microbiological symbiosis (Hartman et al. 2023; Goswami et al. 2017). Hence, quantifiable and subjective changes in the composition and organization of microbial communities in the soil may act as important and delicate markers of both immediate and ongoing modifications to soil health (D. S. M. Amol D Gholap 2011; S. B. Dighe and Amol D Gholap 2012b; Dattatray Tupe and Amol D Gholap* 2012; S. B. Dighe and Amol D Gholap 2012c; Amol D Gholap and Dattatray Tupe 2012a; Tushar V Magar and Amol D Gholap 2011; Amol D Gholap and V. G. Rajurkar 2012a, 2012b). Assessment of microbial

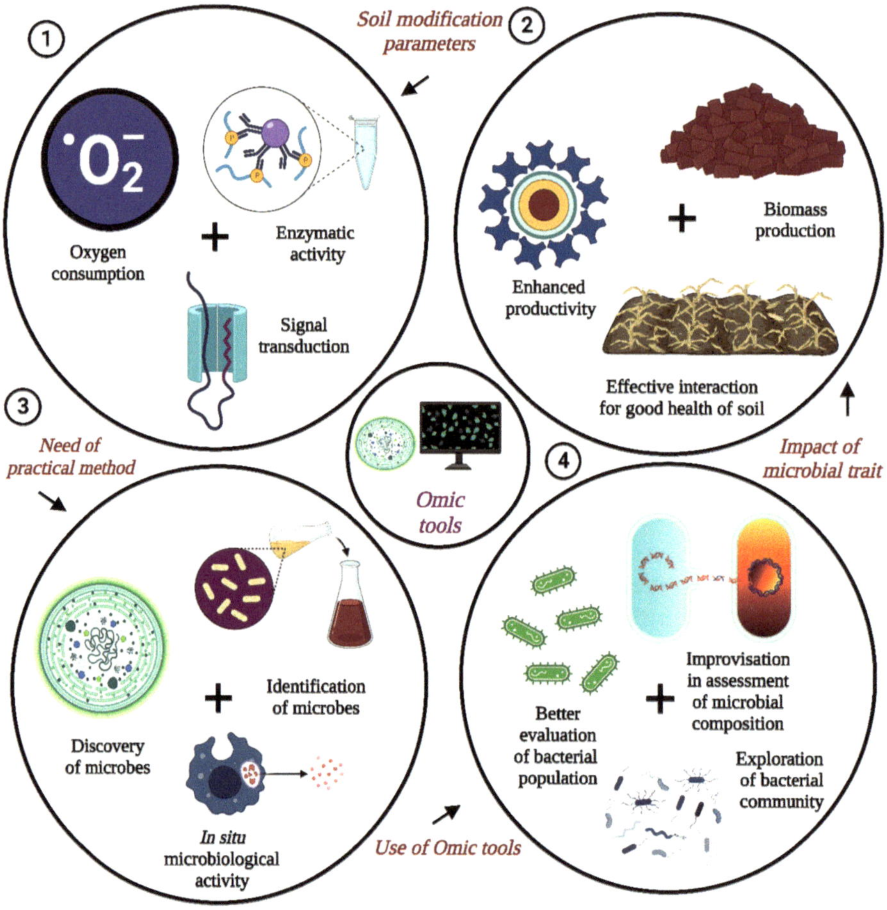

FIGURE 3.1 The need for omics technology is illustrated as 1. the impact of microbial traits; 2. soil modification parameters; 3. the need for practical methods; 4. the use of omics tools.

activity, dispersion, function, and the kinds of communication among diverse species should also be included in the examination of soil microbiological organization, in addition to determining microbial population and variety. The development of practical methods to (a) discover and identify which microbes are found in the soil and (b) to detect microbiological activity in situ has been a long-standing issue in microbial soil ecology. Such difficulties have been compounded by microbe isolation from soil matrix and plant parts, morphological similarities among numerous soil-dwelling bacteria, and shifting microbial taxonomies. Moreover, direct imaging is more difficult with soil microbes than with other microorganisms due to their small size (Santoyo 2022; Berg et al. 2020). The methods used to comprehend and recognize the microbial soil composition and associated interrelationships have dramatically altered over past decades. Soil microbiologists can now use a wide range of modern techniques and methodologies that allow them utilize to more microbes in soil and improve evaluations of bacterial populations. In this chapter, we succinctly go through some of the key methods for examining the microbiological architecture of soil. In contrast to the traditional culture-based approaches for measuring microbial variety, the chapter will discuss omics techniques (B. Zhang et al. 2022; Veresoglou, Halley, and Rillig 2015; Timmis and Ramos 2021).

3.2 REASONS FOR STUDYING MICROBIAL DIVERSITY

There is a lot of genetic data "waiting" to be found in natural microbial communities. It's been noted that endophytic bacteria only make up a small portion of the current microbial species (Veresoglou, Halley, and Rillig 2015; Timmis and Ramos 2021; Morales-Cedeño et al. 2021). Therefore it's crucial to keep working across both culturable as well as nonculturable microorganisms from various settings. Research on variation is crucial for comparing populations. The fact that we don't know enough about living and extinct microbes is yet another crucial reason to investigate bacterial ecology. Scientists disagree on the number of species in the world, the usefulness of the majority of them, and the pace at which they are either vanishing or appearing (Kapinusova, Lopez Marin, and Uhlik 2023; Singh and Dubey 2018; Z. Li et al. 2022). The capacity of an environment to tolerate significant disturbances or stressful conditions can partially depend just on the variety within the system. Diversity studies are crucial and important to:

- Increase awareness of the variety of biological resources and understand the spread of microorganisms.
- Increase awareness of the real-world significance of microbial culture diversity.
- Identify and classify diverse variances that are difficult to manage.
- Understand and identify the management and control of biodiversity.
- Be aware of the significance and effects of biodiversity. (Salamon, Mikołajczak, and Błaszczyk 2023; Subramani and Aalbersberg 2013; Harvey et al. 2023)

3.3 FACTORS GOVERNING MICROBIAL DIVERSITY

Several distinct species will participate in the same activities and may occupy the very same habitats in a bacterial population. Both biotic and abiotic variables can be categorized as the main two processes that affect microbial diversity (Harvey et al. 2023; Hoegh-Guldberg et al. 2017). Abiotic variables can be either chemical or physical in nature and are often known as physicochemical parameters. Examples include the presence of water, salinity, pressure, oxic/anoxic environments, temperature, organic contaminants, pH, heavy metals, pesticides, allelopathy causes, and antibiotics (Moënne-Loccoz et al. 2015; I. Sharma 2021; Hanashiro et al. 2019). In general, all environmental changes have different effects and to different degrees, which causes a change in the biodiversity spectrum. Biotic components, which have a significant effect on microbial variety, provide other living bacteria, plasmids, bacteriophages, proteins, cell-free nucleic acids, and transposons, which are forms of auxiliary DNA that affect the genetic traits and, in just about all situations, the morphologies of their host (Ajay Khedkar, Dattatray Tupe, and Amol Gholap 2012a, 2012b; Amol D. Gholap, Vikas G. Rajurkar 2012; Amol D. Gholap, Vikas G Rajurkar 2012b; Vilas Bhagat 2012; V. G. R. Amol D. Gholap, n.d.; Vikas G. Rajurkar 2012; Gholap, Amol D., and Vikas G. Rajurkar 2012; Gholap, Hatvate et al. 2023; Gholap, Amol D. and Mahajan 2011). Moreover, a number of protozoa have an impact on the variety of microbes because they consume them (Zwanzig 2021; Michaelis and Grohmann 2023; S. Sharma et al. 2023).

3.4 CULTURE TECHNIQUES IN ECO-MICROBIOLOGY

Constrictions and usage standard microbiological methods employ widely used growing mediums, including agar plate, potato dextrose agar, tryptic soy agar, and Luria-Bertani medium, to enhance, identify, and describe microbes. More than 99% of microorganisms in just about any soil environment can be seen under a microscope but not by typical cultivating methods; this is the main obstacle to traditional civilization procedures (Mitra et al. 2021; Tu et al. 2021; Mandal et al. 2020). To increase the proportion of bacterial diversity that is suitable for cultivation, a number of enhanced cultivation techniques, as well as pure culture, have indeed been developed that imitate natural settings in terms of nutrient content (concentration and composition), oxygenation differential, ionic strength, etc. (F. Ajilogba et al. 2022; Kuo, Chiu, and Chou 2023; Fakruddin, Mannan, and Andrews 2013). Throughout this study done by Fakruddin et al., (2013) , a method for huge microorganism cultivation under lower nutritional flow rate conditions, which implicated encapsulating cell lines in gel small particles, was developed for inoculating the same uncultivated microbes from a wide range of conditions, such as depth saltwater, ocean waters, and undiscovered soil ecosystems. However, not all "uncultivated" species can be grown, and many species continue to be as such. Despite being capable of surviving in their natural habitats, these microorganisms stay in a "vital but nonculturable" (VBNC) phase in a laboratory environment. These VBNC species may constitute entirely new groupings, may be numerous or extremely active, and remain unexplored by conventional culture conditions. Potential microbial subdivisions such as OP11, OP10, SC3, WS3, BRC1, WS2, and TM7 are only recognized from their chemical sequencing, according to molecular microbiological studies relying on genes encoded by 16S rRNA. Given the absence of cultivated representatives, such segment clades, like OP11, are thought of as "potential subdivisions" to represent our restricted understanding. They are very diverse and extensively spread in various settings. According to research, there are at least 50 diverse microbial phyla, 50% of which are totally characterized by genetic sequences (Ramamurthy et al. 2014; Laam Li et al. 2014). The majority of the time, organisms recovered using conventional culture techniques are neither quantitatively abundant nor measure the effect inside the habitat from which they were cultivated. Such cultivated bacteria, which make up less than 1% of all microbes, are regarded as the "weed species" of the microbiome. For instance, the majority of species isolated and grown from collected samples correspond to some of the "big four" phyla: Firmicutes, Proteobacteria, Actinobacteria, and Bacteroidetes. This is mostly because these phyla are the easiest to grow in a laboratory environment. However, though Actinobacteria often make up 20% of the microbial flora in soil, such organisms are challenging to cultivate and only represent a small variety of species. Our results suggest that for comprehensive characterization of ambient bacterial diversity, genomic approaches that do not require separation and culturing are extremely desired (Fakruddin, Mannan, and Andrews 2013; Pan and Ren 2022; Pienaar, Singh, and Barnard 2016).

3.5 OMICS TECHNOLOGY

Metagenomics is the study of aggregate genome sequences that are readily extracted from ambient samples. Alternative titles for metagenomics include ecological informatics, communal genome sequencing, and bacterial ecogenomics (Pienaar, Singh, and Barnard 2016). Basically, approaches that exclusively analyze polymerase chain reaction (reverse transcription—RT—versions of chosen genes, such as DNA fingerprint tools) really aren't considered part of metagenomics since they do not offer data on genetic variation outside of the amplified regions (Hiraoka, Yang, and Iwasaki 2016). Metagenomic approaches, as explained in the previous chapter, are rooted in the notion that the complete genetic makeup of surrounding bacterial diversity might be transcribed and examined in a manner similar to what occurs in a bacterial culture society's total genome (Garza and Dutilh 2015; Eisen 2007).

In order to comprehend the metabolic functions of uncultured microbes, including their interactions with certain other both biotic and abiotic factors, metagenomics is therefore essential. Novel bacterial catalysts and medicines with possible uses in biotech, healthcare, and commerce have been found in abundance in ecological metagenomics (Vidal et al. 2022; Devarapalli and Kumavath 2015). Creating a genomic collection involves these actions: whole DNA is isolated out of an organic substance, fragmented nucleotide sequences are cloned through an appropriate carrier, the copies are transformed into bacterial host, and finally, the clones are screened for systems that enhance. Compared to those with bigger segments, metagenomic banks with short plasmid DNA in the range of 2–3 kb offer better coverage of an environment's metagenome. As per estimates, at most, 1011 genome copies will be needed to recover the genes from uncommon microorganisms (Reddy et al. 2013).

When rebuilding metagenomes for genomic research and screening for traits that are coded by genetic markers, DNA molecule banks are very helpful. When examining multi-gene biological reactions, large sequencing collections (100–200 kb) are preferred (Reddy et al. 2013; Thomas, Gilbert, and Meyer 2012). Operational testing of produced phenotypes or set of images in metagenomic research, which includes massive high-throughput scanning, may be employed to filter genome collections. Sets of images in the entire metagenomic sequencing reveal a variety of significant genomic properties, including genomic organizations, the duplication of activities in a population, and traits that are gained by lateral gene exchanges between different species (Handelsman 2004). Libraries are evaluated in operational genome organization (metagenomic analysis driven by function) depending on transcription of a chosen phenotype on the particular medium. Atmospheric metagenomic libraries have uncovered a broad range of metabolic activities. For instance, new antibiotics (such as terragine and turbomycin), bacterial catalysts (e.g., amylases, lipases, cellulases), and other proteins have been found in soil metagenomic libraries, including antiporters (Navgire et al. 2022; de Boer et al. 2021).

The effective production of a target gene in a foreign host, such as *E. coli*, is a prerequisite for metagenomic techniques. As a result, most of external proteins in E. coli express themselves at quite reduced numbers or not at all. In some instances, metagenomic information can be transformed into number of other substitute hosts, such as *Streptomyces*, *Bacillus*, *Agrobacterium*, and *Pseudomonas*, to increase gene transcription. Very desirable methods include techniques that may improve the foreign translation of unidentified genomes within a human host. For illustration, cloned plasmids with powerful promoters that can provide additional signaling molecules or genetic modification of E. coli that can enable the translation and production of a broad variety of genes are both extremely coveted (Fakruddin et al. 2013; Handelsman 2004; Land et al. 2015). The proportion of functional DNA variants producing a trait in a genomic library is often quite minimal. For instance, only one of 730,000 clones in a soil-based ambient metagenomic collection demonstrated lipase activity. A unique pancreatic lipase clone's genome and deduced primary structure showed only a modest identity (50%) with recognized lipases, suggesting that it might come from an uncultivated species. Enhanced testing and identification techniques are required due to the low frequency of continuously producing copies in metagenomics databases. In the separation of genes in collected samples, library building, replication, mining, and their possible use in biotechnology, the aforementioned data should be kept in mind (Peterson and Kaur 2018; Ratcliffe et al. 2022; Vassallo et al. 2022).

3.5.1 Proteogenomics

Only if a gene sequence has a high amount of similarity to molecules that are already present in databases that are currently accessible can it be confidently identified in metaproteomics. Yet, the majority of ambient proteome polling shows only sporadic relationships between proteins and recognized library patterns (Ratcliffe et al. 2022; Peterson and Kaur 2018). Thus, it seems that most small coding regions obtained through metaproteomes go unrecognized and are unable to be attributed to the genealogical and behavioral properties. These restrictions have made the metagenomic component of the proteogenomic technique extremely important to improve the recognition of homologous proteins by genome organization compared to the identical material from which the peptides were isolated. In a study by Delmotte et al., the proteogenomics method was used to interpret phyllosphere microbial species, as presented in Figure 3.2. Proteins were recovered from bacterial biomass that was taken from the surface of the leaf of soybean, Arabidopsis, and clover. These restrictions have been circumvented, though, by integrating the metagenomic and metaproteomic techniques under the umbrella term "proteogenomics". In communal proteogenomics, complete genomes and proteomes are isolated from a single sample, allowing for a more confident connection between physiological systems and genealogical identity (Blakeley-Ruiz and Kleiner 2022; Vassallo et al. 2022). Tryptic digestion, liquid chromatography-based particle sorting, and spectrometry analysis came next. As a result, 2,883 distinct proteins were found among almost 500,000 spectra. The metagenomic information recorded from genomes isolated from an identical stream of microbial biomass greatly enhanced (up to 74%) the number of enzymes that could be identified, showing that the bulk of microbial flora found inside the rhizosphere region were physically distinct from those documented in databases. The bacterial growth species *Pseudomonas*, *Methylobacterium*, and *Sphingomonas*were responsible for the majority of the proteins that were

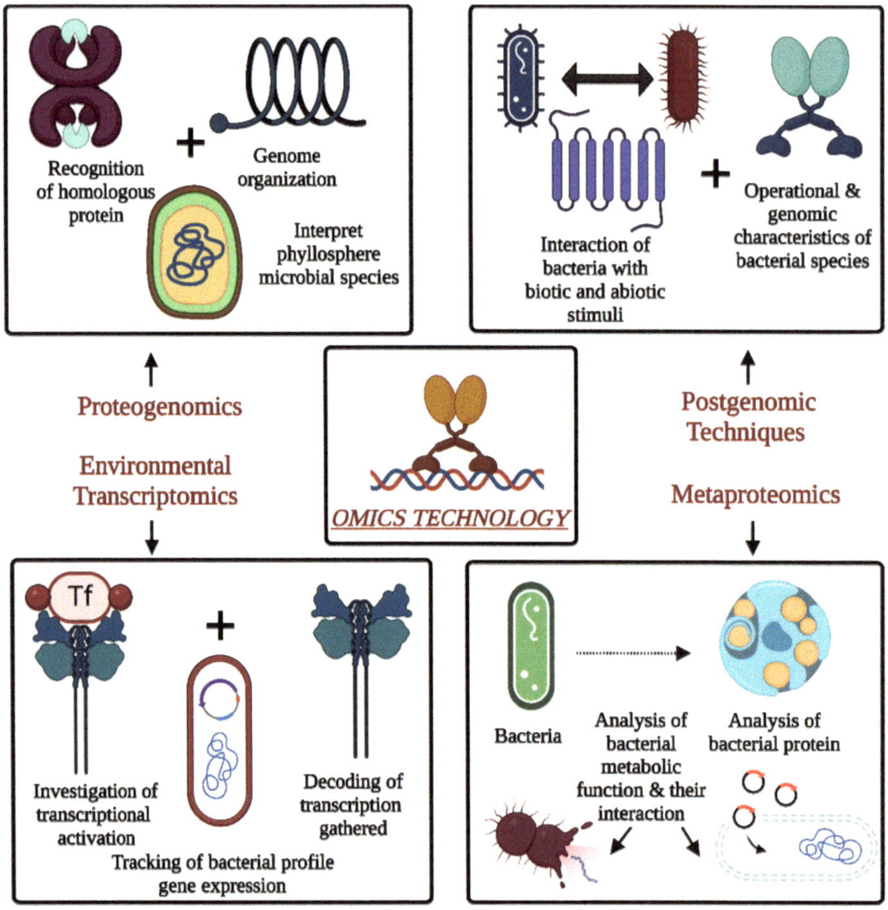

FIGURE 3.2 Summary of omics technology involving proteogenomics, environmental transcriptomics, postgenomic techniques, and metaproteomics.

discovered in the phyllosphere proteomics. *Methylobacterium* taxa capable of utilizing methanol as both a carbon and energy source have been given a huge number of proteins with a role in biodiesel synthesis (Sheynkman et al. 2016; Delmotte et al. 2009; Knief et al. 2012).

3.5.2 Environmental Transcriptomics

This method enables the investigation of worldwide transcriptional activation by randomized decoding of transcription gathered from bacterial diversity at a specific time and location, allowing the tracking of bacterial profiles of gene expression in natural settings. The measurement of modifications to gene production and their control in response to shifting climate factors is especially suited to metatranscriptomics. The primary difficulty in metatranscriptomics is that bacterial bacterial mRNA translations are not poly-A-tailed, making it difficult to acquire complementary DNA (cDNA) (Knief et al. 2012; Zhao et al. 2023). As a result, coextraction of the whole RNA pool's relatively frequent rRNA molecules occurs, which can result in an excessive amount of noisy genomes in a sizable sequencing study. In order to analyze the genetic components of populations of freshwater and marine numerous bacteria, a technique for preferentially enhancing mRNA by layer-by-layer hybridization of rRNA was created and assessed (Gałązka et al. 2018; Lloyd, Lapham, and Teske 2006). This showed the existence of several genes that were connected to biogeochemical activities including sulfur oxidation (soxA), absorption of C1 chemicals (fdh1B), and nitrogen absorption through polyamine breakdown (aphA). Since a society's total RNA pool is inherently rich in both physiologically and phylogenetically relevant components, such as mRNA and rRNA, an analysis technique known as "double-RNA" has been developed more recently. This gave researchers a way to look at microorganisms' structural and metabolic activity in a controlled trial. The research done by José F. Siqueira et al., (2012) used extensive pyrosequencing and transcriptome profiling to generate 193,219 rRNA labels plus 21,133 mRNA labels using sandy soil specimens that had low nutrient content and stable pH. Proteobacteria and Actinobacteria were the most prevalent microorganisms in samples collected, according to the rRNA tags that provided information on the phylogenetic structure of those communities. Crenarchaeota were less prevalent. The in situ transcription of numerous important enzymes involved, including nitrite reductase (nirk) and ammonia monooxygenase (amoA and amoC) implicated in ammonification, could be seen thanks to the mRNA labels. Moreover, bacterial gene sequences for the catalysts 4-hydroxybutyryl-CoA dehydratase and methylmalonyl-CoA mutase, which are involved in Crenarchaeota's carbon fixation processes, were found (Dalmaso, Ferreira, and Vermelho 2015; Suominen et al. 2022). The lack of poly-A ends in bacterial mRNA sequences, which makes it challenging to produce cDNA, is a significant drawback of this method. Current developments in metatranscriptomics have resulted in the development of a "double RNA" technique, in which the whole phylogenetic pool of the taxonomic groups and important biological microbial population is analyzed using the commercially rich RNA, i.e., both mRNA and rRNA. Hence, the area of microbiological identification and its intrinsic recognition has been perfectly enhanced as a consequence of recent advancements in molecular techniques. Both the functional and structural properties of microbes have been shown to be quite diverse. The quantitative evaluation of the enormous diversity of bacterial diversity found in our environment presents a significant difficulty in this research (Miller et al. 2007; Benammar et al. 2020; Ramin and Allison 2019).

Each methodology for selective and entire community analyses has its own benefits and restrictions. Combining the necessary methodologies may provide a better investigation of the biological diversities connected to various groupings in the microscopic ecology (Hugerth and Andersson 2017). This is also why, in presenting a comprehensive picture of mycology, an additional step should be used with various omics methods. In paddy rice soils, the potential for bilateral cross-species transfer of electrons between Methanothrix and Geobacter species in methanogenic terrestrial habitats was investigated (Lodge et al. 2014). Methanothrix is a significant bacterial factor in worldwide methane production, but little is understood about its biology and behavior. Methanothrix may contribute to methane emissions in ecological systems in ways other than just converting acetate into methane. This highlights the value of DIET as a crucial electron supplier for Methanothrix in paddy fields. With the aid of metatranscriptomics, competitive nitrogen conversion gene expressions were identified in wet paddy soil. In soggy agricultural soil, severe anoxic areas help the bacteria aggressively produce RNT, or reduced nitrogen transfer. Underwatering and the presence of strong anaerobic biochemical cycles in agricultural soil are the main hallmarks of periodic anaerobic environments (Müller et al. 2023; Vandegrift et al. 2023).

With no prior PCR processing, an RNA shotgun sequencing study was used to look into the bacterial population linked with RNT. Rice farming is threatened globally as a result of significant soil salinization brought on by coastal erosion in coastal places (Thiam et al. 2021). Even though the negative effects on rice cultivation and development have been thoroughly studied, little is understood about how severely salinity affects the makeup and functioning of microbiological rice ground communities. Thus, utilizing rice husks as a source of carbon, controlled lab studies were carried out to ascertain how they responded to full- and half-strength marine salinity. Both severe and moderate salinity stress reduced biogas and acetate synthesis by 50% and 70%, respectively, when compared with the control group. Large differences in the relative transcript quantity corresponding to neighborhood mRNA expression were seen among some of the family units (Browning and Sawyer 2021; Tarolli et al. 2023).

By examining the pool of particular eukaryotic polyadenylated mRNA that was physically separated from samples collected, researchers were able to examine the distinct activities that eukaryotes in soil could produce. Other extraction procedures can be used to recover sample soil RNA of the necessary grade. Whole soil mRNA or RNA can be converted into cDNA using elevated sequencing. By transferring polyadenylated mRNA full-length cDNAs onto vector expression systems, cDNA banks can be created. As a result, cDNA pools can be used to examine different functional genomics groups. The essential parts, including exoskeletons of residing peat arthropods, including Sphagnum-derived trash made of chitin, xylan, cellulose, and pectin, were additionally utilized to determine the biocatalysts and communities. This was done using Illumina RNA-seq of total RNA and SSU rRNA genes. On adding biomaterials to peat, microbial cell counts increased by a magnitude of five or three. Xylan and pectin were shown to have an enhanced frequency of transcribed genes, which facilitated low-carbon energy sources and energy metabolism. The action of microbial pathogens multiplied in accordance with the presence of a certain polymer. Consequently, metatranscriptomics aids in analyzing the variety of soil microbes as well as population makeup, both of which have remained largely hidden (Mettel et al. 2010; de Boer et al. 2021).

3.5.3 Postgenomic Techniques

Genetic material molecular approaches do not offer sufficient details concerning the expression of genes under in situ circumstances. As such, by utilizing metagenomic datasets, which also contain genomes from species that are uncultivated, additional insight into variability of the microbes can be discovered (Nwachukwu and Babalola 2022; Vieites et al. 2009; Nowinski et al. 2023). Consequently, postgenomic methods such as metaproteomics, metatranscriptomics, and proteogenomics can be used to determine a connection between operational and genomic characteristics of various bacterial diversity. Metagenomics, environmental genetics, communal genomics, and metagenomics all directly gather microbe genes from samples collected without the need for prior understanding of bacterial diversity. Metagenomics analyzes the entire genome of bacterial diversity that has been collected from the environment, much like when a bacterium's whole genome is sequenced (Robinson, Piel, and Sunagawa 2021; Gutleben et al. 2018; Handelsman 2004). Metagenomics can therefore be used to identify how microbes interact with both biotic and abiotic stimuli, uncultivated species, and their metabolic functions. Hence, other active compounds can be produced from functional metagenomic libraries, such as antibacterial agents (terragine) and microbiological catalysts (lipases, amylases, and cellulases). The first step in creating a metagenome pool is isolating the genes from the biological sample, which are then used as carriers for shotgun replication or the randomized micropropagation of DNA molecules. After converting the copies into a host bacterium, the viable clones are then subjected to testing (Purushothaman, Meola, and Egli 2022; Bai et al. 2021; Sangwan, Xia, and Gilbert 2016).

The activated genes' regular low-level activity is a significant drawback of this method. Consequently, for an accurate characterization of the microbiological sample and its equivalents, enhanced detection techniques and high-throughput screening are required (Rabausch et al. 2013; Ngara and Zhang 2018). Lately, metagenome scanning has been utilized to describe the functional and structural genes that the bacterial population produces in response to aerobic circumstances in acidic sulfate environments. After being incubated under aerobic environments, major changes were seen in both the source material and the topsoil. Despite a noticeable decline in archaea, sulfur cycling genes increased dramatically in parent material. At the biological level, it is yet unknown how taxonomy diversification and population makeup are related. The basic aspects of bacterial diversity in the soil are significantly affected by temperature, soil pH, and other pertinent environmental factors. To evaluate the genetic capabilities of the microbial flora, entire sequence metagenomics could be applied (L. Zhang et al. 2021; Yun and Ryu 2005).

3.5.4 Metaproteomics

Ambient proteomics, also known as metaproteomics, is the analysis of proteins at a specific moment on a vast scale as exhibited by a microbial community. Metaproteomics can be used to analyze different microbes quantitatively and qualitatively. Relative to other macromolecules like nucleotides or fats, peptides provide a wider perspective or a sharper image (Leyuan Li and Figeys 2020). Stresses cause the formation of proteofingerprints that reveal microbiological populations' functional ability. Firstly, the colonies are ultracentrifuged out from the bacterial matrix in order to separate the rich proteins, after which the cells are lysed. The creation of population proteofingerprints is then is accomplished using one- or two-dimensional electrophoresis. Subsequently, several analytical methods are used to locate the protein spots (Shrestha et al. 2021; Petriz and Franco 2017). Mass spectrometry and liquid chromatography work together to quickly identify proteins. The precise transcription factor within the soil ecosystem may be assessed extremely effectively with metaproteomics. Metaproteomics is currently being used to evaluate many environmental materials, including soil, sedimentary, water, and saltwater systems (Cortes et al. 2019).

Metaproteomics has been used to study countless different metabolic functions, and it is an effective tool for interpreting the interactions between rigid procedures. Using this technique, the enormous number of expression levels by many organisms in an environment at a certain moment can be investigated. This creates a strong functional relationship between the physiological and phylogenetic information of the microbial population (Cortes et al. 2019; Petriz and Franco 2017; Karaduta, Dvanajscak, and Zybailov 2021; Stamboulian, Canderan, and Ye 2022). On the other hand, the aforesaid procedure is adversely affected by the large diversity of microbial organisms, the diversity of the soil, and inadequate metagenomic data. The recovery of peptides from soil is seriously hampered by complex polysaccharides, phenol, and lipid components, as well as inorganic materials including silt and clay minerals. Humic substances also pose a significant problem. Peptide recognition can be a significant barrier when analyzing metaproteomics data since it depends on the strength, capacity, and database management system of the protein selected for investigation (Salvato, Hettich, and Kleiner 2021; Kleiner 2019).

Today's metaproteomics can disclose every activity in an environmentally stable setting, making it a reliable method for analyzing the surroundings. The various metabolic processes of microbes related to plants acting as metal hyperaccumulators trying to grow in serpentine land, normally polluted with cobalt, chromium, and nickel, as well as rhizospheres of high heavy metal plants, were interpreted using a column chromatography, a relatively high-resolution mass spectrometry-based metaproteomic strategy (Trindade et al. 2021). A unique specific proteins database was established based on previous identifications by 16S DNA profiling. This was subsequently utilized to identify proteins using the LC-MS results, information that when combined allows for the discovery of many proteins that match roles assigned in candidate gene categories. Proteins present in nutrition and metal movement, as well as stimulus response, show a continuity of microorganisms capable of adapting to various microenvironmental situations (Mattarozzi et al. 2017).

Microbial biocatalysts play a crucial part in the transformation to a post-petroleum microbial economy. The capacity of biocatalysts is constrained by constantly searching for rewarding activities, including evaluating genetic data. Thus, a mixture of genomic studies and metatranscriptomics makes it easier to find various biocatalysts directly in environmental sample material. Operational metaproteomics is a useful technique for analyzing proteome biocatalyst activity inside an organic substance (Warnecke and Hess 2009; Cano-Flores et al. 2020). Quantitative measurements of biomass or cell counts for different demographics also aid in the accurate identification of a microbiological group's makeup. As a result, one way to determine the makeup of microbial populations is to apply the quantity of peptides as a measurement of biomass for diverse demographics. There are currently techniques for counting cells, including illumination, in vitro experiments, and 16S rRNA gene amplicon decoding. This technique was utilized to examine microorganisms from two distinct settings, soda springs and saliva from various people. In order to comprehend the interactions between various species in an environmentally robust setting, metaproteomics is a crucial analytical technique (de Sousa et al. 2023; Lin and Tao 2017).

3.6 CONCLUSION

There is a wide variety of microbes in natural settings. Diversity can be examined at several levels, including global, local, and demographic levels, using a variety of methodologies. The field of microbial ecology has a new future thanks to the molecular viewpoint, which provides us with more than simply a view of evolutionary history. Organisms can be recognized without even being grown since molecular-phylogenetic identifiers are predicated on sequencing rather than metabolic characteristics. Hence, the investigation into organic bacterial communities can make use of a full set of images provided by molecular genetics tools. These techniques provide a deeper understanding of bacterial populations by characterizing bacterial activity. These methods can be applied long term to quantitatively study bacterial ecology and deepen our understanding of ecosystem processes. There are benefits and drawbacks to each molecular biology technique that can be used to analyze population form and function, but none of them gives complete visibility to the genomic or operational variety of intricate bacterial communities. To examine the variety, purpose, and biology of microbes, it is best to use a variety of methodologies. It is best to think of civilization and culture-independent molecular methods as complementary rather than antagonistic or exclusive. To offer fresh perspectives on microbiology, a multidisciplinary system incorporating various "omics" technologies to uncover the relationships between genomes, peptides, and external conditions will be necessary. In the future, the advancement of the multi-"omics" method will be a top focus for researchers.

REFERENCES

Bai, Xiangning, Aswathy Narayanan, Piotr Nowak, Shilpa Ray, Ujjwal Neogi, and Anders Sönnerborg. 2021. "Whole-Genome Metagenomic Analysis of the Gut Microbiome in HIV-1-Infected Individuals on Antiretroviral Therapy." *Frontiers in Microbiology* 12 (June). doi:10.3389/fmicb.2021.667718.

Benammar, L., K. İnan Bektaş, T. Menasria, A. O. Beldüz, H. I. Güler, I. K. Bedaida, J. M. Gonzalez, and A. Ayachi. 2020. "Diversity and Enzymatic Potential of Thermophilic Bacteria Associated with Terrestrial Hot Springs in Algeria." *Brazilian Journal of Microbiology* 51 (4): 1987–2007. doi:10.1007/s42770-020-00376-0.

Berg, Gabriele, Daria Rybakova, Doreen Fischer, Tomislav Cernava, Marie-Christine Champomier Vergès, Trevor Charles, Xiaoyulong Chen, et al. 2020. "Microbiome Definition Re-Visited: Old Concepts and New Challenges." *Microbiome* 8 (1): 103. doi:10.1186/s40168-020-00875-0.

Bertola, Marta, Andrea Ferrarini, and Giovanna Visioli. 2021. "Improvement of Soil Microbial Diversity through Sustainable Agricultural Practices and Its Evaluation by -Omics Approaches: A Perspective for the Environment, Food Quality and Human Safety." *Microorganisms* 9 (7): 1400. doi:10.3390/microorganisms9071400.

Bhagat, Vilas, and Amol Gholap. 2012. "Gold Nanorods: A Promising Drug Delivery Vehicle." *Inventi Rapid: NDDS [ISSN 0976-3864]* 4: 1–4.

Blakeley-Ruiz, J. Alfredo, and Manuel Kleiner. 2022. "Considerations for Constructing a Protein Sequence Database for Metaproteomics." *Computational and Structural Biotechnology Journal* 20: 937–52. doi:10.1016/j.csbj.2022.01.018.

Boer, Hugo de, Marcella Orwick Rydmark, Brecht Verstraete, and Barbara Gravendeel. 2021. *Molecular Identification of Plants: From Sequence to Species.* Pensoft Publishers. doi:10.3897/ab.e98875.

Browning, Trevor N., and Derek E. Sawyer. 2021. "Vulnerability to Watershed Erosion and Coastal Deposition in the Tropics." *Scientific Reports* 11 (1): 885. doi:10.1038/s41598-020-79402-y.

Cano-Flores, Arturo, Javier Gómez, Iker S. Escalona-Torres, and Benjamín Velasco-Bejarano. 2020. "Microorganisms as Biocatalysts and Enzyme Sources." In *Microorganisms.* IntechOpen. doi:10.5772/intechopen.90338.

Corato, Ugo De. 2020. "Soil Microbiota Manipulation and Its Role in Suppressing Soil-Borne Plant Pathogens in Organic Farming Systems under the Light of Microbiome-Assisted Strategies." *Chemical and Biological Technologies in Agriculture* 7 (1): 17. doi:10.1186/s40538-020-00183-7.

Cortes, Laetitia, Harm Wopereis, Aude Tartiere, Julie Piquenot, Joost Gouw, Sebastian Tims, Jan Knol, and Daniel Chelsky. 2019. "Metaproteomic and 16S RRNA Gene Sequencing Analysis of the Infant Fecal Microbiome." *International Journal of Molecular Sciences* 20 (6): 1430. doi:10.3390/ijms20061430.

Dalmaso, Gabriel, Davis Ferreira, and Alane Vermelho. 2015. "Marine Extremophiles: A Source of Hydrolases for Biotechnological Applications." *Marine Drugs* 13 (4): 1925–65. doi:10.3390/md13041925.

Dattatray, Tupe, and Amol D. Gholap*. 2012. "The Threat of Photoaging: Clinical Conditions, Evaluation and Treatment." *Inventi Rapid: Cosmeceuticals [ISSN 0976-3864]* 4: 1–3.

Delmotte, Nathanaël, Claudia Knief, Samuel Chaffron, Gerd Innerebner, Bernd Roschitzki, Ralph Schlapbach, Christian von Mering, and Julia A. Vorholt. 2009. "Community Proteogenomics Reveals Insights into the Physiology of Phyllosphere Bacteria." *Proceedings of the National Academy of Sciences* 106 (38): 16428–33. doi:10.1073/pnas.0905240106.

Devarapalli, Pratap, and Ranjith N. Kumavath. 2015. "Metagenomics — A Technological Drift in Bioremediation." In *Advances in Bioremediation of Wastewater and Polluted Soil.* InTech. doi:10.5772/60749.

Eisen, Jonathan A. 2007. "Environmental Shotgun Sequencing: Its Potential and Challenges for Studying the Hidden World of Microbes." Edited by Simon Levin. *PLoS Biology* 5 (3): e82. doi:10.1371/journal.pbio.0050082.

F. Ajilogba, Caroline, Olubukola O. Babalola, Patrick Adebola, and Rasheed Adeleke. 2022. "Bambara Groundnut Rhizobacteria Antimicrobial and Biofertilization Potential." *Frontiers in Plant Science* 13 (July). doi:10.3389/fpls.2022.854937.

Fakruddin, Md., Khanjada Shahnewaj Bin Mannan, and Stewart Andrews. 2013. "Viable but Nonculturable Bacteria: Food Safety and Public Health Perspective." *ISRN Microbiology* 2013 (September): 1–6. doi:10.1155/2013/703813.

Fakruddin, Md., Reaz Mohammad Mazumdar, Khanjada Shahnewaj Bin Mannan, Abhijit Chowdhury, and Md. Nur Hossain. 2013. "Critical Factors Affecting the Success of Cloning, Expression, and Mass Production of Enzymes by Recombinant E. Coli." *ISRN Biotechnology* 2013 (September): 1–7. doi:10.5402/2013/590587.

Gałązka, Anna, Jarosław Grządziel, Rafał Gałązka, Aleksandra Ukalska-Jaruga, Joanna Strzelecka, and Bożena Smreczak. 2018. "Genetic and Functional Diversity of Bacterial Microbiome in Soils With Long Term Impacts of Petroleum Hydrocarbons." *Frontiers in Microbiology* 9 (August). doi:10.3389/fmicb.2018.01923.

Garza, Daniel R., and Bas E. Dutilh. 2015. "From Cultured to Uncultured Genome Sequences: Metagenomics and Modeling Microbial Ecosystems." *Cellular and Molecular Life Sciences* 72 (22): 4287–4308. doi:10.1007/s00018-015-2004-1.

Gholap, Amol D. 2012. "Translational Nanomedicine: Status Assessment and Opportunities." *International Journal of Research in Pharmaceutical and Biomedical Sciences* 3 (4): 1454–63.

Gholap, Amol D., Deblina D. Bhowmik, Aditya Y. Deshmukh, and Navnath T. Hatvate. 2023. "Quintessential Impact of Dendrimer Bioconjugates in Targeted Drug Delivery." In . doi:10.1016/bs.coac.2023.02.009.

Gholap, Amol D., Santosh S. Borude, and Anand M. Mahajan. 2011. "Smart Crystals Technology: A Review." *Pharmacologyonline* 3: 238–43.

Gholap, Amol D., and Santosh B. Dighe. 2012a. "Dandruff: Pathogenesis And Treatment." *Inventi Rapid: Cosmeceuticals [ISSN 0976-3864]* 4: 1–4.

Gholap, Amol D., and Santosh B. Dighe. 2012b. "Role of Sebum in Human Health & Acne." *Inventi Rapid: Cosmeceuticals [ISSN 0976-3864]* 4: 1–3.

Gholap, Amol D., Juhi S. Gupta, Pallavi A. Kamandar, Gauri V. Banchhod, and Navnath T. Hatvate. 2023. "Antibody-Drug Conjugates for Cancer Therapy: An up-to-Date Review on the Chemistry and Pharmacology." In *Bioconjugated Materials Part 2 Applications in Drug Delivery, Vaccine Formulations and Important Conjugates for Cancer Therapy.* doi:10.1016/bs.coac.2023.02.007.

Gholap, Amol D., Navnath T. Hatvate, Pankaj R. Khuspe, Trushali A. Mandhare, Pooja Kashid, and Vinay D. Gaikwad. 2023. "Proteomics in Oncology: Retrospect and Prospects." In *Novel Technologies in Biosystems, Biomedical & Drug Delivery*, 243–69. Singapore: Springer Nature Singapore. doi:10.1007/978-981-99-5281-6_10.

Gholap, Amol D., and Deepak S. Musmade. 2011. "Advances in the Field of Nanooncology: An Overview." *Inventi Impact: NDDS [ISSN 0976-3791]* 4: 1–3.

Gholap, Amol D., and Dattatray Tupe. 2012. "Potential Applications of Botanical Extract." *Inventi Rapid: Ethnopharmacology [ISSN 0976-3805]* 4: 1–4.

Gholap, Amol D., and Vikas G. Rajurkar. n.d. "Potential Applications of Capillary Electromigration Methods For The Improvement Of Separation Efficiency." *Inventi Rapid: Pharm Analysis & Quality Assurance [ISSN 0976-3813]* 2012 (4): 1–3.

Gholap, Amol D., and Vikas G. Rajurkar. 2012a. "New Threat of Clinical Toxicities Evoked from Nanocarrier Systems." *Inventi Rapid: NDDS [ISSN 0976-3791]* 4: 1–4.

Gholap, Amol D., and Vikas G. Rajurkar. 2012b. "Potential Applications of Superparmagnetic Nanoparticles Technology." *Inventi Rapid: NDDS [ISSN 0976-3791]* 4: 1–3.

Gholap, Amol D., Vikas G. Rajurkar, and Santosh B. Dighe. 2012. "Phonophoresis: Ultrasound-Enhanced Transdermal Drug Delivery." *Inventi Rapid: NDDS [ISSN 0976-3791]* 4: 1–3.

Gholap, Amol D., Rushikesh P. Said, Rutuja D. Pawar, Gauri S. Ambore, and Navnath T. Hatvate. 2023. "Importance of Carbohydrate-Drug Conjugates in Vaccine Development: A Detailed Review." In *Bioconjugated Materials Part 2 Applications in Drug Delivery, Vaccine Formulations and Important Conjugates for Cancer Therapy.* doi:10.1016/bs.coac.2023.03.002.

Giovannetti, Marco, Alessandra Salvioli di Fossalunga, Ioannis A. Stringlis, Silvia Proietti, and Valentina Fiorilli. 2023. "Unearthing Soil-Plant-Microbiota Crosstalk: Looking Back to Move Forward." *Frontiers in Plant Science* 13 (January). doi:10.3389/fpls.2022.1082752.

Goswami, Madhurankhi, Purnita Bhattacharyya, Indranil Mukherjee, and Prosun Tribedi. 2017. "Functional Diversity: An Important Measure of Ecosystem Functioning." *Advances in Microbiology* 07 (01): 82–93. doi:10.4236/aim.2017.71007.

Gutleben, Johanna, Maryam Chaib De Mares, Jan Dirk van Elsas, Hauke Smidt, Jörg Overmann, and Detmer Sipkema. 2018. "The Multi-Omics Promise in Context: From Sequence to Microbial Isolate." *Critical Reviews in Microbiology* 44 (2): 212–29. doi:10.1080/1040841X.2017.1332003.

Hanashiro, Fabio Toshiro T., Shinjini Mukherjee, Caroline Souffreau, Jessie Engelen, Kristien I. Brans, Pieter Busschaert, and Luc De Meester. 2019. "Freshwater Bacterioplankton Metacommunity Structure Along Urbanization Gradients in Belgium." *Frontiers in Microbiology* 10 (April). doi:10.3389/fmicb.2019.00743.

Handelsman, Jo. 2004. "Metagenomics: Application of Genomics to Uncultured Microorganisms." *Microbiology and Molecular Biology Reviews* 68 (4): 669–85. doi:10.1128/MMBR.68.4.669-685.2004.

Hartman, Kyle, Marc W. Schmid, Natacha Bodenhausen, S. Franz Bender, Alain Y. Valzano-Held, Klaus Schlaeppi, and Marcel G. A. van der Heijden. 2023. "A Symbiotic Footprint in the Plant Root Microbiome." *Environmental Microbiome* 18 (1): 65. doi:10.1186/s40793-023-00521-w.

Hartman, Kyle, Marcel G. A. van der Heijden, Raphaël A. Wittwer, Samiran Banerjee, Jean-Claude Walser, and Klaus Schlaeppi. 2018. "Cropping Practices Manipulate Abundance Patterns of Root and Soil Microbiome Members Paving the Way to Smart Farming." *Microbiome* 6 (1): 14. doi:10.1186/s40168-017-0389-9.

Harvey, Jeffrey A., Kévin Tougeron, Rieta Gols, Robin Heinen, Mariana Abarca, Paul K. Abram, Yves Basset, et al. 2023. "Scientists' Warning on Climate Change and Insects." *Ecological Monographs* 93 (1). doi:10.1002/ecm.1553.

Hiraoka, Satoshi, Ching-chia Yang, and Wataru Iwasaki. 2016. "Metagenomics and Bioinformatics in Microbial Ecology: Current Status and Beyond." *Microbes and Environments* 31 (3): 204–12. doi:10.1264/jsme2.ME16024.

Hoegh-Guldberg, Ove, Elvira S. Poloczanska, William Skirving, and Sophie Dove. 2017. "Coral Reef Ecosystems under Climate Change and Ocean Acidification." *Frontiers in Marine Science* 4 (May). doi:10.3389/fmars.2017.00158.

Hugerth, Luisa W., and Anders F. Andersson. 2017. "Analysing Microbial Community Composition through Amplicon Sequencing: From Sampling to Hypothesis Testing." *Frontiers in Microbiology* 8 (September). doi:10.3389/fmicb.2017.01561.

Kapinusova, Gabriela, Marco A. Lopez Marin, and Ondrej Uhlik. 2023. "Reaching Unreachables: Obstacles and Successes of Microbial Cultivation and Their Reasons." *Frontiers in Microbiology* 14 (March). doi:10.3389/fmicb.2023.1089630.

Karaduta, Oleg, Zeljko Dvanajscak, and Boris Zybailov. 2021. "Metaproteomics—An Advantageous Option in Studies of Host-Microbiota Interaction." *Microorganisms* 9 (5): 980. doi:10.3390/microorganisms9050980.

Khedkar, Ajay, Dattatray Tupe, and Amol Gholap. 2012a. "Lipid Nanocapsules: A Novel Promising Tool as Drug Delivery System." *Inventi Rapid: NDDS [ISSN 0976-3791]* 4: 1–4.

Khedkar, Ajay, Dattatray Tupe, and Amol Gholap. 2012b. "Pharmacosomes: An Emerging Vesicular System." *Inventi Rapid: NDDS [ISSN 0976-3791]* 4: 1–3.

Khedkar, Ajay, Dattatray Tupe, and Amol Gholap. 2012c. "Cubosomes: Emerging Drug Delivery System." *Inventi Rapid: NDDS [ISSN 0976-3791]* 4: 1–3.

Kleiner, Manuel. 2019. "Metaproteomics: Much More than Measuring Gene Expression in Microbial Communities." *MSystems* 4 (3). doi:10.1128/mSystems.00115-19.

Knief, Claudia, Nathanaël Delmotte, Samuel Chaffron, Manuel Stark, Gerd Innerebner, Reiner Wassmann, Christian von Mering, and Julia A Vorholt. 2012. "Metaproteogenomic Analysis of Microbial Communities in the Phyllosphere and Rhizosphere of Rice." *The ISME Journal* 6 (7): 1378–90. doi:10.1038/ismej.2011.192.

Kuo, Hao-Yu, Ming-Chung Chiu, and Jui-Yu Chou. 2023. "Bacillus Predominates in the Ophiocordyceps Pseudolloydii-Infected Ants, and It Potentially Improves Protection and Utilization of the Host Cadavers." *Archives of Microbiology* 205 (1): 53. doi:10.1007/s00203-022-03385-9.

Land, Miriam, Loren Hauser, Se-Ran Jun, Intawat Nookaew, Michael R. Leuze, Tae-Hyuk Ahn, Tatiana Karpinets, et al. 2015. "Insights from 20 Years of Bacterial Genome Sequencing." *Functional & Integrative Genomics* 15 (2): 141–61. doi:10.1007/s10142-015-0433-4.

Li, Laam, Nilmini Mendis, Hana Trigui, James D. Oliver, and Sebastien P. Faucher. 2014. "The Importance of the Viable but Non-Culturable State in Human Bacterial Pathogens." *Frontiers in Microbiology* 5 (June). doi:10.3389/fmicb.2014.00258.

Li, Leyuan, and Daniel Figeys. 2020. "Proteomics and Metaproteomics Add Functional, Taxonomic and Biomass Dimensions to Modeling the Ecosystem at the Mucosal-Luminal Interface." *Molecular & Cellular Proteomics* 19 (9): 1409–17. doi:10.1074/mcp.R120.002051.

Li, Zhaogao, Weie Wen, Ming Qin, Yuqi He, Delin Xu, and Lin Li. 2022. "Biosynthetic Mechanisms of Secondary Metabolites Promoted by the Interaction Between Endophytes and Plant Hosts." *Frontiers in Microbiology* 13 (July). doi:10.3389/fmicb.2022.928967.

Lin, Baixue, and Yong Tao. 2017. "Whole-Cell Biocatalysts by Design." *Microbial Cell Factories* 16 (1): 106. doi:10.1186/s12934-017-0724-7.

Lloyd, Karen G., Laura Lapham, and Andreas Teske. 2006. "An Anaerobic Methane-Oxidizing Community of ANME-1b Archaea in Hypersaline Gulf of Mexico Sediments." *Applied and Environmental Microbiology* 72 (11): 7218–30. doi:10.1128/AEM.00886-06.

Lodge, D. Jean, Mahajabeen Padamsee, P. Brandon Matheny, M. Catherine Aime, Sharon A. Cantrell, David Boertmann, Alexander Kovalenko, et al. 2014. "Molecular Phylogeny, Morphology, Pigment Chemistry and Ecology in Hygrophoraceae (Agaricales)." *Fungal Diversity* 64 (1): 1–99. doi:10.1007/s13225-013-0259-0.

Magar, Tushar V., Amol D. Gholap, and Deepak S. Musmade. 2011. "Elastic Vesicles as Drugs Carriers Through The Skin." *Inventi Rapid: NDDS [ISSN 0976-3791]* 4: 1–2.

Mandal, Ronit, Xanyar Mohammadi, Artur Wiktor, Anika Singh, and Anubhav Pratap Singh. 2020. "Applications of Pulsed Light Decontamination Technology in Food Processing: An Overview." *Applied Sciences* 10 (10): 3606. doi:10.3390/app10103606.

Mattarozzi, Monica, Marcello Manfredi, Barbara Montanini, Fabio Gosetti, Anna Maria Sanangelantoni, Emilio Marengo, Maria Careri, and Giovanna Visioli. 2017. "A Metaproteomic Approach Dissecting Major Bacterial Functions in the Rhizosphere of Plants Living in Serpentine Soil." *Analytical and Bioanalytical Chemistry* 409 (9): 2327–39. doi:10.1007/s00216-016-0175-8.

Mettel, Carsten, Yongkyu Kim, Pravin Malla Shrestha, and Werner Liesack. 2010. "Extraction of MRNA from Soil." *Applied and Environmental Microbiology* 76 (17): 5995–6000. doi:10.1128/AEM.03047-09.

Michaelis, Claudia, and Elisabeth Grohmann. 2023. "Horizontal Gene Transfer of Antibiotic Resistance Genes in Biofilms." *Antibiotics* 12 (2): 328. doi:10.3390/antibiotics12020328.

Miller, William G., Craig T. Parker, Marc Rubenfield, George L. Mendz, Marc M. S. M. Wösten, David W. Ussery, John F. Stolz, et al. 2007. "The Complete Genome Sequence and Analysis of the Epsilonproteobacterium Arcobacter Butzleri." Edited by Cecile Fairhead. *PLoS One* 2 (12): e1358. doi:10.1371/journal.pone.0001358.

Mishra, Apurva, Lal Singh, and Dharmesh Singh. 2023. "Unboxing the Black Box—One Step Forward to Understand the Soil Microbiome: A Systematic Review." *Microbial Ecology* 85 (2): 669–83. doi:10.1007/s00248-022-01962-5.

Mitra, Mautusi, Kevin Manoap-Anh-Khoa Nguyen, Taylor Wayland Box, Taylor Lynne Berry, and Megumi Fujita. 2021. "Isolation and Characterization of a Heavy Metal- and Antibiotic-Tolerant Novel Bacterial Strain from a Contaminated Culture Plate of Chlamydomonas Reinhardtii, a Green Micro-Alga." *F1000Research* 10 (September): 533. doi:10.12688/f1000research.53779.2.

Moënne-Loccoz, Yvan, Patrick Mavingui, Claude Combes, Philippe Normand, and Christian Steinberg. 2015. "Microorganisms and Biotic Interactions." In *Environmental Microbiology: Fundamentals and Applications*, 395–444. Dordrecht: Springer Netherlands. doi:10.1007/978-94-017-9118-2_11.

Morales-Cedeño, Luzmaria R., Ma. del Carmen Orozco-Mosqueda, Pedro D. Loeza-Lara, Fannie I. Parra-Cota, Sergio de los Santos-Villalobos, and Gustavo Santoyo. 2021. "Plant Growth-Promoting Bacterial Endophytes as Biocontrol Agents of Pre- and Post-Harvest Diseases: Fundamentals, Methods of Application and Future Perspectives." *Microbiological Research* 242 (January): 126612. doi:10.1016/j.micres.2020.126612.

Müller, Markus, Ursula Kües, Katharina B. Budde, and Oliver Gailing. 2023. "Applying Molecular and Genetic Methods to Trees and Their Fungal Communities." *Applied Microbiology and Biotechnology* 107 (9): 2783–2830. doi:10.1007/s00253-023-12480-w.

Navgire, Gauri S., Neha Goel, Gifty Sawhney, Mohit Sharma, Prashant Kaushik, Yugal Kishore Mohanta, Tapan Kumar Mohanta, and Ahmed Al-Harrasi. 2022. "Analysis and Interpretation of Metagenomics Data: An Approach." *Biological Procedures Online* 24 (1): 18. doi:10.1186/s12575-022-00179-7.

Ngara, Tanyaradzwa Rodgers, and Houjin Zhang. 2018. "Recent Advances in Function-Based Metagenomic Screening." *Genomics, Proteomics & Bioinformatics* 16 (6): 405–15. doi:10.1016/j.gpb.2018.01.002.

Nowinski, Brent, Xiaoyuan Feng, Christina M. Preston, James M. Birch, Haiwei Luo, William B. Whitman, and Mary Ann Moran. 2023. "Ecological Divergence of Syntopic Marine Bacterial Species Is Shaped by Gene Content and Expression." *The ISME Journal* 17 (6): 813–22. doi:10.1038/s41396-023-01390-4.

Nwachukwu, Blessing Chidinma, and Olubukola Oluranti Babalola. 2022. "Metagenomics: A Tool for Exploring Key Microbiome With the Potentials for Improving Sustainable Agriculture." *Frontiers in Sustainable Food Systems* 6 (June). doi:10.3389/fsufs.2022.886987.

Pan, Hanxu, and Qing Ren. 2022. "Wake Up! Resuscitation of Viable but Nonculturable Bacteria: Mechanism and Potential Application." *Foods* 12 (1): 82. doi:10.3390/foods12010082.

Peterson, Elizabeth, and Parjit Kaur. 2018. "Antibiotic Resistance Mechanisms in Bacteria: Relationships Between Resistance Determinants of Antibiotic Producers, Environmental Bacteria, and Clinical Pathogens." *Frontiers in Microbiology* 9 (November). doi:10.3389/fmicb.2018.02928.

Petriz, Bernardo A., and Octávio L. Franco. 2017. "Metaproteomics as a Complementary Approach to Gut Microbiota in Health and Disease." *Frontiers in Chemistry* 5 (January). doi:10.3389/fchem.2017.00004.

Pienaar, Jennifer A., Atheesha Singh, and Tobias G. Barnard. 2016. "The Viable but Non-Culturable State in Pathogenic Escherichia Coli: A General Review." *African Journal of Laboratory Medicine* 5 (1). doi:10.4102/ajlm.v5i1.368.

Purushothaman, Srinithi, Marco Meola, and Adrian Egli. 2022. "Combination of Whole Genome Sequencing and Metagenomics for Microbiological Diagnostics." *International Journal of Molecular Sciences* 23 (17): 9834. doi:10.3390/ijms23179834.

Rabausch, U., J. Juergensen, N. Ilmberger, S. Böhnke, S. Fischer, B. Schubach, M. Schulte, and W. R. Streit. 2013. "Functional Screening of Metagenome and Genome Libraries for Detection of Novel Flavonoid-Modifying Enzymes." *Applied and Environmental Microbiology* 79 (15): 4551–63. doi:10.1128/AEM.01077-13.

Rajurkar, Vikas G., and Amol D Gholap. 2012. "Recent Technologies in Nanomedicine for Treatment of Cancer." *Inventi Rapid: Pharm Tech [ISSN 0976-3783]* 4: 1–3.

Ramamurthy, Thandavarayan, Amit Ghosh, Gururaja P. Pazhani, and Sumio Shinoda. 2014. "Current Perspectives on Viable but Non-Culturable (VBNC) Pathogenic Bacteria." *Frontiers in Public Health* 2 (July). doi:10.3389/fpubh.2014.00103.

Ramin, Kelly I., and Steven D. Allison. 2019. "Bacterial Tradeoffs in Growth Rate and Extracellular Enzymes." *Frontiers in Microbiology* 10 (December). doi:10.3389/fmicb.2019.02956.

Ratcliffe, Norman A., João P. Furtado Pacheco, Paul Dyson, Helena Carla Castro, Marcelo S. Gonzalez, Patricia Azambuja, and Cicero B. Mello. 2022. "Overview of Paratransgenesis as a Strategy to Control Pathogen Transmission by Insect Vectors." *Parasites & Vectors* 15 (1): 112. doi:10.1186/s13071-021-05132-3.

Reddy, Amitha P., Christopher W. Simmons, Patrik D'haeseleer, Jane Khudyakov, Helcio Burd, Masood Hadi, Blake A. Simmons, Steven W. Singer, Michael P. Thelen, and Jean S. VanderGheynst. 2013. "Discovery of Microorganisms and Enzymes Involved in High-Solids Decomposition of Rice Straw Using Metagenomic Analyses." Edited by Mark R. Liles. *PLoS One* 8 (10): e77985. doi:10.1371/journal.pone.0077985.

Robinson, Serina L., Jörn Piel, and Shinichi Sunagawa. 2021. "A Roadmap for Metagenomic Enzyme Discovery." *Natural Product Reports* 38 (11): 1994–2023. doi:10.1039/D1NP00006C.

Salamon, Sylwia, Katarzyna Mikołajczak, and Lidia Błaszczyk. 2023. "Constellation of the Endophytic Mycobiome in Spring and Winter Wheat Cultivars Grown under Various Conditions." *Scientific Reports* 13 (1): 6089. doi:10.1038/s41598-023-33195-y.

Salvato, Fernanda, Robert L. Hettich, and Manuel Kleiner. 2021. "Five Key Aspects of Metaproteomics as a Tool to Understand Functional Interactions in Host-Associated Microbiomes." Edited by Deborah A. Hogan. *PLoS Pathogens* 17 (2): e1009245. doi:10.1371/journal.ppat.1009245.

Sangwan, Naseer, Fangfang Xia, and Jack A. Gilbert. 2016. "Recovering Complete and Draft Population Genomes from Metagenome Datasets." *Microbiome* 4 (1): 8. doi:10.1186/s40168-016-0154-5.

Santoyo, Gustavo. 2022. "How Plants Recruit Their Microbiome? New Insights into Beneficial Interactions." *Journal of Advanced Research* 40 (September): 45–58. doi:10.1016/j.jare.2021.11.020.

Sharma, Indu. 2021. "Bioremediation Techniques for Polluted Environment: Concept, Advantages, Limitations, and Prospects." In *Trace Metals in the Environment - New Approaches and Recent Advances*. IntechOpen. doi:10.5772/intechopen.90453.

Sharma, Satish, James Mohler, Supriya D. Mahajan, Stanley A. Schwartz, Liana Bruggemann, and Ravikumar Aalinkeel. 2023. "Microbial Biofilm: A Review on Formation, Infection, Antibiotic Resistance, Control Measures, and Innovative Treatment." *Microorganisms* 11 (6): 1614. doi:10.3390/microorganisms11061614.

Sheynkman, Gloria M., Michael R. Shortreed, Anthony J. Cesnik, and Lloyd M. Smith. 2016. "Proteogenomics: Integrating Next-Generation Sequencing and Mass Spectrometry to Characterize Human Proteomic Variation." *Annual Review of Analytical Chemistry* 9 (1): 521–45. doi:10.1146/annurev-anchem-071015-041722.

Shrestha, Him K., Manasa R. Appidi, Manuel I. Villalobos Solis, Jia Wang, Dana L. Carper, Leah Burdick, Dale A. Pelletier, Mitchel J. Doktycz, Robert L. Hettich, and Paul E. Abraham. 2021. "Metaproteomics Reveals Insights into Microbial Structure, Interactions, and Dynamic Regulation in Defined Communities as They Respond to Environmental Disturbance." *BMC Microbiology* 21 (1): 308. doi:10.1186/s12866-021-02370-4.

Singh, Radha, and Ashok K. Dubey. 2018. "Diversity and Applications of Endophytic Actinobacteria of Plants in Special and Other Ecological Niches." *Frontiers in Microbiology* 9 (August). doi:10.3389/fmicb.2018.01767.

Siqueira JF Jr, Fouad AF, Rôças IN. 2012. "Pyrosequencing as a tool for better understanding of human microbiomes." *J Oral Microbiol* 4. doi: 10.3402/jom.v4i0.10743.

Sousa, Ronaldo Rodrigues de, Rui de Paula Vieira de Castro, Nadinne Medeiros Assis, Ayla Sant'Ana da Silva, Denise Maria Guimarães Freire, Roberto Fernandez-Lafuente, and Viridiana Santana Ferreira-Leitão. 2023. "Technical–Economic Assessment—The Missing Piece for Increasing the Attractiveness of Applied Biocatalysis in Ester Syntheses?" *Catalysts* 13 (2): 223. doi:10.3390/catal13020223.

Stamboulian, Moses, Jamie Canderan, and Yuzhen Ye. 2022. "Metaproteomics as a Tool for Studying the Protein Landscape of Human-Gut Bacterial Species." Edited by Luis Pedro Coelho. *PLoS Computational Biology* 18 (3): e1009397. doi:10.1371/journal.pcbi.1009397.

Subramani, Ramesh, and William Aalbersberg. 2013. "Culturable Rare Actinomycetes: Diversity, Isolation and Marine Natural Product Discovery." *Applied Microbiology and Biotechnology* 97 (21): 9291–9321. doi:10.1007/s00253-013-5229-7.

Suominen, Saara, Gonzalo V. Gomez-Saez, Thorsten Dittmar, Jaap S. Sinninghe Damsté, and Laura Villanueva. 2022. "Interplay between Microbial Community Composition and Chemodiversity of Dissolved Organic Matter throughout the Black Sea Water Column Redox Gradient." *Limnology and Oceanography* 67 (2): 329–47. doi:10.1002/lno.11995.

Tarolli, Paolo, Jian Luo, Eugenio Straffelini, Yuei-An Liou, Kim-Anh Nguyen, Rodolfo Laurenti, Roberta Masin, and Vincenzo D'Agostino. 2023. "Saltwater Intrusion and Climate Change Impact on Coastal Agriculture." Edited by Debora Walker. *PLoS Water* 2 (4): e0000121. doi:10.1371/journal.pwat.0000121.

Thiam, Sophie, Grace B. Villamor, Laurice C. Faye, Jean Henri Bienvenue Sène, Badabate Diwediga, and Nicholas Kyei-Baffour. 2021. "Monitoring Land Use and Soil Salinity Changes in Coastal Landscape: A Case Study from Senegal." *Environmental Monitoring and Assessment* 193 (5): 259. doi:10.1007/s10661-021-08958-7.

Thomas, Torsten, Jack Gilbert, and Folker Meyer. 2012. "Metagenomics - a Guide from Sampling to Data Analysis." *Microbial Informatics and Experimentation* 2 (1): 3. doi:10.1186/2042-5783-2-3.

Timmis, Kenneth, and Juan Luis Ramos. 2021. "The Soil Crisis: The Need to Treat as a Global Health Problem and the Pivotal Role of Microbes in Prophylaxis and Therapy." *Microbial Biotechnology* 14 (3): 769–97. doi:10.1111/1751-7915.13771.

Trindade, Felipe Costa, Markus Gastauer, Silvio Junio Ramos, Cecílio Frois Caldeira, Josiney Farias de Araújo, Guilherme Oliveira, and Rafael Borges da Silva Valadares. 2021. "Soil Metaproteomics as a Tool for Environmental Monitoring of Minelands." *Forests* 12 (9): 1158. doi:10.3390/f12091158.

Tu, Kai-Wen, Ming-Chung Chiu, Wei-Jiun Lin, Yen-Ping Hsueh, Chung-Chi Lin, and Jui-Yu Chou. 2021. "Possible Impacts of the Predominant Bacillus Bacteria on the Ophiocordyceps Unilateralis s. l. in Its Infected Ant Cadavers." *Scientific Reports* 11 (1): 22695. doi:10.1038/s41598-021-02094-5.

Vandegrift, R., D. S. Newman, B. T. M. Dentinger, R. Batallas-Molina, N. Dueñas, J. Flores, P. Goyes, et al. 2023. "Richer than Gold: The Fungal Biodiversity of Reserva Los Cedros, a Threatened Andean Cloud Forest." *Botanical Studies* 64 (1): 17. doi:10.1186/s40529-023-00390-z.

Vassallo, Christopher N., Christopher R. Doering, Megan L. Littlehale, Gabriella I. C. Teodoro, and Michael T. Laub. 2022. "A Functional Selection Reveals Previously Undetected Anti-Phage Defence Systems in the E. Coli Pangenome." *Nature Microbiology* 7 (10): 1568–79. doi:10.1038/s41564-022-01219-4.

Veresoglou, Stavros D., John M. Halley, and Matthias C. Rillig. 2015. "Extinction Risk of Soil Biota." *Nature Communications* 6 (1): 8862. doi:10.1038/ncomms9862.

Vidal, Paula, Mónica Martínez-Martínez, Laura Fernandez-Lopez, Sergi Roda, Celia Méndez-García, Olga V. Golyshina, Víctor Guallar, Ana I. Peláez, and Manuel Ferrer. 2022. "Metagenomic Mining for Esterases in the Microbial Community of Los Rueldos Acid Mine Drainage Formation." *Frontiers in Microbiology* 13 (May). doi:10.3389/fmicb.2022.868839.

Vieites, José M., María-Eugenia Guazzaroni, Ana Beloqui, Peter N. Golyshin, and Manuel Ferrer. 2009. "Metagenomics Approaches in Systems Microbiology." *FEMS Microbiology Reviews* 33 (1): 236–55. doi:10.1111/j.1574-6976.2008.00152.x.

Walder, Florian, Marc W. Schmid, Judith Riedo, Alain Y. Valzano-Held, Samiran Banerjee, Lucie Büchi, Thomas D. Bucheli, and Marcel G. A. van der Heijden. 2022. "Soil Microbiome Signatures Are Associated with Pesticide Residues in Arable Landscapes." *Soil Biology and Biochemistry* 174 (November): 108830. doi:10.1016/j.soilbio.2022.108830.

Warnecke, Falk, and Matthias Hess. 2009. "A Perspective: Metatranscriptomics as a Tool for the Discovery of Novel Biocatalysts." *Journal of Biotechnology* 142 (1): 91–95. doi:10.1016/j.jbiotec.2009.03.022.

Yun, Jiae, and Sangryeol Ryu. 2005. "Screening for Novel Enzymes from Metagenome and SIGEX, as a Way to Improve It." *Microbial Cell Factories* 4 (1): 8. doi:10.1186/1475-2859-4-8.

Zhang, Biao, Kai Xue, Shutong Zhou, Kui Wang, Wenjing Liu, Cong Xu, Lizhen Cui, et al. 2022. "Environmental Selection Overturns the Decay Relationship of Soil Prokaryotic Community over Geographic Distance across Grassland Biotas." *ELife* 11 (January). doi:10.7554/eLife.70164.

Zhang, Lu, FengXin Chen, Zhan Zeng, Mengjiao Xu, Fangfang Sun, Liu Yang, Xiaoyue Bi, et al. 2021. "Advances in Metagenomics and Its Application in Environmental Microorganisms." *Frontiers in Microbiology* 12 (December). doi:10.3389/fmicb.2021.766364.

Zhao, Mingxia, Zhao Peng, Yang Qin, Tej Man Tamang, Ling Zhang, Bin Tian, Yueying Chen, et al. 2023. "Bacterium-Enabled Transient Gene Activation by Artificial Transcription Factors for Resolving Gene Regulation in Maize." *The Plant Cell* 35 (8): 2736–49. doi:10.1093/plcell/koad155.

Zwanzig, Martin. 2021. "The Ecology of Plasmid-Coded Antibiotic Resistance: A Basic Framework for Experimental Research and Modeling." *Computational and Structural Biotechnology Journal* 19: 586–99. doi:10.1016/j.csbj.2020.12.027.

4 Fecal Microbiota Transplantation as a Perspective of Microbiome Engineering

A Vijaya Chitra
Sri Ramakrishna College of Arts and Science for Women
Coimbatore

S Selvajeyanthi
Theeran Chinnamalai College of Arts and Science for Women
Tirupur

K Karthikeyan
Gujarat Institute of Desert Ecology
Gujarat

K Nanthakumar
The Energy and Resources Institute (TERI)
New Delhi

P M Ayyasamy
Periyar University, Salem

4.1 INTRODUCTION

Microbial communities colonize all over the human body, especially in the distal gut, where trillions of microorganisms are present. These indigenous microorganisms, called microbiota, are referred to as a hidden metabolic organ, as they have a direct and indirect influence on host metabolism, physiology, and immunity (Ley *et al.,* 2008). The gut microbiome plays a vital role in the physiology and metabolism of humans. Numerous studies have provided insight into the connection between the gut microbiome and its benefits. Projects such as the Human Microbiome Project and the MetaHit project also aim to elucidate the role of microbiomes in human health and their impact in causing diseases. The normal gut microbiome of healthy individuals predominantly harbors bacteria *Bacteroidetes*, *Firmicutes*, *Actinobacteria*, *Proteobacteria*,and *Verrucomicrobiota* (Jimenez-Avalos *et al.*, 2021) as well as other microorganisms such as fungi, viruses, and archaea (Soo *et al.*, 2020). Though a conserved core gut microbiota exists, each individual will harbor a unique

DOI: 10.1201/9781003394662-4

assembly of the microbial community depending on the environment and the host. This eubiosis state provides a harmonicharmonious and mutual relationship between the microbiome and the host, favoring human health.

In recent years, the gut microbiome has been altered due to many factors. A significant rationale is the inappropriate and frequent usage of antibiotics by humans and in animal feeds. This has led to an increase in antibiotic resistance and has also disrupted the natural gut microbiota of humans. Other major factors that contribute to the alteration of the gut microbiome include lifestyle and food habits. A change from eubiotic microbial ecosystems results in a condition called dysbiosis, which results in the enrichment of a few microorganisms and the reduction of others, functional imbalance, and a decline in activity exceeding its resistance and resilience capabilities. Perturbation of the gut microbiota is linked to various diseases such as constipation, recurrent *Clostridioides difficile* infection (rCDI), inflammatory bowel disease (IBD), neurological disorders, irritable bowel syndrome (IBS), hepatic encephalopathy, metabolic syndrome, autoimmunity, asthma, and allergic diseases (Jimenez-Avalos *et al.*, 2021).

4.2 FUNCTIONS OF THE GUT MICROBIOME

The human gut microbiome is an intricate ecology that serves vital purposes for its host. The gut microbiome plays a major role in metabolism, immunity, and protection against pathogens. The structural integrity of the gut mucosal barrier is maintained by the gut microbiome, and the microbiome has a symbiotic relationship with the gut mucosa. The gut microbiota of each person has a unique network of microorganisms and initially depends on the mode of delivery followed by breast milk feeding. In later stages, diet, environmental exposure, and use of antibiotics determine the microbiota network of an individual.

Recent extensive metagenomic research has shed light on its structure and potential functions. The intricate communities of microbes that live in the human gastrointestinal system are becoming important regulators of human health and illness. The fact that the human host receives a number of crucial services from the gut microbiota attests to its significance. The microbiota offers many benefits to the host through a range of physiological functions such as strengthening gut integrity and shaping the intestinal epithelium (Natividad and Verdu, 2013), harvesting energy (den Besten *et al.*, 2013), protecting against pathogens (Baumler and Sperandio, 2016), and regulating host immunity (Gensollen *et al.*, 2016). The gut microbiota imparts substantial metabolic, immunological, and gut protective functions in a healthy individual. The gut microbiota, which derives its nutrients from host dietary components and shed epithelial cells, is an organ by itself with extensive metabolic capability and substantial functional plasticity (Sonnenburg *et al.*, 2005)

The gut bacteria can produce a wide range of vitamins, amino acids, and can perform bile biotransformation (Vyas and Ranganathan, 2012). The microbiome also offers crucial biochemical pathways for the metabolism of indigestible carbohydrates, such as large polysaccharides like resistant starches, cellulose, hemicellulose, pectins, and gums; some oligosaccharides that are indigestible; unabsorbed sugars and alcohols from the diet (Cummings *et al.*, 1987); and host-derived mucins (Koropatkin *et al.*, 2012). As a result of this capacity, the host can recover energy and absorbable substrates.

The gut microbiota acquires nutrients from the host carbohydrate and breaks down these with its enzymes. During this process, the indigestible oligosaccharides are converted to short-chain fatty acids (SCFAs) such as butyrate, acetate, and propionate (Macfarlane and Macfarlane, 2003; Sartor, 2008), which are utilized by the human body. Butyrate prevents the buildup of toxic products in the body, and the oxalate that is produced during carbohydrate fermentation is countered by the gut microbiota, preventing oxalate stones in the kidney (Magwira *et al.*, 2012).

Microbes swap from energy-rich carbohydrates to energetically less favorable sources such as amino acids or dietary fats (Cummings and Macfarlane, 1991; Wall *et al.*, 2009) during the demand for carbohydrates, resulting in reduced fermentative activity of the microbiota and SCFAs as minor end products (Russell *et al.*, 2011). The gut microbiota has a key role in lipid metabolism, the synthesis of vitamin K and components of vitamin B, and is also enriched with productive protein metabolizing machinery. Some gut microbiota produce compounds such as 3-indolepropionic acid (IPA), a strong antioxidant that protects the nervous system from damage (Bendheim *et al.*, 2002).

Host cells in the gut wall have attachment sites that pathogenic bacteria can use to enter the epithelial cells, and Guarner and Malagelada (2003) described an effect known as the barrier or competitive-exclusion effect, wherein intestinal bacteria produce antimicrobial compounds and compete for nutrients and sites of attachment in the gut lining, preventing colonization by pathogens. Additionally, the researchers noted that many gastrointestinal bacteria possess the potential to synthesize antibiotic compounds known as bacteriocins, which can prevent the growth of rival bacteria.

Having such multifaceted perspectives, the human gut microbiota changes over the course of a lifetime and is thought to be crucial to both health and disease. In a healthy state, the gut microbiota performs a wide range of beneficial tasks, including energy recovery through the metabolism of nondigestible dietary components, host defense against pathogenic invasion, and immune system modulation. Considering all these facts, it is clear that the ability of the gastrointestinal tract depends on the health of the gut microbial flora that is involved in the digestion of nutrients and metabolites such as fatty acids, bile acids, and amino acids (Rinninella *et al.*, 2019).

4.3 DYSBIOSIS AND STRATEGIES FOR FECAL TRANSPLANTATION

The term "microbiome" refers to the collection of genes found in all microorganisms that occupy practically every component of the human body. The microbiome is viewed as a second genome that is symbiotic with the host. These connections and microbiome interactions contribute to playing a crucial role in human health, whether positive, negative, pathogenic, or neutral. The intricate and diverse microbiome functions as a functional expansion of the host genome, with an estimated 50–100 times increase in gene content (Hooper *et al.*, 2001).

The microbial diversity of the gut microbiota is crucial for multiple functions, and any imbalance in microbial diversity, functions, and changes in metabolic products will result in a condition called dysbiosis. This is associated with inflammatory bowel diseases such as Crohn's disease and ulcerative colitis, type 2 diabetes (T2D), metabolic disorders, obesity, celiac disease, cardiovascular disease, and colorectal cancer (CRC). During dysbiosis, microbial composition alteration occurs with an ascending trend of proinflammatory species and a descending rate of gut microbial diversity. This changes the influence of inflammatory pathways in the elderly and contributes to chronic inflammation. Long-term antibiotic usage alters the normal healthy gut microbiota, resulting in dysbiosis and also producing multidrug-resistant organisms through horizontal transfer of resistant genes. With multiple negative effects correlated with dysbiosis, appropriate and effective methods are mandatory for efficient treatment. Restoring a balanced microbial population is a possible treatment approach for disorders associated with gut dysbiosis. Fecal microbiota transplantation (FMT), which involves introducing feces from a healthy donor into a patient's digestive tract to provide therapeutic effects, is anticipated to be a key factor in the widespread therapeutic restoration of the gut microbiota. The process involves infusion or engraftment of liquid filtrate feces from a healthy donor into the gut of a recipient but is not restricted to this method, as many other modes of FMT are also available. In the following section, we will briefly review the strategies of FMT that are currently used.

4.4 STRATEGIES FOR FECAL TRANSPLANTATION

FMT is a medical procedure that involves the transfer of fecal material from a healthy donor to a recipient's gastrointestinal tract to restore the recipient's gut microbiome. FMT has shown promising results in treating a range of gastrointestinal disorders, including *Clostridium difficile* infection (CDI), inflammatory bowel disease, and irritable bowel syndrome. However, there are some challenges associated with FMT, including the risk of infectious disease transmission and the variability in the composition of the donor microbiota. As a result, there is ongoing research to develop engineered fecal transplants that address these challenges. The various strategies being explored for engineering fecal transplants are discussed in the following sections.

4.4.1 Synthetic Fecal Microbiota

One of the strategies being explored for engineering fecal transplants is the development of synthetic fecal microbiota. Synthetic microbiota are designed by selecting specific bacteria, viruses, and other microorganisms that are known to be beneficial for the treatment of specific diseases. Synthetic fecal microbiota provide controlled and consistent microbiota composition compared to traditional fecal transplants and can be tailored to meet individual needs. Synthetic microbiota reduce the risk of infectious disease transmission, as they are designed with sterile laboratory cultures.

Synthetic microbiota can be designed using a bottom-up approach or a top-down approach. The top-down approach involves designing the microbiota based on metabolic activity. The microbiota of a healthy individual is analyzed, and the microbial species and metabolic pathways that are beneficial for health are identified. The preliminary process of the top-down approach includes inoculum and microbial composition profiling of treated or untreated fecal material, followed by the selection of microbes. The selected bacteria are added to a murine model, isolated, and sequenced for further iteration cycles for the complete interpretation and description of adapted synthetic microbial communities (SynComs) (Atarashi *et al.*, 2013). Optimization of synthetic interspecific relationships can be achieved with omics data, such as amplicons, and metagenomics that identify molecular interaction networks on the basis of correlated molecular behavior, and end level optimization is accomplished by the genome-scale metabolic model (GEM), a computational method. Prior knowledge of the community's functioning is not required to implement top-down attempts to engineer biological systems studies (Leon and Nogales, 2022).

In the bottom-up approach, individual microorganisms are selected based on their metabolic activity, colonization, and immune-modulatory properties, and then combined to create a synthetic microbiota. Unlike the top-down approach, prior knowledge of individual component performance is critical for bottom-up approaches. The bottom-up approach uses existing knowledge of the metagenome, abundances, and growth parameters of candidate microorganisms. SynCom or the candidate microorganisms may be designed to have specific functions and can be altered depending on need (Pim *et al.*, 2023). The designed SynCom is tested in an *in vitro* setting with iteration cycles until the composition of the culture's microbiome is understood, followed by *in vivo* testing in animal models. The SynCom is then extracted from the model and sequenced after an incubation or experimental period, ensuring the functional profile of the SynCom. If the functional profile is defective, iteration cycles are performed again to establish the expected output.

Researchers are also developing bioreactors or models that simulate the colon environment for therapeutic applications. A single-stage bioreactor system called the "Robogut" is designed to duplicate the conditions of the human colon. The Robogut protocol includes the preparation of growth medium, building the bioreactor, preparing human fecal material for inoculating the bioreactor, sampling, and harvesting the contents of the bioreactor (Gianetto-Hill *et al.*, 2023). A mini-colon model with a bench-top bioreactor that simulates an environment similar to that of a colon was developed by Jin *et al.* (2022). A multicompartmental *in vitro* model that simulates the dynamic events occurring within the lumen of the gastrointestinal tract of human and monogastric animals was designed by Minekus *et al.* (1995).

4.4.2 Probiotic Mixtures

Combinations of different probiotics can be used for therapeutic applications to obtain a synergistic effect. Kim *et al.* (2010) have reported that the probiotic mixture of *Bifidobacterium bifidum*, *Bifidobacterium lactis*, and *Lactobacillus acidophilus* is helpful for infants prone to a high risk of allergy in the prevention of the development of eczema. A probiotic mixture of *Lactobacillus rhamnosus species GG (LGG)* and *LC705*, *Bifidobacterium breve 99 (Bb99)*, *Propionibacterium freudenreichii Shermanii JS (PJS)*, and *Escherichia coli Nissle 1917 (EcN)* was found to have profound effects on immunomodulatory properties (Fong *et al.*, 2022). The effect of probiotic agents *Lactobacillus acidophilus, L. rhamnosus, L. paracasei, Pediococcus pentosaceus, Bifidobacterium lactis, and B. breve* on visceral fat area (VFA) and intrahepatic fat (IHF) fraction in nonalcoholic fatty liver disease (NAFLD) was tested, and the study found that treatment with probiotics resulted in a significant reduction in IHF and body weight (Ahn *et al.*, 2019). Probiotic mixtures PM-1 (*Lactobacillus acidophilus* and *Bifidobacterium lactis*), a mixture of two probiotics, and probiotic mixture PM-2 (*Lactobacillus acidophilus*, *Lactobacillus paracasei*, *Lactobacillus rhamnosus*, and *Bifidobacterium lactis*), a mixture of four probiotics, acted as potential therapeutic agents against intestinal mucositis (Quaresma *et al.*, 2020). *L. paracasei* and *L. plantarum* have therapeutic potential in the suppression of experimental autoimmune encephalomyelitis (EAE).

4.4.3 Prebiotic-Based Fecal Microbiota

Prebiotics are dietary fibers that promote the growth of beneficial gut bacteria. Prebiotic-based fecal microbiota transplants involve providing prebiotic substrates to the transplanted microbiota to enhance their colonization and therapeutic efficacy. The goal of prebiotic-based fecal transplants is to improve the stability and function of the transplanted microbiota in the recipient's gut.

Most prebiotics are carbohydrate groups and are mostly oligosaccharide carbohydrates (OSCs), which include fructooligosaccharides (FOS), inulin, galactooligosaccharides (GOS), resistant starch, starch and glucose-derived oligosaccharides, pectic oligosaccharide (POS), and non-carbohydrate oligosaccharides. FOS production occurs with the help of the enzyme machinery glycosidase and glycosyl-transferase, while GOS formation is a result of galactosyl-transferase and galactosidase

TABLE 4.1
Probiotic Organisms and Their Application in Treatment of Diseases

Therapeutically useful organisms	Treatment	Reference
Lactobacillus rhamnosus	Acute diarrhea	Canani *et al.*, 2007
Lactobacillus helveticus	Immunomodulatory	LeBlanc *et al.*, 2002
Lactobacillus rhamnosus GG	Food allergy	Berni *et al.*, 2017
Lactobacillus casei, Bifidobacterium lactis		Dupont *et al.*, 2015
Lactobacillus sp.	Urinary tract infections	Lisa hanson *et al.*, 2016
Akkermansia muciniphila	Obesity	Plovier *et al.*, 2017
Lactobacillus plantarum	Cardiovascular disease	Costabile *et al.*, 2017
Faecalibacterium prausnitzii	Obesity-associated gut disorders	Maioli *et al.*, 2021
Saccharomyces boulardii	Prevention of antibiotic-associated diarrhea	La Rosa *et al.*, 2003
Bifidobacteriumlongum, *Collinsella aerofaciens*, and *Enterococcus faecium*	Hepatocellular carcinoma	Zheng *et al.*, 2019

(Davani-Davari *et al.*, 2019). The prebiotics act as an energy source for the gut microbiota, and the products of prebiotic degradation are mainly short-chain fatty acids (SCFAs) that diffuse through gut enterocytes and enter blood circulation. The SCFAs can act as substrates or signals in the host organs. Prebiotics have an important role in the gastrointestinal tract as well as in other distant site organs and systems (den Besten *et al.*, 2013)

Prebiotic-based fecal transplants can be achieved through several techniques, including prebiotic supplementation, prebiotic gene expression, and the use of encapsulated prebiotics. Prebiotic supplementation involves administering prebiotic fibers to the donor microbiota before transplantation, which enhances the growth and colonization of the transplanted microbiota in the recipient's gut. Prebiotic gene expression involves genetically modifying the donor microbiota to produce and secrete prebiotic fibers directly in the recipient's gut. Encapsulated prebiotics are prebiotic fibers that are coated with a protective layer, which allows them to survive the acidic environment of the stomach and reach the colon intact.

Prebiotic-based fecal transplants have several advantages over traditional fecal transplants. They provide a more stable and long-lasting microbiota composition, as the prebiotic fibers promote the growth and colonization of the transplanted microbiota in the recipient's gut. Prebiotic-based fecal transplants also reduce the risk of infectious disease transmission, as they do not require the transfer of live microorganisms.

4.4.4 Genetically Modified Fecal Microbiota

Another approach being explored for engineering fecal transplants is the genetic modification of the fecal microbiota. Genetically modified microbiota are designed by adding or removing genes from the microbiota to alter their metabolic activity, colonization, and immune-modulatory properties. The goal of genetically modified microbiota is to enhance their therapeutic potential and improve their colonization in the recipient's gut.

Genetically engineered bacterial construction mainly includes upstream (functional gene acquisition) and downstream (heterogeneous expression) stages. In the upstream stage, the focus is the acquisition or modification of target genes or large gene clusters in constructing a genetically engineered strain (Cobb and Zhao, 2012). General or long PCR methods or direct DNA synthesis and restriction of enzyme digestion are executed for small gene fragments of <10 kb, whereas recombination methods are used for gene sizes above 50 kb. Genetic modification can be achieved through several techniques, including gene editing, gene knockdown, and gene insertion. Gene knockdown involves reducing the expression of specific genes using RNA interference or antisense technology. Gene insertion involves adding new genes to the microbiota using gene transfer techniques, such as bacterial conjugation or transfection. Gene editing involves using molecular scissors, such as clustered regularly interspaced short palindromic repeats (CRISPRs) and CRISPR-associated (Cas) proteins (CRISPR/Cas9), to cut and modify specific genes or eliminate the microbiota. CRISPR–Cas9 can edit bacterial DNA fragments up to 100 kb in a single step in which RNA-guided Cas9 nuclease targets and cleaves DNA fragments, and the final large gene fragments are assembled via Gibson assembly (Jiang *et al.*, 2015). CRISPR-Cas systems can control gene expression without modification of the genome.

CRISPR has facilitated genome engineering in many organisms, including those genomes that are considered more difficult to manipulate. The CRISPR-directed homologous recombination tool provides the possibility to characterize the gene function of microbiome-related phenotypes by engineering the microbiome and commensal microbes. CRISPR tools are now available for an extensive range of commensal bacteria such as *E. coli*, *Lactobacillus*, *Clostridium*, *Bacteroides*, *Staphylococcus*, and *Bacillus*, and yeasts *Saccharomyces* and *Candida* for site-directed mutagenesis and gene deletion/insertion (Agarwal *et al.*, 2023). *Bacteroides thetaiotaomicron*, a prevalent and stable resident of the human gut, is engineered through CRISPR interference (CRISPRi) for regulated knockdown of recombinant and endogenous gene expression, which alters the metabolic capacity of *B. thetaiotaomicron* and its resistance to antimicrobial peptides. Inducible CRISPRi and recombinase systems act as a resource to modify *Bacteroides* for therapeutic delivery to the gut microbiome (Mimee *et al.*, 2015). CRISPR editing in *B. animalis* subsp. *lactis* demonstrated that genomic and epigenetic landscape factors contribute to the editing efficiency and genome accessibility. The gut microbiome, each with its own unique genome, may require individual CRISPR-based genome engineering approaches for developing future probiotics (Pan *et al.*, 2022). Phage-delivered CRISPR-Cas9 systems have also been developed for modifying the gut microbiome in many studies. They enable strain-specific depletion and genomic deletions in the gut microbiome. The deployment of engineered M13 bacteriophage to deliver DNA to *E. coli* cells in a mouse gut, specifically targeting the strain in the gut of the mammalian host, was reported. The M13-delivered CRISPR-Cas9 system induced chromosomal deletions of fluorescent protein genes *in vitro* and *in vivo* (Kathy *et al.*, 2021), and the method could be adopted for microbiome editing as well as for targeting therapeutic applications.

DNA conjugation is an *in situ* engineering mechanism that transfers genetic material to the target microbiota through donor bacteria. In bacterial conjugation, the DNA is transferred from donor to recipient through a type IV secretion system (T4SS) (Hospenthal *et al.*, 2017). Several researchers have reported the efficiency of conjugation in modifying bacteria for suitable purposes. A strain of *B. subtilis* named XPORT, containing integrative and conjugative elements, was designed to engineer bacteria from the gut microbiome and skin, the undomesticated bacteria. Successful results were obtained in conjugating the undomesticated bacteria in both laboratory and field conditions (Brophy *et al.*, 2018). Jin *et al.* (2022) established a genetic manipulation (GM) pipeline to identify a gene transfer methodology for non-model human gut commensals through multifactorial optimization of their conjugation/transformation conditions, followed by setting up gene manipulation tools such as CRISPRi or Group II intron-based genetic tools to modulate a specific pathway. Metagenomic alteration of the gut microbiome by *in situ* conjugation (MAGIC) was adopted by Ronda *et al.* (2019) to modify gut microbiota through the transfer of mobile plasmids. In this method, an *E. coli* donor strain was engineered to transfer mobile plasmids into recipient Gram-positive and Gram-negative strains through a replicative or integrative process containing broad host-range origin of replication and a transposable Himar cassette and transposase (Tnase) respectively. Transconjugant bacteria were detected on the basis of expression of green fluorescent protein (GFP) and an antibiotic-resistance gene (AbR), and were found to persist only for 72 hours post-gavage. This unstable plasmid maintenance highlights the difficulty in maintaining the engineered plasmids in complex microbial systems and the need for improvements in designing for a long time span. Apart from engineering the microbiome, many factors aid in the conjugation process of gut microbiota. Studies by Neil *et al.* (2020) reported differences in transfer rates during *in situ* and *in vitro* conditions. The Incl2 conjugative plasmid TP114 showed a modest transfer rate *in vitro* but highly efficient gene transfer in the mouse gut microbiota. Analysis of the above signifies the importance of mating pair stabilization, T4SS compatibility, replication compatibility, and other factors such as restriction sites, exclusion systems, and toxin/antitoxin systems for effective conjugation and its maintenance (Neil *et al.*, 2021). Bacterial density, pH, oxygen, bile, and microbial supernatants are other key factors for conjugation in *in vivo* and *in vitro* approaches, respectively (Kessler *et al.*, 2023)

Study on live bacterial therapeutics using native *E. coli* strains has shown persistently improved glucose tolerance in an obesity model study conducted in mice. Native E. coli strains isolated from mouse stool were genetically engineered, and the study demonstrated the ability of the engineered native bacteria to serve as chassis for transgene delivery with significant changes in the physiology of the wild-type host's phenotype (Baylee *et al.*, 2022). Genetically modified microbiota have several potential advantages over traditional fecal transplants. They can be designed to produce specific therapeutic molecules, such as anti-inflammatory cytokines or antimicrobial peptides. They can also be engineered to enhance their colonization in the recipient's gut by altering their adhesion properties, resistance to immune attack, or sensitivity to antibiotics.

Engineering donor microbiota based on their compatibility with the recipient's gut microbiota is another approach being explored for engineering fecal transplants. The goal of this is to provide a more personalized and effective fecal transplant by selecting donor microbiota that are more likely to colonize and function effectively in the recipient's gut. Using several techniques, including microbiome analysis, *in vitro* screening, and computational modeling, this could be accomplished. Analyzing the microbiota composition of the recipient and the potential donor candidates, and selecting the donor with the most similar microbiota composition to the recipient, is one possibility in engineering fecal transplants. The microbiome analysis may be carried out using culturing techniques or advanced molecular techniques.

The dynamics of gut microbiota, their composition, interaction, and metabolic activity are crucial for understanding the function of each taxon. Computational modeling is one of the means that employ mathematical models to predict the colonization and function of the donor microbiota in the recipient's gut based on their metabolic activity and interactions. Computational and mathematical modeling act as complementary tools to generate and test hypotheses on microbial community behavior at the taxon and community levels. One of the computational analysis models is the genome-scale models (GEMs) of metabolism, which predict the metabolic machinery of the microbial community at taxon and community levels. *coda4microbiome* is another computational tool used for analyzing microbiome data within the compositional data analysis (CoDA) framework for cross-sectional and longitudinal studies. Some of the other model methods used in microbiome analysis include ALDEx2, ANCOM, ANCOM-BC, fastANCOM, LinDA, and LOOCM (Calle *et al.*, 2023). With the help of computation, the dynamics of gut microbiota, their function, behavior modification, functionality change, and metabolite influence on dysbiosis can be studied, and an area of focus for treatment can be identified.

In vitro screening involves culturing the donor microbiota in conditions that mimic the recipient's gut environment and selecting the microbiota that show the best colonization and therapeutic efficacy. Culture-based approaches are implemented in deciphering the host-microbiome interactions, but deficits pertain to culturing the total gut microbiota. This makes the process of understanding host-microbiome interactions difficult. Hence, laboratory-scale continuous culture systems are used to simulate environmental, spatial, and temporal features similar to those in the gastrointestinal environment. Models with single or multiple batch vessels with host-free systems are employed to study microbial perturbations from exogenous stimuli. Currently, the most advanced systems to study gut microbiota-host interactions include the host-microbiota interaction module (HMI), the simulator of the human intestinal microbial ecosystem (SHIME), the dynamic gastrointestinal simulator (SIMGI), and TNO gastro-intestinal model (TIM). Other models that mimic host-microbe interactions are the Transwell "apical anaerobic model of the intestinal epithelial barrier", the human oxygen-bacteria anaerobic (HoxBan) system, the human gut-on-a-chip, and the HuMiX model (von Martels *et al.*, 2017)

Significant projects such as the American Gut, the Asian Microbiome Project, the European Microbiome Project, and the Human Microbiome Project have contributed significantly to understanding the interactions and functions of host-associated microbiota. These projects utilize metagenomics, metabolomics, metaproteomics, metatranscriptomics, and other techniques to elucidate the interaction and function of the host-associated microbiota (Turnbaugh *et al.*, 2007; Kisuse and Nakayama, 2017)

Engineered donor microbiota have several advantages over traditional fecal transplants. They provide a more personalized and effective microbiota composition, which can improve the therapeutic efficacy of the transplant. Engineered donor microbiota also reduce the risk of infectious disease transmission, as they are selected based on their compatibility with the recipient's gut microbiota.

4.5 STEPS INVOLVED IN THE FMT PROCESS

The five important concerns dealt with in FMT approaches are donor selection, fecal substance collection and preparation, recipient preparation, route of administration, and monitoring of recovery (Figure 4.1).

4.5.1 Donor Selection and Screening

The first step is to identify a suitable donor who is healthy and has a diverse gut microbiome. The donor undergoes a thorough screening process that includes medical history, physical examination, blood tests, and stool tests to ensure that they do not have any infectious diseases or conditions that could harm the recipient. The medical history of the donor, relating to age, disease history, obesity, any metabolic syndrome, allergies, and antibiotic intake history, can be collected with questionnaires. The donor should be subjected to blood screening for infectious agents such as HIV, hepatitis, history of irritable bowel syndrome, and cancer markers (Kelly *et al.*, 2015). Stool testing is very important to avoid risk, and it involves primary fecal macroscopic observation and occult blood testing, microscopic observation for pathogenic parasites, and enzyme-linked immunoassay tests for rotavirus, *Clostridium difficile* toxin identification, *Helicobacter pylori* (Chang *et al.*, 1999), and *Giardia* (Youn S *et al.*, 2009). Stool culture methods are performed to identify lower GI tract infection-causing bacteria.

4.5.2 Fecal Material Preparation

The fecal sample is collected from the donor using a sterile container and stored in a sterile environment to maintain its microbiome composition. The sample is usually collected on the same day it will be transplanted or frozen for later use. To prepare the fecal material, it is then processed to create a liquid suspension that can be administered to the recipient. The fecal material processing

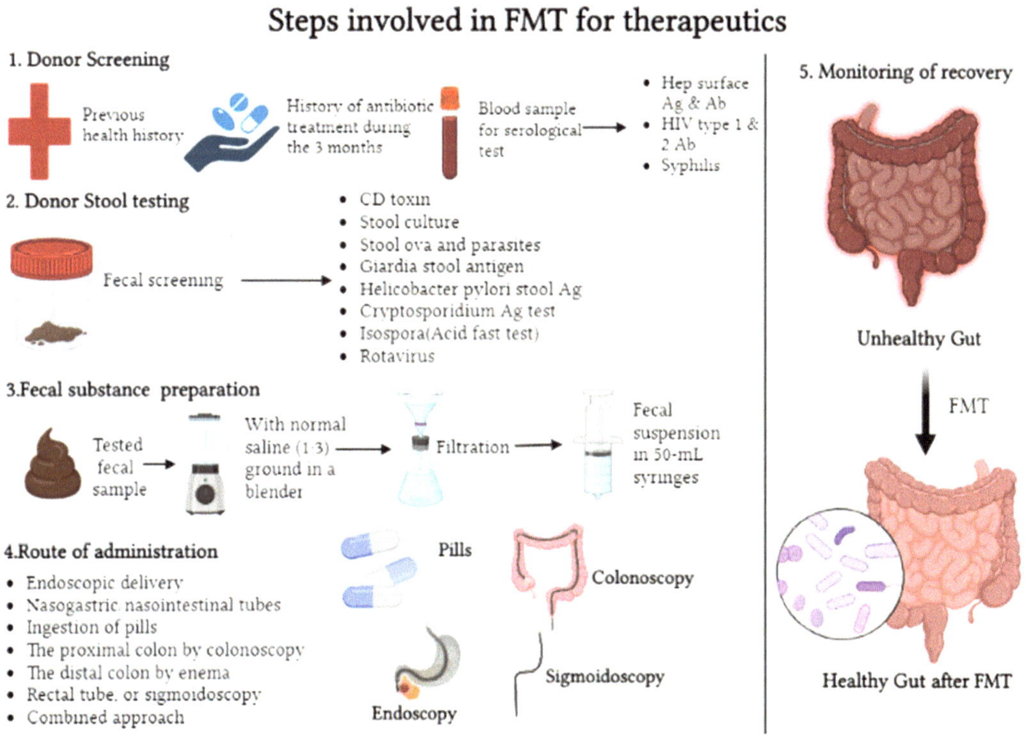

FIGURE 4.1 Steps in FMT for therapeutic application.

involves homogenization with sterile saline or other solutions by a blending method, followed by filtration to remove any solid particles.

4.5.3 Route of Administration

Before the fecal transplantation, the patient undergoes bowel preparation to empty their colon and ensure that the fecal material can reach the intended site in the digestive system. This can involve fasting, taking laxatives, or undergoing enemas. FMT can be administered to the patient via a variety of techniques, including a fecal enema, duodenal tract, nasogastric tract, colonoscopy, rectal enema, and nasojejunal tract (Rohlke and Stollman, 2012). Administration methods through the lower GI tract include colonoscopy, sigmoidoscopy, rectal tube, or retention enema, and upper GI route administration includes nasogastric/nasointestinal tubes or gastroduodenoscopy. Colonoscopic FMT is considered safe, well-tolerated, and easy to perform (Brandt LJ and Aroniadis OC, 2013). According to Kao *et al.* (2017), FMT delivered through oral capsules and colonoscopy endow therapeutic rates greater than 95% and are more accepted treatments by patients.

4.5.4 Post-Transplant Monitoring

After the fecal transplantation, the patient is monitored for any adverse reactions or side effects. This includes monitoring for fever, abdominal pain, diarrhea, or other symptoms. If necessary, additional FMT sessions may be conducted to achieve the desired therapeutic effect.

4.6 EMERGING FMT APPLICATIONS

FMT is an effective therapeutic approach for a variety of disorders linked to altered gut microbiota, including hundreds of randomized controlled trials (RCTs) conducted to treat gastrointestinal diseases and non-gastrointestinal diseases. Numerous gastrointestinal diseases have been treated by FMT, such as CDI, IBD (including Crohn's disease and ulcerative colitis), IBS, chronic constipation (CC), hepatic encephalopathy, metabolic syndrome, and graft-versus-host illness. Additionally, non-GI diseases such as autoimmune diseases, neurologic disorders, obesity, chronic fatigue syndrome, and autism have also been treated by FMT (Aroniadis *et al.*, 2013). A few case studies using FMT treatment are discussed below.

CDI is a common nosocomial infection in the USA, affecting many people and leading to morbidity and mortality even after hospitalized treatment. Treatment for CDI recurrences typically includes extra courses of oral vancomycin, prolonged vancomycin, or metronidazole. A case study of 87 CDI patients from an institution in northern Minnesota treated with 89 FMT courses was studied and analyzed by Rubin *et al.* (2013). The study revealed that FMT administration via the upper GI tract provided clinical resolution to diarrhea. Chehri *et al.* (2018) of the Department of Infectious Diseases, Copenhagen University Hospital Hvidovre, conducted a study on patients who had experienced at least three previous episodes of CDI. Women patients were the majority, at 67.8%, and having an age range of 25–86 years. The patients were offered either rectal bacteriotherapy (RBT), FMT enema (from a single donor), or FMT capsules (mixed from four donors). After treatment and a 180-day follow-up, it was found that the patients had recovered from CDI. Inflammatory bowel diseases are chronic inflammatory conditions of the gastrointestinal system defined by two pathologic effects: ulcerative colitis (UC) affecting the colon and Crohn's disease (CD) affecting any part of the GI. Twelve patients with ulcerative colitis (UC) received multiple FMT treatments for 52 weeks. The results showed a significant effect on UC, and the patients were responsive to FMT (Dang *et al.*, 2020).

FMT is found to have profound effects on the treatment of neurological disorders. A case study on FMT in a 71-year-old male patient with Parkinson's disease and intractable constipation was reported. The FMT treatment reduced tremor in both lower limbs and led to quicker defecation. FMT raised the α-diversity of the gut microbiota, and the relative abundance of *Lachnoclostridium*, *Dialister*, *Alistipes*, and unidentified *Ruminococcaceae* increased after one week; that of Megamonas increased after one month, and that of *Akkermansia* and *Faecalibacterium* increased after three months (Huang *et al.*, 2019). As a potential modulator of human biology, gut microbiota can directly or indirectly regulate brain neurochemistry through immune, metabolic, or endocrine mechanisms (Yoo and Mazmanian, 2017). Meta-analysis of the gut microbiome of Parkinson's patients provides mounting evidence of the role of gut microbiota in motor deficits and neuroinflammation (Romano *et al.*, 2021). A 28-year-old male with bipolar II disorder, after initiating FMT, ceased taking the drug valproate, and after 12 months of initiation of FMT, reduced the intake concentration of the drug quetiapine, had virtually no bipolar symptoms, no distinct highs, improved ADHD symptoms, and had not taken any mood stabilizer for 15 months.

A study reported an 87-year-old male with recurrent ascending cholangitis, treated with multiple broad-spectrum antibiotics, who developed complications during antibiotic treatment. The treatments were unsuccessful and led to bacteremia with multi-drug resistant microorganisms. FMT was performed during each hospitalization and was done three times at different time intervals. After FMT, a decline in the frequency of MDR systemic infections was observed (Gouveia *et al.*, 2020). A published study of 18 children with autism, with FMT administration daily for seven to eight weeks, demonstrated improvement in both gastrointestinal and behavioral symptoms, and a persistent improvement was observed when FMT was followed for two years (Kang *et al.*, 2017, 2019). In a study on 48 patients, treatment with a randomized control trial of FMT appeared to show improvement in gastrointestinal and behavioral symptoms (Zhao *et al.*, 2019).

4.7 CURRENT CHALLENGES AND LIMITATIONS OF MICROBIOME ENGINEERING

Microbiome engineering is a rapidly evolving frontier for solutions to improve human health, agricultural productivity, and the environment. Inoculating new microorganisms intended to enhance function in an already-existing microbial population is a common modification technique. Microbiome engineering seeks to improve the function of an ecosystem by manipulating the composition of microbes. Under this scenario, there are two major challenges for successful microbiome engineering: (1) the design of a microbiome with improved function and (2) the establishment of an improved microbiome in a recipient system of interest (Albright *et al.*, 2022).

The potential for microbiome engineering is also constrained by our incomplete understanding of the human microbiome and its interactions with the host and environment. Simple sequencing has been replaced with a methodical examination of the microbial metagenome, proteome, metabolome, and transcriptome in the study of the microbiome. This has produced a huge amount of data that is challenging to organize, interpret, and reproduce (Gilbert *et al.*, 2010). To solve this challenge, newer infrastructures, software, and models are being developed.

In addition, engineering the microbiota will have a profound effect on human health. It is more likely to influence a person's life duration, eating patterns, and even social behavior. Its safety can be called into doubt as a result. Over time, the functional integrity of one's own immune system will weaken due to an overreliance on the microbiome for protection against invasive infections, toxic antigens, and malignancies. Furthermore, spontaneous mutations are a natural part of every living cell. This frequently results in cancer development, microbial drug resistance, and the elevation of pathogenic prototypes. Therefore, it is possible for advantageous germs to mutate and appear as dangerous or cancer-causing microbes. Before implementing widespread microbiome engineering, a way to remove the designed bacteria is required (Sonnenburg, 2015).

4.8 CONCLUSION

The development of engineered fecal transplants is an exciting area of research that has the potential to revolutionize the treatment of gastrointestinal disorders. The various strategies being explored, including synthetic microbiota, genetically modified microbiota, prebiotic-based fecal transplants, and engineered donor microbiota, offer unique advantages over traditional fecal transplants and may provide more effective and personalized treatments for patients. FMT is now recognized as a promising novel therapeutic option for a wide range of additional illnesses linked to intestinal dysbiosis due to high success rates and short-term safety documented for recurrent CDI. However, more research is needed to determine the safety and efficacy of these approaches and to optimize their use in clinical practice.

REFERENCES

Aggarwal N, Kitano S, Puah GRY, Kittelmann S, Hwang IY, Chang MW. 2023. Microbiome and human health: Current understanding, engineering, and enabling technologies. *Chem Rev.*, Jan 11;123(1):31–72. PMID: 36317983; PMCID: PMC9837825.

Ahn SB, Jun DW, Kang BK, Lim JH, Lim S, Chung MJ. 2019. Randomized, double-blind, placebo-controlled study of a multispecies probiotic mixture in nonalcoholic fatty liver disease. *Sci Rep.*, Apr 5;9(1):5688. PMID: 30952918; PMCID: PMC6450966.

Albright MBN, Louca S, Winkler DE, Feeser KL, Haig SJ, Whiteson KL, Emerson JB, Dunbar J. 2022. Solutions in microbiome engineering: Prioritizing barriers to organism establishment. *ISME J.*, 16(2):331–338. PMID: 34420034; PMCID: PMC8776856.

Aroniadis OC, Brandt LJ. 2013. Fecal microbiota transplantation: past, present and future. Curr Opin Gastroenterol. 2013 Jan;29(1):79-84. PMID: 23041678.

Atarashi K, Tanoue T, Oshima K, Suda W, Nagano Y, Nishikawa H, Fukuda S, Saito T, Narushima S, Hase K, Kim S, Fritz JV, Wilmes P, Ueha S, Matsushima K, Ohno H, Olle B, Sakaguchi S, Taniguchi T, Morita H, Hattori M, Honda K. 2013. Treg induction by a rationally selected mixture of Clostridia strains from the human microbiota. *Nature.*, Aug 8;500(7461):232–236. PMID: 23842501.

Baylee J. Russell, Steven D. Brown, Nicole Siguenza, Irene Mai, Anand R. Saran, Amulya Lingaraju, Erica S. Maissy, Ana C. Dantas Machado, Antonio F.M. Pinto, Concepcion Sanchez, Leigh-Ana Rossitto, Yukiko Miyamoto, R. Alexander Richter, Samuel B. Ho, Lars Eckmann, Jeff Hasty, David J. Gonzalez, Alan Saghatelian, Rob Knight, Amir Zarrinpar, 2022. Intestinal transgene delivery with native E. coli chassis allows persistent physiological changes. *Cell.*, 185(17):3263–3277.e15.

Bendheim PE, Poeggeler B, Neria E, Ziv V, Pappolla MA, Chain DG. 2002. Development of indole-3-propionic acid (OXIGON) for Alzheimer's disease. *J Mol Neurosci.*, 19(1–2):213–217. PMID: 12212784.

Berni Canani R, Di Costanzo M, Bedogni G, Amoroso A, Cosenza L, Di Scala C, et al. 2017. Extensively hydrolyzed casein formula containing *Lactobacillus rhamnosus* GG reduces the occurrence of other allergic manifestations in children with cow's milk allergy: 3-year randomized controlled trial. *J Allergy Clin Immunol.*, 139(6):1906–1913.

Brandt LJ, Aroniadis OC. 2013. An overview of fecal microbiota transplantation: techniques, indications, and outcomes. *Gastrointest Endosc.*, 78(2):240–249. PMID: 23642791.

Brandt LJ, Aroniadis OC, Mellow M, Kanatzar A, Kelly C, Park T, Stollman N, Rohlke F, Surawicz C. 2012. Long-term follow-up of colonoscopic fecal microbiota transplant for recurrent Clostridium difficile infection. *Am J Gastroenterol.*, 107(7):1079–1087. PMID: 22450732.

Brophy JAN, Triassi AJ, Adams BL, Renberg RL, Stratis-Cullum DN, Grossman AD, Voigt CA 2018. Engineered integrative and conjugative elements for efficient and inducible DNA transfer to undomesticated bacteria. *Nat Microbiol.*, 3(9):1043–1053. PMID: 30127494.

Calle ML, Pujolassos M, Susin A. 2023. *coda4microbiome*: Compositional data analysis for microbiome cross-sectional and longitudinal studies. *BMC Bioinformatics.*, Mar 6;24(1):82.

Canani RB, Cirillo P, Terrin G, Cesarano L, Spagnuolo MI, De Vincenzo A, Albano F, Passariello A, De Marco G, Manguso F, Guarino A. 2007. Probiotics for treatment of acute diarrhoea in children: randomised clinical trial of five different preparations. *BMJ.*, Aug 18;335(7615):340.

Chang MC, Wu MS, Wang HH, Wang HP, Lin JT. 1999. Helicobacter pylori stool antigen (HpSA) test--a simple, accurate and non-invasive test for detection of Helicobacter pylori infection. *Hepatogastroenterology.*, Jan-Feb; 46(25):299-302. PMID: 10228811.

Chehri M, Christensen AH, Halkjær SI, Günther S, Petersen AM, Helms M. 2018. Case series of successful treatment with fecal microbiota transplant (FMT) oral capsules mixed from multiple donors even in patients previously treated with FMT enemas for recurrent Clostridium difficile infection. *Medicine (Baltimore).*, 97(31):e11706. PMID: 30075573; PMCID: PMC6081131.

Cobb R. E., Zhao H. 2012. Direct cloning of large genomic sequences. *Nat. Biotechnol.,* 30, 405–406.

Costabile A, Buttarazzi I, Kolida S, Quercia S, Baldini J, Swann JR, Brigidi P, Gibson GR. 2017. An in vivo assessment of the cholesterol-lowering efficacy of Lactobacillus plantarum ECGC 13110402 in normal to mildly hypercholesterolaemic adults. *PLoS One.*, Dec 11;12(12):e0187964.

Cummings JH, Pomare EW, Branch WJ, Naylor CP, Macfarlane GT. 1987. Short chain fatty acids in human large intestine, portal, hepatic and venous blood. Gut. Oct 28(10):1221-7.

Cummings, JH, Macfarlane GT. 1991. The control and consequences of bacterial fermentation in the human colon. *Appl. Bacteriol.*, 70(6): 443–459.

Dang XF, Qing-Xi Wang, Yin Z, Sun L, Yang WH. 2020. Recurrence of moderate to severe ulcerative colitis after fecal microbiota transplantation treatment and the efficacy of re-FMT: A case series. *BMC Gastroenterol.*, Nov 26;20(1):401.

Davani-Davari D, Negahdaripour M, Karimzadeh I, Seifan M, Mohkam M, Masoumi SJ, Berenjian A, Ghasemi Y. 2019. Prebiotics: Definition, types, sources, mechanisms, and clinical applications. *Foods.*, Mar 9;8(3):92.

den Besten G, van Eunen K, Groen AK, Venema K, Reijngoud DJ, Bakker BM, 2013. The role of short-chain fatty acids in the interplay between diet, gut microbiota, and host energy metabolism. *J. Lipid Res.*, 54:2325–2340.

Dupont C, Hol J, Nieuwenhuis EES. 2015. An extensively hydrolysed casein-based formula for infants with cows' milk protein allergy: Tolerance/hypo-allergenicity and growth catch-up. *Br J Nutr.*, 113(7):1102–1112.

Fahnøe U., Bukh J. 2019. Full-length open reading frame amplification of hepatitis C Virus. Methods *Mol. Biol.,*1911, 85–91.

Fong FLY, El-Nezami H, Mykkänen O, Kirjavainen PV. 2022. The effects of single strains and mixtures of probiotic bacteria on immune profile in liver, spleen, and peripheral blood. *Front Nutr.*, Apr 12;9:773298.

Gensollen T, Iyer SS, Kasper DL, Blumberg RS. 2016. How colonization by microbiota in early life shapes the immune system. *Science.,* Apr 29;352(6285):539-44.

Gianetto-Hill CM, Vancuren SJ, Daisley B, Renwick S, Wilde J, Schroeter K, Daigneault MC, Allen-Vercoe E. 2023. The Robogut: A bioreactor model of the human colon for evaluation of gut microbial community ecology and function. *Curr Protoc.*, 3(4):e737.

Gilbert JA, Meyer F, Antonopoulos D, Balaji P, Brown CT, Brown CT, Desai N, Eisen JA, Evers D, Field D, Feng W, Huson D, Jansson J, Knight R, Knight J, Kolker E, Konstantindis K, Kostka J, Kyrpides N, Mackelprang R, McHardy A, Quince C, Raes J, Sczyrba A, Shade A, Stevens R. 2010. Meeting report: The terabase metagenomics workshop and the vision of an Earth microbiome project. *Stand Genomic Sci.*, May 25;3(3):243–248.

Gouveia C, Palos C, Pereira P, Roque Ramos L, Cravo M. 2020. Fecal microbiota transplant in a patient infected with multidrug-resistant bacteria: A case report. *GE Port J Gastroenterol.*, 28(1):56–61.

Guarner F, Malagelada JR. 2003. Gut flora in health and disease. *Lancet.*, Feb 8;361(9356):512–519.

Hanson L, VandeVusse L, Jermé M, Abad CL, Safdar N. 2016. Probiotics for treatment and prevention of urogenital infections in women: A systematic review. *J Midwifery & Women's Health.*, 61:339–355.

Hooper, L. V.; Gordon, J. I. 2001. Commensal Host-Bacterial Relationships in the *Gut. Science.,* 292, 1115–1118.

Hospenthal MK, Costa TRD, Waksman G. 2017. A comprehensive guide to pilus biogenesis in gram-negative bacteria. *Nat. Rev. Microbiol.*, 15(6):365–379.

Huang H, Xu H, Luo Q, He J, Li M, Chen H, Tang W, Nie Y, Zhou Y. 2019. Fecal microbiota transplantation to treat Parkinson's disease with constipation: A case report. *Medicine (Baltimore).*, 98(26):e16163.

Jiang H, Ling Z, Zhang Y, Mao H, Ma Z, Yin Y, Wang W, Tang W, Tan Z, Shi J, Li L, Ruan B. 2015. Altered fecal microbiota composition in patients with major depressive disorder. *Brain Behav Immun.*, 48:186–194.

Jiménez-Avalos JA, Arrevillaga-Boni G, González-López L, García-Carvajal ZY, González-Avila M. 2021. Classical methods and perspectives for manipulating the human gut microbial ecosystem. *Crit Rev Food Sci Nutr.*, 61(2):234–258.

Jin WB, Li TT, Huo D, Qu S, Li XV, Arifuzzaman M, Lima SF, Shi HQ, Wang A, Putzel GG, Longman RS, Artis D, Guo CJ. 2022. Genetic manipulation of gut microbes enables single-gene interrogation in a complex microbiome. *Cell.*, Feb 3;185(3):547–562.e22.

Kang DW, Adams JB, Coleman DM, Pollard EL, Maldonado J, McDonough-Means S, Caporaso JG, Krajmalnik-Brown R. 2019. Long-term benefit of Microbiota transfer therapy on autism symptoms and gut microbiota. *Sci Rep.*, Apr 9;9(1):5821.

Kang DW, Adams JB, Gregory AC, Borody T, Chittick L, Fasano A, Khoruts A, Geis E, Maldonado J, McDonough-Means S, Pollard EL, Roux S, Sadowsky MJ, Lipson KS, Sullivan MB, Caporaso JG, Krajmalnik-Brown R. 2017. Microbiota transfer therapy alters gut ecosystem and improves gastrointestinal and autism symptoms: An open-label study. *Microbiome.*, Jan 23;5(1):10.

Kao D, Roach B, Silva M, Beck P, Rioux K, Kaplan GG, Chang HJ, Coward S, Goodman KJ, Xu H, Madsen K, Mason A, Wong GK, Jovel J, Patterson J, Louie T. 2017. Effect of oral capsule- vs colonoscopy-delivered fecal microbiota transplantation on recurrent clostridium difficile infection: A randomized clinical trial. *JAMA.*, Nov 28;318(20):1985–1993.

Kathy N. Lam, Peter Spanogiannopoulos, Paola Soto-Perez, Margaret Alexander, Matthew J. Nalley, Jordan E. Bisanz, Renuka R. Nayak, Allison M. Weakley, Feiqiao B. Yu, Peter J. Turnbaugh, 2021. Phage-delivered CRISPR-Cas9 for strain-specific depletion and genomic deletions in the gut microbiome, *Cell Reports.,* 37(5):109930.

Kelly JR, Kennedy PJ, Cryan JF, Dinan TG, Clarke G, Hyland NP. 2015. Breaking down the barriers: the gut microbiome, intestinal permeability and stress-related psychiatric disorders. *Front Cell Neurosci.*, Oct 14;9:392.

Kessler C, Hou J, Neo O, Buckner MMC. 2023. In situ, in vivo, and in vitro approaches for studying AMR plasmid conjugation in the gut microbiome. *FEMS Microbiol Rev.*, Jan 16;47(1):fuac044.

Kim JY, Kwon JH, Ahn SH, Lee SI, Han YS, Choi YO, Lee SY, Ahn KM, Ji GE. 2010. Effect of probiotic mix (*Bifidobacterium bifidum, Bifidobacterium lactis, Lactobacillus acidophilus*) in the primary prevention of eczema: A double-blind, randomized, placebo-controlled trial. *Pediatr Allergy Immunol.*, 21(2 Pt 2):e386–393.

Kisuse J, Nakayama J. 2017. 16S rRNA Metagenomics of Asian gut microbiota. In *Understanding Host-Microbiome Interactions - An Omics Approach: Omics of Host-Microbiome Association.* Springer Singapore:71–81.

Koropatkin NM, Cameron EA, Martens EC. 2012. How glycan metabolism shapes the human gut microbiota. *Nat Rev Microbiol.*, Apr 11;10(5):323–335.

La Rosa M, Bottaro G, Gulino N, Gambuzza F, Di Forti F, Inì G, Tornambè E. 2003. Prevention of antibiotic-associated diarrhea with *Lactobacillus sporogens* and fructo-oligosaccharides in children. A multicentric double-blind vs placebo study. *Minerva Pediatr.*, 55(5):447–452.

LeBlanc JG, Matar C, Valdéz JC, LeBlanc J, Perdigon G. 2002. Immunomodulating effects of peptidic fractions issued from milk fermented with Lactobacillus helveticus. *J Dairy Sci.*, 85(11):2733–2742.

Ley RE, Hamady M, Lozupone C, Turnbaugh PJ, Ramey RR, Bircher JS, Schlegel ML, Tucker TA, Schrenzel MD, Knight R, Gordon JI. 2008. Evolution of mammals and their gut microbes. *Science.*, Jun 20;320(5883):1647–1651.

Macfarlane S, Macfarlane GT. 2003. Regulation of short-chain fatty acid production. *Proc Nutr Soc.*, 62:67–72.

Magwira CA, Kullin B, Lewandowski S, Rodgers A, Reid SJ, Abratt VR. 2012. Diversity of faecal oxalate-degrading bacteria in black and white South African study groups: Insights into understanding the rarity of urolithiasis in the black group. *J Appl Microbiol.*, 113(2):418–428.

Maioli TU, Borras-Nogues E, Torres L, Barbosa SC, Martins VD, Langella P, Azevedo VA, Chatel JM. 2021. Possible benefits of *Faecalibacterium prausnitzii* for obesity-associated gut disorders. *Front Pharmacol.*, Dec 2;12:740636.

Mimee M, Tucker AC, Voigt CA, Lu TK. 2015. Programming a human commensal bacterium, *bacteroides thetaiotaomicron*, to sense and respond to stimuli in the murine gut Microbiota. *Cell Syst.*, Jul 29;1(1):62–71.

Minekus M, Phillipe M, Robert H, Jos HJ, Huis in'T V. 1995. A multicompartmental dynamic computer-controlled model simulating the stomach and small intestine. *Altern Lab Anim.*, 23(2):197–209.

Natividad JM, Verdu EF. 2013. Modulation of intestinal barrier by intestinal microbiota: pathological and therapeutic implications. *Pharmacol Res.*, 69(1):42–51.

Neil K, Allard N, Grenier F, Burrus V, Rodrigue S. 2020. Highly efficient gene transfer in the mouse gut microbiota is enabled by the Incl2 conjugative plasmid TP114. *Commun Biol.*, Sep 22;3(1):523.

Neil K, Allard N, Rodrigue S. 2021. Molecular mechanisms influencing bacterial conjugation in the intestinal Microbiota. *Front Microbiol.*, Jun 4;12:673260.

Pan M, Morovic W, Hidalgo-Cantabrana C, Roberts A, Walden KKO, Goh YJ, Barrangou R. 2022. Genomic and epigenetic landscapes drive CRISPR-based genome editing in Bifidobacterium. *Proc Natl Acad Sci U S A.*, Jul 26;119(30):e2205068119.

Pim T van Leeuwen, Stanley Brul, Jianbo Zhang, Meike T Wortel, 2023. Synthetic microbial communities (SynComs) of the human gut: design, assembly, and applications, *FEMS Microbiology Reviews*, 47(2):fuad012.

Plovier H, Everard A, Druart C, Depommier C, Van Hul M, Geurts L, Chilloux J, Ottman N, Duparc T, Lichtenstein L, Myridakis A, Delzenne NM, Klievink J, Bhattacharjee A, van der Ark KC, Aalvink S, Martinez LO, Dumas ME, Maiter D, Loumaye A, Hermans MP, Thissen JP, Belzer C, de Vos WM, Cani PD. 2017. A purified membrane protein from Akkermansia muciniphila or the pasteurized bacterium improves metabolism in obese and diabetic mice. *Nat Med.,* Jan;23(1):107-113.

Quaresma M, Damasceno S, Monteiro C, Lima F, Mendes T, Lima M, Justino P, Barbosa A, Souza M, Souza E, Soares P. 2020. Probiotic mixture containing *Lactobacillus* spp. and Bifidobacterium spp. attenuates 5-fluorouracil-induced intestinal mucositis in mice. *Nutr Cancer.*, 72(8):1355–1365.

Rinninella E, Raoul P, Cintoni M, Franceschi F, Miggiano GAD, Gasbarrini A, Mele MC. 2019. What is the healthy gut microbiota composition? A changing ecosystem across age, environment, diet, and diseases. *Microorganisms.*, Jan 10;7(1):14.

Rohlke F, Stollman N. 2012. Fecal microbiota transplantation in relapsing Clostridium difficile infection. *Therap Adv Gastroenterol.*, 5(6):403–420.

Romano S, Savva GM, Bedarf JR, Charles IG, Hildebrand F, Narbad A. 2021. Meta-analysis of the Parkinson's disease gut microbiome suggests alterations linked to intestinal inflammation. *NPJ Parkinsons Dis.*, Mar 10;7(1):27.

Ronda C, Chen SP, Cabral V, Yaung SJ, Wang HH. 2019. Metagenomic engineering of the mammalian gut microbiome in situ. *Nat Methods.*, 16(2):167–170.

Rubin TA, Gessert CE, Aas J, Bakken JS. 2013. Fecal microbiome transplantation for recurrent Clostridium difficile infection: Report on a case series. *Anaerobe.*, 19:22–26.

Russell WR, Gratz SW, Duncan SH, Holtrop G, Ince J, Scobbie L, Duncan G, Johnstone AM, Lobley GE, Wallace RJ, Duthie GG, Flint HJ. 2011. High-protein, reduced-carbohydrate weight-loss diets promote metabolite profiles likely to be detrimental to colonic health. *Am J Clin Nutr.*, 93(5):1062–1072.

San León D, Nogales J. 2022. Toward merging bottom-up and top-down model-based designing of synthetic microbial communities. *Curr Opin Microbiol.*, 69:102169.

Sartor RB. 2008. Microbial influences in inflammatory bowel diseases. *Gastroenterology.*, 134(2):577–594.

Sonnenburg JL. 2015. Microbiome engineering. *Nature.* Feb 26;518(7540):S10.

Sonnenburg JL, Xu J, Leip DD, Chen CH, Westover BP, Weatherford J, Buhler JD, Gordon JI. 2005. Glycan foraging in vivo by an intestine-adapted bacterial symbiont. *Science.*, Mar 25;307(5717):1955–1959.

Soo WT, Bryant RV, Costello SP. 2020. Faecal microbiota transplantation: indications, evidence and safety. *Aust Prescr.,* 2020 Apr 43(2):36-38.

Turnbaugh PJ, Ley RE, Hamady M, Fraser-Liggett CM, Knight R, Gordon JI. 2007. The human microbiome project. *Nature.*, Oct 18;449(7164):804–810.

von Martels JZH, Sadaghian Sadabad M, Bourgonje AR, Blokzijl T, Dijkstra G, Faber KN, Harmsen HJM. 2017. The role of gut microbiota in health and disease: In vitro modeling of host-microbe interactions at the aerobe-anaerobe interphase of the human gut. *Anaerobe.*, 44:3–12.

Vyas U, Ranganathan N. 2012. Probiotics, prebiotics, and synbiotics: gut and beyond. Gastroenterol Res Pract. 2012:872716.

Wall R, Ross RP, Shanahan F, O'Mahony L, O'Mahony C, Coakley M, Hart O, Lawlor P, Quigley EM, Kiely B, Fitzgerald GF, Stanton C. 2009. Metabolic activity of the enteric microbiota influences the fatty acid composition of murine and porcine liver and adipose tissues. *Am J Clin Nutr.*, 89(5):1393–1401.

Yoo BB, Mazmanian SK. 2017. The enteric network: Interactions between the immune and nervous systems of the gut. *Immunity.*, Jun 20;46(6):910–926.

Youn, S., Kabir, M., Haque, R., & Petri, W. A., Jr, 2009. Evaluation of a screening test for detection of giardia and cryptosporidium parasites. Journal of clinical microbiology, 47(2), 451–452.

Zhang X, Shen D, Fang Z, Jie Z, Qiu X, Zhang C, Chen Y, Ji L. 2013. Human gut microbiota changes reveal the progression of glucose intolerance. *PLoS One.*, Aug 27;8(8):e71108.

Zhao H, Gao X, Xi L, Shi Y, Peng L, Wang C, Zou L, Yang Y. 2019. Mo1667 fecal microbiota transplantation for children with autism spectrum disorder. *Gastrointestinal Endoscopy.*, 89(6):12–13.

Zheng Y, Wang T, Tu X, Huang Y, Zhang H, Tan D, Jiang W, Cai S, Zhao P, Song R. 2019. Gut microbiome affects the response to anti-PD-1 immunotherapy in patients with hepatocellular carcinoma. *J Immunother Cancer.*, 7:1–7.

5 Role of Prebiotics and Probiotics in Human Microbiome Engineering

Shaimaa H. Negm
Port Said University
Egypt

Hany G. EL-Masry
Helwan University
Egypt

Walaa A. Eid
New Valley University
Egypt

5.1 INTRODUCTION

Probiotics are live bacterial species that are able to survive and thrive in the acidic gastric environment and provide beneficial effects to the health of the host by reestablishing or improving the gut microbiota (Gareau et al. 2010). The introduction of probiotics into the gastrointestinal tract might serve as a promising strategy to restore the balance of the gut ecosystem and prevent or treat illness. Great progress has been made in the employment of a living product rather than a chemical to diagnose or treat diseases including diabetes, phenylketonuria, HIV, and inflammatory bowel disease (IBD) (Duan et al. 2015; Chua et al. 2017). However, there is no one-size-fits-all probiotic that works well for everyone, as the gut microbiome is different in each individual. With the development of metabolic engineering and synthetic biology, the engineering of probiotics opens up possibilities of designing microbes to target specific tissues and cells rather than the whole body and the creation of novel probiotics with desired characteristics and functionalities (Ramírez et al. 2017).

5.2 DEFINITION AND GENERAL PRINCIPLES OF PROBIOTICS

Probiotics are defined as "live microorganisms which when administered in adequate amounts confer a health benefit on the host" (Canfora et al. 2015). The evaluation guidelines of these probiotics include microbial identification through phenotypic and genotypic methods, *in vitro* and *in vivo* studies, and double-blind randomized and placebo-controlled human trials. The two major genera of probiotics are *Lactobacillus* and *Bifidobacterium*, where the former is non-spore forming, Gram-positive, rod-shaped, aerotolerant or anaerobic, and belongs to the lactic acid group of bacteria (Olivares et al. 2006) while the latter is non-spore forming and mostly anaerobic. Both genera offer modulation of intestinal crypt dynamics and protection against pathogens such as rotavirus (Thomas and Versalovic 2010).

 DOI: 10.1201/9781003394662-5

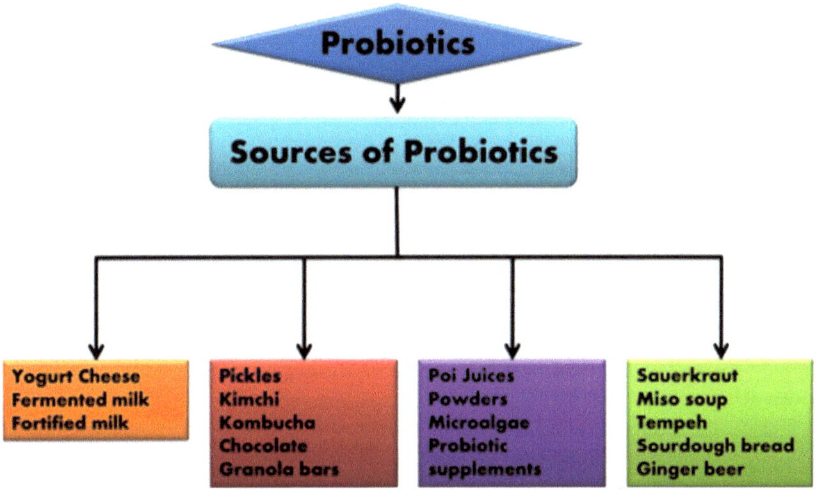

FIGURE 5.1 Different sources of probiotics.

True probiotics are in general of human origin and should be free from any vectors or other factors that can transfer antibiotic resistance. A probiotic should also show antagonism to pathogens, stimulate the immune system, and demonstrate beneficial effects on the host (Gómez-Llorente et al. 2010). Overall, there are four characteristics of effective probiotics: (a) probiotics should be able to survive the passage through the digestive system, (b) probiotics should be able to attach to the intestinal epithelia and colonize, (c) probiotics should be able to maintain good viability upon successful passage through the intestine, (d) probiotics should be able to utilize the nutrients and substrates in a normal diet, and (e) probiotics should be free from pathogens and should not contain any microbial or nonmicrobial toxic products (Gómez-Llorente et al. 2010; Wells 2011). A probiotic should also show antagonism to pathogens and stimulate the immune system and demonstrate beneficial effects on the host (Gómez-Llorente et al. 2010). Figure 5.1: Different sources of probiotics.

5.3 PROBIOTICS, PREBIOTICS, POSTBIOTICS, AND SYNBIOTICS

There have been several definitions that have been postulated for probiotics, prebiotics, and synbiotics; however, they can be best explained as a microbe or a group of microbes that inhabit the gut and nourish the host body internally (Hamasalim 2016). They are commonly consumed as preparations with active live cultures and contain bacteria, such as *Lactobacillus*, *Lactococci*, or *Bifidobacteria*, which have been isolated from natural environments (Bongaerts and Severijnen 2016). Because the diverse characteristics of probiotics have been recognized as key health promoters, research in recent years has mainly focused on investigating the culture conditions and viability of probiotic strains during processing and storage; sensitivity to low pH values, gastric fluid, bile, pancreatic and intestinal fluids, and intestinal or respiratory mucus; adherence to isolated cells or cell cultures; and interactions with other (pathogenic) microorganisms.

5.3.1 Postbiotics

Recent data suggests that bacterial products, in the absence of viable organisms, may have similar effects on signaling pathways and barrier function. These bacterial products are broadly characterized as postbiotics and can be defined as non-viable bacterial products or metabolic byproducts from probiotic microorganisms that have biologic activity in the host (Patel and Denning 2013).

Generally, postbiotics include bacterial metabolic byproducts, such as bacteriocins, organic acids, ethanol, diacetyl, acetaldehydes, and hydrogen peroxide, but it is also found that certain heat-killed probiotics can retain important bacterial structures that may exert biological activity in the host (Islam 2016). Research shows that these metabolic products have a broad inhibitory property towards pathogenic microbes and, therefore, can be used as an alternative to antibiotics (Ooi et al. 2015). Postbiotics are non-toxic, non-pathogenic, and resistant to hydrolysis by mammalian enzymes, as these are non-viable bacterial products or metabolic byproducts from probiotics. In some instances, postbiotics can also enhance barrier function against species like *Saccharomyces boulardii* and improve angiogenesis *in vitro* and *in vivo* in epithelial cells by activation of $\alpha 2\beta 1$ integrin collagen receptors (Giorgetti et al. 2015). Similar properties have also been identified in several other probiotic species such as *Bifidobacterium breve*, *Bifidobacterium lactis*, *Bifidobacterium infantis*, *Bacteroides fragilis*, *Lactobacillus*, *Escherichia coli*, and *Faecalibacterium prausnitzii* (Cicenia et al. 2014).

5.3.2 Prebiotics

Further exploration of probiotics has led to the development of prebiotics, which are specific nutrients that modify the gut microbial flora. Although not easily digested by humans, they have a selective role in stimulating the growth or activity of beneficial bacterial species in the gut (Rastall and Gibson 2015; Thomas 2016). Some of the commonly known prebiotics include the bifidogenic properties of inulin, oligofructose, and fructo-oligosaccharides (FOS) synthetically produced from sucrose, as well as galactose-containing and xylose-containing oligosaccharides (Hutkins et al. 2016). The fermentation of carbohydrates represents a major source of energy for epithelial cells in the colon, and prebiotics can readily fulfill these requirements as a result of their fermentation by gut microbiota, such as *Bifidobacteria*. Besides *Bifidobacteria*, there are several other gut microorganisms that play a significant role in fermenting these non-digestible oligosaccharides. Prebiotics can be obtained naturally from sources like vegetables, fruits, and grains consumed in our daily lives. Prebiotics not only serve as an energy source but also have several health benefits, such as reducing the prevalence and duration of diarrhea, providing relief from inflammation and other symptoms associated with intestinal bowel disorders, and exerting protective effects to prevent colon cancer (Peña 2007). Prebiotics are also implicated in enhancing the bioavailability and uptake of minerals, lowering some risk factors for cardiovascular disease, and promoting satiety and weight loss (Pokusaeva et al. 2011). Despite their vast nutritional and medicinal benefits, research regarding screening new versatile prebiotics is scarce. Therefore, more research should be focused on identifying new health supplements, with screening novel prebiotics being a primary concern.

5.3.3 Synbiotics

Development in microbial research has led to the formation of synbiotics, which are a fusion of probiotics and prebiotics products and help in enhancing the survival and the implantation of live microbial dietary supplements in the gut. The synergistic benefits are more efficiently promoted when both the probiotic and prebiotic work together in the living system. There is mounting scientific evidence that the symbiotic relationship between prebiotics and probiotics contributes significantly to health. Commercial interest in functional foods containing synbiotics has consistently increased due to the awareness of the benefits for gut health, disease prevention, and therapy. Research in this area is currently focused on developing new health-promoting foods and selecting new cultures demonstrating an enhanced ability to colonize the human gut along with an ability to digest new forms of prebiotics (Tufarelli and Laudadio 2016).

5.4 PROBIOTIC MECHANISMS OF ACTION

The overall mechanisms of action for probiotics are targeted towards two major aspects: (a) changing the composition and function of the gut microbiome and (b) promoting intestinal mucosal physiology and immunobiology to facilitate an anti-inflammatory response to promote wound healing. Probiotic mechanisms of action thus include, among others, enhancing the epithelial barrier, increasing the adhesion of healthy probiotic microbes to the intestinal mucosa, inhibiting pathogen adhesion to mucosal surfaces, competitive exclusion of pathogenic microorganisms, production of antimicrobial substances, and modulation of immune functions. For example, certain probiotics produce antimicrobial metabolites that suppress the growth of other microorganisms (Tursi et al. 2010), while others (*Lactobacillus*) promote intestinal integrity and barrier functions, thereby facilitating immune tolerance with decreased bacterial translocation across the mucosal barrier (Gionchetti et al. 2003).

5.4.1 Reestablishment of Intestinal Symbiosis through the Application of Probiotics

5.4.1.1 Competitive Exclusion of Pathogens

Probiotics facilitate microbial diversity through competitive exclusion, where bacteria from probiotic formulations compete with dysbiotic and pathogenic species for receptors present in the gastrointestinal tract (Bermudez-Brito et al. 2012). Though regulatory elements and associated specific pathways underlying these effects are largely unknown, some of the mechanisms proposed include creating an acidic environment to reduce competition, competing for nutritional sources, and the production of bacteriocin or bacteriocin-like substances for the competitive exclusion of pathogens (Royan 2019).

Several mechanisms have been reported through which probiotics facilitate the reestablishment of intestinal symbiosis. For example, it was reported that some probiotics produce antimicrobial peptides or metabolites that suppress the growth of other microorganisms (O'Shea et al. 2012). One clinical study demonstrated that a four-week treatment with *Lactobacillus plantarum* DSM 9843 per day resulted in reduced pain and flatulence in patients with IBS (Nobaek et al. 2000). Both rectal biopsies and fecal analysis indicated that the improvement in the IBS symptoms was associated with increased *L. plantarum* in biopsies and decreased enterococci in fecal specimens. Similar results of IBS improvement have also been reported using a probiotic mix of *Bifidobacterium* strains, *Lactobacillus* strains, and *Streptococcus thermophilus* (Cha et al. 2012).

5.4.1.2 Probiotics in Intestinal Enzymatic Activities and Volatile Fatty Acid Signaling

5.4.1.2.1 Probiotics and Intestinal Enzymatic Activities

Probiotics facilitate beneficial effects through microbial enzymatic activities. Of particular interest is the inhibition of β -glucuronidase activities coming from harmful bacteria. These bacteria hydrolyze glucuronidated metabolites through β -glucuronidase activity in the intestinal lumen, resulting in toxic metabolite formation and intestinal damage. Studies suggest probiotic supplements containing *B. longum* facilitate a decrease in β -glucuronidase activity and associated aberrant crypt formation and colon carcinogenesis (Kulkarni and Reddy 1994). Probiotic interactions with bile acids in the luminal compartment of the intestine promote bile acid metabolism and cholesterol absorption. Bile salt hydrolase (BSH) produced by the probiotics participates in the first reaction of the deconjugation of biliary salts (Pavlović et al. 2012). Due to these beneficial effects of probiotic bacteria, BSH activity has been included in FAO/WHO guidelines for the evaluation of probiotics for food use (Plaza-Diaz et al. 2019). Enzymatic deconjugation of bile acids by BSH from probiotics has been one of the main mechanisms of the hypocholesterolemia effect attributed to probiotics (Pavlović et al. 2012; Kumar et al. 2011).

5.4.1.2.2 Probiotics and Intestinal Volatile Fatty Acid Signaling

Short-chain fatty acids (SCFAs) are not only the key signaling molecules in enterocytes but also an important source of their energy. Together, they maintain gut health. Once absorbed, SCFAs also stimulate cell surface receptors in several other tissues *via* entering systemic circulation. For example, studies using animal models and *in vitro* studies showed SCFAs promote intestinal secretion of polypeptide YY and glucagon-like peptide 1 and enhance satiety through interaction with G protein-coupled receptors (GPR) 41 and GPR43, respectively (Canfora et al. 2015). Studies suggest probiotics promote increased production of SCFAs as determined through increased fecal acetate, propionate, and butyrate in a probiotics-treated group compared to a non-treated group. Studies also show continued production of some of these fatty acids in probiotic-treated groups even after the cessation of probiotic treatment when compared with a nontreated group (Olivares et al. 2006).

5.5 CLINICAL SIGNIFICANCE OF PROBIOTICS AND POTENTIAL APPLICATIONS

The use of probiotics for clinical health benefits is a fascinating area of research that the present era has yet to explore. Some of the elite properties of probiotics, such as anti-pathogenicity, anti-diabetic, anti-obesity, anti-inflammatory, anti-cancer, anti-allergic, and angiogenic activities, and their effect on the brain and central nervous system (CNS) are briefly discussed below and also depicted in Figure 5.2.

5.5.1 Anti-Pathogenic Activity of Probiotics

Anti-pathogenic activity is regarded as one of the most beneficial effects of probiotics because, unlike classic antibiotics, disturbances or alterations in the composition of the complex population of the gut microbiota are inhibited. There has been considerable research on the anti-pathogenic activity of probiotics or a probiotic mixture. Tejero-Sarinena et al. (2013) postulated that probiotics inhibit pathogens by the production of SCFAs, such as acetic, propionic, butyric, and lactic acids. SCFAs help to maintain an appropriate pH in the colonic lumen, which is imperative in the expression of numerous bacterial enzymes and in the metabolism of foreign compounds and carcinogens in the gut (Kareem et al. 2014). Islam (2016) also suggested that a wide variety of anti-pathogenic compounds, such as bacteriocins, ethanol, organic acids, diacetyl, acetaldehydes, hydrogen peroxide (H_2O_2), and peptides are produced by many probiotics. Among these compounds, peptides and

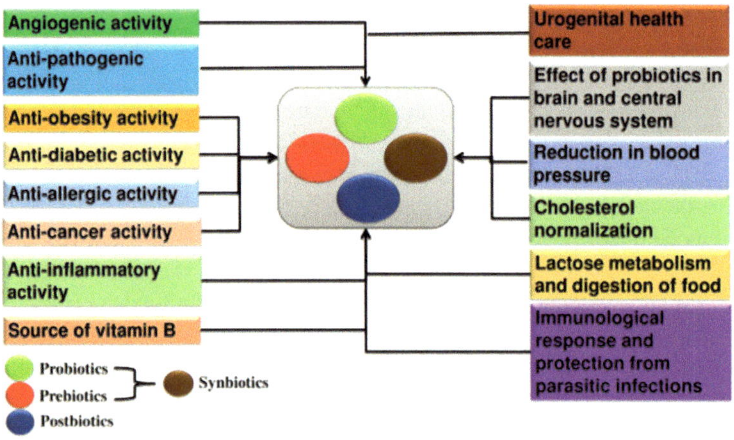

FIGURE 5.2 Applications of probiotics and their modes of action.

bacteriocins, in particular, are mostly involved in increasing the membrane permeability of the target cells, which leads to the depolarization of the membrane potential and, ultimately, cell death (Simova et al. 2009). In addition to producing anti-pathogenic bioactive compounds that directly affect pathogens, probiotics also stimulate host anti-pathogenic defense pathways, such as stimulating or activating the pathway involved in the production of defensins that are cationic anti-microbial peptides produced in several cell types, including Paneth cells in the crypts of the small intestine and intestinal epithelial cells (Figueroa-González et al. 2011). Another mechanism by which probiotics exert anti-pathogenic activity is by competing for pathogen binding and receptor sites, as well as for available nutrients and growth.

5.5.2 Urogenital Healthcare

It is well-known that there is an association between abnormal vaginal microbial flora and an increased incidence of urinary tract infection (UTI). There are about 50 different species inhabiting the vagina, such as *Lactobacillus* species *Lactobacillus brevis*, *Lactobacillus casei*, *Lactobacillus vaginalis*, *Lactobacillus delbrueckii*, *Lactobacillus salivarius*, *Lactobacillus reuteri*, and *Lactobacillus rhamnosus*, which are regarded as the main regulators of the vaginal microenvironment. Imbalance in the microbial composition greatly influences the health of the vaginal microenvironment, potentially leading to the compromised states of bacterial vaginosis (BV) and UTI. These compromised states can be reassured by balancing the number of *Lactobacillus* sp. *via* the supplementation of probiotics (Waigankar et al. 2011).

5.5.3 Anti-Diabetic Activities of Probiotics

Enhancing beneficial microbiota by the use of probiotics is expected to play a significant role in the neutralization of diabetes (Larsen et al. 2010). Gram-negative Bacteroidetes and Gram-positive Firmicutes are two specific bacterial phyla that dominate the gut microenvironment. Recent research has proven that obesity is associated with increased Bacteroidetes over time, concurrent with a reduction in Firmicutes (Kobyliak et al. 2016). More specifically, patients with type-2 diabetes have significantly reduced numbers of Firmicutes species, such that the Bacteroidetes/Firmicutes ratio has increased, which positively correlates with plasma glucose concentration (Barrett et al. 2012). A similar pattern has been implicated in the development of autoimmune diseases, such as type-1 diabetes (Hu et al. 2015). Alterations in the microbiome also increase the invasion of opportunistic pathogens, which are resistant to oxidative stress and simultaneously capable of reducing sulfates and inhibiting the growth of butyrate-producing bacteria (Hartstra et al. 2015). Management of type-2 diabetes by modulating gut hormones, such as gastric inhibitory polypeptide and glucagon-like peptide-1 *via* probiotic and prebiotic interventions is another convincing strategy. In this context, hormones play an implicated role in glucose homeostasis, which results in neutralizing the disorder caused by peripheral insulin resistance or the failure of β-cells to produce insulin (Grover et al. 2012). Currently, research is focused on generating new prebiotics, such as arabinoxylan and arabinoxylan oligosaccharides, which show promising results in counteracting related metabolic disorders, because both carbohydrates have been linked to adiposity reduction.

5.5.4 Anti-Obesity Activity of Probiotics

Probiotics possess physiological functions that contribute to the health of the host environment regulating microbes. In most instances, weight loss is facilitated by thermogenic and lipolytic responses through stimulating the sympathetic nervous system (Karimi et al. 2015). Probiotic strains of *Lactobacillus gasseri* BNR17 have shown properties of inhibiting the increase in adipocyte tissue that is the main source of leptin and adiponectin, thereby limiting leptin secretion (Kang et al.

2013). Other probiotic microbes such as *L. casei*, *Lactobacillus acidophilus*, and *Bifidobacterium longum* have also been reported to have hypocholesterolemic effects (Karimi et al. 2015).

5.5.5 Anti-Inflammatory Activity of Probiotics

Crohn's disease (CD) and ulcerative colitis (UC) are among the most chronic inflammatory diseases of the gastro-intestinal tract (GIT) and are collectively referred to as inflammatory bowel diseases (IBDs) (Iqbal et al. 2014). Research has shown that an imbalance in the gut microbiota plays an important pathophysiological role in the positive regulation of IBD. It is also understood that the disorder could possibly be altered by supplementation with probiotics, prebiotics, and synbiotics (Cammarota et al. 2016). IBD is associated with impaired production of SCFAs, particularly acetate, butyrate, and propionate. Moreover, these SCFAs have been known to play a key role in maintaining colonic homeostasis. They also possess anti-inflammatory effects and improve the propulsive colonic function (Currò et al. 2017). Therefore, it is reasonable to consider that supplementation with indigestible carbohydrates and fiber (prebiotic) alone, or in combination with probiotics to increase the production of SCFAs, could be useful therapeutic approaches. Presently, progress in the field is mostly concerned with developing genetically engineered probiotic bacterial strains that are able to produce and discharge immunomodulators, such as interleukin-10, trefoil factors (compact proteins co-expressed with mucins in the GIT), or lipoteichoic acid (a major constituent of the cell wall of Gram-positive bacteria) that can impact the host immune system, resulting in the restoration of the level of protective commensal bacterial species (Shahverdi 2016). *Lactobacillus*, *Bifidobacterium*, *Enterobacter*, and *E. coli* are the most widely used probiotics in foods. Apart from these organisms, new or genetically modified strains should be developed to counteract IBD (Gowri et al. 2016).

5.5.6 Anti-Cancer Activity of Probiotics

Natural sources that confer anti-carcinogenic effects, such as probiotics, have been receiving prime focus in recent years (Gayathri and Rashmi 2016). *In vitro* studies have demonstrated that probiotic strains *Lactobacillus fermentum* NCIMB-5221 and -8829 are highly potent in suppressing colorectal cancer cells and promoting normal epithelial colon cell growth through the production of SCFAs (ferulic acid). This ability was also compared with other probiotics, namely *L. acidophilus* ATCC 314 and *L. rhamnosus* ATCC 51303, both of which were previously characterized with anti-tumorigenic activity (Kahouli et al. 2015). Again, two different probiotic strains, *L. acidophilus* LA102 and *L. casei* LC232, have been found to show pronounced cytotoxic activities, with *in vitro* anti-proliferative activity against two colorectal cancer cell lines (Caco-2 and HRT-18) (Awaisheh et al. 2016).

5.5.7 Anti-Allergic Activity of Probiotics

In recent times, the beneficial role of probiotics in the protection and management of allergic diseases has advanced the understanding of their cause and prevention. *In vitro* studies of certain probiotics, such as *Lactobacillus plantarum* L67, have shown the potential to prevent allergy-associated disorders with the production of interleukin-12 and interferon-γ in their host (Song et al. 2016). In another study, *L. plantarum* 06CC2 significantly alleviated allergic symptoms and reduced the levels of total immunoglobulin E, ovalbumin-specific immunoglobulin E, and histamine in the sera of ovalbumin-sensitized mice. In spleen cells of the mice, *L. plantarum* 06CC2 is known to significantly enhance the secretions of interferon-γ and interleukin-4, which are responsible for alleviating allergic symptoms (Takeda et al. 2014).

5.5.8 Effect of Probiotics on Brain and CNS

The effect of probiotics on the CNS has been mainly studied in clinical trials, where it has been evident that gut microbiota influence human brain development function (Tillisch 2014). In children with autism spectrum disorder, a daily dose of *L. plantarum* WCFS1 (4.5×10^{10} CFU/day) led to an improvement in their school records and attitude towards food (Umbrello and Esposito 2016). Messaoudi et al. (2011) discussed reduced psychological distress in a randomized trial involving healthy volunteers treated with an oral administration of *Lactobacillus helveticus* R0052 and *B. longum* R0175. In another clinical trial, Rao et al. (2009) showed a decrease in anxiety symptoms by the administration of *L. casei* strain Shirota to patients suffering from chronic fatigue syndrome. However, despite an increase in the *Lactobacillus* and *Bifidobacteria* levels, the bowel functions were not studied. Hence, it is feasible that the reduced anxiety was due to improved bowel function. Szajewska (2016) reported that autism spectrum and attention-deficit/hyperactivity disorders in children could be prevented by *L. rhamnosus* administration to the mother at four weeks from expected delivery. It has been observed that many gut bacteria synthesize neuroactive compounds similar to those produced in the host brain. Human intestinally derived strains of *L. brevis* DPC6108 and *Bifidobacterium dentium* were reported to produce large amounts of γ-aminobutyric acid, a brain neurotransmitter that helps humans to suppress anxiety and depression (Barrett et al. 2012). Doses of a multispecies probiotic containing *L. brevis* W, *B. lactis* W, *L. acidophilus* W37, *Bifidobacterium bifidum* W2, *L. salivarius* W2, *L. casei* W5, and *Lactococcus lactis* (W19 and W58) to healthy humans showed a significant overall reduction in cognitive reactivity to sad mood (Daliri et al. 2016).

5.5.9 Engineering Probiotics for Amelioration of Metabolic Disorders

Enzymes play a central role in cellular metabolism and catalyze complicated biological processes to maintain life. The steady state of the metabolism depends upon multiple enzymatic reactions which can be interrupted by enzyme deficiency (Yen et al. 2009). The missing or defective enzyme results in metabolic disorders in which toxic metabolites may accumulate, or essential products may not be produced (Picca and Dionisi-Vici 2019). Taking enzyme replacements to restore metabolism and removing toxic products or inhibiting their synthesis are promising treatments for metabolic disorders. Many researchers have suggested that engineered probiotics harboring specific enzymes or pathways are able to relieve metabolic disorders in organisms (Isabella et al. 2018; Kurtz et al. 2019).

5.5.10 Modulation of Host Immunobiology by Probiotics

Studies suggest that probiotics suppress intestinal inflammation through several TLR signaling-targeted mechanisms, including through downregulation of TLR expression and the secretion of metabolites capable of inhibiting extracellular proinflammatory cytokines such as TNF-α, or through intracellular inhibition of the transcription factor NF-κB responsible for the expression of inflammatory cytokines in enterocytes (Wells 2011). Other mechanisms through which probiotics suppress inflammation involve suppression of IL-12 production by host cells (Shida et al. 2009) and reinforcing epithelial barrier functions, promoting tight-junction formation (Karczewski et al. 2010). Probiotics also facilitate TLR-mediated T-helper 1 cell differentiation, thereby augmenting antibody production, increasing phagocytic and natural killer cell activity, and upregulating the anti-inflammatory cytokine interleukin- (IL-) 10 and transforming growth factor beta while concurrently downregulating proinflammatory cytokines such as TNF-α , IFN-γ , and IL-8 (Shanahan and Quigley 2014; Naseer et al. 2020). Probiotic *Bifidobacterium longum* spp. showed reduced C-reactive protein levels in patients with ulcerative colitis (Groeger et al. 2013). Indeed, the probiotic also showed dampening of the inflammatory response towards common pathogens such as

Salmonella typhimurium, where the exopolysaccharide coat was shown to have anti-inflammatory effects (Altmann et al. 2016). In addition, the probiotic strain *Faecalibacterium prausnitzii* was reported to have protective effects in patients with IBD through the induction of IL-10, impacting both human and murine dendritic cells, thereby preventing the onset of chronic inflammation (Cao et al. 2014).

5.5.11 Probiotics in Ulcerative Colitis

Probiotics are speculated to work through changing the existing microbiota by replacing the aggressive bacteria with nonaggressive bacteria, thereby mitigating inflammatory responses (Fedorak 2010). There is also a lack of convincing data supporting the efficacy of some probiotics, such as *Lactobacillus*, in the prevention and/or treatment of UC in humans. However, *Escherichia coli* Nissle is one of the promising probiotics that is recommended by the ECCO guidelines as a maintenance therapy for UC. Moreover, the use of *Escherichia coli* Nissle may also worsen the remission of ulcerative colitis in some individuals (Petersen et al. 2014). A comparison of probiotic milk (*Bifidobacterium breve*, *Bifidobacterium bifidum*, and *Lactobacillus acidophilus)* versus placebo is reported to have significantly improved the change in clinical activity index and histological score in the probiotic group during the evaluation of the induction of remission (Kato et al. 2004). The mechanism of action for these effects includes, among others, a reduction of TNF-α , interferon-γ , improvement of the colonic barrier function, conversion of linoleic acid into conjugated linoleic acid, inhibition of TNF-α -induced IL-8 secretion, and upregulation of mucin expression (Gionchetti et al. 2003).

5.6 FINAL CONSIDERATIONS AND FUTURE DIRECTION

Probiotic engineering is the frontier of synthetic biology, tailored with natural science, with great promise to treat previously untreatable chronic inflammatory conditions such as IBD. Emerging technologies such as CRISPR-Cas system and other genetic engineering tools have unlimited potential that can also be utilized to manipulate probiotic bacterial systems to produce useful therapeutic biologics with the goal to either ameliorate and/or treat previously untreatable chronic inflammatory conditions. As new findings increase our understanding of viable bacterial strains and synthetic biology tools, in the foreseeable future, it is highly possible to identify and characterize additional probiotic bacterial strains as potential candidates for probiotic engineering. As the gut microbiome is diverse and varies among different ethnic, geographic, and cultural groups, it will be important to develop a vision for personalized probiotic treatments and develop the technology to make personalized probiotic options a reality. Researchers must also recognize certain performance metrics to ensure that these genetically engineered bacteria are noninvasive and have a site-specific, localized action. With this in mind, it will not be too long before probiotics can be used as a safe, long-term efficacy alternative to treat IBD through the restoration of the normal gut microbiota.

5.7 CONCLUSIONS

Utilization of probiotics as dietary management for the treatment of diseases has gained much attention in the food and medicine industries. The development of metabolic engineering and synthetic biology enables disclosing their mechanisms of action and creating novel probiotic strains with desired functions. Engineering of probiotic strains for the detection and treatment of metabolic disorders, inflammations, and pathogen infections has made great progress in recent years. These advancements enable us to provide insights into the principles and limitations for the development of the perfect probiotics for human health. We believe that the increasing knowledge of the human microbiome and the mechanisms of disease, along with the fast development of metabolic engineering, synthetic biology, and other disciplines, will facilitate optimization of strategies in the design and construction of robust and effective probiotics to prevent disease and improve human health.

REFERENCES

Altmann, F., Kosma, P., O'Callaghan, A., Leahy, S., Bottacini, F., Molloy, E., & O'Mahony, L. (2016). Genome analysis and characterisation of the exopolysaccharide produced by Bifidobacterium longum subsp. longum 35624™. *PloS one, 11*(9), e0162983.

Awaisheh, S. S., Obeidat, M. M., Al-Tamimi, H. J., Assaf, A. M., El-Qudah, J. M., & Rahahleh, R. J. (2016). In vitro cytotoxic activity of probiotic bacterial cell extracts against Caco-2 and HRT-18 colorectal cancer cells. *Milk Science International-Milchwissenschaft, 69*(7), 33–37.

Barrett, E., Ross, R. P., O'Toole, P. W., Fitzgerald, G. F., & Stanton, C. (2012). γ-Aminobutyric acid production by culturable bacteria from the human intestine. *Journal of Applied Microbiology, 113*(2), 411–417.

Barrett, H. L., Callaway, L. K., & Nitert, M. D. (2012). Probiotics: A potential role in the prevention of gestational diabetes?. *Acta diabetologica, 49*, 1–13.

Bermudez-Brito, M., Plaza-Díaz, J., Muñoz-Quezada, S., Gómez-Llorente, C., & Gil, A. (2012). Probiotic mechanisms of action. *Annals of Nutrition and Metabolism, 61*(2), 160–174.

Bongaerts, G. P., & Severijnen, R. S. (2016). A reassessment of the PROPATRIA study and its implications for probiotic therapy. *Nature Biotechnology, 34*(1), 55–63.

Cammarota, G., Pecere, S., Ianiro, G., Masucci, L., & Currò, D. (2016). Principles of DNA-based gut microbiota assessment and therapeutic efficacy of fecal microbiota transplantation in gastrointestinal diseases. *Digestive Diseases, 34*(3), 279–285.

Canfora, E. E., Jocken, J. W., & Blaak, E. E. (2015). Short-chain fatty acids in control of body weight and insulin sensitivity. *Nature Reviews Endocrinology, 11*(10), 577–591.

Cao, Y., Shen, J., & Ran, Z. H. (2014). Association between Faecalibacterium prausnitzii reduction and inflammatory bowel disease: A meta-analysis and systematic review of the literature. *Gastroenterology Research and Practice.*

Cha, B. K., Jung, S. M., Choi, C. H., Song, I. D., Lee, H. W., Kim, H. J., ... Seo, J. G. (2012). The effect of a multispecies probiotic mixture on the symptoms and fecal microbiota in diarrhea-dominant irritable bowel syndrome: A randomized, double-blind, placebo-controlled trial. *Journal of Clinical Gastroenterology, 46*(3), 220–227.

Chua, K. J., Kwok, W. C., Aggarwal, N., Sun, T., & Chang, M. W. (2017). Designer probiotics for the prevention and treatment of human diseases. *Current Opinion in Chemical Biology, 40*, 8–16.

Cicenia, A., Scirocco, A., Carabotti, M., Pallotta, L., Marignani, M., & Severi, C. (2014). Postbiotic activities of lactobacilli-derived factors. *Journal of Clinical Gastroenterology, 48*, S18–S22.

Currò, D., Ianiro, G., Pecere, S., Bibbò, S., & Cammarota, G. (2017). Probiotics, fibre and herbal medicinal products for functional and inflammatory bowel disorders. *British Journal of Pharmacology, 174*(11), 1426–1449.

Daliri, E. B. M., Oh, D. H., & Lee, B. H. (2016). Psychobiotics; a promise for neurodevelopmental therapy. *Journal of Probiotics & Health, 4*, 1e4.

Duan, F. F., Liu, J. H., & March, J. C.(2015). Engineered commensal bacteria reprogram intestinal cells into glucose-responsive insulin-secreting cells for the treatment of diabetes. *Diabetes, 64*, 1794–1803.

Fedorak, R. N. (2010). Probiotics in the management of ulcerative colitis. *Gastroenterology & Hepatology, 6*(11), 688.

Figueroa-González, I., Cruz-Guerrero, A., & Quijano, G. (2011). The benefits of probiotics on human health. *Journal of Biochemical & Microbial Technology S, 1*, 1948–5948.

Gareau, M. G., Sherman, P. M., & Walker, W. A. (2010). Probiotics and the gut microbiota in intestinal health and disease. *Nature Reviews Gastroenterology & Hepatology, 7*, 503.

Gayathri, D., & Rashmi, B. S. (2016). Anti-cancer properties of probiotics: a natural strategy for cancer prevention. *EC Nutrition, 5*(4), 1191–1202.

Gionchetti, P., Rizzello, F., Helwig, U., Venturi, A., Lammers, K. M., Brigidi, P., & Campieri, M. (2003). Prophylaxis of pouchitis onset with probiotic therapy: A double-blind, placebo-controlled trial. *Gastroenterology, 124*(5), 1202–1209.

Giorgetti, G., Brandimarte, G., Fabiocchi, F., Ricci, S., Flamini, P., Sandri, G., ... Tursi, A. (2015). Interactions between innate immunity, microbiota, and probiotics. *Journal of Immunology Research.*

Gómez-Llorente, C., Munoz, S., & Gil, A. (2010). Role of Toll-like receptors in the development of immunotolerance mediated by probiotics. *Proceedings of the Nutrition Society*, *69*(3), 381–389.
Gowri, R. S., Meenambigai, P., Prabhavathi, P., Rajeswari, P. R., & Yesudoss, L. A. (2016). Probiotics and its effects on human health-A review. *International Journal of Current Microbiology and Applied*, *5*(4), 384–392.
Groeger, D., O'Mahony, L., Murphy, E. F., Bourke, J. F., Dinan, T. G., Kiely, B., & Quigley, E. M. (2013). *Bifidobacterium* infantis 35624 modulates host inflammatory processes beyond the gut. *Gut Microbes*, *4*(4), 325–339.
Grover, S., Rashmi, H. M., Srivastava, A. K., & Batish, V. K. (2012). Probiotics for human health–new innovations and emerging trends. *Gut Pathogens*, *4*(1), 1–14.
Hamasalim, H. J. (2016). Synbiotic as feed additives relating to animal health and performance. *Advances in Microbiology*, *6*(4), 288–302.
Hartstra, A. V., Bouter, K. E., Bäckhed, F., & Nieuwdorp, M. (2015). Insights into the role of the microbiome in obesity and type 2 diabetes. *Diabetes Care*, *38*(1), 159–165.
Hu, C., Wong, F. S., & Wen, L. (2015). Type 1 diabetes and gut microbiota: Friend or foe?. *Pharmacological Research*, *98*, 9–15.
Hutkins, R. W., Krumbeck, J. A., Bindels, L. B., Cani, P. D., Fahey Jr, G., Goh, Y. J., ... Sanders, M. E. (2016). Prebiotics: why definitions matter. *Current Opinion in Biotechnology*, *37*, 1–7.
Iqbal, M. Z., Qadir, M. I., Hussain, T., Janbaz, K. H., Khan, Y. H., & Ahmad, B. (2014). Probiotics and their beneficial effects against various diseases. *Pakistan journal of Pharmaceutical Sciences*, *27*(2): 405-15.
Isabella, V. M., Ha, B. N., Castillo, M. J., Lubkowicz, D. J., Rowe, S. E., Millet, Y. A., Anderson, C. L., Li, N., Fisher, A. B., West, K. A., et al. (2018). Development of a synthetic live bacterial therapeutic for the human metabolic disease phenylketonuria. *Nature Biotechnology*. *36*, 857–864.
Islam, S. U. (2016). Clinical uses of probiotics. *Medicine*, *95*(5), ISO 690.
Kahouli, I., Malhotra, M., Alaoui-Jamali, M., & Prakash, S. (2015). In-vitro characterization of the anticancer activity of the probiotic bacterium Lactobacillus fermentum NCIMB 5221 and potential against colorectal cancer. *Journal of Cancer Science and Therapy*, *7*(7), 224–235.
Kang, J. H., Yun, S. I., Park, M. H., Park, J. H., Jeong, S. Y., & Park, H. O. (2013). Anti-obesity effect of Lactobacillus gasseri BNR17 in high-sucrose diet-induced obese mice. *PloS One*, *8*(1), e54617.
Karczewski, J., Troost, F. J., Konings, I., Dekker, J., Kleerebezem, M., Brummer, R. J. M., & Wells, J. M. (2010). Regulation of human epithelial tight junction proteins by Lactobacillus plantarum in vivo and protective effects on the epithelial barrier. *American Journal of Physiology-Gastrointestinal and Liver Physiology*, *298*(6), G851–G859.
Kareem, K. Y., Hooi Ling, F., Teck Chwen, L., May Foong, O., & Anjas Asmara, S. (2014). Inhibitory activity of postbiotic produced by strains of Lactobacillus plantarum using reconstituted media supplemented with inulin. *Gut Pathogens*, *6*(1), 1–7.
Karimi, G., Sabran, M. R., Jamaluddin, R., Parvaneh, K., Mohtarrudin, N., Ahmad, Z., ... Khodavandi, A. (2015). The anti-obesity effects of Lactobacillus casei strain Shirota versus Orlistat on high fat diet-induced obese rats. *Food & Nutrition Research*, *59*(1), 29273.
Kato, K., Mizuno, S., Umesaki, Y., Ishii, Y., Sugitani, M., Imaoka, A., & Arakawa, Y. (2004). Randomized placebo-controlled trial assessing the effect of bifidobacteria-fermented milk on active ulcerative colitis. *Alimentary Pharmacology & Therapeutics*, *20*(10), 1133–1141.
Kobyliak, N., Conte, C., Cammarota, G., Haley, A. P., Styriak, I., Gaspar, L., ... & Kruzliak, P. (2016). Probiotics in prevention and treatment of obesity: a critical view. *Nutrition & Metabolism*, *13*(1), 1–13.
Kulkarni, N., & Reddy, B. S. (1994). Inhibitory effect of Bifidobacterium iongum cultures on the azoxymethane-induced aberrant crypt foci formation and fecal bacterial β-glucuronidase. *Proceedings of the Society for Experimental Biology and Medicine*, *207*(3), 278–283.
Kumar, R., Grover, S., & Batish, V. K. (2011). Hypocholesterolaemic effect of dietary inclusion of two putative probiotic bile salt hydrolase-producing Lactobacillus plantarum strains in Sprague–Dawley rats. *British Journal of Nutrition*, *105*(4), 561–573.
Kurtz, C. B., Millet, Y. A., Puurunen, M. K., Perreault, M., Charbonneau, M. R., Isabella, V. M., Kotula, J. W., Antipov, E., Dagon, Y., & Denney, W. S. (2019). An engineered *E. coli* Nissle improves hyperammonemia and survival in mice and shows dose-dependent exposure in healthy humans. *Science Translational Medicine*, *11*, eaau7975.
Larsen, N., Vogensen, F. K., Van Den Berg, F. W., Nielsen, D. S., Andreasen, A. S., Pedersen, B. K., ... Jakobsen, M. (2010). Gut microbiota in human adults with type 2 diabetes differs from non-diabetic adults. *PLoS One*, *5*(2), e9085.

Messaoudi, M., Lalonde, R., Violle, N., Javelot, H., Desor, D., Nejdi, A., ... Cazaubiel, J. M. (2011). Assessment of psychotropic-like properties of a probiotic formulation (Lactobacillus helveticus R0052 and Bifidobacterium longum R0175) in rats and human subjects. *British Journal of Nutrition*, *105*(5), 755–764.

Naseer, M., Poola, S., Ali, S., Samiullah, S., & Tahan, V. (2020). Prebiotics and Probiotics in Inflammatory Bowel Disease: Where are we now and where are we going?. *Current Clinical Pharmacology*, *15*(3), 216–233.

Nobaek, S., Johansson, M. L., Molin, G., Ahrné, S., & Jeppsson, B. (2000). Alteration of intestinal microflora is associated with reduction in abdominal bloating and pain in patients with irritable bowel syndrome. *The American Journal of Gastroenterology*, *95*(5), 1231–1238.

Olivares, M., Díaz-Ropero, M. P., Gómez, N., Lara-Villoslada, F., Sierra, S., Maldonado, J. A., & Xaus, J. (2006). Oral administration of two probiotic strains, Lactobacillus gasseri CECT5714 and Lactobacillus coryniformis CECT5711, enhances the intestinal function of healthy adults. *International Journal of Food Microbiology*, *107*(2), 104–111.

Ooi, M. F., Mazlan, N., Foo, H. L., Loh, T. C., Mohamad, R., Rahim, R. A., & Ariff, A. (2015). Effects of carbon and nitrogen sources on bacteriocin-inhibitory activity of postbiotic metabolites produced by Lactobacillus plantarum I-UL4. *Malaysian Journal of Microbiology*, 176–184.

O'Shea, E. F., Cotter, P. D., Stanton, C., Ross, R. P., & Hill, C. (2012). Production of bioactive substances by intestinal bacteria as a basis for explaining probiotic mechanisms: bacteriocins and conjugated linoleic acid. *International Journal of Food Microbiology*, *152*(3), 189–205.

Patel, R. M., & Denning, P. W. (2013). Therapeutic use of prebiotics, probiotics, and postbiotics to prevent necrotizing enterocolitis: What is the current evidence?. *Clinics in Perinatology*, *40*(1), 11–25.

Pavlović, N., Stankov, K., & Mikov, M. (2012). Probiotics-interactions with bile acids and impact on cholesterol metabolism. *Applied Biochemistry and Biotechnology*, *168*, 1880–1895.

Peña, A. S. (2007). Intestinal flora, probiotics, prebiotics, synbiotics and novel foods. *Revista Española de Enfermedades Digestivas*, *99*(11), 653.

Petersen, A. M., Mirsepasi, H., Halkjær, S. I., Mortensen, E. M., Nordgaard-Lassen, I., & Krogfelt, K. A. (2014). Ciprofloxacin and probiotic Escherichia coli Nissle add-on treatment in active ulcerative colitis: A double-blind randomized placebo controlled clinical trial. *Journal of Crohn's and Colitis*, *8*(11), 1498–1505.

Picca, S., & Dionisi-Vici, C.(2019). *Hyperammonemia and metabolic diseases. Critical care pediatric nephrology and dialysis: A practical handbook* (pp. 311–323). Singapore: Springer.

Plaza-Diaz, J., Ruiz-Ojeda, F. J., Gil-Campos, M., & Gil, A. (2019). Mechanisms of action of probiotics. *Advances in Nutrition*, *10*(suppl_1), S49–S66.

Pokusaeva, K., Fitzgerald, G. F., & van Sinderen, D. (2011). Carbohydrate metabolism in Bifidobacteria. *Genes & Nutrition*, *6*, 285–306.

Ramírez, A. M, Rodriguez-López, A., Ardila, A., Beltran, L., Patarroyo, C. A., Melendez, A. D. P., Sánchez, O. F., Alméciga-Díaz, C. J. (2017). Production of human recombinant phenylalanine hydroxylase in *Lactobacillus plantarum* for gastrointestinal delivery. *European Journal of Pharmaceutical Sciences*, *109*, 48–55.

Rao, A. V., Bested, A. C., Beaulne, T. M., Katzman, M. A., Iorio, C., Berardi, J. M., & Logan, A. C. (2009). A randomized, double-blind, placebo-controlled pilot study of a probiotic in emotional symptoms of chronic fatigue syndrome. *Gut Pathogens*, *1*(1), 1–6.

Rastall, R. A., & Gibson, G. R. (2015). Recent developments in prebiotics to selectively impact beneficial microbes and promote intestinal health. *Current Opinion in Biotechnology*, *32*, 42–46.

Royan, M. (2019). Mechanisms of probiotic action in the honeybee. *Critical Reviews™ in Eukaryotic Gene Expression*, *29*(2): 95-103.

Shahverdi, E. (2016). Probiotics and gastrointestinal diseases. *International Journal of Digestive Diseases*, *2*, 1–2.

Shanahan, F., & Quigley, E. M. (2014). Manipulation of the microbiota for treatment of IBS and IBD—challenges and controversies. *Gastroenterology*, *146*(6), 1554–1563.

Shida, K., Kiyoshima-Shibata, J., Kaji, R., Nagaoka, M., & Nanno, M. (2009). Peptidoglycan from lactobacilli inhibits interleukin-12 production by macrophages induced by Lactobacillus casei through Toll-like receptor 2-dependent and independent mechanisms. *Immunology*, *128*(1pt2), e858–e869.

Simova, E. D., Beshkova, D. B., & Dimitrov, Z. P. (2009). Characterization and antimicrobial spectrum of bacteriocins produced by lactic acid bacteria isolated from traditional Bulgarian dairy products. *Journal of Applied Microbiology*, *106*(2), 692–701.

Song, S., Lee, S. J., Park, D. J., Oh, S., & Lim, K. T. (2016). The anti-allergic activity of Lactobacillus plantarum L67 and its application to yogurt. *Journal of Dairy Science*, *99*(12), 9372–9382.

Szajewska, H. (2016). What are the indications for using probiotics in children?. *Archives of Disease in Childhood*, *101*(4), 398–403.

Takeda, S., Hidaka, M., Yoshida, H., Takeshita, M., Kikuchi, Y., Tsend-Ayush, C., ... Kurokawa, M. (2014). Antiallergic activity of probiotics from Mongolian dairy products on type I allergy in mice and mode of antiallergic action. *Journal of Functional Foods*, *9*, 60–69.

Tejero-Sariñena, S., Barlow, J., Costabile, A., Gibson, G. R., & Rowland, I. (2013). Antipathogenic activity of probiotics against Salmonella Typhimurium and Clostridium difficile in anaerobic batch culture systems: is it due to synergies in probiotic mixtures or the specificity of single strains?. *Anaerobe*, *24*, 60–65.

Thomas C. and Versalovic J. (2010) Probiotics-host communication: modulation of signaling pathways in the intestine. *Gut Microbes* 1: 148–163.

Thomas, L. V. (2016). Probiotics–the journey continues. *International Journal of Dairy Technology*, *69*(4), 469–480.

Tillisch, K. (2014). The effects of gut microbiota on CNS function in humans. *Gut Microbes*, *5*(3), 404–410.

Tufarelli, V., & Laudadio, V. (2016). An overview on the functional food concept: Prospectives and applied researches in probiotics, prebiotics and synbiotics. *Journal of Experimental Biology and Agricultural Sciences*, *4*(3, Suppl.), 273–278.

Tursi, A., Brandimarte, G., Papa, A., Giglio, A., Elisei, W., Giorgetti, G. M., & Gasbarrini, A. (2010). Treatment of relapsing mild-to-moderate ulcerative colitis with the probiotic VSL# 3 as adjunctive to a standard pharmaceutical treatment: a double-blind, randomized, placebo-controlled study. *The American Journal of Gastroenterology*, *105*(10), 2218.

Umbrello, G., & Esposito, S. (2016). Microbiota and neurologic diseases: Potential effects of probiotics. *Journal of Translational Medicine*, *14*(1), 1–11.

Waigankar, Santosh S.; Patel, Vimal.(2011). Role of probiotics in urogenital healthcare. *Journal of Mid-life Health*, *2*(1), 5–10.

Wells, J. M. (2011). Immunomodulatory mechanisms of lactobacilli. In *Microbial cell factories* (Vol. 10, No. 1, pp. 1–15). BioMed Central.

Yen, C.-L. E., Cheong, M.-L., Grueter, C., Zhou, P., Moriwaki, J., Wong, J. S., Hubbard, B., Marmor, S., Farese, R. V. Jr. (2009). Deficiency of the intestinal enzyme acyl CoA: Monoacylglycerol acyltransferase-2 protects mice from metabolic disorders induced by high-fat feeding. *Nature Medicine*, *15*, 442.

6 The Role of Inulin in Altering the Animal Microbiome

A Comprehensive Review

Amol D. Gholap, Sampada D. Sawant and Sagar R. Pardeshi*
Department of Pharmaceutics, St. John
Institute of Pharmacy and Research
Palghar Maharashtra, India

Pankaj R. Khuspe
Department of Pharmaceutics, Shriram Shikshan
Sanstha's College of Pharmacy
Paniv, Solapur, Maharashtra, India

Sadikali F. Sayyad and Machindra Chavan
Amrutvahini College of Pharmacy
Sangamner Maharashtra, India

Navnath T. Hatvate
Institute of Chemical Technology Mumbai, Marathwada Campus
Jalna, Maharashtra, India

Md Faiyazuddin
School of Pharmacy, Al-Karim University
Katihar, Bihar, India

Md Jasim Uddin
Department of Pharmaceutical Technology,
Faculty of Pharmacy, Universiti Malaya
Kuala Lumpur, Malaysia

6.1 INTRODUCTION

During the previous decade, the globe experienced enormous progress in dietary research that birthed the notion of a healthy diet founded on the recognition of important elements for sustaining growth, maturation, and productivity, as well as the treatment of deficiency diseases. Throughout

* Corresponding author. Email: amolgholap16@gmail.com

DOI: 10.1201/9781003394662-6

the last few decades, research activities have led to the creation of dietary guidelines for numerous organisms, varieties, and productivities from around the world in order to improve efficiency and provide greater yields from cattle companies (Cena and Calder 2020; Popkin, Adair, and Ng 2012). In the mid-twentieth century, among the significant contributions of nutrition research is indeed the principle of the "healthy diet", which can be described (relevant to ruminant animals) as "a suitable combination of fodder and concentrate in the type of mixture or block proficient to satisfy the criteria of nutrient content required for the repairs, progress, and manufacturing of animal life". (Mozaffarian et al. 2018; Cencic and Chingwaru 2010). A "proper diet" and "nutritional quality" were essential themes at the turn of the twenty-first century. But, as lifestyles evolve in the twenty-first century, buyer needs and requirements have completely shifted (Gholap and Deepak, 2011; Gholap and Santhosh, 2012). Prebiotics have subsequently been proposed as a new idea in functional food science because of worries about antibiotics or hormone residues, increasing awareness, and safety considerations. Given the foregoing, animal health studies are currently focusing on feeding security, excellence, designing meat products, nutritious meat products, environmental therapy for gastrointestinal issues, and sustainable animal agriculture, as presented in Figure 6.1. It appears that prebiotics, notably inulin, have the ability to alleviate those difficulties while providing the optimum result for animal productivity. The inclusion of prebiotics in animal nutrition is a comparatively recent venture, and the first results are promising. Even though most food products

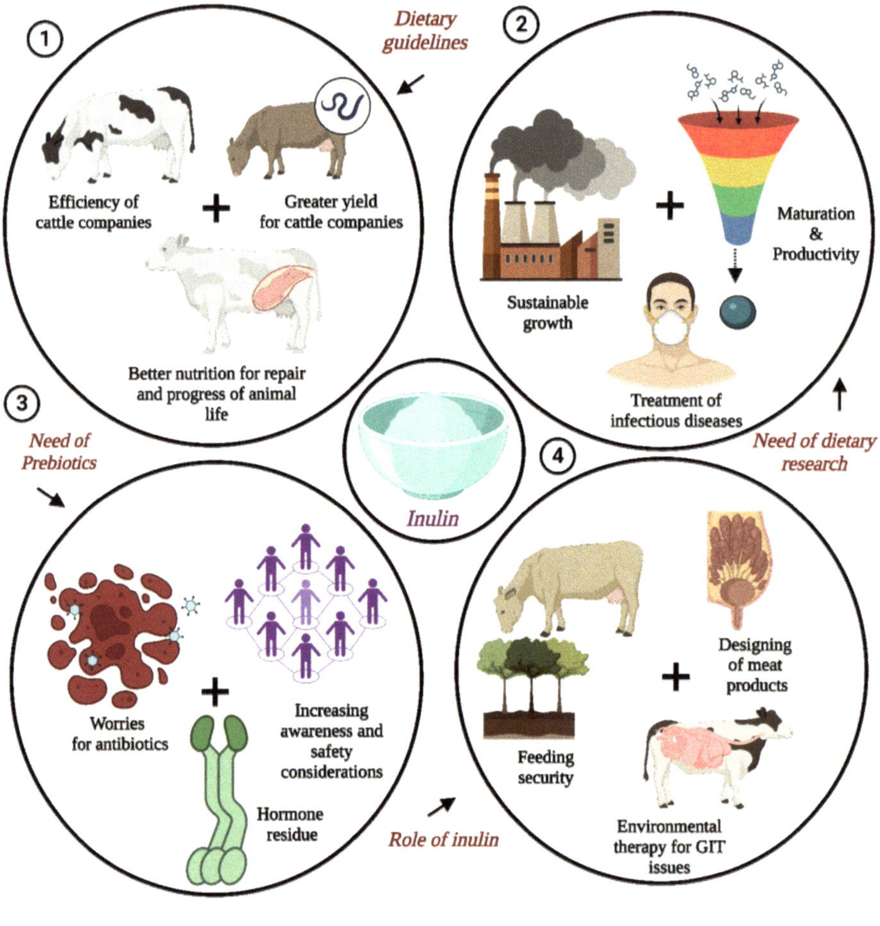

FIGURE 6.1 The summary of prebiotics' status in the twenty-first century includes 1. the need for dietary research, 2. dietary guidelines, 3. the need for prebiotics, and 4. the role of inulin.

address just one or a few activities, prebiotics address a broad spectrum of bodily functions, from stool motility and colon microflora activities to mineral uptake, lipid balance, and immunology (Cena and Calder 2020; Bassaganya-Riera et al. 2021; Cena and Calder 2020).

6.2 GASTROINTESTINAL MICROFLORA

Domestic animals are grouped into two categories based on the presence of bacterial habitation in the intestinal mucosa: single gut fermentors—non-ruminants (chicken and turkey, pig, horse, rabbit, donkey, etc.)—and multi-gut fermenters—ruminants, including pseudoruminants (cattle, camels, sheep, buffalo, mithuns, goats, etc.). Ruminant gastrointestinal pathways contain two bacterial ecosystems: the foregut (intestine, omasum, and reticulum) and the hindgut (cecum) (O'Donnell et al. 2017; Stevens and Hume 1998). Non-ruminants, on the other hand, are identified by the existence of a singular bacterial habitat, namely the hindgut. This same existence of a foregut, including an immense ability to shelter diversified classifications of intestinal flora, helps give ruminant creatures an advantage compared to other living creatures in regards to diet and reliance on lignocellulosic organic matter to fulfill energy demands, such as through volatile fats (butyrate, acetate, propionate, etc.) produced from the deterioration as well as anaerobic digestion of natural fibers (Niba et al. 2009; Xu et al. 2021). The solitary fermenting location, which is located in the hindgut of non-ruminant mammals, results in a complete reliance on volatile fats for energy needs. Rumen liquids are estimated to include (number/mL) 108 to 109 archaea, 102 to 104 fungi, 1010 to 1012 microbes, 104 to 106 protozoa, phages, yeast, and other microbes (Matthews et al. 2019; Cammack et al. 2018; Xu et al. 2021). Likewise, millions more microorganisms, archaea, fungi, and yeast strains live in the hindgut of non-ruminants. Because of the microenvironment accessible in the gastrointestinal system in terms of food, acidity, interaction, temperature, and interaction, all helpful as well as detrimental microorganisms find a place for development as well as reproduction. The difficulty in clarifying some aspects after the conclusion of the human genomic program compels investigators to reconsider gut microflora in order to better understand as well as address unresolved issues. In this aspect, intestinal microbiota are generally believed to develop alongside evolutionary history, while people have indeed been proposed to just be "metaorganisms", with ten-times more germs than animal tissue. Human "metaorganisms" encompass roughly 1014 bacterial microorganisms with a bulk greater than 1 kg and are thus often considered a concealed organ with the capability of a kidney (Samanta et al. 2013; Xu et al. 2021). Full growth of the gastrointestinal system and immune response occurs only after the formation of the intestinal microbiota, suggesting an absence of signaling inside the genetic code to occur through the complete development of the intestinal system or even the immune response (Gholap, Hatvate et al. 2023; Gholap et al. 2024; Gholap, Sayyad et al. 2023; Vilas Bhagat 2012; Rajurkar and Gholap 2012; Anand M. Mahajan and Amol D. Gholap 2011; D. S. M. Amol D. Gholap 2011; Anand M. Mahajan and Amol D. Gholap n.d.; Gholap, Amol D. and Mahajan 2011). The contribution and relevance of microorganisms populating the foregut and hindgut of ruminants, in addition to the foregut of non-ruminants, can be deduced from the example above (Matthews et al. 2019; Vemuri et al. 2020).

6.2.1 Classification of Gut Microflora

Researchers have used traditional methods of cultivation to group intestinal flora depending on strains, form, biochemical properties, food usage, and so on. They have since assumed that only 10% to 20% of microorganisms from the habitat might be suitable for cultivation using conventional cultivation techniques (Ito et al. 2019; Davis 2014). To tackle the problem of degraded land flora, gut environmentalists are currently studying the bacteria using 16S rRNA-based methods that address suitable for cultivation as well as uncultivable intestinal flora. Disregarding the particular substrate usage sequence (hemicellulose, protein, cellulose, pectin, lipid, etc.) as well as biological characteristics, the digestive tract microbial community might be widely divided into two primary groups: infective or harmful or deleterious bacterial flora, and health-promoting or beneficial or bacterium-repressive bacterial flora (Fijan 2014; Berg et al. 2020). Harmful bacteria include *E. coli*, *Salmonella enterica*, *Streptococcus faecalis*, and *Clostridium perfringenss*. *Bifidobacteria* and *Lactobacillus* are thought to be the most important members of the helpful or wellness class (Atanasov et al. 2021; Usta 2013). The importance of prebiotic usage stems from their capacity to preferentially enhance the development and multiplication of indigenous microbes like *Bifidobacteria* and *Lactobacillus* inside the hindgut, which in turn represses the action of fermentative or detrimental gut bacteria, thereby lowering the density of poisonous fermentation (Tsvetkov et al. 2018; Markowiak and Śliżewska 2017; Macfarlane and Cummings 1999).

6.3 HARMFUL FERMENTATION METABOLITES

The large intestine, particularly the cecum, is indeed the principal place for millions of microorganisms from various groups in non-ruminants (chicken, pig, horse, and dog). With ruminants (livestock, bison, sheep, and goat), digested nutrients are subjected to extended activity by a broad set of microorganisms and their vast array of hydrolytic enzymes until they are used by the animal's own stomach or pancreas enzyme ("Diseases of the Alimentary Tract" 2017; Mutungwazi et al. 2022). The intestinal movement involves the passage of processed substances into the large intestine, exposing the consumed contents to the activity of biocatalysts once again. As a consequence, many metabolic events take place in the vicinity of bacterial homes, like the gut or intestine, leading to the creation of large metabolites with either positive or negative activity. During the bacterial fermentation of foodstuffs in the intestinal tract, a variety of hazardous compounds are produced (MOON et al. 2010; Hao et al. 2021). Amines, ammonia, nitrosamines, secondary bile acid, phenols, indole, cresols, oestrogens, skatole, aglycone, and other toxins are significant toxic metabolites of the hindgut. Such compounds can cause liver injury (nitrogen, imidazole, etc.), carcinogenic effects (heterocyclic amines, phenolic compounds, cresols, etc.), genotoxicity (aglycone), and other negative effects. *Clostridium* spp. and *Escherichia coli* (amines, ammonia, phenol, nitrosamines, secondary bile acids, indole, and aglycone), *Bacteroides* spp., *Streptococcus faecalis* (aglycones, secondary bile acids, and nitrosamines), and *Proteus* spp. (amines, ammonia, and indoles) produce these harmful metabolites. Azoreductase is the main enzyme responsible for the creation of carcinogens from procarcinogens inside the intestinal mucosa. *Clostridium* spp. has the highest standard of azoreductase enzymatic activity, followed by *Eubacterium* spp. and *Peptostreptococcus* spp. and *Bifidobacteria* spp., which seems to have no activity. Many gastrointestinal illnesses have been seen as a result of the breakdown of intestinal microbiota equilibrium, in addition to the creation of harmful compounds or the production of enzymes implicated in the manufacture of harmful products (L. Wang et al. 2019; Jakszyn et al. 2004).

6.4 PREBIOTICS

Fermented foods are a class of proteins distinguished for their capacity to encourage the development or proliferation of specific desirable intestinal bacteria. The prohibition of using antibiotics and hormones as food supplements, customer perception, and tight glycemic methods of control are deciding forces in the fields of food products, particularly prebiotic oligosaccharides (Dimidi et al. 2019; Leeuwendaal et al. 2022). Even though the idea of functional ingredients was first invented by the brilliant ancient Greek scholar Hippocrates, who said, "Let meal be thy treatment, and drug be thy meal", empirical evidence has just recently started to support that concept by employing prebiotics for attenuating intestinal flora. The word "prebiotics" was developed recently by Gibson and Roberfroid, who modified "pro" into "pre", meaning "before" or "for" (Vinayamohan et al. 2023; Dey et al. 2023). Prebiotics are poorly absorbed dietary elements that benefit the host by specifically promoting the development or action of a few or a restricted group of intestinal bacteria.

The modified version of probiotic bacteria includes "preferentially cultured ingredients that enable specific modifications in terms of makeup and/or function inside the gut microbiome that impart advantages to human well-being" because they correspond with certain elements of dietary fiber. Later, the Food and Agriculture Organization (FAO) described the prebiotic as "a non-viable food component that confers a benefit to the host's health associated with the manipulation of microorganisms". Even though the impact of most functional meals is restricted to one or a few activities, prebiotics address a variety of biological methods, including improved gut health, increased nutrient uptake, lipid lowering, immune activation, and microbial rejection. As a result, probiotic bacteria are distinguished by their non-digestibility at the stomach level, specific activation of the helpful gut microflora, and natural origin, including, of course, the lack of residual concerns (Devirgiliis, Barile, and Perozzi 2011; Dahiya and Nigam 2022a; X. Wang, Zhang, and Zhang 2021).

6.4.1 Evolution of Prebiotic Concept

Essentially, the innovative notion of prebiotics was developed with the goal of its applicability to public health (Gholap et al., 2011). In the past few decades, there's been a growing fascination with the connection between gastrointestinal flora and various physiological functions of humans, with particular emphasis on symbiotic carbohydrates, which are not digested or used by enzymes produced by multicellulars but possess a glandular structure and are fully available to a small subset of gastrointestinal bacterial diversity, i.e., beneficial intestinal microbiomes . Carbohydrates occur in the form of fructan in over 36,000 distinct species all around the world. Due to early predecessors' inefficient storage of phytoconstituents, physical proof of the consumption of prebiotic-rich crops is virtually non-existent . So far, it can be assumed that history's probiotic bacteria, such as inulin and oligofructose, were consumed by Pliocene and Pleistocene progenitors a million years ago through varied plant materials. Symbiotic products gradually formed a component of human meals in some locations in previous times, and it is believed that the amount of precursor intake may well have exceeded the level consumed by the population today. Prebiotics were defined scientifically in the mid-1990s, recognizing the importance of employing tree compounds to enhance gut health as well as performance. Due to their significance, prebiotics are already thought to be a pioneer in several nutraceuticals utilized for routinely maintaining health in addition to environmental therapy for digestive tract disorders (Bevilacqua et al., 2024; Habib et al., 2024).

6.4.2 Characteristics of Prebiotics

Prebiotics were first proposed when considering human gut ecology. As a result, the parameters used to designate a biomolecule as a prebiotic can vary from those established in previous studies. The following conditions are required for a chemical to become a prebiotic in cattle: its inability to be digested by the animal's natural gastrointestinal or pancreatic enzymes; used exclusively by so-called healthy intestinal microorganisms (except several rumen bacteria ferment prebiotics) (Markowiak and Śliżewska 2017; George Kerry et al. 2018); made from plants or by enzyme production; non-absorption first from the digestive tract's epithelial surface; maintenance of both functional and structural integrity when traveling through the digestive tract at either an alkaline or acidic pH; demonstrates effectiveness even at small doses; maintains integrity during mechanical digestion phase activities such as mastication, biting, combining with liquids; and is not carcinogenic. Its digestion must not result in the production of hazardous metabolites. It is simple to combine with other feed additives or micronutrient blends (Davani-Davari et al. 2019; Sionek et al. 2023; R. P. Singh, Shadan, and Ma 2022).

6.4.3 Inulin-Type Prebiotics

Notwithstanding the reality that the initial idea of a biostimulant was developed with nature and the impact of lactose on the human gastrointestinal tract in mind, only a small amount of prebiotics has emerged that are currently on their way to establishing their significance in the marketplace of healthy ingredients. However, inulin ranks first on the list of prebiotics due to its extensive accessibility from a variety of sources at a low cost, and it consists of a set of biomolecules known as inulin, oligofructose, and fructooligosaccharides (R. P. Singh, Shadan, and Ma 2022; Davani-Davari et al. 2019; Sionek et al. 2023). Inulin and its various forms were found as endogenous storage carbohydrates in a wide range of plants. It was identified as an "unusual material" in 1804 by German scientist Rose from a warm-water extraction from the roots of a plant (*Inula helenium*) of the Compositae family. The term "inulin" was later invented by Thomson in 1818. Molecularly, it is a fructan, composed of sugar units joined by a 2,1 bond, usually terminated by a single glucose unit. Inulin is depicted by the structural formula GFn, while oligofructose is depicted by Fm and/or GFn; "G" stands for glucosyl unit, "F" stands for fructosyl unit, "n" stands for the number of fructose units joined to the end glucose units, and "m" stands for the number of fructose units covalently joined to one another along the carbohydrate chain. Of these, "n" is larger than two and much lower than 60 in the case of inulin. Oligofructose, which is generated by hydrolytic cleavage of inulin, may consist of a combination of both GFn and Fm units, with degree of polymerization (DP) varying between 2 and 7, or 8 and 10, according to the manufacturer (Andersson et al. 2001; Sanjuán et al. 2023). However, only GFn-form oligofructose can be generated from inulin by the hydrolysis process or by the transfer tosylation of sugar monomers in the absence of fructosyl transferase. The very first extension of sugar, including one fructose monomer, generates fructooligosaccharides known as 1-kestose, and subsequent elongation produces 1-nystose, and so on. DP is the number of monomer units (fructose) contained within prebiotic inulin or oligofructose and is affected by plant origin, stage of growth, meteorological circumstances, post-harvest treatment, and other factors. Inulin can additionally contain tiny levels of Fmfructans (F2); however, GFnfructans having DP between 2 and 60 are common for the same (Davani-Davari et al. 2019; Firrman et al. 2023; Nath et al. 2018).

6.4.4 Effect of Inulin

6.4.4.1 Livestock

Over the last 15 years, there have been signs of great interest from across the scientific community in ensuring the multifunctional usage of prebiotics to improve the health of humans (Samanta et al. 2013). As a result, prebiotics have a place in enhancing digestive health and functioning, regularizing fecal output in older people, acting as sweeteners for diabetic patients, environmental therapy for digestive diseases, including bowel inflammation, and so forth. While the first studies on prebiotic application in animal feeds have shown positive results, the quantity of data on cattle is limited. Attempts have been made in this study to integrate that fragmentary material into useful objectives (Cunningham et al. 2021; Teferra 2021).

6.4.4.2 Swine

Swine are subjected to numerous stimuli during their lives, the most critical of which is weaning. Variations in gut microflora composition frequently cause a rupture in gut microbial equilibrium throughout this period. Antibiotics are currently prohibited in food in the European Union, and some nations are contemplating prohibiting them as food supplements. In such cases, prebiotics sourced from plants or produced by bacterial enzymes promise to fill the antibiotic void in order to improve cattle output while keeping the drawbacks of antibiotic use in mind (Cunningham et al. 2021; Zeineldin, Aldridge, and Lowe 2019; Davani-Davari et al. 2019; Sanjuán et al. 2023). On the sixth day of intake, newborn piglets fed oligofructose had a larger quantity of *Bifidobacteria* (1.68 x 1010 vs. 4.85 x 109 CFU/g fecal contents). Following daily administration of 9.5 g of oligofructose, the community of lactobacilli was considerably higher for both cecum (9.3 vs. 7.7 log10 CFU per gram) and colon (9.3 vs. 8.3 log10 CFU per gram) in an effort to determine the potentiality of prebiotic inulin on healing from diarrhea caused by cholera enterotoxin at the age of three weeks. In the same period, the overall community of dangerous bacteria, *Enterobacteria*, was greatly diminished in oligofructose pigs at both the cecum (6.2 vs. 7.4 log10 CFU per gram) and the colon (6.2 vs. 8.5 log10 CFU per gram) compared to pigs who were not given the prebiotics. With prebiotic as well as probiotic supplementation, we observed a substantial rise in the number of overall anaerobic bacteria, *Bifidobacteria* and *Lactobacillus*, with a concurrent reduction in the community of *Enterococci* and *Clostridium*. A few studies, however, failed to find a positive effect of prebiotics on mammals. Switching the cellulose portion with oligofructose in pig meals seemed to have no effect on the intestine community of total anaerobic bacteria, or *Bacillus subtilis*. The same scientists found a non-significant increase in the number of *Bifidobacteria* in the intestinal absorption secretions of feeding piglets. Pig manure has a foul odor because of the presence of microbiota such as phenolic acids, naturally occurring substances, sulfur compounds, essential fats, and so on (Guevarra et al. 2019; Zeineldin, Aldridge, and Lowe 2019). The amount of ammonia in the pigs' intestines as well as their bowels was greatly reduced after chicory inulin was added to their meals. Prebiotics improved the ileum's ability to digest both lower-moisture and organic material. Swine using inulin as feedstuff had significantly lower parasite loads, showing the ability of prebiotics to reduce parasite disease. Weaning swine that ingested oligofructose on a regular basis along with milk substitute material were able to mitigate the negative effects of *E. coli* and showed greater longevity, higher fecal matter percentages of *Bifidobacteria*, and reduced levels of fecal *E. coli*. Due to its prebiotic capability, inulin showed a beneficial change in the gut microbiome, which in turn improved animal performance (Lauridsen et al. 2021; Nath et al. 2018; Firrman et al. 2023; Andersson et al. 2001). Prior to the notion of prebiotics, attempts had been undertaken to clarify how oligofructose affected food consumption and live body weight gain in weaning swine. Due to insufficient amounts of prebiotics, neither oligofructose nor Jerusalem artichoke powder had a discernible impact on swine's feed consumption, everyday growth, or the effectiveness of feed conversion. Subsequent investigations showed that supplementing developing swine either with oligofructose or trans-galactooligosaccharides decreased dry mass consumption, improved everyday mass growth, and improved the effectiveness of feed conversion (Knecht et al. 2020; Guevarra et al. 2019).

6.4.4.3 Poultry

To provide a nutritional habitat for millions of microbiota (1011 CFU/g corresponding to 200 or more types), as well as to take advantage of the synergistic effect of bacterial fermentation, poultry, like other animals, possess a specialized organ (ceca) near the terminal part of the gastrointestinal tract (Marcolla, Ju, and Willing 2023; Shang et al. 2018). The bacteria of the poultry intestine are roughly categorized as follows: dominant, sub-dominant, and transitory groups. The main species present are *Enterococcus*, *Lactobacillus*, and *Lactobacillus Enterobacteriaceae*, which are present in the cecum at the start of a chick's development. After the second week of the chick's development, *Bacteroides* and *Eubacterium* spp. become dominant. The microbiota in the intestine of chickens is thought to have a variety of functions, including promoting digestion, metabolic activity, rejecting parasites, boosting the immune system, and promoting vitamin production (Siddique et al. 2021; Dapkevicius et al. 2021; Hanchi et al. 2018).

It has been observed that in an in vitro culture system, Salmonella serotype was not able to develop when oligofructose served as the only energy source, whereas so-called advantageous microbiota, such as *Lactobacillus lactis*, *Pediococcus*, and *Enterococcus faecium*, could develop well because they were able to precipitate powerful enzymes in terms of dissolving the osidic bonds existing in prebiotics. Including inulin as a prebiotic is thus significant for preventing *Salmonella* infections in hens (Tinrat et al. 2018; King, Adeola, and Asem 2000). Oligofructose was added to the meals of chicks, which greatly reduced the amount of *Salmonella* infections in the gastrointestinal tract. The performance of growth was improved when inulin was added as prebiotics to the diet of layers as well as broiler birds. Contrarily, several studies failed to detect an improvement in growth and yield after inulin administration in birds. The degree of incorporation, type of feed, animal uniqueness, degree of cleanliness, etc. may all play a role in the variation in inulin's efficacy. The relative mass and length of the bird's intestines and lining of the stomach, intestine, and anal canal were unaffected by the dietary addition of progressive quantities of inulin. It's possible that inulin's rising water absorption as well as its limited ability to raise the fluidity of digestion are also responsible for its failure to successfully affect the overall intestinal conformation. The mass and length of the tibia were unaffected by the addition of prebiotic inulin to broilers fed; however, the apparent accumulation of Ca, Zn, and Cu was linearly improved. Research showed that prebiotics can promote mineral absorption and bone mineralization, improving the ash and calcium contents of the tibia in broiler birds receiving inulin (Figure 6.2). Prebiotic inulin is used by a specific set of microflora, *Bifidobacteria* and *Lactobacillus*, in the intestinal tract to produce a variety of short-chain fatty acids and lactic acids instead of being digested by the bird's natural stomach or pancreas enzymes (Samanta et al. 2013; Piskin et al. 2022). Lowered luminal pH is a result of the large intestine's synthesis of fatty acids. It has been discovered that an acidic pH promotes increased absorption of nutrients as well as increases mineral bioavailability. In this regard, butyrate outperforms acetate in terms of ensuring greater absorption. Lowered luminal pH is a result of the large intestine's synthesis of fatty acids. It has been discovered that an acidic pH promotes increased absorption of nutrients as well as increases mineral bioavailability. In this regard, butyrate outperforms acetate in

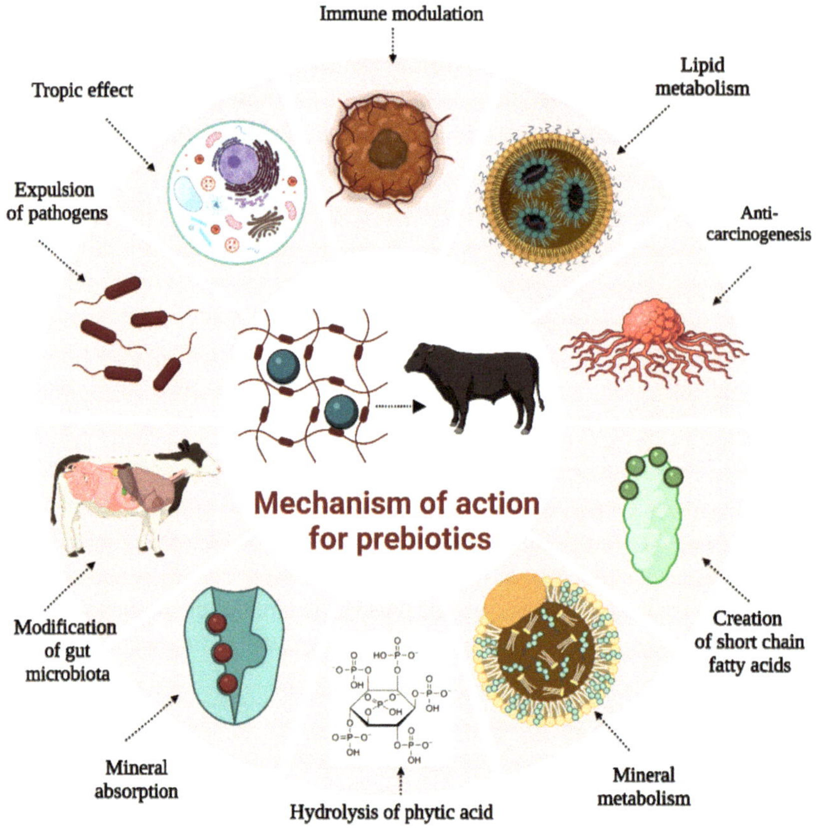

FIGURE 6.2 The mechanism of action for prebiotics consists of trophic effect, hydrolysis of phytic acid, modulation of gut microflora, immune modulation, lipid metabolism, mineral metabolism, and anti-carcinogenesis, etc.

terms of guaranteeing greater Ca uptake (Silva, Bernardi, and Frozza 2020; Xiao, Metzler-Zebeli, and Zebeli 2015).

6.4.4.4 Dogs

German Shepherd dogs with intestinal bacterial overgrowth were given oligofructose for about 50 days at a dose of 1% of their diet (w/w). This considerably decreased the number of total gut bacteria, facultative aerobes in the duodenal fluid, and mucosa (Samanta et al. 2013; Sasso et al. 2023). It is thought that extra oligofructose is selectively digested and used by enteric bacteria as well as bacterial strains, preventing harmful bacteria from adhering to and surviving inside the intestines of beagles. When female hounds consume graded amounts of oligofructose, the quantity of *Clostridium perfringens* in their feces significantly decreases, but the population of *Bifidobacteria* and overall aerobe concentrations increases (Topping and Clifton 2001; Cryan et al. 2019; Minamoto et al. 2019). Oligofructose ferments quickly in the presence of dog feces microflora in vitro, producing short-chain fats including acetate, butyrate, and propionate. The colonic epithelium is thought to respond favorably to short-chain fatty acids. Oligofructose-consuming dogs had longer, heavier small intestines with more absorption surface area available (Propst et al. 2003; Twomey et al. 2003). After oligofructose supplementation, fecal concentrations of ammonia, cadaverine, putrescine, and total phenols were observed to be lower. Inulin consumption in dogs at a concentration of 7% of food (w/w) had no influence on overall perceived dry mass digestibility, but it did cause organic compounds, protein content, and fat digestibility to be at lower levels than the control group. Contrary to what was stated above, no differences in the perceived dry mass, organic compounds, protein content, or fat digestibility were seen after giving the healthy dogs varying amounts of oligofructose as a supplement. Following regular oligofructose ingestion over a period of six weeks, dogs' plasma cholesterol levels drastically decreased (Flickinger et al. 2003; A. M. Pereira et al. 2020).

6.4.4.5 Ruminant Livestock

The cecum serves as the principal location for decomposition in rabbits, which is carried out by a number of anaerobic species. The diets of rabbits are designed to be high in protein and carbohydrates but low in nutritional value. In order to maintain the homeostasis of the gut microbiota in relation to its functionality, it appears that the diet of rabbits should be provided with sufficient amounts of fermentable carbohydrates (prebiotic) (S. Li, Tang, and Le 2023; Combes et al. 2013). Oligofructose is commonly added to European feeds in amounts of 1 to 3 g/kg to promote gut health and lower death rates. In contrast to control animals, fewer rabbits showed clinical signs of infection in humans during *E. coli*-related diarrhea after consuming oligofructose. Furthermore, prebiotic oligosaccharide-consuming rabbits gained more weight on average each day and had greater body weights. As a result of the formation of short-chain fatty acids from the fermentation of prebiotics, a relatively low amount of oligofructose in the food of rabbits is capable of lowering the gastrointestinal pH. Ruminant animals have millions of different kinds of microflora, including bacteria, fungi, yeasts, phage particles, archaea, etc., with the exception of protozoa, which are thought to be present solely in the foregut and absent in the hindgut (Huang et al. 2021; Samanta et al. 2013). Many rumen bacteria ferment prebiotics to use them as an energy source. The positive effects of prebiotics, particularly on the ecology of gut microbes, have motivated ruminant researchers to investigate their potential for use with various livestock species, such as cattle, sheep, and buffalo. Ruminants are subjected to many types of stress on various occasions, such as weaning and transport, which negatively affects the well-being of the cattle and causes poor intestinal morphology, off-feeding, growth depression, diarrhea, etc. Prebiotic ecological treatment may offer prospective solutions in such circumstances to help animals with digestive issues (Xu et al. 2021; Sanjorjo et al. 2023). There is sparse information on the effects of prebiotics on ruminant animals. It is possible for all rumen hemicellulolytic bacteria to use xylooligosaccharides as a growth substrate. These include *Ruminococcus albus*, *Eubacterium ruminantium*, and *Butyrovibrio fibrisolvens* (Izuddin et al. 2018). When prebiotics are given to Holstein cows kept on orchard grass hay or alfalfa silage, rumen pH stays the same. Other studies discovered that the pH of prebiotic-supplemented rumen of steers did not change considerably, but they did record significantly increased reduction-oxidation capacity. Due to the rumen's use of ammonium during microbial synthesis of proteins, Holstein cows and steers treated with prebiotics had slightly lower rumen ammonia nitrogen concentrations (Weimer et al. 1999; Zhu et al. 2020). Prebiotic dosing also led to decreased rumen nitrate ion content in sheep, which may be related to the inhibition of ammonia-producing microorganisms. Prebiotic supplementation has little effect on the digestibility of nutrients from dry matter intake; however, it did result in higher nitrogen retention because of higher bacterial synthesis of proteins in the rumen. Pre-ruminant calves' live weight increases are greatly enhanced when inulin is added to their milk replacer, and their stools have higher reliability. Increased fermentation in the small intestine, enhanced circulation for microbial nitrogen in the intestinal tract, and stable microflora makeup in the rumen, small intestine, and intestine of calves are thought to be the causes of the weight gain. Rumen inoculums from sheep fed a forage-only diet accelerate the digestion of inulin at pH 6.0 compared to neutral pH. Oligofructose-supplemented diets in calves reduced the amount of fecal *E. coli* as well as the overall anaerobic microbiota, but *Bifidobacteria* populations showed sharp increases. This may be explained by the positive effects produced by prebiotic ingestion and subsequent fermentation in calves' hindguts. Oligofructose was added to the milk substitute material feeding calves, which led to increased body weight growth, enhanced feed conversion efficiency, and harder feces. Due to the alteration of the rumen metabolic profile, the addition of oligosaccharides at a level of 0.5% to 1% of the whole mixed diet (w/w) considerably increased the organic compounds as well as dried substance digestibility of the total mixed ration (van Niekerk et al. 2020).

6.4.5 Mechanism of Action of Prebiotics

Roberfroid and Gibson introduced the idea of prebiotics and outlined the criteria for a chemical to be classified as a prebiotic: it has to be inedible by microorganisms and enzymes but definitely used by a certain set of good gut bacteria (Davani-Davari et al., 2019; Gibson and Roberfroid, 1995). Prebiotics display multi-dimensional activities, in contrast to other functional meals, starting with modification of gut microbiota and moving on to pathogen exclusion, mineral absorption, etc., as illustrated in Figure 6.2. Thus, the channel of activities should be extensively distributed between several systems, including gastrointestinal, immune regulatory, and skeletal organs, among others, in order that the full beneficial impact on the system of cattle is realized (Piccioni et al., 2023; Salminen et al., 2021).

According to the literature, prebiotics are finding a place in daily consumption for maintaining gut health and curing gastrointestinal illnesses. Of course, mice or rats are used as laboratory animals for the majority of research aimed at understanding the process of prebiotic activities, with human volunteers or in vitro research providing further support. Under these conditions, some features of prebiotic activity are considered, which are discussed in the following sections (Megur et al. 2022; Song, Jeon, and Unno 2022; Roberfroid et al. 2010).

6.4.5.1 Trophic Effect

Piglets were maintained on a diet that included 1.4 grams of oligofructose every day in order to better understand the trophic function of the sugar. Both the number of tagged cells (8.2 vs. 9.6 cells/crypt) and the density of cecum mucosal cells (40.4 vs. 44.7 cells/crypt) considerably increased. In comparison to control piglets, oligofructose supplementation was shown to result in an increase in the distal colonic mucous crypt depth, labeled cells, the proliferative zone, and the labeling index (Samanta et al. 2013; Roberfroid et al. 2010). A similar increase in mucosa cell density, crypt depth, labeling index, number of labeled cells, and proliferative zone was observed at the distal colon following consumption of oligofructose permeate. It appears that consuming the prebiotic oligofructose improves markers of epithelial cell growth along the whole length of the piglets' large intestine. Later, the rat model's colonic mucosa's direct trophic effect was confirmed by adding chicory inulin, oligofructose, or pectin to the animal's diet. Increased regions of absorption of nutrients, particularly those for minerals and micronutrients, were followed by increases in epithelial cell numbers and secretory capabilities as well as the width and length of colon crypts (Grancieri et al. 2023; Cummins, Jones, and Thompson 2006). Following consumption of prebiotics including FOS and inulin, additional experiments in the colonic mucosa and rat cecum observed a 2.2-times rise in macroscopic areas and a 2.4-times rise in tissue wall mass. Male Sprague Dawley rats receiving prebiotics had a substantial increase in crypt depths (p 0.001) and goblet cell counts (p 0.001) at the cecum epithelium. Research on male Wistar rats supplemented with yacon flour revealed a rise in the amount of bifurcating crypts for the animals' large intestines to have a larger absorbent surface. This could be as a result of increased crypt fission and epithelial cell proliferation. Animals' lower guts create varying amounts of short-chain fatty acids (butyrate, acetate, and propionate) as well as lactic acid through the hydrolysis and consumption of prebiotics, which lowers the pH of the lumen. Among these, butyric acid and fatty acids are said to have potent trophic effects that stimulate the proliferation of mucosa. Yet, the rat cecum mucosa was capable of being stimulated by lactic acid to undergo mitosis (Lobo et al. 2007; Grancieri et al. 2017). When germ-free animals failed to show trophic effects in the large intestine after consuming prebiotics, the above notion of trophic activities was further supported. Supporting the microbial production of fatty acids using prebiotics is essential to proving the above functionality. Tight connections between neighboring epithelial cells and sizable protein complexes are necessary for the paracellular pathway's permeability (Honoré et al. 2018).

Because prebiotics may improve the permeability of paracellular channels, there was a sharp increase in paracellular ion transport after the introduction of non-digestible oligosaccharides within a model of cell line caco-2 intestine. The mucus produced by the goblet cells in the stomach epithelial lining is a cysteine-rich, high molecular weight glycoprotein (Croix et al. 2011). The primary purpose of mucus is to create a slimy film along the whole mucosal surface. It is commonly considered the intestine's first line of defense. Although mucus thickness differs along the extent of the gastrointestinal system, the distal section of the gut has the thickest mucus. Despite being a great source of energy and nitrogen, few intestinal microfloras have the ability to secrete the enzymes required for the breakdown of mucus. The bacterial flora has a big impact on how various mucus types are distributed (Grondin et al. 2020; Teigen et al. 2019). Neutral and sulfomucin concentrations are higher while sialomucin concentrations are lower inside the small intestine. In contrast, sialomucin levels are higher in the colonic mucosa and cecum but sulfated mucin content has significantly decreased there. Prebiotic administration resulted in thickening of the mucus layer and enhanced goblet cell production (Alves Júnior et al. 2017).

6.4.5.2 Hydrolysis of Phytic Acid

Plants can contain phosphorus in the form of phytic or phytate acid. In contrast to monogastric animals, phytate or phytic acid from plants is typically prone to microbial breakdown for the additional generation of phosphorus inside the reticulo-rumen (Alves Júnior et al. 2017; Bohn, Meyer, and Rasmussen 2008). In the latter species, phytate binds to minerals such as Ca, Zn, and Fe to prevent the absorption of those substances. Similarly, phytic acid has the ability to decrease both the absorption of Fe and its availability in the gastrointestinal system. Prebiotic consumption is said to have stimulating effects on the absorption of iron by raising the amount of soluble iron in the large intestine (B. Singh and Satyanarayana 2011; Rizwanuddin et al. 2023). Fructooligosaccharide addition improves zinc bioavailability in rats and humans, restoring zinc absorption. Prebiotic supplementation significantly lowers the phytate concentration in rat feces because prebiotics grow and change the gut microbiota, which results in larger concentrations of phytate being hydrolyzed. By forming soluble ligands with cations, short chain fatty acids generated by gut microflora stop the development of the insoluble mineral-phytic acid complex (Raboy 2020; Cangussu et al. 2018).

6.4.5.3 Modulation of Gut Microflora

By definition, consuming prebiotics alters the composition of the gut flora in favor of the growth and proliferation of good bacteria *Lactobacillus* and *Bifidobacteria* (Davani-Davari et al. 2019). As they can ferment indigestible carbohydrates, lower gut pH, create short-chain fatty acids, aid in the competitive inhibition of pathogens, boost the production of immunoglobulin, and other beneficial properties, these bacteria are frequently referred to as beneficial bacteria (Guarino et al. 2020; Yaqoob et al. 2021). In the large intestine, prebiotics are converted into short-chain fatty acids such as butyrate and acetate, which serve as the host's main source of energy (Ashaolu, Ashaolu, and Adeyeye 2021).

Bifidobacteria encourage the growth of bacteria that produce butyrate, such as *Eubacteria*, which are found in the gastrointestinal system, as *Bifidobacteria* cannot make butyrate through their own fermentation process of metabolism (Rivière et al. 2016; Parada Venegas et al. 2019). The significance of short-chain fatty acids is multifaceted and includes cellular development and proliferation, lipid metabolism, colonic movement of epithelial cells, and glucose metabolism, among other things. Prebiotics, on the other hand, have the power to lower levels of toxic gut flora such as *Bacteroides*, *Enterococcus*, *Clostridium*, and *Enterobacteriaceae*. Following the consumption of prebiotics, the percentage of harmful metabolites—indoles, ammonia, thiols, phenols, etc.—produced by those dangerous bacteria is significantly reduced (Silva, Bernardi, and Frozza 2020; LeBlanc et al. 2017). These harmful metabolites can cause cancer in some people. Hence, consuming prebiotics lowers the number of dangerous microorganisms in the gut, which in turn lowers the synthesis of hazardous microbial metabolites and reduces the risk of colorectal cancer. Not all prebiotics equally promote the multiplication and growth of the entire collection of beneficial gut microflora, as each strain of gut flora uses substrates in distinct ways. The level of polymerization, the presence of other carbohydrates, the composition, the amount consumed, and other factors all affect the prebiotic potential. The number of *Eubacteria* and *Bifidobacteria* in the mucosa increased with the consumption of fructooligosaccharides and inulin. Similarly, fructooligosaccharides from chicory roots were discovered to significantly boost *Enterococcus faecalis* and *Lactobacillus plantarum* growth in an in vitro system. Recently, it was discovered that prebiotics are also broken down by other commensal gut flora, such as *Faecalibacterium prausnitzii*, with butyrate being the result of fermentation (You et al. 2022; Engel and Moran 2013; Sasso et al. 2023).

6.4.5.4 Immune Modulation

Dietary prebiotics can potentially exclude gut pathogens by competitive inhibition, altering the host defense system and immune system, and modifying the makeup and activity of microflora (Bothiraja et al. 2014; Tushar et al., 2011; Ajay Khedkar and Dattatray Tupe 2012a; Gholap 2012; V. G. R. Amol D Gholap 2012c, 2012a; Gholap, Bhowmik et al. 2023; AMOL D GHOLAP 2012; Gholap, Gupta et al. 2023; Dattatray Tupe and Amol D Gholap* 2012; S. B. D. Amol D Gholap 2012c; Ajay Khedkar and Dattatray Tupe 2012b; Gholap, Said et al. 2023; Gholap et al. 2012; V. G. R. Amol D Gholap 2012b; D. T. Amol D Gholap 2012a; V. G. R. Amol D Gholap, n.d.; Vikas G Rajurkar 2012; Gholap et al. 2011). The exact mechanism causing this effect is still unclear, and it is uncertain whether it could be either the indirect or direct result of triggering the creation of short-chain fatty acids with short chains or immune-modulating microflora (Markowiak and Śliżewska 2018; Callaway et al. 2008). Short-chain fatty acids connect to certain receptors, such as Gpr41 (G-protein coupled receptor 41) of immune cells situated at the gut-associated lymphoid tissue (GALT), to activate the goblet cells to increase mucin secretion and decrease translocation. In addition, the intestinal mucosa and spleen exhibit modified cytokine formation and produce more mucosal immunoglobulin (Sekirov et al. 2010; Kreuzer and Hardt 2020; Di Gioia and Biavati 2018).

6.4.5.5 Lipid Metabolism

Prebiotics are said to provide benefits for the cardiovascular system and lipidemia in both humans and animals, as they lower triglyceride and cholesterol levels. Rats supplemented with oligofructose had significantly lower levels of serum triglycerides and phospholipids, particularly VLDL (very low density lipoprotein) (Ooi and Liong 2010; D. I. A. Pereira and Gibson 2002; Nie and Luo 2021). Reduction in the activity of malic enzymes, hepatic lipogenic enzymes FAS, acetyl-CoA carboxylase, ATP citrate lyase, fatty acid synthase mRNA, and glucose-6-phosphate 1-dehydrogenase are the mechanisms through which this occurs. Prebiotics may alter the expression of genes of the lipogenic enzymes, according to some theories. Prebiotics control the free cholesterol levels of plasma in rats fed a high-fat diet, showing extra-hepatic control of lipid metabolism. Prebiotic effects appear to be related to insulin, which amplifies the impact on gene regulation. Prebiotic consumption led to the formation of short-chain fatty acids in the gut lumen, followed by an increase in their concentration (Zhang et al. 2022; Saad et al. 2013; den Besten et al. 2013). Fatty acid production is reported to be inhibited by propionate in vivo. For the enterocytes that make up the intestinal epithelium, butyrate is the preferred energy source. Acetate enters the hepatocyte after being absorbed to participate in lipogenesis. The protein that controls the entry of acetate into hepatocyte cells is in competition with propionate. As a result, prebiotics control lipid metabolism by regulating the acetate to a propionate proportion in cells (Lange, Proczko-Stepaniak, and Mika 2023; McNabney and Henagan 2017).

6.4.5.6 Mineral Metabolism

Prebiotics increase calcium, magnesium, iron, and zinc absorption in the gastrointestinal tract as well as bone mineralization in animals, which demonstrates its beneficial effects. Inulin has been found to have a favorable impact on mineral imbalance, especially Ca, P, and Zn, in both broilers and layers. Higher pH levels increase Ca solubility (Samanta et al. 2013; Davani-Davari et al. 2019; Macfarlane and Cummings 1999). Prebiotic treatment in broilers may lead to greater Ca retention because the production of beneficial gut bacteria from prebiotic breakdown lowers the mucosal pH at the lower section of the large intestine. Enhanced solubility of Ca and a higher differential across the epithelium caused by the acidic pH produced by prebiotic fermentation in the lower portion of the large intestine further encourage inactive absorption of the element. Higher solubility of Ca and an enhanced gradient throughout the epithelium caused by the acidic pH produced by prebiotic fermentation in the lower portion of the large intestine further encourage passive intake of the element. Nonetheless, other schools of thought have suggested that prebiotics cause increased Ca retention through a divalent ion carrier system (You et al. 2022; Dahiya and Nigam 2022b). The permeability and electrical resistance of the rat mucosal epithelium are increased and decreased, respectively, by the introduction of short-chain fatty acids to the diet. In order to increase the porosity for the entry of minerals, it is theorized that prebiotics interact with strong epithelial junctions. The positive effect of inulin on mineral consumption is mediated by a complicated process which could result in increased mineral solubility due to increased synthesis of short-chain fatty acids that result from inulin breakdown. Increased expression of calcium-binding proteins, the release of bone-stimulating factors, the breakdown of phytates, enhanced absorption surface areas, altered intestinal mucosa due to the positive effects of bacterial byproducts on enterocyte proliferation, and an overall improvement in gut health are other factors that contribute to higher mineral absorption (Han et al. 2021; Budhwar, Sethi, and Chakraborty 2020). Another theory for increased mineral absorption takes into account the co-absorption of short-chain fatty acids and Ca in the colon. Prebiotic fermentation also controls the production of proteins involved in mucosal ion binding and sequestration. It's important to note that the amount of enhanced mineral absorption is unaffected by the prebiotic fructan's chain length or kind of branching. The gut's lactate pool and the pH of the lumen have a big impact on how much magnesium is absorbed. A different school of thinking supported the use of short-chain fatty acids in order to activate the Mg2+/2H+ antiport and improve Mg absorption. Butyrate is the most powerful stimulator for the magnesium flow from the mucous to serosal direction among the several short-chain fatty acids (Byrne et al. 2015; den Besten et al. 2013).

6.4.5.7 Anti-Carcinogenesis

The usual lifespan of animals and fowl does not allow sufficient opportunity for the growth of colon cancer. Other types of tumors are occasionally found in domestic animals, though. The microbial decomposition of plant proteins in the gastrointestinal system leads to the creation of many mutagenic chemicals, as well as the secretion of enzymes that convert pro-carcinogens into carcinogens, according to research on gut ecology (Yang et al. 2009; Śliżewska, Markowiak-Kopeć, and Śliżewska 2020; Rebersek 2021). Among the metabolites produced by gut microbes that have been linked to cancer are skatole, phenols, cresols, nitrosamines, indole, oestrogens, secondary bile acids, aglycones, and others. *Clostridium* spp., *E. coli*, *Bacteroides* spp., and *Proteus* spp. are examples of the intestinal flora that actively contribute to the creation of those carcinogenic substances. Prebiotics are added to animal diets to influence the gut microbiota towards higher populations of *Lactobacillus* and *Bifidobacteria*, and that in turn decreases the population of those deadly bacteria and provides an environment that is unsuitable for their growth (X. Wang, Zhang, and Zhang 2021; de Vos et al. 2022).

Quick oleic acid-generating microbes exist in larger quantities, and microbe-generated oleic acid concentrations are greater in stools if inulin supplements are used. In a study, intestinal scarring decreased in a mouse model of long-term irradiation enteropathy by transplanting intestinal flora and compounds generated from oligosaccharides (Guo et al. 2021; Charlet et al. 2022). In the human embryonic neuroblastoma cell type NIH/3T3, therapy with compounds from gut bacteria which are produced by inulin significantly reduced the levels of genes related to scarring and collagen synthesis. Using metabolites formed from inulin also decreased the levels of genes involved in the intercellular synthesis process in NIH/3T3 cell cultures. Fiber content has well-known health impacts (Shrifteylik et al. 2023).

The inclusion of 10% inulin, a fermentable fiber, to food modified the gut microbiome and reduced systemic probiotic renal tubular poisons, phosphate, and parathyroid hormone in a mouse model. Modifications in bone formation parameters, reduced endothelial dysfunction, and lower interest enlargement, stiffness, and fibrosis were all related to these changes. When paired, nutritional digestion fibers could serve as a cutting-edge alternative to existing CKD-MBD therapy (Zupcic, Slezak, and Radloff 2023).

The greatest biological health risk for vascular dementia is the apolipoprotein 4 gene (APOE4) (AD). Before developing signs of Alzheimer's disease (AD), APOE4 patients exhibited participatory management metabolism failure. There is increasing proof that physiological instability expedites the onset of AD, which comprises increased α-synuclein (A) storage, neuronal injury, and mental impairment (C.-C. Liu et al. 2013; Kim, Basak, and Holtzman 2009). As a conclusion, initial metabolism activity retention may be important for lowering the risk of AD. In this, it has been proven that lignin enhances digestion in the intestine, external, and nervous systems, as proved by rises in the levels of short-chain fatty acids, chymotrypsin metabolism, ketone bodies, gluconeogenic metabolites, and scyllo-inositol. Lignin raises useful microbes and reduces dangerous microbes in the excrement of youths in a subclinical APOE4 transgenic (E4FAD) mouse model. It has also been shown that lignin decreases the production of top player genes in the brain (Husain, Laurent, and Plourde 2021; Raulin et al. 2022).

This data could be utilized especially for developing personalized diet-appropriate interventions that improve overall metabolism with a bioactive meal, potentially decreasing the likelihood of AD in undiagnosed APOE4 bearers. Insulin's impact on intestinal microbiota, microbial functions, and lipid levels was investigated in a study in particular microbe chicks afflicted by *S. enteritidis* (SE) (B. Li, Tang, and Le 2023; Croci et al. 2021). Specific pathogen-free Orchard Acre-site poultry (n = 240, one day old) were divided into five groups: a treatment group fed a standard diet without lignan intake or SE inflammation; three groups given a standard diet supplemented with lactose up to 28 days old; and a group supplied a basal diet until being exposed to SE contest at 28 days old. A day after infection, small intestinal short-chain fatty acid levels, microbiome composition, and activity were investigated (Hoffman et al. 2019; Mazhar, Zhu, and Qin 2023).

According to the findings, SE virus greatly diminished the levels of small intestinal benzoate compared to the control treatment (p0.05), but lignan dietary supplementation recovered these changes relative to an SE control (p=0.05). In contrast to the SE control, lignan administration at 1% raised the relative abundance of b *Staphylococcus* while reducing overall relative abundance of *Sellimonas* and *Subdoligranulum* (p=0.05) (B. Li, Tang, and Le 2023; Croci et al. 2021). According to the metagenomic analysis, the operational patterns of bacterial diversity showed that SE contamination dramatically improved the relative abundance of processes connected to glucose metabolism, the metabolism of amino acids, mitochondrial biogenesis, metabolic activity of coenzymes and micronutrients, and carbohydrate synthesis and digestion (p=0.05), and drastically decreased the relative abundance of processes related to purine metabolic activities, translation, duplication, and repair relative to control circumstances. As a consequence of altering short-chain fatty acids synthesis and bacterial activity patterns, pectin regulated the microbiome produced by SE disease (Hoffman et al. 2019; Mazhar, Zhu, and Qin 2023).

6.5 CONCLUSION

For many years, therapeutic doses of antibiotics and chemotherapy agents have been added to animal feed to expand production and ensure greater financial returns. Cattle supervisors are now leaning towards environmental feed ingredients as a result of a ban on using antibiotics in the feed of cattle, as well as increasing concerns about residues. Irrespective of the promising results of the latest studies, the use of synbiotics in livestock diets is still a more recent effort. Synbiotics hold very good potential, and consumer perception of dietary and health problems is increasing simultaneously with time. Synbiotics remain quite costly as a medicine, making them expensive for most livestock farmers. In fact, there is an immediate need for research studies to assess the efficacy of synbiotics in animal meals on a daily basis and to develop advanced methods for precursor production on an industrial level. Close contact between the offspring and its surroundings, method of distribution, hygiene conditions, medication, type of nutrition, etc. results in infection with different factions of microorganisms in the gastrointestinal tract of newborn cattle or hatched chicks. It has regularly been observed that more than 90% of a healthy human's total cell count is made up of bacteria, the largest portion being located in the intestinal mucosa. In terms of outlook, future studies on synbiotics as nutritional components should focus on (i) improving ways to generate nutrients at a low cost, (ii) ascertaining action mechanisms, and (iii) creating research evidence that supports the potential effects of synbiotics in poultry diets, especially the amount and quality of manufacturing. Synbiotics might be the right approach from the choice of useful nutrients, as they exhibit value in a number of ways. The effect of synbiotics on microflora needs to be further studied using reduction tools with next-generation genomics. Additional research regarding the use of synbiotics in cattle must also include biological factors.

REFERENCES

"Diseases of the Alimentary Tract." 2017. In *Veterinary Medicine*, 175–435. Elsevier. doi:10.1016/B978-0-7020-5246-0.00007-3.

Andersson, Henrik, Nils-Georg Asp, Åke Bruce, Stefan Roos, Torkel Wadström, and Agnes E. Wold. 2001. "Health Effects of Probiotics and Prebiotics A Literature Review on Human Studies." *Näringsforskning* 45 (1): 58–75. doi:10.3402/fnr.v45i0.1790.

Ashaolu, T.J., J.O. Ashaolu, and S.A.O. Adeyeye. 2021. "Fermentation of Prebiotics by Human Colonic Microbiota in Vitro and Short-chain Fatty Acids Production: A Critical Review." *Journal of Applied Microbiology* 130 (3): 677–687. doi:10.1111/jam.14843.

Atanasov, Atanas G., Sergey B. Zotchev, Verena M. Dirsch, and Claudiu T. Supuran. 2021. "Natural Products in Drug Discovery: Advances and Opportunities." *Nature Reviews Drug Discovery* 20 (3): 200–216. doi:10.1038/s41573-020-00114-z.

Bassaganya-Riera, Josep, Elliot M. Berry, Ellen E. Blaak, Barbara Burlingame, Johannes le Coutre, Willem van Eden, Ahmed El-Sohemy, et al. 2021. "Goals in Nutrition Science 2020-2025." *Frontiers in Nutrition* 7 (February). doi:10.3389/fnut.2020.606378.

Berg, Gabriele, Daria Rybakova, Doreen Fischer, Tomislav Cernava, Marie-Christine Champomier Vergès, Trevor Charles, Xiaoyulong Chen, et al. 2020. "MICROBIOME DEFINITION RE-VISITED: Old Concepts and New Challenges." *Microbiome* 8 (1): 103. doi:10.1186/s40168-020-00875-0.

Besten, Gijs den, Karen van Eunen, Albert K. Groen, Koen Venema, Dirk-Jan Reijngoud, and Barbara M. Bakker. 2013. "The Role of Short-Chain Fatty Acids in the Interplay between Diet, Gut Microbiota, and Host Energy Metabolism." *Journal of Lipid Research* 54 (9): 2325–2340. doi:10.1194/jlr.R036012.

Bevilacqua, A., D. Campaniello, B. Speranza, A. Racioppo, M. Sinigaglia, and M.R. Corbo. 2024. "An Update on Prebiotics and on Their Health Effects." *Foods* 13, 446. doi:10.3390/foods13030446.

Bhagat, Vilas, and Amol Gholap. 2012. "Gold Nanorods: A Promising Drug Delivery Vehicle." *Inventi Rapid: NDDS [ISSN 0976-3864]* 4: 1–4.

Bohn, Lisbeth, Anne S. Meyer, and Søren K. Rasmussen. 2008. "Phytate: Impact on Environment and Human Nutrition. A Challenge for Molecular Breeding." *Journal of Zhejiang University SCIENCE B* 9 (3): 165–191. doi:10.1631/jzus.B0710640.

Bothiraja, Chellampillai, Amol D. Gholap, Karimunnisa S. Shaikh, and Atmaram P. Pawar. 2014. "Investigation of Ethyl Cellulose Microsponge Gel for Topical Delivery of Eberconazole Nitrate for Fungal Therapy." *Therapeutic Delivery* 5 (7): 781–94. doi:10.4155/tde.14.43.

Budhwar, Savita, Kashika Sethi, and Manali Chakraborty. 2020. "Efficacy of Germination and Probiotic Fermentation on Underutilized Cereal and Millet Grains." *Food Production, Processing and Nutrition* 2 (1): 12. doi:10.1186/s43014-020-00026-w.

Byrne, C. S., E. S. Chambers, D. J. Morrison, and G. Frost. 2015. "The Role of Short Chain Fatty Acids in Appetite Regulation and Energy Homeostasis." *International Journal of Obesity* 39 (9): 1331–1338. doi:10.1038/ijo.2015.84.

Callaway, T. R., T. S. Edrington, R. C. Anderson, R. B. Harvey, K. J. Genovese, C. N. Kennedy, D. W. Venn, and D. J. Nisbet. 2008. "Probiotics, Prebiotics and Competitive Exclusion for Prophylaxis against Bacterial Disease." *Animal Health Research Reviews* 9 (2): 217–225. doi:10.1017/S1466252308001540.

Cammack, K. M., K. J. Austin, W. R. Lamberson, G. C. Conant, and H. C. Cunningham. 2018. "Tiny but Mighty: The Role of the Rumen Microbes in Livestock Production." *Journal of Animal Science* (January). doi:10.1093/jas/skx053.

Cangussu, Alex Sander Rodrigues, Deborah Aires Almeida, Raimundo Wagner de Souza Aguiar, Sidnei Emilio Bordignon-Junior, Kelvinson Fernandes Viana, Luiz Carlos Bertucci Barbosa, Edson Wagner da Silva Cangussu, et al. 2018. "Characterization of the Catalytic Structure of Plant Phytase, Protein Tyrosine Phosphatase-Like Phytase, and Histidine Acid Phytases and Their Biotechnological Applications." *Enzyme Research* 2018 (March): 1–12. doi:10.1155/2018/8240698.

Cena, Hellas, and Philip C. Calder. 2020. "Defining a Healthy Diet: Evidence for the Role of Contemporary Dietary Patterns in Health and Disease." *Nutrients* 12 (2): 334. doi:10.3390/nu12020334.

Cencic, Avrelija, and Walter Chingwaru. 2010. "The Role of Functional Foods, Nutraceuticals, and Food Supplements in Intestinal Health." *Nutrients* 2 (6): 611–625. doi:10.3390/nu2060611.

Charlet, Rogatien, Chrystelle Le Danvic, Boualem Sendid, Patricia Nagnan-Le Meillour, and Samir Jawhara. 2022. "Oleic Acid and Palmitic Acid from Bacteroides Thetaiotaomicron and Lactobacillus Johnsonii Exhibit Anti-Inflammatory and Antifungal Properties." *Microorganisms* 10 (9): 1803. doi:10.3390/microorganisms10091803.

Combes, S., L. Fortun-Lamothe, L. Cauquil, and T. Gidenne. 2013. "Engineering the Rabbit Digestive Ecosystem to Improve Digestive Health and Efficacy." *Animal* 7 (9): 1429–1439. doi:10.1017/S1751731113001079.

Croci, Sonia, Lina Ilaras D'Apolito, Valeria Gasperi, Maria Valeria Catani, and Isabella Savini. 2021. "Dietary Strategies for Management of Metabolic Syndrome: Role of Gut Microbiota Metabolites." *Nutrients* 13 (5): 1389. doi:10.3390/nu13051389.

Croix, Jennifer A., Franck Carbonero, Gerardo M. Nava, Mark Russell, Eugene Greenberg, and H. Rex Gaskins. 2011. "On the Relationship between Sialomucin and Sulfomucin Expression and Hydrogenotrophic Microbes in the Human Colonic Mucosa." Edited by Stefan Bereswill. *PLoS One* 6 (9): e24447. doi:10.1371/journal.pone.0024447.

Cryan, John F., Kenneth J. O'Riordan, Caitlin S. M. Cowan, Kiran V. Sandhu, Thomaz F. S. Bastiaanssen, Marcus Boehme, Martin G. Codagnone, et al. 2019. "The Microbiota-Gut-Brain Axis." *Physiological Reviews* 99 (4): 1877–2013. doi:10.1152/physrev.00018.2018.

Cummins, Adrian G., Ben J. Jones, and Fiona M. Thompson. 2006. "Postnatal Epithelial Growth of the Small Intestine in the Rat Occurs by Both Crypt Fission and Crypt Hyperplasia." *Digestive Diseases and Sciences* 51 (4): 718–723. doi:10.1007/s10620-006-3197-9.

Cunningham, Marla, M. Andrea Azcarate-Peril, Alan Barnard, Valerie Benoit, Roberta Grimaldi, Denis Guyonnet, Hannah D. Holscher, et al. 2021. "Shaping the Future of Probiotics and Prebiotics." *Trends in Microbiology* 29 (8): 667–85. doi:10.1016/j.tim.2021.01.003.

Dahiya, Divakar, and Poonam Singh Nigam. 2022a. "The Gut Microbiota Influenced by the Intake of Probiotics and Functional Foods with Prebiotics Can Sustain Wellness and Alleviate Certain Ailments like Gut-Inflammation and Colon-Cancer." *Microorganisms* 10 (3): 665. doi:10.3390/microorganisms10030665.

Dahiya, Divakar, and Poonam Singh Nigam. 2022b. "Probiotics, Prebiotics, Synbiotics, and Fermented Foods as Potential Biotics in Nutrition Improving Health via Microbiome-Gut-Brain Axis." *Fermentation* 8 (7): 303. doi:10.3390/fermentation8070303.

Dapkevicius, Maria de Lurdes Enes, Bruna Sgardioli, Sandra P. A. Câmara, Patrícia Poeta, and Francisco Xavier Malcata. 2021. "Current Trends of Enterococci in Dairy Products: A Comprehensive Review of Their Multiple Roles." *Foods* 10 (4): 821. doi:10.3390/foods10040821.

Dattatray, Tupe, and Amol D. Gholap*. 2012. "The Threat of Photoaging: Clinical Conditions, Evaluation and Treatment." *Inventi Rapid: Cosmeceuticals [ISSN 0976-3864]* 4: 1–3.

Davani-Davari, Dorna, Manica Negahdaripour, Iman Karimzadeh, Mostafa Seifan, Milad Mohkam, Seyed Masoumi, Aydin Berenjian, and Younes Ghasemi. 2019. "Prebiotics: Definition, Types, Sources, Mechanisms, and Clinical Applications." *Foods* 8 (3): 92. doi:10.3390/foods8030092.

Davis, Catherine. 2014. "Enumeration of Probiotic Strains: Review of Culture-Dependent and Alternative Techniques to Quantify Viable Bacteria." *Journal of Microbiological Methods* 103 (August): 9–17. doi:10.1016/j.mimet.2014.04.012.

Devirgiliis, Chiara, Simona Barile, and Giuditta Perozzi. 2011. "Antibiotic Resistance Determinants in the Interplay between Food and Gut Microbiota." *Genes & Nutrition* 6 (3): 275–284. doi:10.1007/s12263-011-0226-x.

Dey, Tushar Kumar, Johanna F. Lindahl, Rajkumari Sanjukta, A. Arun Prince Milton, Samir Das, Porteen Kannan, Åke Lundkvist, Arnab Sen, and Sandeep Ghatak. 2023. "Characterization of Lactic Acid Bacteria and Pathogens Isolated from Traditionally Fermented Foods, In Relation to Food Safety and Antimicrobial Resistance in Tribal Hill Areas of Northeast India." Edited by Ali Akbar. *Journal of Food Quality* 2023 (May): 1–12. doi:10.1155/2023/6687015.

Dimidi, Eirini, Selina Cox, Megan Rossi, and Kevin Whelan. 2019. "Fermented Foods: Definitions and Characteristics, Impact on the Gut Microbiota and Effects on Gastrointestinal Health and Disease." *Nutrients* 11 (8): 1806. doi:10.3390/nu11081806.

Engel, Philipp, and Nancy A. Moran. 2013. "The Gut Microbiota of Insects – Diversity in Structure and Function." *FEMS Microbiology Reviews* 37 (5): 699–735. doi:10.1111/1574-6976.12025.

Fijan, Sabina. 2014. "Microorganisms with Claimed Probiotic Properties: An Overview of Recent Literature." *International Journal of Environmental Research and Public Health* 11 (5): 4745–4767. doi:10.3390/ijerph110504745.

Firrman, Jenni, LinShu Liu, Karley Mahalak, Weiming Hu, Kyle Bittinger, Ahmed Moustafa, Steven M. Jones, Adrienne Narrowe, and Peggy Tomasula. 2023. "An in Vitro Analysis of How Lactose Modifies the Gut Microbiota Structure and Function of Adults in a Donor-Independent Manner." *Frontiers in Nutrition* 9 (January). doi:10.3389/fnut.2022.1040744.

Flickinger, E. A., E. M. W. C. Schreijen, A. R. Patil, H. S. Hussein, C. M. Grieshop, N. R. Merchen, and G. C. Fahey. 2003. "Nutrient Digestibilities, Microbial Populations, and Protein Catabolites as Affected by Fructan Supplementation of Dog Diets." *Journal of Animal Science* 81 (8): 2008–2018. doi:10.2527/2003.8182008x.

George Kerry, Rout, Jayanta Kumar Patra, Sushanto Gouda, Yooheon Park, Han-Seung Shin, and Gitishree Das. 2018. "Benefaction of Probiotics for Human Health: A Review." *Journal of Food and Drug Analysis* 26 (3): 927–939. doi:10.1016/j.jfda.2018.01.002.

Gholap, Amol D. 2012. "Translational Nanomedicine: Status Assessment and Opportunities." *International Journal of Research in Pharmaceutical and Biomedical Sciences* 3 (4): 1454–63.

Gholap, Amol D., Deblina D. Bhowmik, Aditya Y. Deshmukh, and Navnath T. Hatvate. 2023. "Quintessential Impact of Dendrimer Bioconjugates in Targeted Drug Delivery." In *Bioconjugated Materials Part 2 Applications in Drug Delivery, Vaccine Formulations and Important Conjugates for Cancer Therapy Comprehensive Analytical Chemistry*. doi:10.1016/bs.coac.2023.02.009.

Gholap, Amol D., Santosh S. Borude, and Anand M. Mahajan. 2011. "Smart Crystals Technology: A Review." *Pharmacologyonline* 3: 238–243.

Gholap, Amol D., Santosh S. Borude, Anand M. Mahajan, and Mr Amol Dilip Gholap. 2011. "Aquasomes: A Potential Drug Delivery Carrier." *Pharmacologyonline* 3: 230–237.

Gholap, Amol D., and Dattatray Tupe. 2012a. "Potential Applications of Botanical Extract." *Inventi Rapid: Ethnopharmacology [ISSN 0976-3805]* 4: 1–4.

Gholap, Amol D., and Dattatray Tupe. 2012b. "Artificial Viral Envelopes: Emerging Vehicle For Gene Delivery." *Pharmatutor-Art-1439:1*: 1–3.

Gholap, Amol D., and Santosh B. Dighe. 2012. "Dandruff: Pathogenesis And Treatment." *Inventi Rapid: Cosmeceuticals [ISSN 0976-3864]* 4: 1–4.

Gholap, Amol D., Juhi S. Gupta, Pallavi A. Kamandar, Gauri V. Banchhod, and Navnath T. Hatvate. 2023. "Antibody-Drug Conjugates for Cancer Therapy: An up-to-Date Review on the Chemistry and Pharmacology." In *Bioconjugated Materials Part 2 Applications in Drug Delivery, Vaccine Formulations and Important Conjugates for Cancer Therapy.* doi:10.1016/bs.coac.2023.02.007.

Gholap, Amol D., and Deepak S. Musmade. 2011. "Advances in the Field of Nanooncology: An Overview." *Inventi Impact: NDDS [ISSN 0976-3791]* 4: 1–3.

Gholap, Amol D., Navnath T. Hatvate, Pankaj R. Khuspe, Trushali A. Mandhare, Pooja Kashid, and Vinay D. Gaikwad. 2023. "Proteomics in Oncology: Retrospect and Prospects." In *Novel Technologies in Biosystems, Biomedical & Drug Delivery*, 243–269. Singapore: Springer Nature Singapore. doi:10.1007/978-981-99-5281-6_10.

Gholap, Amol D., and Vikas G. Rajurkar. 2012a. "Lipid Nanocapsules: A Novel Promising Tool as Drug Delivery System." *Inventi Rapid: NDDS [ISSN 0976-3791]* 4: 1–4.

Gholap, Amol D., and Vikas G. Rajurkar. 2012b. "New Threat of Clinical Toxicities Evoked from Nanocarrier Systems." *Inventi Rapid: NDDS [ISSN 0976-3791]* 4: 1–4.

Gholap, Amol D., and Vikas G. Rajurkar. 2012c. "Potential Applications of Superparmagnetic Nanoparticles Technology." *Inventi Rapid: NDDS [ISSN 0976-3791]* 4: 1–3.

Gholap, Amol D., and Vikas G. Rajurkar. n.d. "Potential Applications of Capillary Electromigration Methods For The Improvement Of Separation Efficiency." *Inventi Rapid: Pharm Analysis & Quality Assurance [ISSN 0976-3813]* 2012 (4): 1–3.

Gholap, Amol D, Vikas G. Rajurkar, Santosh B. Dighe. 2012. "Phonophoresis: Ultrasound-enhanced Transdermal Drug Delivery." *Inventi Rapid: NDDS* 2012(4): 1–3.

Gholap, Amol D., Rushikesh P. Said, Rutuja D. Pawar, Gauri S. Ambore, and Navnath T. Hatvate. 2023. "Importance of Carbohydrate-Drug Conjugates in Vaccine Development: A Detailed Review." In *Bioconjugated Materials Part 2 Applications in Drug Delivery, Vaccine Formulations and Important Conjugates for Cancer Therapy.* doi:10.1016/bs.coac.2023.03.002.

Gholap, Amol D., Sadikali F. Sayyad, Navnath T. Hatvate, Vilas V. Dhumal, Sagar R. Pardeshi, Vivek P. Chavda, and Lalitkumar K. Vora. 2023. "Drug Delivery Strategies for Avobenzone: A Case Study of Photostabilization." *Pharmaceutics* 15 (3): 1008. doi:10.3390/pharmaceutics15031008.

Gholap, Amol D., Satish Rojekar, Harshad S. Kapare, Nikhar Vishwakarma, Sarjana Raikwar, Atul Garkal, Tejal A. Mehta, Harsh Jadhav, Mahendra Kumar Prajapati, and Uday Annapure. 2024. "Chitosan Scaffolds: Expanding Horizons in Biomedical Applications." *Carbohydrate Polymers* 323 (January): 121394. doi:10.1016/j.carbpol.2023.121394.

Gibson, Glenn R., and Marcel B. Roberfroid. 1995. "Dietary Modulation of the Human Colonic Microbiota: Introducing the Concept of Prebiotics." *The Journal of Nutrition* 125 (6): 1401–1412. doi:10.1093/jn/125.6.1401.

Gioia, Diana Di, and Bruno Biavati, eds. 2018. *Probiotics and Prebiotics in Animal Health and Food Safety.* Cham: Springer International Publishing. doi:10.1007/978-3-319-71950-4.

Grancieri, Mariana, Neuza Maria Brunoro Costa, Maria das Graças Vaz Tostes, Daniela Silva de Oliveira, Louisiane de Carvalho Nunes, Letícia de Nadai Marcon, Thaísa Agrizzi Veridiano, and Mirelle Lomar Viana. 2017. "Yacon Flour (Smallanthus Sonchifolius) Attenuates Intestinal Morbidity in Rats with Colon Cancer." *Journal of Functional Foods* 37 (October): 666–675. doi:10.1016/j.jff.2017.08.039.

Grancieri, Mariana, Mirelle Lomar Viana, Daniela Furtado de Oliveira, Maria das Graças Vaz Tostes, Mariana Drummond Costa Ignacchiti, André Gustavo Vasconcelos Costa, and Neuza Maria Brunoro Costa. 2023. "Yacon (Smallanthus Sonchifolius) Flour Reduces Inflammation and Had No Effects on Oxidative Stress and Endotoxemia in Wistar Rats with Induced Colorectal Carcinogenesis." *Nutrients* 15 (14): 3281. doi:10.3390/nu15143281.

Grondin, Jensine A., Yun Han Kwon, Parsa Mehraban Far, Sabah Haq, and Waliul I. Khan. 2020. "Mucins in Intestinal Mucosal Defense and Inflammation: Learning From Clinical and Experimental Studies." *Frontiers in Immunology* 11 (September). doi:10.3389/fimmu.2020.02054.

Guarino, Michele, Annamaria Altomare, Sara Emerenziani, Claudia Di Rosa, Mentore Ribolsi, Paola Balestrieri, Paola Iovino, Giulia Rocchi, and Michele Cicala. 2020. "Mechanisms of Action of Prebiotics and Their Effects on Gastro-Intestinal Disorders in Adults." *Nutrients* 12 (4): 1037. doi:10.3390/nu12041037.

Guevarra, Robin B., Jun Hyung Lee, Sun Hee Lee, Min-Jae Seok, Doo Wan Kim, Bit Na Kang, Timothy J. Johnson, Richard E. Isaacson, and Hyeun Bum Kim. 2019. "Piglet Gut Microbial Shifts Early in Life: Causes and Effects." *Journal of Animal Science and Biotechnology* 10 (1): 1. doi:10.1186/s40104-018-0308-3.

Guo, Yingjie, Yanan Yu, Hailong Li, Xueli Ding, Xiaoyu Li, Xue Jing, Jianwei Chen, et al. 2021. "Inulin Supplementation Ameliorates Hyperuricemia and Modulates Gut Microbiota in Uox-Knockout Mice." *European Journal of Nutrition* 60 (4): 2217–2230. doi:10.1007/s00394-020-02414-x.

Habib, M., I. Qureshi, K. Jan, and K. Bashir. 2024. "Prebiotics, Probiotics, Synbiotics, and Nutraceuticals." In: Bashir, K., Jan, K., and Ahmad, F.J. (eds) *Functional Foods and Nutraceuticals: Chemistry, Health Benefits and the Way Forward*. Cham: Springer. doi: 10.1007/978-3-031-59365-9_9.

Han, Shengyi, Yanmeng Lu, Jiaojiao Xie, Yiqiu Fei, Guiwen Zheng, Ziyuan Wang, Jie Liu, et al. 2021. "Probiotic Gastrointestinal Transit and Colonization After Oral Administration: A Long Journey." *Frontiers in Cellular and Infection Microbiology* 11 (March). doi:10.3389/fcimb.2021.609722.

Hanchi, Hasna, Walid Mottawea, Khaled Sebei, and Riadh Hammami. 2018. "The Genus Enterococcus: Between Probiotic Potential and Safety Concerns—An Update." *Frontiers in Microbiology* 9 (August). doi:10.3389/fmicb.2018.01791.

Hao, Yangyi, Chunyan Guo, Yue Gong, Xiaoge Sun, Wei Wang, Yajing Wang, Hongjian Yang, Zhijun Cao, and Shengli Li. 2021. "Rumen Fermentation, Digestive Enzyme Activity, and Bacteria Composition between Pre-Weaning and Post-Weaning Dairy Calves." *Animals* 11 (9): 2527. doi:10.3390/ani11092527.

Hoffman, Jared D., Lucille M. Yanckello, George Chlipala, Tyler C. Hammond, Scott D. McCulloch, Ishita Parikh, Sydney Sun, Josh M. Morganti, Stefan J. Green, and Ai-Ling Lin. 2019. "Dietary Inulin Alters the Gut Microbiome, Enhances Systemic Metabolism and Reduces Neuroinflammation in an APOE4 Mouse Model." Edited by Florian Reichmann. *PLoS One* 14 (8): e0221828. doi:10.1371/journal.pone.0221828.

Honoré, Stella Maris, Maria Virginia Grande, Jorge Gomez Rojas, and Sara Serafina Sánchez. 2018. "Smallanthus Sonchifolius (Yacon) Flour Improves Visceral Adiposity and Metabolic Parameters in High-Fat-Diet-Fed Rats." *Journal of Obesity* 2018 (October): 1–15. doi:10.1155/2018/5341384.

Huang, Pan, Xuemei Cui, Zhipeng Wang, Chenwen Xiao, Quanan Ji, Qiang Wei, Yee Huang, Guolian Bao, and Yan Liu. 2021. "Effects of Clostridium Butyricum and a Bacteriophage Cocktail on Growth Performance, Serum Biochemistry, Digestive Enzyme Activities, Intestinal Morphology, Immune Responses, and the Intestinal Microbiota in Rabbits." *Antibiotics* 10 (11): 1347. doi:10.3390/antibiotics10111347.

Husain, Mohammed Amir, Benoit Laurent, and Mélanie Plourde. 2021. "APOE and Alzheimer's Disease: From Lipid Transport to Physiopathology and Therapeutics." *Frontiers in Neuroscience* 15 (February). doi:10.3389/fnins.2021.630502.

Ito, Tamaki, Tsuyoshi Sekizuka, Norimi Kishi, Akifumi Yamashita, and Makoto Kuroda. 2019. "Conventional Culture Methods with Commercially Available Media Unveil the Presence of Novel Culturable Bacteria." *Gut Microbes* 10 (1): 77–91. doi:10.1080/19490976.2018.1491265.

Izuddin, Wan Ibrahim, Teck Chwen Loh, Anjas Asmara Samsudin, and Hooi Ling Foo. 2018. "In Vitro Study of Postbiotics from Lactobacillus Plantarum RG14 on Rumen Fermentation and Microbial Population." *Revista Brasileira de Zootecnia* 47 (November). doi:10.1590/rbz4720170255.

Jakszyn, Paula, Antonio Agudo, Raquel Ibáñez, Reina García-Closas, Guillem Pera, Carlos A. González, and Pilar Amiano. 2004. "Development of a Food Database of Nitrosamines, Heterocyclic Amines, and Polycyclic Aromatic Hydrocarbons." *The Journal of Nutrition* 134 (8): 2011–2014. doi:10.1093/jn/134.8.2011.

Júnior, Alves, Antonio José Tiburcio, José Aires Pereira, Adrieli Heloísa Campardo Pansani, Daniela Oliveira Magro, Cláudio Saddy Rodrigues Coy, and Carlos Augusto Real Martinez. 2017. "Tissue Sulfomucin and Sialomucin Content in Colon Mucosa without Intestinal Transit Subjected to Intervention with Curcuma Longa (Curcumin)." *Acta Cirurgica Brasileira* 32 (3): 182–193. doi:10.1590/s0102-865020170030000002.

Khedkar, Ajay, Dattatray Tupe, and Amol Gholap. 2012. "Cubosomes: Emerging Drug Delivery System." *Inventi Rapid: NDDS [ISSN 0976-3791]* 4: 1–3.

Khedkar, Ajay, Dattatray Tupe, and Amol Gholap. 2012b "Pharmacosomes: an emerging vesicular system." *Inventi Rapid: NDDS.*

Kim, Jungsu, Jacob M. Basak, and David M. Holtzman. 2009. "The Role of Apolipoprotein E in Alzheimer's Disease." *Neuron* 63 (3): 287–303. doi:10.1016/j.neuron.2009.06.026.

King, Dale E., Olayiwola Adeola, and Elikplimi K. Asem. 2000. "Ontogenetic Development of Intestinal Digestive Functions in White Pekin Ducks." *The Journal of Nutrition* 130 (1): 57–62. doi:10.1093/jn/130.1.57.

Knecht, Damian, Paulina Cholewińska, Anna Jankowska-Mąkosa, and Katarzyna Czyż. 2020. "Development of Swine's Digestive Tract Microbiota and Its Relation to Production Indices—A Review." *Animals* 10 (3): 527. doi:10.3390/ani10030527.

Kreuzer, Markus, and Wolf-Dietrich Hardt. 2020. "How Food Affects Colonization Resistance Against Enteropathogenic Bacteria." *Annual Review of Microbiology* 74 (1): 787–813. doi:10.1146/annurev-micro-020420-013457.

Lange, Oliwia, Monika Proczko-Stepaniak, and Adriana Mika. 2023. "Short-Chain Fatty Acids—A Product of the Microbiome and Its Participation in Two-Way Communication on the Microbiome-Host Mammal Line." *Current Obesity Reports* 12 (2): 108–126. doi:10.1007/s13679-023-00503-6.

Lauridsen, Charlotte, Jean Jacques Matte, Martin Lessard, Pietro Celi, and Gilberto Litta. 2021. "Role of Vitamins for Gastro-Intestinal Functionality and Health of Pigs." *Animal Feed Science and Technology* 273 (March): 114823. doi:10.1016/j.anifeedsci.2021.114823.

LeBlanc, Jean Guy, Florian Chain, Rebeca Martín, Luis G. Bermúdez-Humarán, Stéphanie Courau, and Philippe Langella. 2017. "Beneficial Effects on Host Energy Metabolism of Short-Chain Fatty Acids and Vitamins Produced by Commensal and Probiotic Bacteria." *Microbial Cell Factories* 16 (1): 79. doi:10.1186/s12934-017-0691-z.

Leeuwendaal, Natasha K., Catherine Stanton, Paul W. O'Toole, and Tom P. Beresford. 2022. "Fermented Foods, Health and the Gut Microbiome." *Nutrients* 14 (7): 1527. doi:10.3390/nu14071527.

Li, Bowen, Xue Tang, and Guowei Le. 2023. "Dietary Habits and Metabolic Health." *Nutrients* 15 (18): 3975. doi:10.3390/nu15183975..

Liu, Chia-Chen, Takahisa Kanekiyo, Huaxi Xu, and Guojun Bu. 2013. "Apolipoprotein E and Alzheimer Disease: Risk, Mechanisms and Therapy." *Nature Reviews Neurology* 9 (2): 106–118. doi:10.1038/nrneurol.2012.263.

Lobo, Alexandre R., Célia Colli, Eliana P. Alvares, and Tullia M. C. C. Filisetti. 2007. "Effects of Fructans-Containing Yacon (Smallanthus Sonchifolius Poepp & Endl.) Flour on Caecum Mucosal Morphometry, Calcium and Magnesium Balance, and Bone Calcium Retention in Growing Rats." *British Journal of Nutrition* 97 (4): 776–785. doi:10.1017/S0007114507336805.

Macfarlane, G.T, and J.H Cummings. 1999. "Probiotics and Prebiotics: Can Regulating the Activities of Intestinal Bacteria Benefit Health?" *BMJ* 318 (7189): 999–1003. doi:10.1136/bmj.318.7189.999.

Magar, Tushar V., Amol D. Gholap, Deepak S. Musmade. 2011. "Elastic Vesicles as Drugs Carriers Through The Skin." *Inventi Rapid: NDDS [ISSN 0976-3791]* 4: 1–2.

Mahajan, Anand M., Amol D. Gholap*, and Santosh S. Borude. 2011. "POEMS Syndrome: A Review." *Pharmacologyonline* 3: 265–274.

Mahajan, Anand M., Amol D. Gholap*, and Santosh S. Borude. n.d. "Recent Updates on Q Fever: Diagnosis And Treatment." *Pharmacologyonline* 3: 254–264.

Marcolla, Camila S., Tingting Ju, and Benjamin P. Willing. 2023. "Cecal Microbiota Development and Physiological Responses of Broilers Following Early Life Microbial Inoculation Using Different Delivery Methods and Microbial Sources." Edited by Christopher A. Elkins. *Applied and Environmental Microbiology* 89 (5). doi:10.1128/aem.00271-23.

Markowiak, Paulina, and Katarzyna Śliżewska. 2017. "Effects of Probiotics, Prebiotics, and Synbiotics on Human Health." *Nutrients* 9 (9): 1021. doi:10.3390/nu9091021.

Markowiak, Paulina, and Katarzyna Śliżewska. 2018. "The Role of Probiotics, Prebiotics and Synbiotics in Animal Nutrition." *Gut Pathogens* 10 (1): 21. doi:10.1186/s13099-018-0250-0.

Matthews, Chloe, Fiona Crispie, Eva Lewis, Michael Reid, Paul W. O'Toole, and Paul D. Cotter. 2019. "The Rumen Microbiome: A Crucial Consideration When Optimising Milk and Meat Production and Nitrogen Utilisation Efficiency." *Gut Microbes* 10 (2): 115–132. doi:10.1080/19490976.2018.1505176.

Mazhar, Muhammad, Yong Zhu, and Likang Qin. 2023. "The Interplay of Dietary Fibers and Intestinal Microbiota Affects Type 2 Diabetes by Generating Short-Chain Fatty Acids." *Foods* 12 (5): 1023. doi:10.3390/foods12051023.

McNabney, Sean, and Tara Henagan. 2017. "Short Chain Fatty Acids in the Colon and Peripheral Tissues: A Focus on Butyrate, Colon Cancer, Obesity and Insulin Resistance." *Nutrients* 9 (12): 1348. doi:10.3390/nu9121348.

Megur, Ashwinipriyadarshini, Eric Banan-Mwine Daliri, Daiva Baltriukienė, and Aurelijus Burokas. 2022. "Prebiotics as a Tool for the Prevention and Treatment of Obesity and Diabetes: Classification and Ability to Modulate the Gut Microbiota." *International Journal of Molecular Sciences* 23 (11): 6097. doi:10.3390/ijms23116097.

Minamoto, Yasushi, Tomomi Minamoto, Anitha Isaiah, Panpicha Sattasathuchana, Agostino Buono, Venkat R. Rangachari, Isaac H. McNeely, Jonathan Lidbury, Jörg M. Steiner, and Jan S. Suchodolski. 2019. "Fecal Short-chain Fatty Acid Concentrations and Dysbiosis in Dogs with Chronic Enteropathy." *Journal of Veterinary Internal Medicine* 33 (4): 1608–1618. doi:10.1111/jvim.15520.

MOON, Yea Hwang, Ji Un OK, Shin Ja LEE, Jong Kyu HA, and Sung Sill LEE. 2010. "A Comparative Study on the Rumen Microbial Populations, Hydrolytic Enzyme Activities and Dry Matter Degradability between Different Species of Ruminant." *Animal Science Journal* 81 (6): 642–647. doi:10.1111/j.1740-0929.2010.00782.x.

Mozaffarian, Dariush, Sonia Y. Angell, Tim Lang, and Juan A. Rivera. 2018. "Role of Government Policy in Nutrition—Barriers to and Opportunities for Healthier Eating." *BMJ* (June): k2426. doi:10.1136/bmj.k2426.

Mutungwazi, Asheal, Grace N. Ijoma, Henry J. O. Ogola, and Tonderayi S. Matambo. 2022. "Physico-Chemical and Metagenomic Profile Analyses of Animal Manures Routinely Used as Inocula in Anaerobic Digestion for Biogas Production." *Microorganisms* 10 (4): 671. doi:10.3390/microorganisms10040671.

Nath, Arijit, Gokce Haktanirlar, Áron Varga, Máté András Molnár, Krisztina Albert, Ildikó Galambos, András Koris, and Gyula Vatai. 2018. "Biological Activities of Lactose-Derived Prebiotics and Symbiotic with Probiotics on Gastrointestinal System." *Medicina* 54 (2): 18. doi:10.3390/medicina54020018.

Niba, A.T., J.D. Beal, A.C. Kudi, and P.H. Brooks. 2009. "Bacterial Fermentation in the Gastrointestinal Tract of Non-Ruminants: Influence of Fermented Feeds and Fermentable Carbohydrates." *Tropical Animal Health and Production* 41 (7): 1393–1407. doi:10.1007/s11250-009-9327-6.

Nie, Ying, and Feijun Luo. 2021. "Dietary Fiber: An Opportunity for a Global Control of Hyperlipidemia." Edited by Wuquan Deng. *Oxidative Medicine and Cellular Longevity* 2021 (April): 1–20. doi:10.1155/2021/5542342.

Niekerk, J.K. van, A.J. Fischer-Tlustos, L.L. Deikun, J.D. Quigley, T.S. Dennis, F.X. Suarez-Mena, T.M. Hill, R.L. Schlotterbeck, L.L. Guan, and M.A. Steele. 2020. "Effect of Amount of Milk Replacer Fed and the Processing of Corn in Starter on Growth Performance, Nutrient Digestibility, and Rumen and Fecal Fibrolytic Bacteria of Dairy Calves." *Journal of Dairy Science* 103 (3): 2186–2199. doi:10.3168/jds.2019-17372.

O' Donnell, Michelle M., Hugh M. B. Harris, R. Paul Ross, and Paul W. O'Toole. 2017. "Core Fecal Microbiota of Domesticated Herbivorous Ruminant, Hindgut Fermenters, and Monogastric Animals." *MicrobiologyOpen* 6 (5). doi:10.1002/mbo3.509.

Ooi, Lay-Gaik, and Min-Tze Liong. 2010. "Cholesterol-Lowering Effects of Probiotics and Prebiotics: A Review of in Vivo and in Vitro Findings." *International Journal of Molecular Sciences* 11 (6): 2499–2522. doi:10.3390/ijms11062499.

Parada Venegas, Daniela, Marjorie K. De la Fuente, Glauben Landskron, María Julieta González, Rodrigo Quera, Gerard Dijkstra, Hermie J.M. Harmsen, Klaas Nico Faber, and Marcela A. Hermoso. 2019. "Short Chain Fatty Acids (SCFAs)-Mediated Gut Epithelial and Immune Regulation and Its Relevance for Inflammatory Bowel Diseases." *Frontiers in Immunology* 10 (March). doi:10.3389/fimmu.2019.00277.

Pereira, Ana Margarida, Margarida Guedes, Elisabete Matos, Edgar Pinto, Agostinho A. Almeida, Marcela A. Segundo, Alexandra Correia, Manuel Vilanova, António J.M. Fonseca, and Ana Rita J. Cabrita. 2020. "Effect of Zinc Source and Exogenous Enzymes Supplementation on Zinc Status in Dogs Fed High Phytate Diets." *Animals* 10 (3): 400. doi:10.3390/ani10030400.

Pereira, Dora I.A., and Glenn R. Gibson. 2002. "Effects of Consumption of Probiotics and Prebiotics on Serum Lipid Levels in Humans." *Critical Reviews in Biochemistry and Molecular Biology* 37 (4): 259–281. doi:10.1080/10409230290771519.

Piccioni, Andrea, Marcello Covino, Marcello Candelli, Veronica Ojetti, Annunziata Capacci, Antonio Gasbarrini, Francesco Franceschi, and Giuseppe Merra. 2023. "How Do Diet Patterns, Single Foods, Prebiotics and Probiotics Impact Gut Microbiota?" *Microbiology Research* 14 (1): 390–408. doi:10.3390/microbiolres14010030.

Piskin, Elif, Danila Cianciosi, Sukru Gulec, Merve Tomas, and Esra Capanoglu. 2022. "Iron Absorption: Factors, Limitations, and Improvement Methods." *ACS Omega* 7 (24): 20441–20456. doi:10.1021/acsomega.2c01833.

Popkin, Barry M., Linda S. Adair, and Shu Wen Ng. 2012. "Global Nutrition Transition and the Pandemic of Obesity in Developing Countries." *Nutrition Reviews* 70 (1): 3–21. doi:10.1111/j.1753-4887.2011.00456.x.

Propst, E.L., E.A. Flickinger, L.L. Bauer, N.R. Merchen, and G.C. Fahey. 2003. "A Dose-Response Experiment Evaluating the Effects of Oligofructose and Inulin on Nutrient Digestibility, Stool Quality, and Fecal Protein Catabolites in Healthy Adult Dogs." *Journal of Animal Science* 81 (12): 3057–3066. doi:10.2527/2003.81123057x.

Raboy, Victor. 2020. "Low Phytic Acid Crops: Observations Based on Four Decades of Research." *Plants* 9 (2): 140. doi:10.3390/plants9020140.

Rajurkar, Vikas G., and Amol D. Gholap. 2012. "Recent Technologies in Nanomedicine for Treatment of Cancer." *Inventi Rapid: Pharm Tech [ISSN 0976-3783]* 4: 1–3.

Raulin, Ana-Caroline, Sydney V. Doss, Zachary A. Trottier, Tadafumi C. Ikezu, Guojun Bu, and Chia-Chen Liu. 2022. "ApoE in Alzheimer's Disease: Pathophysiology and Therapeutic Strategies." *Molecular Neurodegeneration* 17 (1): 72. doi:10.1186/s13024-022-00574-4.

Rebersek, Martina. 2021. "Gut Microbiome and Its Role in Colorectal Cancer." *BMC Cancer* 21 (1): 1325. doi:10.1186/s12885-021-09054-2.

Rivière, Audrey, Marija Selak, David Lantin, Frédéric Leroy, and Luc De Vuyst. 2016. "Bifidobacteria and Butyrate-Producing Colon Bacteria: Importance and Strategies for Their Stimulation in the Human Gut." *Frontiers in Microbiology* 7 (June). doi:10.3389/fmicb.2016.00979.

Rizwanuddin, Sheikh, Vijay Kumar, Pallavi Singh, Bindu Naik, Sadhna Mishra, Mansi Chauhan, Per Erik Joakim Saris, Ankit Verma, and Vivek Kumar. 2023. "Insight into Phytase-Producing Microorganisms for Phytate Solubilization and Soil Sustainability." *Frontiers in Microbiology* 14 (April). doi:10.3389/fmicb.2023.1127249.

Roberfroid, Marcel, Glenn R. Gibson, Lesley Hoyles, Anne L. McCartney, Robert Rastall, Ian Rowland, Danielle Wolvers, et al. 2010. "Prebiotic Effects: Metabolic and Health Benefits." *British Journal of Nutrition* 104 (S2): S1–63. doi:10.1017/S0007114510003363.

Saad, N., C. Delattre, M. Urdaci, J.M. Schmitter, and P. Bressollier. 2013. "An Overview of the Last Advances in Probiotic and Prebiotic Field." *LWT - Food Science and Technology* 50 (1): 1–16. doi:10.1016/j.lwt.2012.05.014.

Salminen, Seppo, Maria Carmen Collado, Akihito Endo, Colin Hill, Sarah Lebeer, Eamonn M. M. Quigley, Mary Ellen Sanders, et al. 2021. "The International Scientific Association of Probiotics and Prebiotics (ISAPP) Consensus Statement on the Definition and Scope of Postbiotics." *Nature Reviews Gastroenterology & Hepatology* 18 (9): 649–667. doi:10.1038/s41575-021-00440-6.

Samanta, A.K., Natasha Jayapal, S. Senani, A.P. Kolte, and Manpal Sridhar. 2013. "Prebiotic Inulin: Useful Dietary Adjuncts to Manipulate the Livestock Gut Microflora." *Brazilian Journal of Microbiology* 44 (1): 1–14. doi:10.1590/S1517-83822013005000023.

Sanjorjo, Rey Anthony, Tenzin Tseten, Min-Kyoung Kang, Moonhyuk Kwon, and Seon-Won Kim. 2023. "In Pursuit of Understanding the Rumen Microbiome." *Fermentation* 9 (2): 114. doi:10.3390/fermentation9020114.

Sanjuán, Juan, Maria Caridad Nápoles, Daniel Pérez-Mendoza, María J. Lorite, and Dulce N. Rodríguez-Navarro. 2023. "Microbials for Agriculture: Why Do They Call Them Biostimulants When They Mean Probiotics?" *Microorganisms* 11 (1): 153. doi:10.3390/microorganisms11010153.

Sasso, Janet M., Ramy M. Ammar, Rumiana Tenchov, Steven Lemmel, Olaf Kelber, Malte Grieswelle, and Qiongqiong Angela Zhou. 2023. "Gut Microbiome–Brain Alliance: A Landscape View into Mental and Gastrointestinal Health and Disorders." *ACS Chemical Neuroscience* 14 (10): 1717–63. doi:10.1021/acschemneuro.3c00127.

Sekirov, Inna, Shannon L. Russell, L. Caetano M. Antunes, and B. Brett Finlay. 2010. "Gut Microbiota in Health and Disease." *Physiological Reviews* 90 (3): 859–904. doi:10.1152/physrev.00045.2009.

Shang, Yue, Sanjay Kumar, Brian Oakley, and Woo Kyun Kim. 2018. "Chicken Gut Microbiota: Importance and Detection Technology." *Frontiers in Veterinary Science* 5 (October). doi:10.3389/fvets.2018.00254.

Shrifteylik, Anna, Morgan Maiolini, Matthew Dufault, Daniel L. Austin, Bobban Subhadra, Purushottam Lamichhane, and Rahul R. Deshmukh. 2023. "A Current Review on the Role of Prebiotics in Colorectal Cancer." *Biologics* 3 (3): 209–31. doi:10.3390/biologics3030012.

Siddique, Abubakar, Sara Azim, Amjad Ali, Fazal Adnan, Maryum Arif, Muhammad Imran, Erika Ganda, and Abdur Rahman. 2021. "Lactobacillus Reuteri and Enterococcus Faecium from Poultry Gut Reduce Mucin Adhesion and Biofilm Formation of Cephalosporin and Fluoroquinolone-Resistant Salmonella Enterica." *Animals* 11 (12): 3435. doi:10.3390/ani11123435.

Silva, Ygor Parladore, Andressa Bernardi, and Rudimar Luiz Frozza. 2020. "The Role of Short-Chain Fatty Acids From Gut Microbiota in Gut-Brain Communication." *Frontiers in Endocrinology* 11 (January). doi:10.3389/fendo.2020.00025.

Singh, Bijender, and T. Satyanarayana. 2011. "Microbial Phytases in Phosphorus Acquisition and Plant Growth Promotion." *Physiology and Molecular Biology of Plants* 17 (2): 93–103. doi:10.1007/s12298-011-0062-x.

Singh, Rajnish Prakash, Afreen Shadan, and Ying Ma. 2022. "Biotechnological Applications of Probiotics: A Multifarious Weapon to Disease and Metabolic Abnormality." *Probiotics and Antimicrobial Proteins* 14 (6): 1184–1210. doi:10.1007/s12602-022-09992-8.

Sionek, Barbara, Aleksandra Szydłowska, Dorota Zielińska, Katarzyna Neffe-Skocińska, and Danuta Kołożyn-Krajewska. 2023. "Beneficial Bacteria Isolated from Food in Relation to the Next Generation of Probiotics." *Microorganisms* 11 (7): 1714. doi:10.3390/microorganisms11071714.

Śliżewska, Katarzyna, Paulina Markowiak-Kopeć, and Weronika Śliżewska. 2020. "The Role of Probiotics in Cancer Prevention." *Cancers* 13 (1): 20. doi:10.3390/cancers13010020.

Song, Hokyung, Dabin Jeon, and Tatsuya Unno. 2022. "Evaluation of Prebiotics through an in Vitro Gastrointestinal Digestion and Fecal Fermentation Experiment: Further Idea on the Implementation of Machine Learning Technique." *Foods* 11 (16): 2490. doi:10.3390/foods11162490.

Stevens, C. Edward, and Ian D. Hume. 1998. "Contributions of Microbes in Vertebrate Gastrointestinal Tract to Production and Conservation of Nutrients." *Physiological Reviews* 78 (2): 393–427. doi:10.1152/physrev.1998.78.2.393.

Teferra, Tadesse F. 2021. "Possible Actions of Inulin as Prebiotic Polysaccharide: A Review." *Food Frontiers* 2 (4): 407–416. doi:10.1002/fft2.92.

Teigen, Levi M., Zhuo Geng, Michael J. Sadowsky, Byron P. Vaughn, Matthew J. Hamilton, and Alexander Khoruts. 2019. "Dietary Factors in Sulfur Metabolism and Pathogenesis of Ulcerative Colitis." *Nutrients* 11 (4): 931. doi:10.3390/nu11040931.

Tinrat, Sirikhwan, Piyatip Khuntayaporn, Krit Thirapanmethee, and Mullika Traidej Chomnawang. 2018. "In Vitro Assessment of Enterococcus Faecalis MTC 1032 as the Potential Probiotic in Food Supplements." *Journal of Food Science and Technology* 55 (7): 2384–2394. doi:10.1007/s13197-018-3155-5.

Topping, David L., and Peter M. Clifton. 2001. "Short-Chain Fatty Acids and Human Colonic Function: Roles of Resistant Starch and Nonstarch Polysaccharides." *Physiological Reviews* 81 (3): 1031–1064. doi:10.1152/physrev.2001.81.3.1031.

Tsvetkov, Ivan, Atanas Atanassov, Mariana Vlahova, Lucien Carlier, Nikolai Christov, Francois Lefort, Krasimir Rusanov, et al. 2018. "Plant Organic Farming Research – Current Status and Opportunities for Future Development." *Biotechnology & Biotechnological Equipment* 32 (2): 241–260. doi:10.1080/13102818.2018.1427509.

Twomey, L.N., J.R. Pluske, J.B. Rowe, M. Choct, W. Brown, and D.W. Pethick. 2003. "The Effects of Added Fructooligosaccharide (Raftilose®P95) and Inulinase on Faecal Quality and Digestibility in Dogs." *Animal Feed Science and Technology* 108 (1–4): 83–93. doi:10.1016/S0377-8401(03)00162-7.

Usta, Canan. 2013. "Microorganisms in Biological Pest Control — A Review (Bacterial Toxin Application and Effect of Environmental Factors)." In *Current Progress in Biological Research*. InTech. doi:10.5772/55786.

Vemuri, Ravichandra, Esaki M. Shankar, Marcello Chieppa, Rajaraman Eri, and Kylie Kavanagh. 2020. "Beyond Just Bacteria: Functional Biomes in the Gut Ecosystem Including Virome, Mycobiome, Archaeome and Helminths." *Microorganisms* 8 (4): 483. doi:10.3390/microorganisms8040483.

Vinayamohan, Poonam Gopika, Leya Susan Viju, Divya Joseph, and Kumar Venkitanarayanan. 2023. "Fermented Foods as a Potential Vehicle of Antimicrobial-Resistant Bacteria and Genes." *Fermentation* 9 (7): 688. doi:10.3390/fermentation9070688.

Vos, Willem M de, Herbert Tilg, Matthias Van Hul, and Patrice D. Cani. 2022. "Gut Microbiome and Health: Mechanistic Insights." *Gut* 71 (5): 1020–1032. doi:10.1136/gutjnl-2021-326789.

Wang, Lizhi, Kaizhen Liu, Zhisheng Wang, Xue Bai, Quanhui Peng, and Lei Jin. 2019. "Bacterial Community Diversity Associated With Different Utilization Efficiencies of Nitrogen in the Gastrointestinal Tract of Goats." *Frontiers in Microbiology* 10 (February). doi:10.3389/fmicb.2019.00239.

Wang, Xinzhou, Peng Zhang, and Xin Zhang. 2021. "Probiotics Regulate Gut Microbiota: An Effective Method to Improve Immunity." *Molecules* 26 (19): 6076. doi:10.3390/molecules26196076.

Weimer, P.J., G.C. Waghorn, C.L. Odt, and D.R. Mertens. 1999. "Effect of Diet on Populations of Three Species of Ruminal Cellulolytic Bacteria in Lactating Dairy Cows." *Journal of Dairy Science* 82 (1): 122–134. doi:10.3168/jds.S0022-0302(99)75216-1.

Xiao, Jin, Barbara Metzler-Zebeli, and Qendrim Zebeli. 2015. "Gut Function-Enhancing Properties and Metabolic Effects of Dietary Indigestible Sugars in Rodents and Rabbits." *Nutrients* 7 (10): 8348–8365. doi:10.3390/nu7105397.

Xu, Qingbiao, Qinqin Qiao, Ya Gao, Jinxiu Hou, Mingyang Hu, Yufeng Du, Ke Zhao, and Xiang Li. 2021. "Gut Microbiota and Their Role in Health and Metabolic Disease of Dairy Cow." *Frontiers in Nutrition* 8 (August). doi:10.3389/fnut.2021.701511.

Yang, Chung S., Xin Wang, Gang Lu, and Sonia C. Picinich. 2009. "Cancer Prevention by Tea: Animal Studies, Molecular Mechanisms and Human Relevance." *Nature Reviews Cancer* 9 (6): 429–439. doi:10.1038/nrc2641.

Yaqoob, M.U., M.E. Abd El-Hack, F. Hassan, M.T. El-Saadony, A.F. Khafaga, G.E. Batiha, N. Yehia, et al. 2021. "The Potential Mechanistic Insights and Future Implications for the Effect of Prebiotics on Poultry Performance, Gut Microbiome, and Intestinal Morphology." *Poultry Science* 100 (7): 101143. doi:10.1016/j.psj.2021.101143.

You, Siyong, Yuchen Ma, Bowen Yan, Wenhui Pei, Qiming Wu, Chao Ding, and Caoxing Huang. 2022. "The Promotion Mechanism of Prebiotics for Probiotics: A Review." *Frontiers in Nutrition* 9 (October). doi:10.3389/fnut.2022.1000517.

Zeineldin, Mohamed, Brian Aldridge, and James Lowe. 2019. "Antimicrobial Effects on Swine Gastrointestinal Microbiota and Their Accompanying Antibiotic Resistome." *Frontiers in Microbiology* 10 (May). doi:10.3389/fmicb.2019.01035.

Zhang, Bai-Hui, Fan Yin, Ya-Nan Qiao, and Shou-Dong Guo. 2022. "Triglyceride and Triglyceride-Rich Lipoproteins in Atherosclerosis." *Frontiers in Molecular Biosciences* 9 (May). doi:10.3389/fmolb.2022.909151.

Zhu, Yuanting, Jinxin Liu, Julian M. Lopez, and David A. Mills. 2020. "Inulin Fermentation by Lactobacilli and Bifidobacteria from Dairy Calves." Edited by Shuang-Jiang Liu. *Applied and Environmental Microbiology* 87 (1). doi:10.1128/AEM.01738-20.

Zupcic, Ana, Paul Slezak, and Judith Radloff. 2023. "The Gastrointestinal Microbiota as a Potential Cause and Target in Chronic Kidney Disease Accentuating Treatment and Intervention Strategies." *Applied Sciences* 13 (5): 3212. doi:10.3390/app13053212.

7 Role of Probiotics on Animal Nutrition and Health

Emi Grace Mary Gowshika Rajendran
(Ph.D. in Community Nutrition), Department of Home Science, Women's Christian College
Chennai, Tamil Nadu, India

7.1 INTRODUCTION

The term "probiotic" was first introduced in 1965 by Lilly and Stillwell to describe microorganisms that produce growth-promoting substances. Probiotics are live microorganisms primarily used to balance the microflora in the gastrointestinal tract of organisms, resulting in beneficial effects on their hosts, such as promoting health and growth (Valdes et al., 2018). Previous studies have shown that probiotics secrete substances like anti-toxins and antimicrobial compounds, which inhibit the activity of pathogens and help regulate the immune system, leading to the production of antibodies (Dankowiakowska, Bednarczyk, and Kozłowska, 2013; Ding et al., 2021). According to the Food and Agriculture Organization of the United Nations, World Health Organization, and the International Scientific Association for Probiotics and Prebiotics, probiotics are defined as "living microorganisms that, when administered in appropriate amounts, provide health benefits to the host". Probiotics are generally recognized as safe (GRAS) and offer a natural approach to prevent harmful bacteria that could cause microbial infections (Markowiak and Slizewska, 2017). However, despite their generally safe reputation, it is essential to verify the practical benefits of probiotics for the host.

During the 20th century, antibiotics were widely used to prevent, control, and treat diseases and infections in animals. They were also found to improve growth, feed efficiency, and reduce mortality (Castanon, 2007). However, studies have revealed that the therapeutic use of antibiotics in animals has contributed to the rise of antimicrobial resistance. This, in turn, can lead to antibiotic resistance in humans, diminishing the effectiveness of these drugs in treating human diseases (Prestinaci, Pezzotti, and Pantosti, 2015). Moreover, animals contaminated with bacteria can cause infections in humans (Heredia and García, 2018). Mead et al. (1999) reported that foodborne pathogens result in five million illnesses, 46,000 hospitalizations, and 1,458 deaths in the United States annually. In the context of livestock, the ongoing increase in antimicrobial resistance is considered a potential source of antimicrobial-resistant genes that could eventually spread to human pathogenic bacteria (Bywater, 2004). To address these concerns, probiotics are being seen as a promising alternative to antimicrobial agents or antibiotics in animal health and livestock production.

Lactic acid bacteria (LAB) are commonly used as commercial probiotics to improve animal health. According to the American Food and Drug Administration (FDA), LAB are abundantly present in the gastrointestinal tract of healthy animals (Parvez et al., 2006; Bhogoju et al., 2018). These bacteria are Gram-positive, facultative, aerobic or anaerobic cocci or rod-shaped organisms. One of their main metabolic products is lactic acid, produced through carbohydrate fermentation. Additionally, LAB aid in food breakdown, leading to the production of hydrogen peroxide, lactic acid, and other substances that create an acidic environment unfavorable for harmful or pathogenic organisms. Due to their widespread occurrence in food and their beneficial contributions to the healthy microbiota of both animal and human mucosal surfaces, LAB are generally recognized as

DOI: 10.1201/9781003394662-7

safe. Thanks to their positive and non-pathogenic effects, LAB are considered promising potential probiotics.

Lactobacillus and *Bifidobacterium* are the most commonly utilized bacterial strains as probiotics. Among the well-documented probiotic species within the LAB group is *Lactobacillus reuteri*, predominantly found in the gut flora of animals and birds (Bhogoju et al., 2018). Besides LAB, there are several other bacterial types that can serve as probiotics. In aquaculture, *Streptomyces* is particularly employed as a probiotic due to its unique ability to produce various antimicrobial agents as secondary metabolites. Studies by Das et al. and Augustine et al. have demonstrated several promising outcomes regarding the use of *Streptomyces* as a probiotic in aquaculture (Das, Ward, and Burke, 2010; Augustine, Jacob, and Philip, 2015).

7.2 PROBIOTIC MODE OF ACTION

Probiotics have various anticipated modes of action, including inhibiting enteric pathogenic microbes and improving animal performance. However, the specific mechanisms by which probiotics exert their biological functions are not fully understood. Two terms, "non-specifically competitive exclusion" and "colonization resistance", are used to describe their modes of action. Oelschlaeger proposed three ways in which probiotics function in the host system (Oelschlaeger, 2010):

i. Probiotics may effectively modify both the innate and acquired immune systems of the host, leading to the prevention of infectious diseases and reducing inflammation in the host's digestive tract.
ii. Probiotics act directly on other microorganisms, preventing and controlling infections, and restoring the microbial balance in the gut.
iii. Microbial products, such as toxins, antimicrobials, and host metabolites, play a crucial role in probiotic actions. Probiotics help inactivate toxins, detoxifying bile salts, improve the digestion of food components, and enhance nutrient absorption in the gut.

According to research conducted by Galdeano et al. (2019), probiotics compete with pathogenic bacteria to attach to mucus, which aids in adjusting the host's immune responses. Additionally, probiotics can stimulate the synthesis of vitamins in the host by providing extra nutrients and digestive enzymes (Wang, 2007; Markowiak and Slizewska, 2017). This helps prevent the growth of harmful bacteria and promotes the production of inhibitory substances like volatile fatty acids and hydrogen peroxide, which enhance the host's ability to resist pathogens (Plaza-Diaz et al., 2019).

Research indicates that yeast-based probiotics increase the population of cellulolytic bacteria in ruminants, which positively affects microbial fermentation. This, in turn, leads to enhanced cellulose degradation and improved production of microbial protein (Amin and Mao, 2021). Both *Lactobacillus* and *Bifidobacterium* play a role in promoting the growth of closely related bacterial species by producing proteins or polypeptide bacteriocins. This helps reduce the number of harmful microbes in the gut. Probiotic species, including LAB, *Bifidobacterium*, and *Bacillus*, can also produce various types of thermostable bacteriocins with antimicrobial properties. These bacteriocins act against a range of pathogens such as *Bacillus*, *Staphylococcus*, *Enterococcus*, *Listeria*, and *Salmonella* species (Piqué, Berlanga, and Miñana-Galbis, 2019).

Havenaar et al. (1992) proposed certain distinctive traits of probiotics based on their beneficial properties. These characteristics include demonstrating favorable effects on the host, being non-pathogenic and nontoxic, having the capability to survive for extended periods with high cell counts, successfully passing through the digestive system, and colonizing the gastrointestinal tract (GIT) by attaching to the intestinal epithelium. Moreover, probiotics should have the ability to produce antimicrobial substances against pathogens, stabilize gut microflora, utilize nutrients from a regular diet, and ultimately provide health benefits to the host.

Several probiotic strains have been reported to possess anti-inflammatory properties, which help maintain the balance between pro- and anti-inflammatory cytokines (Pagnini et al., 2010; Cristofori et al., 2021). Additionally, probiotics produce antimicrobial substances like bacteriocins, hydrogen peroxide, and volatile fatty acids (Vieco-Saiz et al., 2019). Studies have demonstrated that organic compounds produced by probiotic bacteria can inhibit pathogenic bacteria, including *H. pylori* (Rezaee, Kermanshahi, and Falsafi, 2019). Jiang et al.'s research indicated that probiotics can strengthen intestinal barrier integrity by enhancing the expression of tight junction proteins through the activation of the p38 and ERK signaling pathways (2011). Moreover, probiotics have shown antiviral properties in animals (Lehtoranta, Latvala, and Lehtinen, 2020).

In cancer research, it is speculated that probiotic cultures have the potential to reduce exposure to chemical carcinogens. They achieve this by: (i) detoxifying consumed carcinogens, (ii) modifying the intestinal environment to reduce populations or activities of carcinogen-producing bacteria, (iii) generating metabolic products (e.g., butyrate) that induce apoptosis, (iv) producing inhibitory compounds to hinder tumor cell growth, or (v) stimulating the immune system to counteract cancer cell propagation, leading to a more effective defense mechanism (Parvez et al., 2006).

Probiotics can provide protection to birds by mitigating the negative impact of enzyme activity and by increasing the activity or production of digestive enzymes. Moreover, probiotics have the potential to produce enzymes that release nutrients in the host's gastrointestinal tract. For instance, Nahashon et al. (1994) observed an increase in phytase activity in the crop of layers fed diets containing *L. acidophilus*, which resulted in improved phosphorus retention.

To gain a comprehensive understanding of how probiotics function, each approach must be studied on a case-by-case basis. The effects of probiotics stem from the interaction between the host and the probiotic microorganisms. Hence, further investigations are necessary to explore the intricacies of this host-microbe interaction, which could shed light on the modes of action of probiotics. In the past, limited research funding and underdeveloped molecular techniques might have hindered in-depth studies on the actions of probiotics. However, with the availability of advanced molecular techniques, researchers now have powerful tools to understand the effects and mechanisms of probiotics. The rapid progress in molecular methods and DNA sequencing can significantly aid in understanding microbial ecology and how probiotics exert their effects.

7.3 EFFECT OF PROBIOTICS IN ANIMAL GROWTH PERFORMANCE

The gut microbial community in animals plays a crucial role in maintaining their well-being and facilitating normal digestive processes. While probiotics were initially used in human nutrition, their application has now extended to animal feeding. Novel animal feeds containing gut microbiota have been formulated to enhance nutrient utilization and gut health, ultimately benefiting the animal production sector. Studies have indicated that probiotics offer advantages to host animals by promoting a balanced composition of intestinal microbes (Valdes et al., 2018). Administering probiotics has demonstrated a heightened impact on the levels of immunoglobulins like M and A, as well as an increase in overall antioxidant capacity in the bloodstream (Wang et al., 2018). Additionally, probiotic microorganisms contribute to regulating the host's immune response by boosting the presence of anti-inflammatory agents such as IL-10 and TGF-β (Kim and Lillehoj, 2019). Existing literature also reveals that administering *Lactobacillus rhamnosus* has a positive effect on intestinal growth, reducing epithelial apoptosis in the intestine, which, in turn, aids in combating gastrointestinal tract diseases (Chen et al., 2019).

7.3.1 Ruminant Animals

Rumen microbial ecology is a highly intricate process in which microorganisms break down the polysaccharides and proteins ingested by the host. This breakdown leads to the synthesis of short-chain fatty acids (SCFAs) and microbial protein, which serve as energy and protein sources for the host. Yeast, particularly *Saccharomyces cerevisiae*, is a commonly used probiotic in ruminants, and it mainly influences the microbial population and nutrient breakdown (Ghazanfar, 2021). Studies have shown that probiotic supplementation in ruminants can have positive effects on milk production and weight gain. For example, a dietary supplementation of 5×10^9 cfu of *E. faecium* and 2×10^9 yeast cells (*S. cerevisiae*) per cow per day resulted in a 2.3 L increase in milk production per cow per day (Nocek and Kautz, 2006). In goats, a probiotic mixture of *E. faecium* DDE 39, *L. alimentarius* DDL 48, *L. reuteri* DDL 19, and *B. bifidium* DDBA, isolated from a healthy goat, led to a 9% improvement in average body weight (Apás et al., 2010). Dairy heifers showed enhanced growth rates when administered *S. cerevisiae* (Ghazanfar et al., 2015). Besides lactic acid bacteria and other non-pathogenic microorganisms with health-promoting characteristics, certain strains of yeast, such as *Saccharomyces boulardii*, and *Escherichia coli*, like *E. coli Nissle* 1917, have been used in livestock feeding and medicine (Raheem et al., 2021).

Research indicates that probiotic microbes like *Bacillus subtilis*, *Saccharomyces cerevisiae*, and *Enterococcus faecalis* can enhance milk secretion in cows, and *Bifidobacterium bifidum* administration inhibits milk allergy reactions (Ma et al., 2019; Jing, Liu, and Wang, 2020). Probiotics like *Bacillus subtilis* and *Bacillus amyloliquefaciens* stimulate GH/IGF-1 hormone, leading to improved intestinal maturation and growth competency (Du et al., 2018). Moreover, *Lactobacillus*-based teat spray improves mammary gland condition and the functions of the teat sphincter (Alawneh et al., 2020). The addition of *Rhodopseudomonas palustris* has been shown to improve microbial fermentation and high rumen microbial growth performance (Chen et al., 2019; Chen et al., 2020).

7.3.2 Monogastric Animals

Probiotics have shown extensive effects in both poultry and humans compared to other monogastric animals. In broiler chickens, probiotics have been found to improve growth performance and control enteric diseases such as salmonellosis, coccidiosis, and necrotic enteritis (M'Sadeq et al., 2015). Additionally, probiotics like lactic acid bacteria, bacillus spores, and yeast have been reported to increase egg production and reduce yolk cholesterol levels in poultry (Panda et al., 2003; Yousefi and Karkoodi, 2006). Guo et al. (2006) demonstrated that *B. subtilis* strain MA139 significantly improved feed conversion ratio (FCR) in poultry. For primiparous sows, supplementing *E. faecium* at 5×10^8 cfu/kg feed resulted in increased feed consumption and enhanced reproductive performance (Bohmer, Kramer, and Roth-Maier, 2006). Le Bon et al. (2010) showed a significant reduction in *E. coli* levels in weaned piglets after four weeks of treatment with *P. acidilactici* and *S. boulardii*. Administration of *S. cerevisiae* to mature horses increased the apparent nutrient digestion rate (Agazzi et al., 2011).

In atopic children, probiotics were found to up-regulate anti-inflammatory cytokines, including interleukin-10, as reported by Pessi et al. (2000). A placebo-controlled trial in children with cystic fibrosis showed reduced pneumonia severity with the administration of *Lactobacillus* GG (Bruzzese et al., 2014). In rats with colon cancer, dietary supplementation of *L. acidophilus* resulted in a significant decrease in cell number in a dose-dependent manner (Drago, 2019).

7.3.3 Poultry

Poultry plays a crucial role as a primary protein source globally and significantly contributes to the economies of various countries, including the United States. Poultry farming has emerged as a cost-effective industry in many nations. Due to their efficiency in feed use, conversion, and overall production, poultry are an economical source of protein. As the human population grows, there is a rapid increase in the consumption and demand for poultry products. The attractiveness of poultry lies in their low input costs, which is partly due to their efficient use of feed and overall production efficiency. Since feeding constitutes a major portion (65–75%) of the total production cost in poultry, there is a strong need to continually improve feed efficiency. Probiotics have significantly contributed to this effort by promoting microbial balance, maintaining a healthy gut, and enhancing nutrient absorption. This has led to increased production efficiency in poultry farming.

The primary functions of the gastrointestinal tract (GIT) are to absorb nutrients from the diet and eliminate waste products. The microbial ecosystem of the gut is influenced by various factors, including diet nutrients, host substrates like mucin and bile acids, and immunological responses (Bywater, 2004; Jha et al., 2020). In commercial poultry production, feed enzymes are commonly used to enhance poultry performance by altering the gut environment and its associated microbiota (Santos, Storch, and Gil-Turnes, 2005). To maintain efficient nutrient utilization, especially in stressful commercial poultry production systems, antibiotics have traditionally been used as antimicrobials and growth promoters. However, the emergence of antimicrobial drug resistance from antibiotic use has prompted the industry to reconsider their use.

Probiotics are live microbial feed supplements or defined microorganisms that significantly improve the intestinal microbial balance of the host animal (Afrc, 1989; Schrezenmeir and De Vrese, 2001). They have a substantial impact on the host's health by influencing digestion, nutrient absorption, intestinal morphology, and defense against infections (Mead, 2000). Some of these microorganisms have been well characterized, and others may have beneficial effects on bird performance. Probiotics, like bacteria interacting with the gastrointestinal mucosa, can communicate with underlying epithelial and mucosal lymphoid elements, stimulating host defenses in the gut, as described by Walker (2008). They play a significant role in maintaining a balanced intestinal microbiota, which, in turn, improves nutrient utilization and helps exclude pathogenic microorganisms, thereby reducing the need for antibiotics in poultry production.

The gastrointestinal microbiota of monogastric animals, including poultry, consists mainly of anaerobic bacteria, especially Gram-positive bacteria, which increase in density from the proximal to the distal gut. Among these microorganisms, potential probiotics can be found that contribute to maintaining gut flora homeostasis and promoting gut health.

Considerable efforts have been devoted to identifying and characterizing beneficial microorganisms or candidate probiotics in the gastrointestinal tract of chickens. The adoption of next-generation sequencing technologies has accelerated the discovery and understanding of these bacteria, moving away from traditional culture methods. The aim is to establish a balanced and stable microbiota that maintains a symbiotic relationship with intestinal functions. Such equilibrium is vital for the well-being and performance of poultry, promoting digestion and absorption, functional immune status, gut mucosa integrity, and neuroendocrine and motor functions of the gut (Celi et al., 2017). Oral administration of lactic acid bacteria (*Lactobacillus casei* 1.2435, *Lactobacillus rhamnosus* 621, and *Lactobacillus rhamnosus* A4) to ducks improved their performance and altered their cecal microbiota interaction pattern, leading to intestinal flora dysbiosis (Shi et al., 2020).

Previous reports have demonstrated that adding probiotics enhances nutrient digestibility and absorption in poultry (Chen et al., 2009). For instance, birds fed with *B. subtilis* showed significantly higher digestibility of crude fiber and crude protein compared to untreated birds (Manafi et al., 2017). The addition of *Pediococcus acidilactici*, with or without a combination of oligosaccharides and butyric acid, restored amylase activity in *Salmonella typhimurium*-challenged broilers (Jazi et al., 2018). Similarly, Jin et al. (2000) found that a mixture of 12 *Lactobacillus* strains increased amylase levels in the small intestine and reduced intestinal and fecal β-glucuronidase and fecal β-glucosidase activities in day-old Arbor Acres chicks. Recent studies showed that supplementing broilers with *L. reuteri*, *B. licheniformis*, and *B. subtilis* improved body weight gain (BWG) and feed conversion ratio (FCR) when compared to non-treated controls (Zaghari, Sarani, and Hajati, 2020; Bhogoju et al., 2021). Additionally, Zhen et al. observed increased BWG and FCR from day 15 to day 21 in Cobb broilers challenged with *Salmonella enteritidis* when supplemented with *Bacillus coagulans* (Zhen et al., 2018).

Poultry raised in large-scale rearing facilities often face stressful conditions, leading to various health problems and economic losses due to diseases and environmental challenges. Historically, the control and prevention of diseases in poultry involved the extensive use of antibiotics and veterinary medicines. However, this practice resulted in the development of antimicrobial resistance, which negatively impacted both the animal and food industries, as this resistance can be transferred to humans. As a result, the poultry industry has been actively seeking alternatives to antibiotics, such as probiotics.

Maintaining a healthy diet and proper use of nutrients and probiotics for poultry growth promotion have garnered significant attention from nutritionists and veterinary experts. The occurrence of enteric diseases like *Salmonella* and *Campylobacter*, as well as productivity loss and mortality, are major concerns for the poultry industry and consumers. The supplementation of *Bacillus subtilis* DSM 32315 in the diet of broiler chickens showed beneficial effects in reducing the negative impacts of a *Clostridium perfringens* challenge on performance and intestinal microbiota (Bortoluzzi et al., 2019).

Whelan et al. (2019) observed positive changes in the gut microbial populations of broilers when administering *B. subtilis* DSM 32315 to the birds. This resulted in improved broiler performance when challenged with necrotic enteritis, suggesting that the probiotic had a mechanism that reduced the pathology of necrotic enteritis. The timing of probiotic administration may be crucial for ensuring its beneficial effects in poultry. Early administration of probiotics to chicks or at a young age can influence the gut microflora profile of the adult bird, as the gastrointestinal tract microbiota develops relatively early in the young chick's life cycle, possibly in vivo (Ilina et al., 2016). Orally administering probiotics to newly hatched chicks was found to rapidly establish an adult-type intestinal microflora, providing immediate resistance to colonization by pathogenic microorganisms like *Salmonella* and *Campylobacter.* The "competitive exclusion" effect has become an accepted strategy for controlling poultry-associated salmonellas in some countries, as reported by Mead (2000).

7.3.3.1 Use of Probiotics in Poultry Production

La Ragione (2001) reported that oral consumption of *Bacillus subtilis* spores could reduce *E. coli* colonization in chickens. Enteric diseases like salmonellosis, campylobacteriosis (a zoonotic disease), necrotic enteritis, and coccidiosis pose significant economic threats to the poultry industry (Skinner et al., 2010). Among these diseases, salmonellosis is a major food safety concern, as it can cause serious foodborne illnesses in humans. Probiotics are considered an alternative approach to controlling *Salmonella* and addressing the challenge of antibiotic-resistant strains. Haghighi et al. (2008) found that probiotics can significantly reduce cecal colonization by *Salmonella*, depending on the probiotic dose. Biloni et al. (2013) demonstrated that the administration of probiotics like *L. salivarius* and *Pediococcus parvulus* can slow down the transmission of *Salmonella* infection within poultry flocks. Giannenas et al. (2012) observed a decline in coccidiosis with the use of probiotics such as *E. faecium*, *B. animalis*, *L. reuteri*, and *B. subtilis*, either individually or in combination. Gaggìa et al. (2010) found that probiotics provided to the host during the early stages of life can induce changes in gene expression in intestinal epithelial cells, creating a favorable environment for the probiotic bacteria and leading to a significant reduction in mortality. The use of *Lactobacillus*-based probiotic products resulted in improved body weight (BW) and reduced feed conversion ratio (FCR) in broilers.

Probiotics also have positive effects on egg production in poultry, as well as growth rate, feed conversion, and feed efficiency. In laying hens, a mixed probiotic culture of *L. acidophilus*, *L. casei*, *B. thermophilus*, and *Enterococcus faecium* led to increased egg size and reduced feed cost (Adugna and Belete, 2020). *Bifidobacterium thermophilus* and *Enterococcus faecium* improved egg quality and production (Lokapirnasari et al., 2019). Long-term feeding of *E. faecium* as probiotics in chickens improved egg-laying intensity and feed conversion efficiency (Panda et al., 2003).

Previous studies (Nahashon et al., 1994) have shown that including 1100 ppm of *L. acidophilus* in the diets of single comb white Leghorn layers resulted in increased appetite and improved egg production, egg mass, egg weight, egg size, and feed conversion ratios. Adding fat to the diets containing *Lactobacillus* reduced daily feed consumption as expected, while still positively impacting bird performance and leading to better feed conversion, egg mass, egg sizes, and body weight gains. The retention of nitrogen and phosphorus also significantly improved in this study ($p < 0.05$).

Feeding *L. acidophilus* to single comb white Leghorn pullets had a lasting and positive effect on bird performance during the laying period (Nahashon et al., 1996). Positive correlations were observed between *Lactobacillus* diets and nitrogen and calcium retention, egg mass, as well as between fat, nitrogen, calcium, and phosphorus retention, body weight gain, calcium, and phosphorus retention, and egg mass, respectively. Layers fed *Lactobacillus* diets showed lower intestinal length and dry weight ($p < 0.05$) compared to the control diet. Daily feed consumption, egg size, nitrogen, and calcium retentions increased, while intestinal length decreased in layers fed *Lactobacillus* diets, and these effects were observed up to 59 weeks of age.

Nahashon et al. (1994) conducted a study to examine the impact of *Lactobacillus* supplementation on layer performance in diets with different ingredient compositions. They found that corn-soy-based diets improved egg weight, egg mass, egg size, and body weight gains, while the addition of barley to the corn-soy diets only improved body weight gains. Regardless of the diet type, the passage rates of digesta through the gastrointestinal tract were significantly increased with *Lactobacillus* supplementation. Moreover, the retention of fat, calcium, phosphorus, copper, and manganese in laying hens was significantly increased with *Lactobacillus* supplementation. Feeding *Lactobacillus* also stimulated the cellularity of Peyer's patches in the ileum, indicating a stimulation of the mucosal immune system that responds to antigenic stimuli by secreting immunoglobulin (IgA).

Probiotics, including *Lactobacillus*, are believed to play a crucial role in regulating the immune response by stimulating various subclasses of immune cells to produce cytokines. In chickens, *Lactobacillus* can modify the antibody response to antigens and also promote the production of transforming growth factor β (Valdes et al., 2018). Probiotics have been shown to prevent gastrointestinal diseases such as salmonellosis, necrotic enteritis, and coccidiosis (El-Sawah et al., 2020; El-Sharkawy et al., 2020; Rajput et al., 2020; Fazelnia et al., 2021), and they have also been found to improve nutrient digestion and enhance cecal microbial populations in broilers (Khalid et al., 2021). Several studies have demonstrated that probiotic administration in poultry enhances immunity (Punetha et al., 2018; Terada et al., 2019; Kong et al., 2020; Bilal et al., 2021). Zhang et al. (2021) observed improved chicken growth rate, immune system, and antioxidant levels with the addition of *Lactobacillus casei* and *Bifidobacterium* in the feed.

7.4 CHALLENGES OF ANIMAL FEEDING USING PROBIOTICS

Probiotic microorganisms commonly used in animal feed are generally considered to be safe. However, there are certain risks associated with their use, including the potential spread of antibiotic resistance due to the presence of communicable antibiotic resistance genes in some probiotic bacteria. Additionally, there is a possibility of infections from probiotic microorganisms, including the occurrence of enterotoxins and emetic toxins. Most research on probiotics focuses on their effectiveness rather than their safety, and the available information about the safety of probiotics is mainly limited to *Lactobacillus* and *Bifidobacterium* (Shanahan, 2012).

While probiotics used in animal feed are generally safe for animals, humans, and the environment, caution should be exercised with regard to unsafe or harmful microorganisms. Potential risks associated with the use of probiotics in animal feed include (Sanders et al., 2010):

i. Gastrointestinal tract infection in the animals that consume the probiotics.
ii. Gastrointestinal tract infections in consumers who consume animal products derived from probiotic-fed animals.
iii. Transmission of antibiotic resistance from probiotics to other pathogenic microbes.
iv. Infections in animal handlers and those involved in handling animal food.
v. Sensitization or irritation of the skin or eyes in individuals administering probiotics.

vi. Production of toxins by probiotics that may cause harmful metabolic or toxic effects on the host.
vii. Hyperstimulation of the immune system in susceptible hosts.

Further research is needed to better understand and ensure the safety of probiotics in animal feed. Precautions should be taken to prevent the potential risks associated with their use.

Prior to considering microorganisms as probiotics for animal feed, it is crucial to evaluate the potential risks associated with them. The identification of microorganisms down to the strain level is essential to assess the specificity of each bacterium and understand its beneficial properties. When evaluating the safety of microorganisms for use as probiotics in animal feed, certain concerns must be addressed. Probiotic microorganisms commonly used, such as *Lactobacillus* and *Bifidobacterium*, are generally regarded as the safest option (Shortt, 1999). These microorganisms have a long history of extensive use in various fermented foods and are naturally abundant in the gastrointestinal tract of both humans and animals. Infections related to these microorganisms are very rare. *L. acidophilus* and *L. bulgaricus*, in particular, have been classified as generally recognized as safe by the US Food and Drug Administration.

Some *Bacillus* species, including *B. subtilis*, *B. megaterium*, *B. licheniformis*, and *B. coagulans*, have been recognized as safe by the European Food Safety Authority (EFSA) due to the absence of toxins. However, when it comes to *Enterococcus* bacteria, while they do offer numerous beneficial effects, there have been a few reported infections in humans, both in community and hospital settings. Therefore, a more rigorous safety evaluation is required before considering the use of *Enterococcus* bacteria as probiotics in animal feed (Arias and Murray, 2012).

7.5 COMPLEMENTARY EFFECTS OF PREBIOTICS ON POULTRY FEEDING

Prebiotics are food components that cannot be digested by the host but have a beneficial impact by selectively stimulating the growth and/or activity of specific beneficial bacteria or probiotics in the colon (Gibson et al., 2004). In monogastric animals like poultry, these non-digestible polysaccharides pass through the small intestine without being hydrolyzed by endogenous digestive enzymes and reach the large intestine and cecum intact. There, they are hydrolyzed by bacteria, leading to the growth and activity of beneficial bacterial populations or probiotics that reside in the poultry's gastrointestinal tract. These indigestible carbohydrates, such as oligo- or polysaccharides, play a significant role in conferring beneficial effects to poultry.

Common prebiotics used in the industry include mannanoligosaccharides (MOS), fructooligosaccharides (FOS), galactooligosaccharides and glucooligosaccharides (GOS), xylooligosaccharides (XOS), oligofructose, mannose, stachyose, and even peptides, proteins, and short-chain fatty acids (SCFAs) (Huang et al., 2006; Jiang et al., 2006). These oligosaccharides are naturally present in various plants and vegetables like bananas, onions, chicory root, Jerusalem artichokes, and bamboo shoots. Prebiotics provide a source of energy and carbon for gut microorganisms, thereby modulating the metabolism of gut flora or probiotics. The composition of gut flora, influenced primarily by the diet, plays a pivotal role in determining the microbial profile and its associated functions (Rehman et al., 2020; Yaqoob et al., 2021).

According to Hajati and Rizaei (2010), lactose can be classified as a prebiotic in chickens due to its composition of glucose and galactose, which exhibit prebiotic effects. Since chickens lack the lactase enzyme, lactose passes through the upper parts of the digestive system and reaches the large intestine and cecum, where it is utilized during microbial fermentation. These prebiotics not only directly benefit gut microorganisms by providing them with nutrients but also benefit the host by producing useful compounds like SCFAs through the breakdown of prebiotics such as MOS, FOS, and GOS. Additionally, they positively impact the development and functionality of the digestive system, influencing mucosal morphology and mucin dynamics. Furthermore, they enhance the immune system by promoting the production of antimicrobial peptides and pro-inflammatory cytokines (Pan and Yu, 2013).

When a product combines both probiotics and prebiotics, creating a synergistic effect that enhances the functionality of the probiotic compound, it is referred to as a synbiotic (Schrezenmeir and De Vrese, 2001). The impact of synbiotics on poultry, including broilers and layers, has been extensively researched (Pan and Yu, 2013; Shi et al., 2019). Prebiotics have also been considered as potential components of free-range broiler feeds (Pan and Yu, 2013; Ricke, 2018; Shi et al., 2019). Through sequencing of the gut microbiome to assess the effect of prebiotics on bird performance, lactic acid bacteria and *Bifidobacterium* were detected, associated with improved bird performance (Shi et al., 2019). Several performance traits of free-range broilers were also evaluated, showing a positive response to prebiotic feeding (Ricke, 2018; Shi et al., 2019).

7.6 FUTURE IMPLICATIONS

Research has provided compelling evidence that probiotics offer substantial health benefits, acting as a source of health-promoting antimicrobials and nutrients in animal production. Probiotics can serve as alternatives to growth-stimulating antibiotics, thereby enhancing the immunological capacity of animals. Despite the current understanding of probiotics' effects on organisms, ongoing research aims to further elucidate their modes of action. The mechanism of action of probiotics is crucial to targeting specific aspects of growth or animal performance in the future. Investigating specific gene expression pathways and metabolic pathways related to the effects of probiotics can unveil numerous applications for probiotics in diagnostic and therapeutic contexts, benefiting both humans and animals.

Advanced molecular techniques, such as transcriptome and metabolomics analyses, provide profound insights into the mechanisms of action of probiotics, shedding light on their beneficial effects and their role in improving animal performance. Further research on specific gene expression pathways, revealed through metabolomics assays associated with probiotics' effects, can unlock various applications of probiotics for specific diagnostic and therapeutic purposes. By targeting specific applications, probiotics can address several disease-related concerns in both human and animal health.

Despite the recognized benefits of probiotics in enhancing animal performance and health, there are limitations associated with their use. Certain probiotic organisms, such as enterococci, may carry transmissible drug-resistant genes, and others, such as *Bacillus cereus*, can produce harmful enterotoxins (Ayichew et al., 2017; Zommiti et al., 2020). Additionally, key challenges in using probiotics include understanding potential interactions with host cells and determining safe dosages for specific recipients. Therefore, it is essential to conduct further studies to establish the more effective and appropriate use of probiotics based on the conditions of the host organisms or target recipients.

7.7 CONCLUSION

Current trends in poultry production require a more robust approach to efficiently producing meat and eggs while ensuring high-quality poultry products for consumers. This can be achieved by reducing production costs through optimized feeding practices and maintaining the health of poultry flocks. Efficiently managed and healthy flocks can lead to lower production costs, making poultry products more affordable and safer for consumers. This review presents compelling evidence for the use of probiotics and prebiotics in poultry production. Studies have shown that probiotics can positively influence and balance the gastrointestinal flora of birds, increasing beneficial microorganisms like *Lactobacillus acidophilus*, *Bifidobacterium*, *L. plantarum*, and *B. subtilis* while reducing pathogenic microorganisms such as *Salmonella* and *Campylobacter*.

Due to the excessive use of antibiotics in the poultry industry leading to antimicrobial resistance, the search for antibiotic alternatives has intensified, and probiotics have emerged as promising candidates. Beyond their potential antimicrobial effects, probiotics also contribute additional nutrients through their metabolites. Therefore, considerable effort should be directed toward developing diets that support the desired gut microbiomes, as diet significantly influences gut flora profiles. Utilizing advanced technologies such as next-generation sequencing, transcriptomics, and metabolomics, further exploration of microorganisms with probiotic potential and properties should be pursued.

The review also highlights the potential of prebiotics such as MOS, FOS, GOS, XOS, and SCFAs in poultry production. These prebiotics serve as an energy source for probiotics, and their byproducts from bacterial fermentation can be absorbed as nutrients by the host. Consequently, combining probiotics and prebiotics shows promise in the poultry industry, especially in enhancing feed efficiency and promoting health through competitive exclusion and the display of antimicrobial properties.

REFERENCES

Adugna, W.; Belete, Y. Review on effects of probiotic in chicken feed. *Adv. Biotechnol. Microbiol.* 2020, 3, 555913.

Afrc, R.F. Probiotics in man and animals. *J. Appl. Bacteriol.* 1989, 66, 365–378.

Agazzi, A.; Ferroni, M.; Fanelli, A., et al. Evaluation of the effects of live yeast supplementation on apparent digestibility of high-fiber diet in mature horses using the acid insoluble ash marker modified method. *J. Equine Vet. Sci.* 2011, 31, 13–18.

Alawneh, J.I.; James, A.S.; Phillips, N., et al. Efficacy of a lactobacillus-based teat spray on udder health in lactating dairy cows. *Front. Vet. Sci.* 2020, 7, 584436.

Amin, A.B.; Mao, S. Influence of yeast on rumen fermentation, growth performance and quality of products in ruminants: A review. *Anim. Nutr.* 2021, 7, 31–41.

Apás, A.L.; Dupraz, J.; Ross, R.; González, S.N.; Arena, M.E. probiotic administration effect on faecal mutagenicity and microflora in the goat's gut. *J. Biosci. Bioeng.* 2010, 110, 537–540.

Arias, C.A.; Murray, B.E. The rise of the *enterococcus*: Beyond vancomycin resistance. *Nat. Rev. Genet.* 2012, 10, 266–278.

Augustine, D.; Jacob, J.C.; Philip, R. Exclusion of Vibrio spp. by an antagonistic marine actinomycete *Streptomyces rubrolavendulae* M56. *Aquac. Res.* 2015, 47, 2951–2960.

Ayichew, T.; Belete, A.; Alebachew, T.; Tsehaye, H.; Berhanu, H.; Minwuyelet, A. bacterial probiotics their importance and limitations: A review. *J. Nutr. Health Sci.* 2017, 4, 202.

Bhogoju, S.; Khwatenge, C.; Taylor-Bowden, T., et al. Effects of *Lactobacillus reuteri* and *Streptomyces coelicolor* on growth performance of broiler chickens. *Microorganisms* 2021, 9, 1341.

Bhogoju, S.; Nahashon, S.; Wang, X.; Darris, C.; Kilonzo-Nthenge, A. A comparative analysis of microbial profile of guinea fowl and chicken using metagenomic approach. *PLoS One* 2018, 13, e0191029.

Bilal, M.; Si, W.; Barbe, F.; Chevaux, E.; Sienkiewicz, O.; Zhao, X. Effects of Novel Probiotic strains of *Bacillus pumilus* and *Bacillus subtilis* on production, gut health, and immunity of broiler chickens raised under suboptimal conditions. *Poult. Sci.* 2021, 100, 100871.

Biloni, A.; Quintana, C.; Menconi, A., et al. Evaluation of effects of early bird associated with FloraMax-B11 on *Salmonella enteritidis*, Intestinal morphology, and performance of broiler chickens. *Poult. Sci.* 2013, 92, 2337–2346.

Bohmer, B.M.; Kramer, W.; Roth-Maier, D.A. Dietary probiotic supplementation and resulting effects on performance, health status, and microbial characteristics of *primiparous sows. J. Anim. Physiol. Anim. Nutr.* 2006, 90, 309–315.

Bortoluzzi, C.; Vieira, B.S.; Dorigam, J.C.D.P.; Menconi, A.; Sokale, A.; Doranalli, K.; Applegate, T.J. *Bacillus subtilis* DSM 32315 Supplementation attenuates the effects of *Clostridium perfringens* challenge on the growth performance and intestinal microbiota of broiler chickens. *Microorganisms* 2019, 7, 71.

Bruzzese, E.; Callegari, M.L.; Raia, V., et al. Disrupted intestinal microbiota and intestinal inflammation in children with cystic fibrosis and its restoration with *Lactobacillus* GG: A randomised clinical trial. *PLoS One* 2014, 9, e87796.

Bywater, R.J. Veterinary use of antimicrobials and emergence of resistance in zoonotic and sentinel bacteria in the EU. *J. Vet. Med. Ser. B* 2004, 51, 361–363.

Castanon, J.I.R. History of the use of antibiotic as growth promoters in European poultry feeds. *Poult. Sci.* 2007, 86, 2466–2471.

Celi, P.; Cowieson, A.; Fru-Nji, F.; Steinert, R.; Kluenter, A.-M.; Verlhac, V. Gastrointestinal functionality in animal nutrition and health: New opportunities for sustainable animal production. *Anim. Feed Sci. Technol.* 2017, 234, 88–100.

Chen, B.; Peng, M.; Tong, W.; Zhang, Q.; Song, Z. The quorum quenching bacterium *Bacillus licheniformis* T-1 Protects Zebrafish against *Aeromonas hydrophila* Infection. *Probiotics Antimicrob. Proteins* 2019, 12, 160–171.

Chen, K.-L.; Kho, W.-L.; You, S.-H.; Yeh, R.-H.; Tang, S.-W.; Hsieh, C.-W. Effects of *Bacillus subtilis* var. natto and *Saccharomyces cerevisiae* mixed fermented feed on the enhanced growth performance of broilers. *Poult. Sci.* 2009, 88, 309–315.

Chen, L.; Li, H.; Li, J.; Chen, Y.; Yang, Y. *Lactobacillus rhamnosus* GG treatment improves intestinal permeability and modulates microbiota dysbiosis in an experimental model of sepsis. *Int. J. Mol. Med.* 2019, 43, 1139–1148.

Chen, Y.; Wang, Y.; Wang, W., et al. Beneficial effect of *Rhodopseudomonas palustris* on in vitro rumen digestion and fermentation. *Benef. Microbes* 2020, 11, 91–99.

Cristofori, F.; Dargenio, V.N.; Dargenio, C.; Miniello, V.L.; Barone, M.; Francavilla, R. Anti-inflammatory and immunomodulatory effects of probiotics in gut inflammation: A door to the body. *Front. Immunol.* 2021, 12, 578386.

Dankowiakowska, A.; Bednarczyk, M.; Kozłowska, I. Probiotics, prebiotics and snybiotics in poultry—Mode of action, limitation, and achievements. *J. Cent. Eur. Agric.* 2013, 14, 467–478.

Das, S.; Ward, L.R.; Burke, C. Screening of marine *Streptomyces* spp. for potential use as probiotics in aquaculture. *Aquaculture* 2010, 305, 32–41.

Ding, S.; Yan, W.; Ma, Y.; Fang, J. The impact of probiotics on gut health via alternation of immune status of monogastric animals. *Anim. Nutr.* 2021, 7, 24–30.

Drago, L. Probiotics and colon cancer. *Microorganisms* 2019, 7, 66.

Du, R.; Jiao, S.; Dai, Y., et al. Probiotic *Bacillus amyloliquefaciens* C-1 improves growth performance, stimulates GH/IGF-1, and regulates the gut microbiota of growth-retarded beef calves. *Front. Microbiol.* 2018, 9, 2006.

El-Sawah, A.A.; Aboelhadid, S.M.; El-Nahass, E.N., et al. Efficacy of probiotic *Enterococcus faecium* in combination with diclazuril against coccidiosis in experimentally infected broilers. *J. Appl. Microbiol.* 2020, 129, 1020–1028.

El-Sharkawy, H.; Tahoun, A.; Rizk, A.M., et al. Evaluation of *Bifidobacterium* and *Lactobacillus* probiotics as alternative therapy for *Salmonella typhimurium* infection in Broiler Chickens. *Animals* 2020, 10, 1023.

Fazelnia, K.; Fakhraei, J.; Yarahmadi, H.M.; Amini, K. Dietary Supplementation of potential probiotics *Bacillus subtilis*, *Bacillus licheniformis*, and *Saccharomyces cerevisiae* and Synbiotic improves growth performance and immune responses by modulation in intestinal system in broiler chicks challenged with Salmonella Typhimurium. *Probiotics Antimicrob. Proteins* 2021, 13, 1081–1092.

Gaggìa, F.; Mattarelli, P.; Biavati, B. Probiotics and prebiotics in animal feeding for safe food production. *Int. J. Food Microbiol.* 2010, 141, S15–S28.

Ghazanfar, S. Understanding the mechanism of action of indigenous target probiotic yeast: Linking the manipulation of gut microbiota and performance in animals. In *Saccharomyces*. InTech: London, UK, 2021.

Ghazanfar, S.; Anjum, M.I.; Azim, A.; Ahmed, I. Effects of dietary supplementation of yeast (*Saccharomyces cerevisiae*) culture on growth performance, blood parameters, nutrient digestibility and faecal flora of dairy heifers. *J. Anim. Plant Sci.* 2015, 25, 53–59.

Giannenas, I.; Papadopoulos, E.; Tsalie, E., et al. Assessment of dietary supplementation with probiotics on performance, intestinal morphology and microflora of chickens infected with *Eimeria tenella. Vet. Parasitol.* 2012, 188, 31–40.

Gibson, G.R.; Probert, H.M.; Van Loo, J., et al. Dietary modulation of the human colonic microbiota: Updating the concept of prebiotics. *Nutr. Res. Rev.* 2004, 17, 259–275.

Guo, X.; Li, D.; Lu, W.; Piao, X.; Chen, X. Screening of *Bacillus* strains as potential probiotics and subsequent confirmation of the in vivo effectiveness of *Bacillus subtilis* MA139 in pigs. *Antonie Leeuwenhoek* 2006, 90, 139–146.

Haghighi, H.R.; Abdul-Careem, M.F.; Dara, R.A.; Chambers, J.R.; Sharif, S. Cytokine gene expression in chicken caecal tonsils following treatment with probiotics and Salmonella infection. *Vet. Microbiol.* 2008, 126, 225–233.

Hajati, H.; Rezaei, M. The application of prebiotics in poultry production. *Int. J. Poult. Sci.* 2010, 9, 298–304.

Havenaar, R.; Brink, B.T.; Veld, J.H.J.H.I. Selection of strains for probiotic use. In *Probiotics*. Springer: Dordrecht, The Netherlands, 1992; pp. 209–224. 22.

Heredia, N.; García, S. Animals as sources of food-borne pathogens: A review. *Anim. Nutr.* 2018, 4, 250–255.

Huang, R.-L.; Deng, Z.-Y.; Yang, C.-B., et al. Dietary oligochitosan supplementation enhances immune status of broilers. *J. Sci. Food Agric.* 2006, 87, 153–159.

Ilina, L.A.; Yildirim, E.A.; Nikonov, I., et al. Metagenomic bacterial community profiles of chicken embryo gastrointestinal tract by using T-RFLP analysis. *Dokl. Biochem. Biophys.* 2016, 466, 47–51.

Jazi, V.; Foroozandeh, A.; Toghyani, M.; Dastar, B.; Koochaksaraie, R.R. Effects of *Pediococcus acidilactici*, mannan-oligosaccharide, butyric acid and their combination on growth performance and intestinal health in young broiler chickens challenged with *Salmonella typhimurium. Poult. Sci.* 2018, 97, 2034–2043.

Jha, R.; Das, R.; Oak, S.; Mishra, P. Probiotics (Direct-Fed Microbials) in poultry nutrition and their effects on nutrient utilization, growth and laying performance, and gut health: A systematic review. *Animals* 2020, 10, 1863.

Jiang, H.Q.; Gong, L.M.; Ma, Y.X.; He, Y.H.; Li, D.F.; Zhai, H.X. Effect of stachyose supplementation on growth performance, nutrient digestibility and caecal fermentation characteristics in broilers. *Br. Poult. Sci.* 2006, 47, 516–522.

Jiang, M.; Dai, C.; Zhao, D.-H. VSL#3 probiotics regulate the intestinal epithelial barrier in vivo and in vitro via the p38 and ERK signalling pathways. *Int. J. Mol. Med.* 2011, 29, 202–208.

Jin, L.Z.; Ho, Y.W.; Abdullah, N.; Jalaludin, S. Digestive and bacterial enzyme activities in broilers fed diets supplemented with *Lactobacillus* cultures. *Poult. Sci.* 2000, 79, 886–891.

Jing, W.; Liu, Q.; Wang, W. *Bifidobacterium bifidum* TMC3115 ameliorates milk protein allergy in by affecting gut microbiota: A randomized double-blind control trial. *J. Food Biochem.* 2020, 44, e13489.

Khalid, A.H.; Ullah, K.S.; Naveed, S., et al. Effects of spray dried yeast (*Saccharomyces cerevisiae*) on growth performance and carcass characteristics, gut health, caecal microbiota profile and apparent ileal digestibility of protein, amino acids and energy in broilers. *Trop. Anim. Health Prod.* 2021, 53, 252.

Kim, W.H.; Lillehoj, H.S. Immunity, immunomodulation, and antibiotic alternatives to maximize the genetic potential of poultry for growth and disease response. *Anim. Feed Sci. Technol.* 2019, 250, 41–50.

Kong, Y.; Gao, C.; Du, X.; Zhao, J.; Li, M.; Shan, X.; Wang, G. Effects of single or conjoint administration of lactic acid bacteria as potential probiotics on growth, immune response and disease resistance of snakehead fish (Channa argus). *Fish Shellfish Immunol.* 2020, 102, 412–421.

La Ragione, R. *Bacillus subtilis* spores competitively exclude *Escherichia coli* O78:K80 in poultry. *Vet. Microbiol.* 2001, 79, 133–142.

Le Bon, M.; Davies, H.E.; Glynn, C., et al. Influence of probiotics on gut health in the weaned pig. *Livest. Sci.* 2010, 133, 179–181.

Lehtoranta, L.; Latvala, S.; Lehtinen, M.J. Role of probiotics in stimulating the immune system in viral respiratory tract infections: A narrative review. *Nutrients* 2020, 12, 3163.

Lilly, D.M.; Stillwell, R.H. Probiotics: Growth-promoting factors produced by microorganisms. *Science* 1965, 147, 747–748.

Lokapirnasari, W.P.; Pribadi, T.B.; Al Arif, A., et al. Potency of probiotics *Bifidobacterium* spp. and *Lactobacillus casei* to improve growth performance and business analysis in organic laying hens. *Vet. World* 2019, 12, 860–867.

M'Sadeq, S.A.; Wu, S.; Swick, R.; Choct, M. Towards the control of necrotic enteritis in broiler chickens with in-feed antibiotics phasing-out worldwide. *Anim. Nutr.* 2015, 1, 1–11.

Ma, Z.; Cheng, Y.; Wang, S.; Ge, J.; Shi, H.; Kou, J. Positive effects of dietary supplementation of three probiotics on milk yield, milk composition and intestinal flora in Sannan dairy goats varied in kind of probiotics. *J. Anim. Physiol. Anim. Nutr.* 2019, 104, 44–55.

Maldonado Galdeano, C.; Cazorla, S.I.; Lemme Dumit, J.M.; Vélez, E.; Perdigón, G. Beneficial effects of probiotic consumption on the immune system. *Ann. Nutr. Metab.* 2019, 74, 115–124.

Manafi, M.; Khalaji, S.; Hedayati, M.; Pirany, N. Efficacy of *Bacillus subtilis* and bacitracin methylene disalicylate on growth performance, digestibility, blood metabolites, immunity, and intestinal microbiota after intramuscular inoculation with *Escherichia coli* in broilers. *Poult. Sci.* 2017, 96, 1174–1183.

Markowiak, P.; Slizewska, K. Effects of probiotics, prebiotics, and synbiotics on human health. *Nutrients* 2017, 9, 1021.

Mead, G. Prospects for 'competitive exclusion' treatment to control salmonellas and other foodborne pathogens in poultry. *Vet. J.* 2000, 159, 111–123.

Mead, P.S.; Slutsker, L.; Dietz, V.; McCaig, L.F., et al. Food-related illness and death in the United States. *Emerg. Infect. Dis.* 1999, 5, 607–625.

Nahashon, S.N.; Nakaue, H.S.; Mirosh, L.W. Performance of single comb white leghorn fed a diet supplemented with a live microbial during the growth and egg laying phases. *Anim. Feed Sci. Technol.* 1996, 57, 25–38.

Nahashon, S.N.; Nakaue, H.S.; Mirosh, L.W. Phytase activity, phosphorus and calcium retention, and performance of single comb white leghorn layers fed diets containing two levels of available phosphorus and supplemented with direct-fed Microbials. *Poult. Sci.* 1994, 73, 1552–1562.
Nocek, J.; Kautz, W. Direct-fed microbial supplementation on ruminal digestion, health, and performance of pre- and postpartum dairy cattle. *J. Dairy Sci.* 2006, 89, 260–266, Erratum in J. Dairy Sci. 2011, 94, 1641.
Oelschlaeger, T.A. Mechanisms of probiotic actions—A review. *Int. J. Med. Microbiol.* 2010, 300, 57–62.
Pagnini, C.; Saeed, R.; Bamias, G.; Arseneau, K.O.; Pizarro, T.T.; Cominelli, F. Probiotics promote gut health through stimulation of epithelial innate immunity. *Proc. Natl. Acad. Sci. USA* 2010, 107, 454–459.
Pan, D.; Yu, Z. Intestinal microbiome of poultry and its interaction with host and diet. *Gut Microbes* 2013, 5, 108–119.
Panda, A.; Reddy, M.; Rao, S.R.; Praharaj, N. Production performance, Serum/Yolk cholesterol and immune competence of white leghorn layers as influenced by dietary supplementation with probiotic. *Trop. Anim. Health Prod.* 2003, 35, 85–94.
Parvez, S.; Malik, K.; Kang, S.A.; Kim, H.-Y. Probiotics and their fermented food products are beneficial for health. *J. Appl. Microbiol.* 2006, 100, 1171–1185.
Pessi, T.; Sütas, Y.; Hurme, M.; Isolauri, E. Interleukin-10 generation in atopic children following oral *Lactobacillus rhamnosus* GG. *Clin. Exp. Allergy* 2000, 30, 1804–1808.
Piqué, N.; Berlanga, M.; Miñana-Galbis, D. Health benefits of heat-killed (Tyndallized) probiotics: An overview. *Int. J. Mol. Sci.* 2019, 20, 2534.
Plaza-Diaz, J.; Ruiz-Ojeda, F.J.; Gil-Campos, M.; Gil, A. Mechanisms of action of probiotics. *Adv. Nutr.* 2019, 10, S49–S66, Correction in Adv. Nutr. 2020, 11, 1054.
Prestinaci, F.; Pezzotti, P.; Pantosti, A. Antimicrobial resistance: A global multifaceted phenomenon. *Pathog. Glob. Health* 2015, 109, 309–318.
Punetha, M.; Roy, A.K.; Ajithakumar, H.M.; Para, I.A.; Gupta, D.; Singh, M.; Bharati, J. Immunomodulatory effects of probiotics and prilled fat supplementation on immune genes expression and lymphocyte proliferation of transition stage Karan Fries cows. *Vet. World* 2018, 11, 209–214.
Raheem, A.; Liang, L.; Zhang, G.; Cui, S. Modulatory effects of probiotics during pathogenic infections with emphasis on immune regulation. *Front. Immunol.* 2021, 12, 616713.
Rajput, D.S.; Zeng, D.; Khalique, A., et al. Pretreatment with probiotics ameliorate gut health and necrotic enteritis in broiler chickens, a substitute to antibiotics. *AMB Express* 2020, 10, 220.
Rehman, A.; Arif, M.; Sajjad, N.; Al-Ghadi, et al. Dietary effect of probiotics and prebiotics on broiler performance, carcass, and immunity. *Poult. Sci.* 2020, 99, 6946–6953.
Rezaee, P.; Kermanshahi, R.K.; Falsafi, T. Antibacterial activity of *lactobacilli* probiotics on clinical strains of *Helicobacter pylori*. *Iran. J. Basic Med. Sci.* 2019, 22, 1118–1124.
Ricke, S.C. Impact of prebiotics on poultry production and food safety. *Yale J. Biol. Med.* 2018, 91, 151–159.
Sanders, M.E.; Akkermans, L.M.; Haller, D.; Hammerman, C.; Heimbach, J.T.; Hörmannsperger, G.; Huys, G. Safety assessment of probiotics for human use. *Gut Microbes* 2010, 1, 164–185.
Santos, J.G.D.L.; Storch, O.; Gil-Turnes, C. *Bacillus cereus* var. *toyoii* and *Saccharomyces boulardii* increased feed efficiency in broilers infected with *Salmonella enteritidis*. *Br. Poult. Sci.* 2005, 46, 494–497.
Schrezenmeir, J.; De Vrese, M. Probiotics, prebiotics, and synbiotics—Approaching a definition. *Am. J. Clin. Nutr.* 2001, 73, 361s–364s.
Shanahan, F. A commentary on the safety of probiotics. Gastroenterol. *Clin. N. Am.* 2012, 41, 869–876.
Shi, S.; Qi, Z.; Jiang, W.; Quan, S.; Sheng, T.; Tu, J.; Shao, Y.; Qi, K. Effects of probiotics on caecal microbiome profile altered by duck *Escherichia coli* 17 infection in Cherry Valley ducks. *Microb. Pathog.* 2020, 138, 103849.
Shi, Z.; Rothrock, M.J., Jr.; Ricke, S.C.; Rothrock, M.J. Applications of microbiome analyses in alternative poultry broiler production systems. *Front. Vet. Sci.* 2019, 6, 00157.
Shortt, C. The probiotic century: Historical and current perspectives. *Trends Food Sci. Technol.* 1999, 10, 411–417.
Skinner, J.T.; Bauer, S.; Young, V.; Pauling, G.; Wilson, J. An economic analysis of the impact of subclinical (Mild) necrotic enteritis in broiler chickens. *Avian Dis.* 2010, 54, 1237–1240.
Terada, T.; Nii, T.; Isobe, N.; Yoshimura, Y. Effects of probiotics *Lactobacillus reuteri* and *Clostridium butyricum* on the expression of toll-like receptors, Pro- and Anti-inflammatory Cytokines, and antimicrobial peptides in broiler chick intestine. *J. Poult. Sci.* 2019, 57, 310–318.
Valdes, A.; Walter, J.; Segal, E.; Spector, T.D. Role of the gut microbiota in nutrition and health. *BMJ* 2018, 361, k2179.

Vieco-Saiz, N.; Belguesmia, Y.; Raspoet, R.; Auclair, E.; Gancel, F.; Kempf, I.; Drider, D. Benefits and inputs from lactic acid bacteria and their bacteriocins as alternatives to antibiotic growth promoters during food-animal production. *Front. Microbiol.* 2019, 10, 57.

Walker, W.A. Mechanisms of action of probiotics. *Clin. Infect. Dis.* 2008, 46, S87–S91.

Wang, Y.-B. Effect of probiotics on growth performance and digestive enzyme activity of the shrimp *Penaeus vannamei*. *Aquaculture* 2007, 269, 259–264.

Wang, Y.B.; Dong, Z.; Song, D., et al. Effects of microencapsulated probiotics and prebiotics on growth performance, antioxidative abilities, immune functions, and caecal microflora in broiler chickens. *Food Agric. Immunol.* 2018, 29, 859–869.

Whelan, R.A.; Doranalli, K.; Rinttilä, T.; Vienola, K.; Jurgens, G.; Apajalahti, J. The impact of *Bacillus subtilis* DSM 32315 on the pathology, performance, and intestinal microbiome of broiler chickens in a necrotic enteritis challenge. *Poult. Sci.* 2019, 98, 3450–3463.

Yaqoob, M.; El-Hack, M.A.; Hassan, F., et al. The potential mechanistic insights and future implications for the effect of prebiotics on poultry performance, gut microbiome, and intestinal morphology. *Poult. Sci.* 2021, 100, 101143.

Yousefi, M.; Karkoodi, K. Effect of Probiotic Thepax® and *Saccharomyces cerevisiae* supplementation on performance and egg quality of laying hens. *Int. J. Poult. Sci.* 2006, 6, 52–54.

Zaghari, M.; Sarani, P.; Hajati, H. Comparison of two probiotic preparations on growth performance, intestinal microbiota, nutrient digestibility and cytokine gene expression in broiler chickens. *J. Appl. Anim. Res.* 2020, 48, 166–175.

Zhang, L.; Zhang, R.; Jia, H.; Zhu, Z.; Li, H.; Ma, Y. Supplementation of probiotics in water beneficial growth performance, carcass traits, immune function, and antioxidant capacity in broiler chickens. *Open Life Sci.* 2021, 16, 311–322.

Zhen, W.; Shao, Y.; Gong, X.; Wu, Y.; Geng, Y.; Wang, Z.; Guo, Y. Effect of dietary *Bacillus coagulans* supplementation on growth performance and immune responses of broiler chickens challenged by *Salmonella enteritidis*. *Poult. Sci.* 2018, 97, 2654–2666.

Zommiti, M.; Chikindas, M.L.; Ferchichi, M. Probiotics—Live biotherapeutics: A story of success, limitations, and future prospects—Not only for humans. *Probiotics Antimicrob. Proteins* 2020, 12, 1266–1289.

8 Microbiome Engineering of Livestock by Using Feed Enzymes

Silva Shihab, Anu Jacob and Jissin Mathew
Department of Biotechnology, Karunya Institute of Technology and Sciences

8.1 INTRODUCTION

The microbiota refers to the complex microbial population that inhabits the gastrointestinal tract of animals, and the microbiome describes the collection of all these microbes' genomes. The interaction between livestock and microbes determines the health, productivity, and meat quality of the livestock (Yeoman and White 2014). The microbiome contributes to the digestion and assimilation of nutrients, the immune system, and meat quality. Safe and quality meat sources are required for human nutrition worldwide (Shang et al. 2018).

The efficiency of food production, including food produced from crops and animals, needs to be increased because future projections indicate that global food demand will increase dramatically. The factors that affect gut microbial composition and function are dietary habits, genetics, age, gender, and location (Fuyong Li et al. 2019); nevertheless, the effect of diet has received the most attention to date. Recently, the challenges related to the microbiota are in focus to improve feed utilization efficiency and product quality, and minimize methane release. The gut microbiota has received significant attention in research addressing animal health and productivity because of the critical role played by microbes in nutrition, fitness, and performance (W. Park 2018). The gut microbiota efficiently impacts health traits, immunity, behavior, and stress response in livestock (such as pigs, chickens, and cows), which offers an innovative way to assess well-being and health (S. Chen, Luo, and Yan 2021).

8.1.1 The Livestock Industry

Globally, the livestock industry plays a vital role in a nation's economy. Most populations depend on it for food security and their means of subsistence. Livestock provides valuable products, such as meat, milk, and wool, for humans, and currently, the industry faces various hurdles. A projected world population of 9.15 billion by 2050 will necessitate a 70% rise in food production, from 2007, in developed countries and possibly a twofold increase in developing countries (Gerber et al. 2013).

Global meat consumption between 1988 and 2018 was estimated by FAOSTAT (2018) to be over 350 million tonnes per year, with an additional quantity requirement of 200 million tonnes by 2050. The population, economic growth, changes in lifestyle and food choices have all been linked to the anticipated extraordinary rise in the meat market (FAO 2018). Further, FAOSTAT (2018) estimates that meat output would need to double from its 2008 level to meet world meat consumption by 2050. Therefore, careful consideration must be given to the interaction between the two, considering the significance of the livestock industry in the global market and the current state of climate change.

DOI: 10.1201/9781003394662-8

Climate change is a concern faced by the livestock industry due to its impacts on the nutritional value of feed and forage, water supply, animals and diseases, reproduction, and biodiversity. However, livestock production accounts for global greenhouse gas (GHG) emissions and accelerates climate change. The developments in animal farming practices have resulted from the growing market for animal products, with a move towards more intensive and industrialized techniques for animal production. Consumers, decision-makers, and stakeholders in the livestock farming sector now seriously consider animal welfare and sustainability. New agricultural practices and technology have emerged, including precision and digital livestock farming, which may improve sustainability and the well-being of animals (Neethirajan 2023).

Worldwide, livestock utilize about 26% of the ice-free land, and a third of the farmland is used for animal feed production (FAO 2021). The livestock industry produces about 40% of the global agricultural gross domestic product (GDP). Livestock provide about 33% of the protein and 17% of the calories consumed globally. The industry generates significant employment possibilities for rural households (Thornton 2010; Rojas-Downing et al. 2017). Furthermore, livestock is a major source of nutrition, nutritional security, and income (Swanepoel, Stroebel, and Moyo 2010). Based on the life cycle assessment, agriculture is responsible for approximately 14% of the world's anthropogenic GHG emissions, while just less than 5% of the total amount emitted is due to direct emissions from livestock (Meyer and Pachauri 2014).

8.2 MICROBIAL COMMUNITIES AND LIVESTOCK

Five livestock species, namely, cows, chickens, goats, pigs, and sheep, produce the vast majority of animal-based products that people consume (meat, milk, eggs) (FAOSTAT 2022). These species differ noticeably, with chickens being birds and the others being mammals. Additionally, there are differences between ruminants and monogastric animals in their physiological traits (Forcina et al. 2022). Monogastrics are distinguished by quicker development and a shorter lifetime, while ruminants benefit from their ability to extract energy from low-quality food and the challenges they encounter in maintaining a balanced and healthy ruminal flora. The digestive tract controls the animal's performance and is the most fundamental element in animal production and the focus of microbiome study, followed by the reproductive tract (Celi et al. 2017). However, a rising number of these studies have also recently focused on environmental and public health issues, particularly those related to GHG emissions (Min et al. 2020), the prevalence of food-borne infections (Black et al. 2021), and the increase of antibiotic resistance (Zalewska et al. 2021).

8.2.1 Function of Microbiota in Animals

Primary microbiota members digest consumed fiber, carbohydrates, protein, and lipids into shorter chains (or oligomers) and monomers (such as glucose, amino acids), which are then utilized as nutrients by different microbes in the microbial community (Millen, Arrigoni, and Pacheco 2016). The main products of rumen microbial fermentation are volatile fatty acids (VFAs), including acetate, propionate, and butyrate, which the host absorbs and uses as an energy source. Since ruminally obtained VFAs can supply as much as 70% of the host's energy requirements, animal performance depends on their generation. The rumen microbiota's ability to break down nitrogen-containing substances, such as peptides, ammonia, and urea, is essential for supplying the host with microbial proteins for the production of milk and muscle (Bach, Calsamiglia, and Stern 2005). Increasing our understanding of the ruminant microbiome has tremendous potential for improving animal health and productivity while reducing environmental pollution (Cholewińska et al. 2020). Bacteria possess enzymes that break down starch, plant cell walls, proteins, and lipids (Huws et al. 2018), while archaea release methane in the rumen through the hydrogenotrophic pathway, and less frequently via the methylotrophic and acetoclastic pathways (Morgavi et al. 2010; De la Fuente et al. 2019). Whenever animal diets are high in readily available starch, protozoa help to maintain the rumen's pH, participate in the early colonization of fiber, which has a favorable impact on fiber degradation, and release H_2, which later acts as a substrate for methanogens to convert CO_2 to methane (Newbold et al. 2015).

The microbial community residing in the foregut of ruminants and the hindgut of monogastric animals improves feed efficiency as it breaks down feedstuff and transforms them into beneficial compounds, such as VFA and water-soluble vitamins, which are then absorbed by the host animal (Lourenco and Welch 2022). It imparts health advantages to the host by forming the intestinal epithelial barrier and battling pathogenic microorganisms to stop their damaging propagation (Yu et al. 2012). Moreover, the microbiome influences the formation and function of the mucosal immune system (Malmuthuge, Griebel, and Guan 2014).

8.2.2 Microbiota Profile

The microbiota composition modulates the metabolomic factors that influence feed utilization and metabolic function of animals. The diversity and population of the gut microbiome are the main factors affecting its health and productivity in livestock. The intricate microbial ecosystem found in the gastrointestinal tract (GIT) is essential for preserving intestinal homeostasis, developing mucosal and lymphoid structures, and activating the host immune cell repertoire (Malmuthuge and Guan 2017). Another crucial component is the microbial fermentation of ingested plant biomass in the rumen, which enables the animal to utilize the nutritional potential of the host-indigestible plant feedstock (Mackie 2002).

The profiling of the microbiota began with advances in the culture techniques of microbes. However, only 8% of the microbes constituted cultivable isolates (Weimer 2015). Further, advances in high-throughput sequencing methods have enabled the study of various gut microbial species (Myer et al. 2016; F Li et al. 2016). Gut microbial profiling is essential for tracking livestock health status and establishing treatments to improve livestock health status and prevent the onset or aggravation of pathological states (Clavijo and Flórez 2018). The bacterial population reaches 10^{12} cells per gram of rumen content, with the diversity exceeding 7,000 species of 19 phyla, constituting Firmicutes (56%), Bacteroidetes (31%), and Proteobacteria (4%) (Chaucheyras-Durand and Ossa 2014; Kim, Morrison, and Yu 2011). The archaea population ranges between 10^8 and 10^{10} gene copies per gram of rumen content (Mosoni et al. 2011), attributed to Methanobrevibacter (>60%), Methanomicrobium (15%), and rumen cluster C (16%) (Kim, Morrison, and Yu 2011). About 10^5 and 10^6 cells per gram of rumen content in the rumen is protozoa, with *Entodinium* as the primary genus (Chaucheyras-Durand and Ossa 2014).

8.2.3 Microbiome Contributes to Health Homeostasis

The sensitive growth equilibrium within the microbial community is vulnerable to incursions, which can result in the loss or gain of various microbial species and a microbial imbalance known as dysbiosis (Wilkins, Monga, and Miller 2019). Xenobiotics seldom alter the host genome. On the contrary, dietary factors, antibiotic use, pathogen infection, and other incidents in life easily influence the microbiome (Honda and Littman 2012).

8.3 MICROBIOME ENGINEERING

The core microbiome in the host influences their characteristics as microbiological communities. Microbiome engineering is a method that enhances the functioning of hosts through artificial selection for microbial communities with particular effects on host wellness. It uses multigenerational artificial selection on hosts with different microbiome compositions, altering the host characteristics (Mueller and Sachs 2015). Microbiome engineering allows the modification of microbial assemblages to enhance host phenotypes and benefit ecosystems (Panzer and Lynch 2015).

Every ecosystem has distinct characteristics as the outcome of complex interactions among the microbial populations and their hosts, and reactions to the metabolites that communities create (Waldor et al. 2015). Therefore, protuberance of the microbiome can result in alterations to the hosts and environment, including decreased host well-being and soil fertility (Navarrete et al. 2015). Microbiome engineering can change microbial populations to enhance host traits and improve ecosystems. As a notion for solving several microbiome-related challenges, various ecological, biochemical, and molecular-based strategies have been developed. Engineering the microbiome may make it possible to research the role of certain bacteria in developing new treatments for diseases such as inflammatory and immunological conditions as well as metabolic conditions such as phenylketonuria and chronic kidney disease (El Aidy and Kleerebezem 2013). To accurately manipulate microbial populations to enhance health and infection outcomes, new methods are still needed (Inda et al. 2019).

8.4 MICROBIOME ENGINEERING APPROACHES IN LIVESTOCK

Antibiotics are often applied in livestock husbandry to inhibit bacterial infections and enhance animal development; however, their usage has led to the creation of antibiotic-resistant bacteria and generated alarms about the presence of leftover antibiotics in its products. To improve livestock health in the farming business, microbiome engineering approaches (Figure 8.1) involving feed enzymes, prebiotics, and probiotics to change the microbiota population are a potential solution to the use of antibiotics (Foo et al. 2017).

Feed enzymes such as phytase, amylase, non-starch polysaccharide (NSP)-digesting enzymes (such as xylanase, β-glucanase, and β-mannanase), proteases, and lysozyme are applied to improve substrate degradation, increase the synthesis of prebiotics from dietary NSPs, and decrease antinutritive elements to support gut health in pigs and poultry (Kiarie, Romero, and Nyachoti 2013).

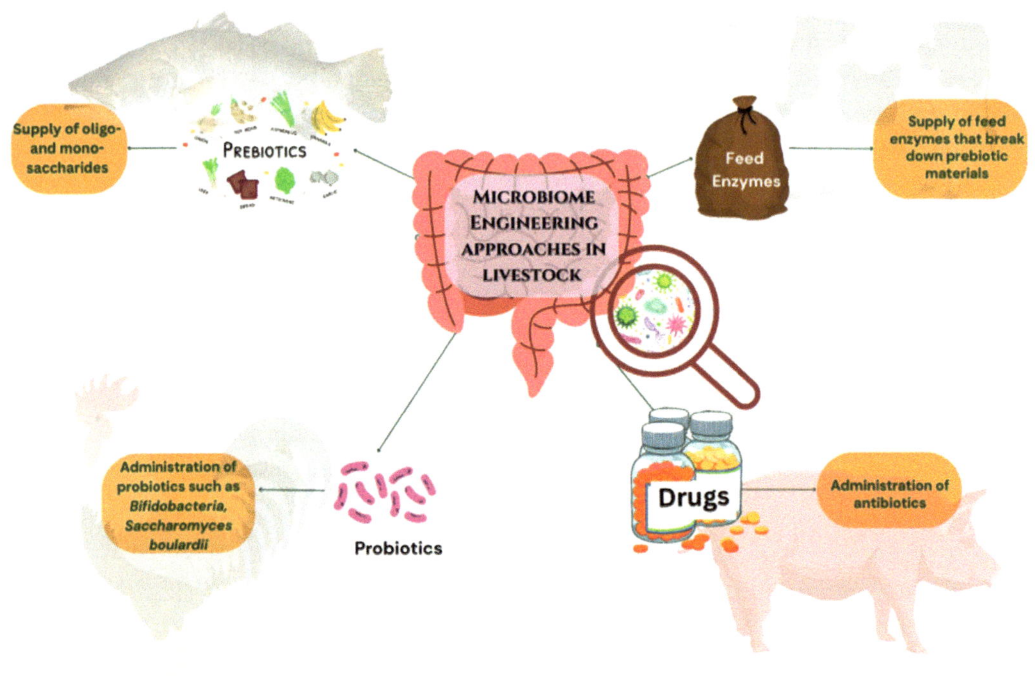

FIGURE 8.1 Microbiome approaches in livestock.

The inclusion of feed enzymes in animal diets is an approach to enhancing animal development and managing infectious illnesses, such as post-weaning colibacillosis in piglets, swine dysentery, salmonellosis, and necrotic enteritis, by improving gut health and nutrient utilization.

The development of animal development and the management of infectious illnesses, such as post-weaning colibacillosis in piglets, swine dysentery, salmonellosis, and necrotic enteritis, have been made possible by feed enzymes. Prebiotic provision permits alteration of the animal gut microbiota and leads to the synthesis of (i) short-chain fatty acids (SCFAs) that decrease luminal pH, serve as energy sources, control metabolism, and regulate immune systems (Den Besten et al. 2013); (ii) factors that prevent pathogens; and (iii) prebiotic derivatives that lessen pathogens' adherence to host cells (Pourabedin and Zhao 2015).

8.5 MICROBIOME ENGINEERING OF LIVESTOCK BY USING FEED ENZYMES

Feed enzymes have developed into a crucial tool for boosting the nutrient content of feed materials, cutting feed costs, protecting the environment, and enhancing animal performance. By definition, enzymes are biological catalysts that accelerate particular chemical reactions. These enzymes are produced efficiently and used in animal feeds. Feed enzymes are the most extensively studied feed additives (Ravindran 2013). Different feed enzymes are in the market against various target substrates in livestock (Table 8.1). Feed costs account for 40–60% of the total production cost in the dairy industry (Bozic et al. 2012), and improvements in the nutritional digestibility of feedstuff increase animal feed utilization and decrease costs. The poultry industry was among the first to demonstrate an interest in the application of feed enzymes in the 1980s (Velázquez-De Lucio et al. 2021). Over time, their use in animal feed grew noticeably, with an estimated market value of 2.88 billion dollars in 2030 (Skyquest 2022). The diverse microbial ecosystem facilitates the digestion of fibrous feedstuff in ruminants. Cell wall components and the cross-linked form of polysaccharides, lignin, and phenolic chemicals constrain the digestion of plant cell wall polysaccharides by ruminal microorganisms. For a while now, monogastric animal production has been effectively using exogenous fibrolytic enzymes (EFE). Animal reactions to EFE feed additives have varied due to variations in the source of the enzyme, application method, the types of foods, and the animal.

Enzymes can improve feed consumption by enhancing the speed and magnitude of pre-ingestive, ruminal, and post-ruminal fiber breakdown, digestion, and decomposition, by improving the ruminal residence, enhancing ruminal microbial composition and attachment, triggering ruminal microbes, and lowering digestive fluid viscosity (Adesogan et al. 2014).

The primary way the enzymes increase production and feed consumption is by breaking down food into molecules that are simple to digest. Amylases, lipases, phytases, galactosidases, glucanases, xylanases, proteases, pepsin, methionine, and polygalacturonase are a few of the significant enzyme products that have been approved for use in animal diets (Niranjan et al. 2023). They are all liquid fermentation byproducts of either bacteria or fungus and are applied as liquids, solids, powders, coated preparation granulates, and micro granulates.

In the present era, the feed industry has a choice of four main types of readily accessible enzyme products: microbial phytases; glycanases, which target viscous cereals (like wheat and barley); enzymes that target non-viscous cereals (like corn and sorghum); and enzymes that target non-cereals (like soybean meal and grain legumes). Most enzyme products, except microbial phytases, comprise a combination of enzymes generated by one or more types. The formulations with diverse enzyme activities offer a practical advantage for enhancing feed consumption in poultry (Cowieson, Hruby, and Pierson 2006; Selle and Ravindran 2007). When using enzymes in combination, the effects on nutrient consumption and animal performance may be additive, subadditive, or synergistic (Cowieson and Adeola 2005; Juanpere et al. 2005). Enzymes in combination, instead of single enzymes, form the next line of feed enzymes because of their extraordinarily complicated structural makeup. Nutrients in feedstuff are not present at the native stage but rather as complexes with varieties linked to protein, lipid, fiber, and complex carbohydrates. In wheat-based diets,

TABLE 8.1
Type of Commercial Feed Enzymes and Target Substrates

Livestock	Enzyme	Feedstuff	Observation	Reference
Broiler chickens	β-glucanase, xylanase, cellulases	Hull-less barley	In finisher diets, hull-less barley combined with an enzyme cocktail can replace up to 75% of the corn without reducing production.	(Teymouri, Zarghi, and Golian 2018)
Broiler chickens	β-mannanase	Galactomannan	β mannanase promoted growth performance, reduced intestinal viscosity, and increased ileal digestible energy.	(Latham et al. 2018)
Lambs	Cellulase and xylanase	High-fiber diet—olive cake	By raising feed intake and improving fiber digestibility, lamb growth performance was improved without having a negative impact on the health of the animals.	(Abid et al. 2020)
Broiler birds	Phytase, nonstarch polysaccharide-degrading enzyme (NSPase)	Corn, soybean meal, and corn distiller dried grains	Phytase supplementation enhanced bone mineralization and mortality-corrected feed conversion ratio (mFCR). NSPase substantially increased total performance, including growth rate, mFCR, and performance index.	(Poernama, Wibowo, and Liu 2021)
Pig	Xylanase and phytase	Nutrient-reduced diets based on wheat and millrun	Increased growth performance and total tract DE content. The upper SI reverted to alkaline conditions more quickly after phytase insertion.	(Nortey, Owusu-Asiedu, and Zijlstra 2015)
Pig	Xylanase, protease, and α-amylase	Corn-soybean-based diet	Greater apparent total tract digestibility of DM, CP, and GE.	(Zhang et al. 2014)
Poultry	Phytase	Corn, soybean meal, and complete corn-soybean meal diet	Phytase breaks down phytate effectively into small IP esters. The myo-inositol level is unaffected by the phytase treatment and is constant between the control and phytase treatments.	(Hirvonen et al. 2019)
Pig	Phytase	Corn-soybean meal-based diets	The addition of phytase improved P digestibility and complete total tract $InsP_6$ with extremely high prececal hydrolysis.	(Rosenfelder-Kuon et al. 2020)
Pigs	Xylanase	Corn distillers' dried grains with solubles	Dietary xylanase supplementation decreased digesta viscosity, improved AID of nutrients, decreased inflammatory response, and altered intestinal architecture, all of which improved ADG in nursery pigs.	(H. Chen, Zhang, and Kim 2020)
Finisher pigs	β-mannanase	Metabolizable energy reduced diets containing xylanase, phytase	Increases growth to feed ratio, energy, protein utilization, and backfat thickness in finisher pigs without causing metabolic and intestinal ecosystem abnormalities.	(Genova et al. 2023)

(Continued)

TABLE 8.1 (CONTINUED)
Type of Commercial Feed Enzymes and Target Substrates

Livestock	Enzyme	Feedstuff	Observation	Reference
Dairy cows	Fibrolytic enzyme (mixture of xylanase and cellulase)	Faba beans (*Vicia faba* L.)	Optimum degree of enzyme addition has the potential to improve substrate fermentation, provide additional energy for animals, and improve animal performance.	(Yang 2021)
Dairy cows	Fibrolytic enzyme (mixture of xylanase and cellulase)	Precalving and high-yield lactation diets	The addition of enzymes to pens of cows in transition while in lactation enhanced milk, fat, and protein production and sped up time to first breeding but did not affect time to pregnancy.	(Golder, Rossow, and Lean 2019)
Goats	Exogenous enzymes, cellulase, and xylanase	Guinea grass ecotype "A" (*Panicum maximum*) and rice straw (*Oryza sativa*).	Exogenous cellulose and xylanase addition to goats' diets boosted average daily gain, consumption, and digestibility of dry matter and neutral detergent fiber.	(Seresinhe, Sujani, and Pathirana 2023)
Sheep	Cellulase	Buckwheat straw and alfalfa	Providing Tan sheep with buckwheat straw and alfalfa treated with enzyme and bacteria enhanced the animals' weight gain, and modified the rumen bacterial population and functional genes in the rumen.	(Jiang et al. 2020)
Beef cattle	Recombinant xylanase	Straw (rice, barley, and wheat) and corn stover	Xylanase addition boosted the effective rumen degradability of feed by about 3% but only when they were processed with ammonia fiber expansion (AFEX) technology.	(Beauchemin et al. 2019)
Dairy cows	Fibrolytic enzymes	Total mixed ration	The findings demonstrated that treatments had no impact on the intake of dry matter. When fibrolytic enzymes were supplied to nursing dairy cows, the quantity of milk increased by 4% and the milk's fat content increased.	(Shadmanesh 2014)
Dairy cows	Exogenous fibrolytic enzymes	Total mixed ration	Exogenous fibrolytic enzymes supplied to lactating dairy cows increased milk output, fat-corrected milk, improved feed efficiency, and energy-corrected milk production.	(Mohamed et al. 2013)
Cows	Exogenous enzymes (ZADO®)	Total mixed rations	Enzyme addition boosted rumen microbial N production, the consumption of dry matter (DM) and organic matter (OM), and it increased the digestibility of all nutrients in the cows' entire tract.	(Gado et al. 2009)
Lactating Holstein cows	Exogenous fibrolytic enzyme additive	Total mixed rations	For dairy cows in early lactation, an enzyme supplement improved the efficacy of fat-corrected milk yield in a dose-dependent manner by up to 11%.	(Chung et al. 2012)

only xylanase-targeting arabinoxylans may not offer complete advantages. In wheat-based poultry foods, the benefits of combining a carbohydrase with xylanase activity and a microbial phytase are observed in terms of protein, energy consumption, and growth performance (Selle, Ravindran, and Partridge 2009). By increasing substrate availability and decreasing its antinutritive elements (NSP and phytate) on feed consumption, the effect of one kind of feed enzyme may be aided by another, perhaps in a mutual manner. The poultry sector is the biggest consumer of these feed enzymes. Because of the chicken industry's high degree of integration, new technologies are adopted more quickly, and adding exogenous enzymes to food is now considered standard practice to increase nutrient consumption and digestibility. Almost all broiler diets in the application include glycanases (xylanases and glucanases). Due to concerns about phosphorus contamination of discharges from rigorous animal operations and the increasing cost of inorganic phosphates, the usage of microbial phytase in chicken foods has expanded over the past ten years. As a result, microbial phytase has supplanted glycanases as the main feed enzyme globally in recent years. Another new development is the availability of enzyme cocktails that target corn-soy diets and comprise amylases, xylanases, lipases, and proteases. Interest in the utilization of such cocktails is growing (Cowieson, Hruby, and Pierson 2006; Cowieson and Adeola 2005; Selle and Ravindran 2007; Slominski 2011).

8.5.1 Modes of Action of Feed Enzymes

Feed enzymes have different modes of action, and the precise mode(s) of action are not yet known, despite their growing adoption as feed additives. Most experts agree that one or more among the following mechanisms account for the perceived benefits (Bedford and Cowieson 2012).

1. Feed enzymes can degrade certain compounds not digested by endogenous enzymes in feed.
2. Degradation of antinutritional substances hinder the direct absorption of nutrients, unintentionally increase intestinal viscosity, or both.
3. Destruction of the endosperm's structural integrity and the discharge of nutrients contained or attached to the cell wall.
4. Switching to digestive sites with greater efficiency.
5. Decrease in endogenous discharges and protein releases from the gut, hence lower maintenance requirements (Cowieson et al. 2009; Cowieson and Ravindran 2007).
6. Change in the morphology of the digestive tract and reduction in its weight (Wu et al. 2004).
7. Changes to the small intestine's microbial profile. The application of enzymes directly affects the populations of bacteria that make up the microfloral populations because enzymes affect the quantity and types of substrates present in the gut (Apajalahti, Kettunen, and Graham 2004; Bedford and Cowieson 2012).
8. Improvement of the poultry's natural digestive enzymes leads to better digestion, particularly applicable for newly born chicks with developing digestive systems.

8.6 CONCLUSION

The vital goal of supplementing enzymes is to promote the digestion of nutritional constituents (protein, amino acids, carbohydrates, lipids, and energy) to improve performance and profitability. However, there are also many other factors that are more significant in upcoming production systems.

1. More adaptable feed formulae: an expanded feedstuff choice by lowering or eliminating the restriction on poorly digested constituents.

2. Across batches, minimum variation in the nutritional quality of products. Decrease particular ingredients' variance between excellent and poor quality samples, and increase the value of components by adding additional enzymes. Hence, a rise in the level of feed formulation precision.
3. Birds fed meals rich in NSP experienced decreased excreta moisture content and decreased damp litter occurrence. Low litter quality can result in, among other things, several welfare problems, such as foot ulcers, hock burns, and carcass degradation (Collett 2012).
4. Enhanced gut health because of better digestion, which results in fewer undigested nutrients reaching the lower intestine, and the composition of the gut flora shifts in favor of healthy bacterial species (Bedford and Cowieson 2012). An associated result is the protected bird's general health, which is partly attributable to the effects of flora on immunological function (Yegani and Korver 2008).
5. Improved homogeneity of animals by promoting the development of animals that perform poorly.

In summation, exogenous enzymes provide advantages beyond just enhancing food utilization and have consequences for the continuing changes in global livestock production regarding the environment, intestinal health, animal welfare, and sustainability.

REFERENCES

Abid, K., J. Jabri, H. Ammar, S. B. Said, H. Yaich, A. Malek, and M. Kamoun. 2020. "Effect of Treating Olive Cake with Fibrolytic Enzymes on Feed Intake, Digestibility and Performance in Growing Lambs." *Animal Feed Science and Technology* 261: 114405.

Adesogan, A. T., Z. X. Ma, J. J. Romero, and K. G. Arriola. 2014. "Ruminant Nutrition Symposium: Improving Cell Wall Digestion and Animal Performance with Fibrolytic Enzymes." *Journal of Animal Science* 92 (4): 1317–30.

Aidy, Sahar El, and Michiel Kleerebezem. 2013. "Molecular Signatures for the Dynamic Process of Establishing Intestinal Host-Microbial Homeostasis: Potential for Disease Diagnostics?" *Current Opinion in Gastroenterology* 29 (6): 621–27. doi:10.1097/MOG.0b013e328365d365.

Apajalahti, J., A. Kettunen, and H. Graham. 2004. "Characteristics of the Gastrointestinal Microbial Communities, with Special Reference to the Chicken." *World's Poultry Science Journal* 60 (2): 223–232+249+252+255+259+262. doi:10.1079/WPS20040017.

Bach, A., S. Calsamiglia, and M. D. Stern. 2005. "Nitrogen Metabolism in the Rumen." *Journal of Dairy Science* 88: E9–21.

Beauchemin, K. A., G. O. Ribeiro, T. Ran, M. R. M. Milani, W. Yang, H. Khanaki, and T. A. McAllister. 2019. "Recombinant Fibrolytic Feed Enzymes and Ammonia Fibre Expansion (AFEX) Pretreatment of Crop Residues to Improve Fibre Degradability in Cattle." *Animal Feed Science and Technology* 256: 114260.

Bedford, M. R., and A. J. Cowieson. 2012. "Exogenous Enzymes and Their Effects on Intestinal Microbiology." *Animal Feed Science and Technology* 173 (1–2). Elsevier B.V.: 76–85. doi:10.1016/j.anifeedsci.2011.12.018.

Besten, Gijs Den, Karen Van Eunen, Albert K. Groen, Koen Venema, Dirk Jan Reijngoud, and Barbara M. Bakker. 2013. "The Role of Short-Chain Fatty Acids in the Interplay between Diet, Gut Microbiota, and Host Energy Metabolism." *Journal of Lipid Research* 54 (9): 2325–40. doi:10.1194/jlr.R036012.

Black, Z., I. Balta, L. Black, P. J. Naughton, J. S. G. Dooley, and N. Corcionivoschi. 2021. "The Fate of Foodborne Pathogens in Manure Treated Soil." *Frontiers in Microbiology* 12: 3873.

Bober, Josef R., Chase L. Beisel, and Nikhil U. Nair. 2018. "Synthetic Biology Approaches to Engineer Probiotics and Members of the Human Microbiota for Biomedical Applications." *Annual Review of Biomedical Engineering* 20 (March): 277–300. doi:10.1146/annurev-bioeng-062117-121019.

Bozic, M., J. Newton, C. S. Thraen, and B. W. Gould. 2012. "Mean-Reversion in Income over Feed Cost Margins: Evidence and Implications for Managing Margin Risk by US Dairy Producers." *Journal of Dairy Science* 95 (12): 7417–28.

Celi, P., A. J. Cowieson, F. Fru-Nji, R. E. Steinert, A. M. Kluenter, and V. Verlhac. 2017. "Gastrointestinal Functionality in Animal Nutrition and Health: New Opportunities for Sustainable Animal Production." *Animal Feed Science and Technology* 234: 88–100.

Chaucheyras-Durand, F., and F. Ossa. 2014. "The Rumen Microbiome: Composition, Abundance, Diversity, and New Investigative Tools." *The Professional Animal Scientist* 30 (1): 1–12.

Chen, H., S. Zhang, and S. W. Kim. 2020. "Effects of Supplemental Xylanase on Health of the Small Intestine in Nursery Pigs Fed Diets with Corn Distillers' Dried Grains with Solubles." *Journal of Animal Science* 98 (6): skaa185.

Chen, Siyu, Shuyan Luo, and Chao Yan. 2021. "Gut Microbiota Implications for Health and Welfare in Farm Animals: A Review." *Animals* 12 (1): 93.

Cholewińska, P., K. Czyż, P. Nowakowski, and A. Wyrostek. 2020. "The Microbiome of the Digestive System of Ruminants–a Review." *Animal Health Research Reviews* 21 (1): 3–14.

Chung, Y. H., M. Zhou, L. Holtshausen, T. W. Alexander, T. A. McAllister, L. L. Guan, and K. A. Beauchemin. 2012. "A Fibrolytic Enzyme Additive for Lactating Holstein Cow Diets: Ruminal Fermentation, Rumen Microbial Populations, and Enteric Methane Emissions." *Journal of Dairy Science* 95 (3): 1419–27.

Clavijo, V., and M. J. V. Flórez. 2018. "The Gastrointestinal Microbiome and Its Association with the Control of Pathogens in Broiler Chicken Production: A Review." *Poultry Science* 97 (3): 1006–21.

Collett, Stephen R. 2012. "Nutrition and Wet Litter Problems in Poultry." *Animal Feed Science and Technology* 173 (1–2). Elsevier B.V.: 65–75. doi:10.1016/j.anifeedsci.2011.12.013.

Cowieson, A. J., and O. Adeola. 2005. "Carbohydrases, Protease, and Phytase Have an Additive Beneficial Effect in Nutritionally Marginal Diets for Broiler Chicks." *Poultry Science* 84 (12): 1860–67. doi:10.1093/ps/84.12.1860.

Cowieson, A. J., M. R. Bedford, P. H. Selle, and V. Ravindran. 2009. "Phytate and Microbial Phytase: Implications for Endogenous Nitrogen Losses and Nutrient Availability." *World's Poultry Science Journal* 65 (3): 401–18. doi:10.1017/S0043933909000294.

Cowieson, A. J., and V. Ravindran. 2007. "Effect of Phytic Acid and Microbial Phytase on the Flow and Amino Acid Composition of Endogenous Protein at the Terminal Ileum of Growing Broiler Chickens." *British Journal of Nutrition* 98 (4): 745–52. doi:10.1017/S0007114507750894.

Cowieson, A. J., M. Hruby, and E. E. M. Pierson. 2006. "Evolving Enzyme Technology: Impact on Commercial Poultry Nutrition." *Nutrition Research Reviews* 19 (1): 90–103. doi:10.1079/nrr2006121.

FAO. 2018. "Shaping the Future of Livestock Sustainably, Responsibly, Efficiently." In *The 10th Global Forum for Food and Agriculture (GFFA)*. FAO, January 18–20.

FAO. 2021. *Livestock and Landscapes: Sustainability Pathways. Food and Agriculture Organizations of the United Nations.* Food and Agriculture Organization of the United Nations. https://www.fao.org/3/ar591e/ar591e.pdf (Accessed on 3 November 2021).

FAOSTAT. 2018. *FAO Statistical Database.* FAOSTAT.

FAOSTAT. 2022. *FAO Statistical Database.* FAOSTAT.

Foo, Jee Loon, Hua Ling, Yung Seng Lee, and Matthew Wook Chang. 2017. "Microbiome Engineering: Current Applications and Its Future." *Biotechnology Journal* 12 (3): 1–11. doi:10.1002/biot.201600099.

Forcina, G., L. Pérez-Pardal, J. Carvalheira, and A. Beja-Pereira. 2022. "Gut Microbiome Studies in Livestock: Achievements, Challenges, and Perspectives." *Animals* 12 (23): 3375.

Gado, H. M., A. Z. M. Salem, P. H. Robinson, and M. Hassan. 2009. "Influence of Exogenous Enzymes on Nutrient Digestibility, Extent of Ruminal Fermentation as Well as Milk Production and Composition in Dairy Cows." *Animal Feed Science and Technology* 154 (1–2): 36–46.

Genova, J. L., P. E. Rupolo, L. B. D. Azevedo, D. Henz, S. T. Carvalho, ... Kipper, M., and P. L. D. O. Carvalho. 2023. "β-Mannanase Supplementation in Diets Reduced in 85 Kcal Metabolizable Energy/Kg Containing Xylanase-Phytase Improves Gain to Feed Ratio, Nutrient Usage, and Backfat Thickness in Finisher Pigs." *Frontiers in Veterinary Science* 10: 1144692.

Gerber, P. J., A. N. Hristov, B. Henderson, H. Makkar, J. Oh, C. Lee, R. Meinen, et al. 2013. "Technical Options for the Mitigation of Direct Methane and Nitrous Oxide Emissions from Livestock: A Review." *Animal* 7 (s2): 220234.

Golder, H. M., H. A. Rossow, and I. J. Lean. 2019. "Effects of In-Feed Enzymes on Milk Production and Components, Reproduction, and Health in Dairy Cows." *Journal of Dairy Science* 102 (9): 8011–26.

Hirvonen, J., J. Liljavirta, M. T. Saarinen, M. J. Lehtinen, I. Ahonen, and P. Nurminen. 2019. "Effect of Phytase on in Vitro Hydrolysis of Phytate and the Formation of Myo-Inositol Phosphate Esters in Various Feed Materials." *Journal of Agricultural and Food Chemistry* 67 (41): 11396–402.

Honda, K., and D. R. Littman. 2012. "The Microbiome in Infectious Disease and Inflammation." *Annual Review of Immunology* 30: 759–95.

Huws, S. A., C. J. Creevey, L. B. Oyama, I. Mizrahi, S. E. Denman, M. Popova, R. Muñoz-Tamayo, et al. 2018. "Addressing Global Ruminant Agricultural Challenges through Understanding the Rumen Microbiome: Past, Present, and Future." *Frontiers in Microbiology* 9: 2161.

Inda, María Eugenia, Esther Broset, Timothy K. Lu, and Cesar de la Fuente-Nunez. 2019. "Emerging Frontiers in Microbiome Engineering." *Trends in Immunology* 40 (10): 952–73. doi:10.1016/j.it.2019.08.007.

Jiang, B., T. Wang, Y. Zhou, and F. Li. 2020. "Effects of Enzyme+ Bacteria Treatment on Growth Performance, Rumen Bacterial Diversity, KEGG Pathways, and the CAZy Spectrum of Tan Sheep." *Bioengineered* 11 (1): 1221–32.

Juanpere, J., A. M. Pérez-Vendrell, E. Angulo, and J. Brufau. 2005. "Assessment of Potential Interactions between Phytase and Glycosidase Enzyme Supplementation on Nutrient Digestibility in Broilers." *Poultry Science* 84 (4): 571–80. doi:10.1093/ps/84.4.571.

Kiarie, Elijah, Luis F. Romero, and Charles M. Nyachoti. 2013. "The Role of Added Feed Enzymes in Promoting Gut Health in Swine and Poultry." *Nutrition Research Reviews* 26 (1): 71–88. doi:10.1017/S0954422413000048.

Kim, M., M. Morrison, and Z. Yu. 2011. "Status of the Phylogenetic Diversity Census of Ruminal Microbiomes." *FEMS Microbiology Ecology* 76 (1): 49–63.

la Fuente, G. De, D. R. Yañez-Ruiz, A. R. Seradj, J. Balcells, and A. Belanche. 2019. "Methanogenesis in Animals with Foregut and Hindgut Fermentation: A Review." *Animal Production Science* 59 (12): 2109–22.

Latham, R. E., M. P. Williams, H. G. Walters, B. Carter, and J. T. Lee. 2018. "Efficacy of β-Mannanase on Broiler Growth Performance and Energy Utilization in the Presence of Increasing Dietary Galactomannan." *Poultry Science* 97 (2): 549–56.

Li, F., G. Henderson, X. Sun, F. Cox, P. H. Janssen, and L. L. Guan. 2016. "Taxonomic Assessment of Rumen Microbiota Using Total RNA and Targeted Amplicon Sequencing Approaches." *Frontiers in Microbiology* 7: 987.

Li Fuyong, Changxi Li, Yanhong Chen, Junhong Liu, Chunyan Zhang, Barry Irving, Carolyn Fitzsimmons, Graham Plastow, and Le Luo Guan. 2019. "Host Genetics Influence the Rumen Microbiota and Heritable Rumen Microbial Features Associate with Feed Efficiency in Cattle." *Microbiome* 7: 1–17.

Lourenco, J. M., and C. B. Welch. 2022. "Using Microbiome Information to Understand and Improve Animal Performance." *Italian Journal of Animal Science* 21 (1): 899–913.

Mackie, Roderick I. 2002. "Mutualistic Fermentative Digestion in the Gastrointestinal Tract: Diversity and Evolution." *Integrative and Comparative Biology* 42 (2): 319–26.

Malmuthuge, N., P. J. Griebel, and L. L. Guan. 2014. "Taxonomic Identification of Commensal Bacteria Associated with the Mucosa and Digesta throughout the Gastrointestinal Tracts of Preweaned Calves." *Applied and Environmental Microbiology* 80 (6): 2021–28.

Malmuthuge, N., and L. L. Guan. 2017. "Understanding Host-Microbial Interactions in Rumen: Searching the Best Opportunity for Microbiota Manipulation." *Journal of Animal Science and Biotechnology* 8 (1): 17.

Meyer, R. K., and L. A. Pachauri. 2014. "IPCC 2014: Climate Change 2014: Synthesis Report."

Millen, D. D., M. D. B. Arrigoni, and R. D. L. Pacheco. 2016. *Rumenology*. Springer.

Min, B. R., S. Solaiman, H. M. Waldrip, D. Parker, R. W. Todd, and D. Brauer. 2020. "Dietary Mitigation of Enteric Methane Emissions from Ruminants: A Review of Plant Tannin Mitigation Options." *Animal Nutrition* 6 (3): 231–46.

Mohamed, D. E. D. A., B. E. Borhami, K. A. El-Shazly, and S. M. Sallam. 2013. "Effect of Dietary Supplementation with Fibrolytic Enzymes on the Productive Performance of Early Lactating Dairy Cows." *Journal of Agriculture Science* 5: 146–55.

Morgavi, D. P., E. Forano, C. Martin, and C. J. Newbold. 2010. "Microbial Ecosystem and Methanogenesis in Ruminants." *Animal* 4 (7): 1024–36.

Mosoni, P., C. Martin, E. Forano, and D. P. Morgavi. 2011. "Long-Term Defaunation Increases the Abundance of Cellulolytic Ruminococci and Methanogens but Does Not Affect the Bacterial and Methanogen Diversity in the Rumen of Sheep." *Journal of Animal Science* 89 (3): 783–91.

Mueller, U. G., and J. L. Sachs. 2015. "Engineering Microbiomes to Improve Plant and Animal Health." *Trends in Microbiology* 23 (10). Elsevier Ltd: 606–17. doi:10.1016/j.tim.2015.07.009.

Myer, P. R., M. Kim, H. C. Freetly, and T. P. Smith. 2016. "Evaluation of 16S RRNA Amplicon Sequencing Using Two Next-Generation Sequencing Technologies for Phylogenetic Analysis of the Rumen Bacterial Community in Steers." *Journal of Microbiological Methods* 127: 132–40.

Navarrete, Acacio A., Siu M. Tsai, Lucas W. Mendes, Karoline Faust, Mattias De Hollander, Noriko A. Cassman, Jeroen Raes, Johannes A. Van Veen, and Eiko E. Kuramae. 2015. "Soil Microbiome Responses to the Short-Term Effects of Amazonian Deforestation." *Molecular Ecology* 24 (10): 2433–48. doi:10.1111/mec.13172.

Neethirajan, S. 2023. "The Significance and Ethics of Digital Livestock Farming." *AgriEngineering* 5 (1): 488–505.

Newbold, C. J., G. De la Fuente, A. Belanche, E. Ramos-Morales, and N. R. McEwan. 2015. "The Role of Ciliate Protozoa in the Rumen." *Frontiers in Microbiology* 6: 1313.
Niranjan, D., N. B. Sridhar, U. S. Chandra, A. Manjunatha, S. S. Borthakur, M. H. Vinuta, and B. R. Mohan. 2023. "Recent Perspectives of Growth Promoters in Livestock: An Overview." *Journal of Livestock Science* 14: 53–64.
Nortey, T. N., A. Owusu-Asiedu, and R. T. Zijlstra. 2015. "Effects of Xylanase and Phytase on Digestion Site of Low-Density Diets Fed to Weaned Pigs." *Livestock Res. Rural Devel* 27: 133.
Panzer, Ariane R., and Susan V. Lynch. 2015. "Influence and Effect of the Human Microbiome in Allergy and Asthma." *Current Opinion in Rheumatology* 27 (4): 373–80. doi:10.1097/BOR.0000000000000191.
Park, Woojun. 2018. "Gut Microbiomes and Their Metabolites Shape Human and Animal Health." *Journal of Microbiology* 56: 151–53.
Poernama, F., T. A. Wibowo, and Y. G. Liu. 2021. "The Effect of Feeding Phytase Alone or in Combination with Nonstarch Polysaccharides-Degrading Enzymes on Broiler Performance, Bone Mineralization, and Carcass Traits." *Journal of Applied Poultry Research* 30 (1): 100134.
Pourabedin, Mohsen, and Xin Zhao. 2015. "Prebiotics and Gut Microbiota in Chickens." *FEMS Microbiology Letters* 362 (15). doi:10.1093/femsle/fnv122.
Ravindran, V. 2013. "Feed Enzymes: The Science, Practice, and Metabolic Realities." *Journal of Applied Poultry Research* 22 (3): 628–36. doi:10.3382/japr.2013-00739.
Rojas-Downing, M. M., A. P. Nejadhashemi, T. Harrigan, and S. A. Woznicki. 2017. "Climate Change and Livestock: Impacts, Adaptation, and Mitigation." *Climate Risk Management* 16: 145–63.
Rosenfelder-Kuon, P., B. Klein, N., Zegowitz, M. Schollenberger, L. Kühn, I., Thuringer and M. Rodehutscord. 2020. "Phytate Degradation Cascade in Pigs as Affected by Phytase Supplementation and Rapeseed Cake Inclusion in Corn–Soybean Meal-Based Diets." *Journal of Animal Science* 98 (3): 053.
Selle, P. H., and V. Ravindran. 2007. "Microbial Phytase in Poultry Nutrition." *Animal Feed Science and Technology* 135 (1–2): 1–41. doi:10.1016/j.anifeedsci.2006.06.010.
Selle, P. H., V. Ravindran, and G. G. Partridge. 2009. "Beneficial Effects of Xylanase and/or Phytase Inclusions on Ileal Amino Acid Digestibility, Energy Utilisation, Mineral Retention and Growth Performance in Wheat-Based Broiler Diets." *Animal Feed Science and Technology* 153 (3–4): 303–13. doi:10.1016/j.anifeedsci.2009.06.011.
Seresinhe, T., S. Sujani, and I. Pathirana. 2023. "Exogenous Fibrolytic Enzymes: For the Better Utilization of Guinea Grass and Rice Straw as Ruminant Feeds." In *Exogenous Enzymes as Feed Additives in Ruminants*, 63–76. Springer International Publishing.
Shadmanesh, A. 2014. "Effect of Dietary Suppplement with Fibrolytic Enzymes on the Productive Performance of Early Lactating Dairy Cows." *Indian Journal of Fundamental and Applied Life Sciences* 4: 396–401.
Shang, Yue, Sanjay Kumar, Brian Oakley, and Woo Kyun Kim. 2018. "Chicken Gut Microbiota: Importance and Detection Technology." *Frontiers in Veterinary Science* 5: 254.
Skyquest. 2022. "Global Feed Enzymes Market Size, Share, Growth Analysis, by Type (Phytase, Carbohydrase), By Application (Poultry, Swinr)- Industry Forecast 2023-2030." *Skyquest*. https://www.skyquestt.com/report/feed-enzymes-market#:~:text=Feed Enzymes Market size was,period (2023-2030).
Slominski, B. A. 2011. "Recent Advances in Research on Enzymes for Poultry Diets." *Poultry Science* 90 (9): 2013–23. doi:10.3382/ps.2011-01372.
Swanepoel, Frans, Aldo Stroebel, and Siboniso Moyo. 2010. *The Role of Livestock in Developing Communities: Enhancing Multifunctionality*. UJ Press.
Teymouri, H., H. Zarghi, and A. Golian. 2018. "Evaluation of Hull-Less Barley with or without Enzyme Cocktail in the Finisher Diets of Broiler Chickens." *Journal of Agricultural Science and Technology* 20 (3): 469–83.
Thornton, Philip K. 2010. "Livestock Production: Recent Trends, Future Prospects." *Philosophical Transactions of the Royal Society B: Biological Sciences* 365 (1554): 2853–67.
Velázquez-De Lucio, B. S., E. M. Hernández-Domínguez, M. Villa-Garcia, G. Diaz-Godinez, V. Mandujano-Gonzalez, B. Mendoza-Mendoza, and J. Alvarez-Cervantes. 2021. "Exogenous Enzymes as Zootechnical Additives in Animal Feed: A Review." *Catalysts* 11: 851.
Waldor, Matthew K., Gene Tyson, Elhanan Borenstein, Howard Ochman, Andrew Moeller, B. Brett Finlay, Heidi H. Kong, et al. 2015. "Where Next for Microbiome Research?" *PLoS Biology* 13 (1): 1–9. doi:10.1371/journal.pbio.1002050.
Weimer, Paul J. 2015. "Redundancy, Resilience, and Host Specificity of the Ruminal Microbiota: Implications for Engineering Improved Ruminal Fermentations." *Frontiers in Microbiology* 6: 296.
Wilkins, L. J., M. Monga, and A. W. Miller. 2019. "Defining Dysbiosis for a Cluster of Chronic Diseases." *Scientific Reports* 9 (1): 110.

Wu, Y. B., V. Ravindran, D. G. Thomas, M. J. Birtles, and W. H. Hendriks. 2004. "Influence of Phytase and Xylanase, Individually or in Combination, on Performance, Apparent Metabolisable Energy, Digestive Tract Measurements and Gut Morphology in Broilers Fed Wheat-Based Diets Containing Adequate Level of Phosphorus." *British Poultry Science* 45 (1): 76–84. doi:10.1080/00071660410001668897.

Yang, J. C. 2021. "Effect of Fibrolytic Enzyme on Lactational Performance, Feeding Behavior and Digestibility in Lactating Dairy Cows Fed a Whole Plant Faba Bean Silage-Based Diet." (Doctoral dissertation, University of Saskatchewan).

Yegani, M., and D. R. Korver. 2008. "Factors Affecting Intestinal Health in Poultry." *Poultry Science* 87 (10): 2052–63. doi:10.3382/ps.2008-00091.

Yeoman, Carl J., and Bryan A. White. 2014. "Gastrointestinal Tract Microbiota and Probiotics in Production Animals." *Annual Review of Animal Biosciences* 2: 469–86.

Yu, L. C. H., J. T. Wang, S. C. Wei, and Y. H. Ni. 2012. "Host-Microbial Interactions and Regulation of Intestinal Epithelial Barrier Function: From Physiology to Pathology." *World Journal of Gastrointestinal Pathophysiology* 3 (1): 27.

Zalewska, M., A. Błażejewska, A. Czapko, and M. Popowska. 2021. "Antibiotics and Antibiotic Resistance Genes in Animal Manure–Consequences of Its Application in Agriculture." *Frontiers in Microbiology* 12: 640.

Zhang, G. G., Z. B. Yang, Y. Wang, W. R. Yang, and H. J. Zhou. 2014. "Effects of Dietary Supplementation of Multi-Enzyme on Growth Performance, Nutrient Digestibility, Small Intestinal Digestive Enzyme Activities, and Large Intestinal Selected Microbiota in Weanling Pigs." *Journal of Animal Science* 92 (5): 2063–69.

9 Manipulating Phageomes for Modulating Bacterial Community in Animal and Human Gut, and the Effects of Microbiome Engineering

Subhrojyoti Ghosh
Department of Biotechnology, IIT Madras
Chennai

Prakriti Seth
Department of Biotechnology, National Institute of Technology
Rourkela, Odisha

Anuvab Dey
Department of Biosciences and Bioengineering,
IIT Guwahati
North Guwahati, Assam

Rajanyaa Chowdhury and Ankita Chowdhury
Heritage Institute of Technology, Chowbaga
Anandapur, Kolkata

9.1 INTRODUCTION

Microorganisms have evolved to create ecological ecosystems on Earth over the course of millions of years of co-evolution. These persistent microbial communities, or microbiomes, can be found in live hosts like people, animals, and plants. Microbes in microbiomes develop synergistic connections through dynamic interactions with their hosts and among themselves. These interactions shape the microbiome's makeup and significantly impact the ecosystems in which they live. They can also have far-reaching impacts on the hosts' physiology and the environment. For instance, the microbiome of the human stomach is related to the physiology and mental health of the host (Grice and Segre 2012). In recent years, interest in the study of microbiomes has risen, and insights into variables that contribute to their differences have been acquired.

The fundamental microbiome of ecosystems determines their traits (Mueller and Sachs 2015). Each ecosystem has a distinct characteristic resulting from a complex web of interactions between

 DOI: 10.1201/9781003394662-9

the microbial populations and their hosts and responses to the microbial communities' metabolites (Waldor et al. 2015). As a result, disruptions to the microbiome can significantly impact the hosts and environment, resulting in decreased host fitness (Round and Mazmanian 2009; Navarrete et al. 2015). Microbiome engineering can change microbial compositions to enhance host phenotypes and benefit ecosystems.

The human microbiome is arguably the one for which microbiome engineering is most frequently used. Given that the human microbiome has been proven to influence the physiology of the host and its composition is related to diseases and disorders like diabetes and cancer, there is a possibility of modifying microbiota in humans to treat diseases (Grice and Segre 2012). Since the Human Microbiome Project (HMP) was established, metagenomic characterization of the human body's microbiota, particularly those of the gut, oral, skin, urogenital, and lung/nasal origin, has advanced significantly. As a result, the HMP has made it easier to identify the characteristics that distinguish healthy from diseased microbiota. The knowledge gathered from the HMP will be extremely helpful for developing techniques to restore microbial balance for therapeutic purposes in microbiome engineering.

Trillions of microorganisms, including bacteria, archaea, bacteriophages, and eukaryotes, colonize the human gastrointestinal tract (GIT), forming a massive ecological community through complex metabolic activities and constant host interactions. The gut microbiota is made up of microorganisms, and its collective genomes are referred to as the gut microbiome (Turnbaugh et al. 2007; Peterson et al. 2009). Bacterial species dominate the gut microbiota, with their genes accounting for approximately 99% of the gut microbiome (Wróblewska 2006). Despite the fact that a healthy person's total bacterial cell mass is estimated to be around 200 g, the bacterial gut metagenome outnumbers the number of distinct genes found in the human genome 150 times (Sender, Fuchs, and Milo 2016; Qin et al. 2010). Thus, the gut microbiota has a significant impact on both normal human physiology and disease susceptibility, including pathogen defense (Kamada et al. 2013; David et al. 2014), nutrient utilization, and peripheral immune system education (Cho and Blaser 2012; Hooper, Littman, and Macpherson 2012). The gut microbiome controls a variety of metabolic pathways, interacting dynamically with the host via metabolites produced. Changes in these metabolic pathways may alter the composition of gut lumen metabolites, disrupting the host system's original balance. The host's responses are dynamic, causing the up- or down-regulation of related metabolic pathways until a new balance is achieved. Obesity, colorectal cancer, cardiovascular disease, and inflammatory bowel disease (IBD) have all been linked to dysbiosis of the host's microbiota (Cho and Blaser 2012).

Like the human microbiome, the animal microbiome comprises a variety of microbial communities, and the microbiota's makeup is correlated with the growth and well-being of the hosts. Understanding the gut microbiota in representative animals such as mice, rats, broilers, swine, and cows has been a major focus of recent research. Despite being the most frequently studied species, mice and rats are mostly employed as animal models for engineering the human microbiome.

9.2 PHAGEOMES

A community of bacteriophages, along with their metagenomes, when localized in a particular environment, is known as a phageome (Ma et al. 2018; Townsend et al. 2021). Bacteriophages are viruses that can infect archaea and bacteria and replicate inside these microorganisms. Phageomes are a subcategory of viromes that includes viruses linked to the environment and the host (Virgin 2014). Phages constitute the majority of viromes and are known to be organisms present in maximum abundance. They have been known for centuries, but such vast diversity has existed for merely two decades (Adriaenssens 2021). Hence, after thorough research, the importance of phage-based processes has been found in different environments like oceans, soils, and even as a part of the human microbiome.

Recently, a change in viral taxonomy was implemented, and higher ranks of viruses were added, referred to as the mega taxonomy of viruses. This provided 15 hierarchical ranks to classify all viruses present in the world (Koonin et al. 2020). Among them, the diversity of phages can be divided into four realms: *Duplodnaviria*, *Monodnaviria, Varidnaviria*, and *Riboviria*, which include six kingdoms and seven phyla. In the class *Caudoviricetes*, we observe the most commonly isolated phages, the dsDNA (double-stranded DNA)-tailed bacteriophages with a major capsid protein like HK97. The family classification level has been mainly used to visualize the metagenomics data, and there has been a rapid revolution in phage taxonomy.

Initially, a morphology-based classification was done, but a shift has been observed favoring a genome-based classification (Turner, Kropinski, and Adriaenssens 2021). With the onset of the age of genomics in the early 2000s, extensive research and sequencing have been done on phage genomes. This revealed higher genomic diversity than what was thought previously. Hence, many new families are being introduced to ensure that the members of the same family share a set of core genes (Adriaenssens 2021). This wasn't the idea behind the classification of families like *Myoviridae*, *Siphoviridae*, and *Podoviridae*, which have now been abolished.

Looking beyond the previously characterized tailed bacteriophages, a similar increase in genomic diversity was observed in other phage clades. This information was mainly driven by metagenome-derived data. Hence, observing the different lineages of bacteriophages, it can be derived that there needs to be a fundamental change to the classification system to address the increasing genomic diversity of phagosomes.

9.3 GUT PHAGEOME ECOLOGY OF ANIMALS AND HUMANS

A complex ecosystem can be observed inside animal and human gastrointestinal tracts (Zuppi et al. 2022). This ecosystem consists of different microbes such as bacteria, viruses, fungi, protozoa, and archaea (Martín et al. 2014). The complexity of this gut ecosystem depends on many factors. These determining factors include peristaltic motion, the anatomy of the gut, the mucus layer present lining the walls and its shedding, compounds secreted by hosts (including gastric juice or bile), the immune system of the host, and the entry of new microbes through food consumed (Thursby and Juge 2017). The communities of microbes in the hosts' gut can regulate various aspects of host well-being by mediating the absorption of nutrients, vitamin and neurotransmitter synthesis, modulation and development of the immune system, and varying resistance among different pathogens. Therefore, it is clear that many diseases can occur due to disturbances caused to the ecosystem of microbes in the gut, starting from gut inflammation to neurological disorders. Many alterations to the gut phageome are observed to be connected to various gastrointestinal tract diseases in multiple cases, which shows how much they contribute to gut health (Norman et al. 2015).

9.3.1 Humans

In order to be healthy, humans require a balanced and stable gut microbiome. This is determined by the bacteriophages that infect the hosts (Arumugam et al. 2011). Studies have concluded that a high extent of heterogeneity among bacteriophages was observed in a healthy human gut microbiome. Extensive research on the available datasets of phageome metagenomics revealed 23 common bacteriophages in over half of the 64 healthy individuals participating in this research from around the world. Individuals with irritable/gastrointestinal bowel disease had a significantly lesser amount of these shared bacteriophages than healthy individuals. From these data, we get an idea of the healthy gut phageome present in humans. A total of 44 bacteriophage groups were observed in the gut of participants, among which only nine (i.e., 20% of the total) were shared in more than half of the 64 individuals studied. Overall, the community of bacteriophages consists of three classes, including core bacteriophages present in more than half of the individuals. A few sets of common bacteriophages were present in just 20–50% of the individuals, while another set were rarely shared or unique to an individual. Only the core and common bacteriophages were distributed globally and comprised the human gut phageome. They play an extremely essential role in maintaining the healthy structure and function of the microbiome, which contributes to human health ("Healthy Human Gut Phageome Manrique, Pilar" 2016).

9.3.2 Animals

Studies revealed a connection between the various microorganisms present in the gut. This was proven through an animal model like gnotobiotic mice, where lysis of the population of bacteria mediated by phages led to a cascade of effects on populations of bacteria that were not even targeted by the predating phages (Hsu et al. 2019). Due to the varying abundance of species, many direct consequences are observed in the ecosystem, depending on the microbe's ecological role. Temperate phages were found to be capable of contributing to the virulence and fitness of the bacterial host by offering virulence genes, including phage-encoded toxins and immune evasion genes. In this way, they exert an inflammatory effect on the gastrointestinal tract, causing changes to the gut ecosystem (Penadés et al. 2015). With the introduction of prophages in hosts, the lytic cycle can be initiated, leading to the death of the host. As a result, it changes the population dynamics and exerts a change in the gastrointestinal tract (Lin and Lin 2019).

9.4 NECESSITY OF GUT MICROBIOTA MODULATION BY MANIPULATING PHAGEOMES

Out of the several kinds of microbiota residing in our body, such as oral, skin, gut, and respiratory microbiota, the ones in the gut, including bacteria, fungi, viruses, and archaea, play a crucial role in regulating human health through vitamin production, immune response stimulation, fermentation of food, etc. (Hillman et al. 2017; Ursell et al. 2014). An imbalance in the gut microbiota, known as microbiota dysbiosis and due to lifestyle habits, health conditions like infection, and environmental factors such as an improper diet with low fiber, etc., can result in several diseases such as cardiopathy, respiratory diseases, and cancers due to metabolic irregularities resulting from gut-barrier disruption (Hrncir 2022). Extensive studies have been conducted on the role of bacteria and fungi in the human and animal gut in regulating health and disease. However, the role of gut viruses or phageomes is still a less explored territory. It should be noted that gut bacteria alone are responsible for many diseases such as respiratory, neurological, and autoimmune domains like irritable bowel syndrome (IBS), atopic asthma, rheumatoid arthritis, cardiovascular diseases (CVDs), type II diabetes, eczema, mental health disorders like Parkinson's disease, Alzheimer's disease, and autism spectrum disorder (ASD). The phageomes that infiltrate the gut bacteria, if modulated properly, can bring about a drastic reduction or improvement in conditions caused by the latter and assist in the development of a healthy human and animal health environment along with agricultural and biotechnological advancements (Vijay and Valdes 2021; Maldonado-Contreras et al. 2020). To delve deeply into this matter, the concern of antibiotic resistance comes into play. Chronic gastrointestinal inflammation due to gut bacteria dysbiosis gives rise to most gastrointestinal malignancies such as colorectal cancer (CRC). To treat the source, mostly generic antibiotics are provided, which fail in most cases due to the increasing percentage of antibiotic-resistant bacteria in the human and animal gut. Moreover, in CRC patients, it is seen that gut barrier-disrupting bacteria like *Fusobacterium* sp. compete with and reduce the population of beneficial *Clostridium butyricum*, which leads to chemoresistance and tumor progression (Yang and Jobin 2017; Zheng et al. 2019). Along with the increasing population of disease-causing bacteria in the above-mentioned cases, the prominence of gut viruses, namely phageomes, has also been noticed and can be studied as a phage biomarker system in the diseased gut. Natural bactericidal activity, target-specific nature against a particular bacterial stratum, and localized proliferation in the gut result in eradicating the concerned bacteria without facing any antimicrobial resistance. In addition, studies have shown the in-vivo use of modulated phages to be free from major side effects, making them safe. In the future, treatment options could be combined with phage-plus-antibiotic-mediated treatment (Gordillo Altamirano and Barr 2019; Sarker et al. 2016; Grubb et al. 2020). This discussion makes it clear that phageome manipulation, if implemented properly, can result in a complete turn of events in gut microbiota modulation and open new avenues in disease regulation as well as biotechnology, completely evading the curse of antimicrobial resistance safely and making this sector of phageome-based technology extremely necessary.

9.5 ROLE OF PHAGEOME AS MICROBIOTA MODULATOR

As mentioned in the previous section, the role of bacteria and fungi in the modulation of animal and human gut microbiota by methods like prebiotic-mediated modulation, and cell-mediated general and targeted modulation through engineered probiotics has already been extensively studied (Lee et al. 2018). This section will deal with phage-mediated modulation, i.e., ways in which the phageome can play a vital role in gut microbiota modulation and serve as a potent method in microbiome engineering. The different established and emerging methods of phageome manipulation, along with case studies, are discussed in the sections that follow (Figure 9.1).

9.5.1 Exogenous Phage Addition for Specific Pathogen Suppression

Addition of bacterio-specific phage strains in the form of oral phage cocktails, sometimes combined with antibiotics, has served to eradicate or at least suppress specific bacterial strains in many cases. A five-phage cocktail system was administered to IBD patients who exhibited a high percentage of *K. pneumoniae* in the intestinal and fecal mucosa. The phages proved to be stable in the mouse as well as human GI tract and significantly reduced the abundance of the mentioned bacterial host,

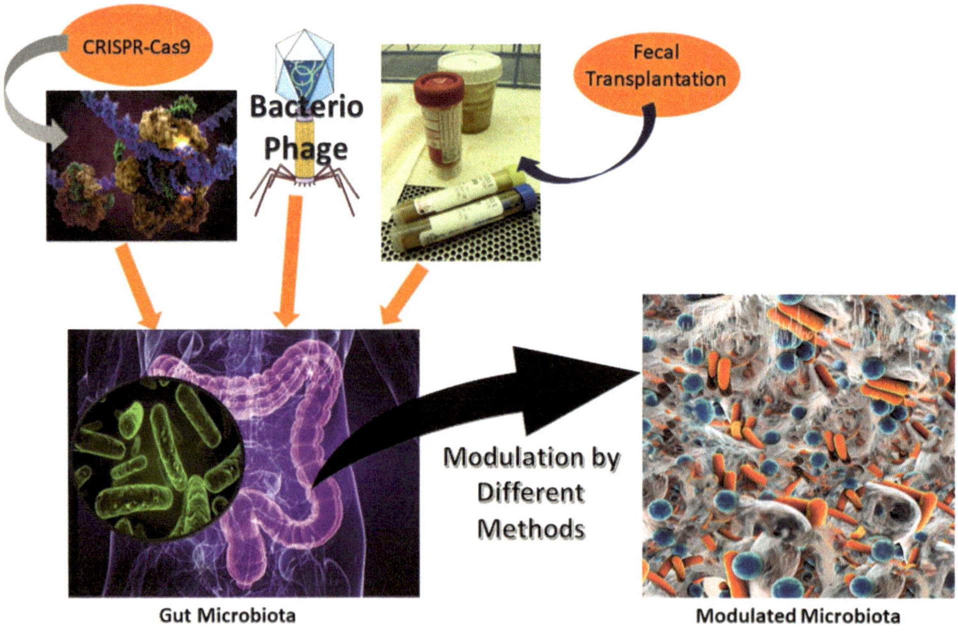

FIGURE 9.1 Modulation methods of gut microbiota involving phageomes.

which automatically reduced the intestinal inflammation in the patients (Federici et al. 2022). In another study, a three-phage cocktail was administered in the intestines of transgenic mice with adherent invasive *Escherichia coli* (AIEC). Results showed drastic downregulation of AIEC colonies in the intestines and feces of the administered mice (Galtier et al. 2017). Next, in mouse models with induced intestinal tumors, phages isolated from IBD patients were exogenously administered and were seen to reduce the AIEC population prominent during tumorigenesis, which resulted in reduction of tumor size, decrease in carcinogenic genes as well as metastasis, and increased survival rate of the mice (Gogokhia et al. 2019). As well as these phage cocktail-based therapies, scientists developed a phage-mediated drug delivery system in colorectal cancer patients, which was seen to eliminate the *Fusobacterium* sp. colonies thriving in the tumors, resulting in an increase in the beneficial *Clostridium butyricum* population, leading to carcinogenic cell suppression and minimized chemoresistance (Zheng et al. 2019). In all these studies, the phage colonies were viable in vivo showing almost no side effects, and they suppressed or completely eliminated the targeted microbiota but, on the other hand, failed to show any modification in the same (Bao et al. 2018).

9.5.2 Filtered Fecal Transplantation for Eradicating Specific Diseases

Filtered fecal transplantation, also known as fecal microbiota transplant (FMT), is an age-old gut microbiota modulation technique that involves transferring a donor's fecal matter to a recipient, leading to the modification of the gut microbiota population of the recipient (Waller, Leong, and Paramsothy 2022). The fecal matter from the donor is sterile-filtered, which leaves the fecal population mainly restricted to phages, and that is how FMT was proven to be a phage-mediated modulation technique rather than being a generic microbiota-mediated method (Ott et al. 2017). The ideology that the efficacy of FMT lies within the phageome and not any other associated nanoparticle was also studied. It involved treating post-antibiotic treatment mice with indigenous FMT and heat-and-nuclease-treated FMT that killed the phageome in the latter. It was seen that the group receiving indigenous FMT had more similarity to the pre-antibiotic treatment healthy gut microbiota, unlike the ones receiving the heat-and-nuclease treated variety, concluding that phages impart the role of restoring microbiota balance in FMT techniques (Draper et al. 2020; Sausset et al. 2020). As a case study, infants born preterm with necrotizing enterocolitis (NEC) were studied. NEC is a lethal bowel disease that exhibits increase in proteobacteria and facultative anaerobic population in the GI tract. By administering FMT, intestinal permeability and abundance of proteobacteria were reduced, and there was an increase in phage density (Brunse et al. 2022). Though FMT has been used for many years to treat chronic IBS and *C. difficile* infection, it has some serious side effects such as the transfer of pathogenic microbes from donor to recipient and a lack of specificity, which need to be regulated to uphold the efficacy of the method (DeFilipp et al. 2019).

9.5.3 CRISPR-Cas9-Mediated Manipulation (Engineered Phages)

The recent advancement of molecular scissor-based genome editing through the CRISPR-Cas9 system can alleviate the drawbacks of the above-mentioned techniques, such as the inability to modify the gut microbiome, lack of efficacy, and pathogenic microbe transfer. Engineered phageomes can be used to target bacteria of the same species but with subgenomic variations, specifically targeting the patient-specific microbiota, leading to increased treatment efficacy on a personalized level (Abedon et al. 2011). CRISPR-Cas9-mediated phageome manipulation helped in the reversal of antibiotic resistance in gut bacteria. Here, CRISPR-Cas9 was used to suppress the beta-lactamase coding ndm-1, reducing broad-spectrum antibiotic sensitivity in the species (Yosef et al. 2015). The unique side of using engineered phageomes is that they can modify and reprogram the gut microbiota in addition to eradicating or suppressing different species like the other methods. Genome editing by site-directed mutagenesis was used to alter the receptor-binding region of a T3 phage for generating engineered phages that suppressed the growth of bacteria and prevented antibiotic resistance (Khan Mirzaei and Deng 2022; Yehl et al. 2019). Furthermore, engineered phageomes can be manipulated to expand the bacterial host range for extensive therapeutic applications, as was done by constructing hybrid phage fragments of T7 phage with different fiber proteins and phage tails transducing a variety of bacterial genera other than the usual *E.coli* target (Yosef et al. 2017). No system is foolproof, and this emerging method of engineered phage systems is no exception. Escape frequency and misplaced gene targeting are some serious concerns clouding this technology, which needs to be addressed with advancements in biotechnology to reduce errors (Bikard et al. 2014).

9.5.4 Culture-Independent Phage Characterization

Even with all the knowledge that we have regarding phages, characteristics of a huge percentage of them still remain unaccounted for, which creates a void in the domain of the phage-bacteria network since most of the classical characterization methods are culture-dependent. This somehow hinders their isolation and subsequent sequencing for further analysis. Thus, to deal away with this obstacle, culture cultivation-independent techniques for phage characterization are emerging that will open a new arena in phage-mediated microbiota modulation (Yarza et al. 2014). A brief description of four such culture cultivation-independent techniques is given in this section. Firstly, microfluidic digital PCR is a technique that uses cellular level marker genes to identify phages along with their hosts simultaneously. To achieve this end, environmental microbes are collected, processed, and loaded into the concerned PCR channel for the digital co-detection of phage and host.

Next, a technique called phage fluorescence in situ hybridization (phageFISH) has been developed, which uses sequence-based fluorescent-labeled probes for the detection of both intra- and extracellular phage DNA. This method has been seen to greatly increase the efficiency of gene detection (Barrero-Canosa and Moraru 2019). Thirdly, viral tagging (VT) is another culture-independent emerging method that uses fluorescence activated cell sorting (FACS) to differentiate phages with infected bacterium from the non-infected ones at the single-cell level as well as the colony level. This helps in a precise screening of the phageome that is not possible by the other methods (Džunková et al. 2019). Lastly, the technique of proximity ligation assay is a multidisciplinary method that combines computational and experimental assays to detect contact points between DNA molecules in a microenvironment using meta3c (metagenomic chromosome conformation) capture. Here, sequences of the microbial population are used to generate a meta3c database that assembles and assigns bacterial sequences based on frequencies of physical contact with the phage DNA (Alam 2018). The only drawback of these emerging techniques is that they are relatively new and need extensive research for proper implementation.

The above-mentioned classical as well as emerging techniques in the field of phageome-mediated microbiota modulation, as shown in Table 9.1, are about to open a new dimension in the field of microbiome engineering if dealt with carefully and with precision.

9.6 EFFECT OF MANIPULATED PHAGEOMES ON CHRONIC INTESTINAL DISORDERS

Chronic gastrointestinal (GI) inflammation is the leading cause of GI cancer development. Patients with inflammatory bowel disease (IBD) and small intestinal bacterial overgrowth (SIBO) have chronic intestinal inflammation and are treated with immunosuppressive medications, which increases their risk of developing intestinal cancer (Stidham and Higgins 2018). CRC, small bowel adenocarcinoma, intestinal lymphoma, anal cancer, cholangiocarcinoma, and extra-intestinal malignancies are among them (Johnson et al. 2013; Axelrad, Lichtiger, and Yajnik 2016; Stidham and Higgins 2018). Furthermore, chronic GI diseases reduce quality of life, and social and occupational life by causing poor GI motility, dysfunctional gut-brain communication, mucosal immune activation, and gut microbiota changes (Holtmann, Ford, and Talley 2016; Singh et al. 2020, 2022).

TABLE 9.1
Advantages and Disadvantages of the Existing and Emerging Phage-mediated Gut Microbiota Modulation Techniques

No.	Type of Technique	Advantages	Disadvantages	
1.	Exogenous phage addition	No in-vivo side effects. Complete eradication of the targeted microbiota.	Only eradication of microbiota, no modification.	(Bao et al. 2018)
2.	Filtered fecal transplantation	Age-old trusted technique to treat chronic ailments	Lack of specificity. Pathogenic microbe transfer from donor to recipient.	(Waller, Leong, and Paramsothy 2022)
3.	Engineered phage administration	Species with sub-genomic variation can also be targeted. Patient-specific microbiota modulation is possible. Efficacy is higher than other methods.	Misplaced gene targeting. The escape frequency of unaltered phage is high.	(Abedon et al. 2011; Bikard et al. 2014)
4.	Culture cultivation-independent methods	A huge population of unaccounted phageome can be characterized and sequenced before manipulation. Computational methods increase efficiency and reduce errors.	Extensive research is needed for proper implementation, as methods are still advancing.	(Yarza et al. 2014; Alam 2018)

Treatments for intestinal inflammation, such as corticosteroids and immunosuppressive agents, are frequently prescribed for long periods of time, focusing primarily on symptom relief rather than addressing the underlying cause, and may promote carcinogenesis (Axelrad, Lichtiger, and Yajnik 2016; Magro et al. 2020). Furthermore, antibiotics lack taxonomic precision and eliminate a large consortium of bacteria, including a beneficial community with anti-inflammatory properties, complicating CRC treatment (Zheng et al. 2019). Recent research has revealed that the viral component of the microbiome, which is primarily composed of bacteriophages, i.e., phages, appears to interact with both the bacteriome and the host, playing a significant role in homeostatic regulation and disease progression in patients with GI inflammation and malignancies (Virgin 2014; Pfeiffer and Virgin 2016; Mirzaei and Maurice 2017; Mukhopadhya et al. 2019).

Phages colonize the skin, oral and nasal cavities, lungs, gut, genital tract, and urinary tract, among other places (Barr 2017; Ganeshan and Hosseinidoust 2019). The gut phage population, also known as the gut phageome, is a significant component of the gut, with an estimated 1012 total phages within the colon of an average adult human (Clokie et al. 2011; Barr 2017; Nguyen et al. 2017). It is thought to play a significant role in shaping the intestinal ecosystem by modulating bacterial diversity via a lytic lifestyle and facilitating horizontal gene transfer while integrated as a prophage (Duerkop et al. 2012; Hsu et al. 2019; Shkoporov et al. 2019; Gutiérrez and Domingo-Calap

2020). Furthermore, phages can physically interact with mucosal surfaces, bypass epithelial cell layers, spread throughout the body, and influence the immune system (Barr 2017; Nguyen et al. 2017). Indeed, the phageome has been shown to directly influence immunity by modulating innate immunity via phagocytosis and cytokine responses and adaptive immunity via effects on antibody production, enabling anti-inflammatory properties (Górski and Weber-Dabrowska 2005; Górski et al. 2018; Van Belleghem et al. 2018). It was found to play a role in lowering the risk of clinical GI complications in stem cell transplant recipients (HCT) in particular (Górski et al. 2018). Interestingly, even when individuals have similar bacterial community structures, there is a significant variation in their intestinal phageome (Van Belleghem et al. 2018). Indeed, when comparing the bacterial and viral populations in CRC patients, the bacterial fecal signature in CRC patients appears to be dominated by a small number of key drivers, such as *Fusobacterium*, whereas the phageome signature appears to be more diverse (Hannigan et al. 2018; Handley and Devkota 2019). Longitudinal metagenomic studies have revealed that the human gut phageome is highly personalized, with a healthy core that is highly prevalent, persistent, and individual-discriminatory, composed primarily of lytic *Caudovirales* phages (crAss-like, *Siphoviridae*, *Myoviridae*, *Podoviridae*), and members of the family *Microviridae*, which are thought to infect the majority of the bacterial microbiota ((Reyes et al. 2010; Barr 2017; Moreno-Gallego et al. 2019; Draper et al. 2020). For at least a year and up to 26 months, the core phageome has been shown to have high levels of inter-personal diversity and within-person conservation ((Minot et al. 2011; Shkoporov et al. 2019). Because the core phageome is so prevalent, it was hypothesized that it could play a role in gut health and cancer progression (Hannigan et al. 2018). In fact, these core phages were found to be significantly lower in people with gastrointestinal diseases like UC and CD, as well as in people with malnutrition, obesity, and GI graft-versus-host disease (GI GvHD) (Norman et al. 2015; Reyes et al. 2015; Kim and Bae 2016; Barr 2017; Legoff et al. 2017; Clooney et al. 2019). Indeed, in mice models of colitis as well as human IBD patients, there was a decrease in the abundance of Clostridiales phages and an increase in the abundance of *Streptococcus* phages, making them potential IBD markers (Norman et al. 2015; Duerkop et al. 2012). IBS is another GI disorder in which the gut microbiome is thought to play a role in pathogenesis. It is distinguished by intermittent abdominal pain and changes in bowel habits (Mihindukulasuriya et al. 2021). One study compared the phageomes of people with diarrhea-predominant IBS and people with constipation-predominant IBS to healthy controls. In diarrhea-predominant IBS, the authors discovered a decrease in the abundance of *Microviridae* targeting *Chlamydia trachomatis* and an increase in *Rhodococcus* Siphoviridae (Mihindukulasuriya et al. 2021). Furthermore, the composition of the gut phageome has been shown to change at various stages of CRC progression, with a high abundance of *Streptococcus* phages at early-stage CRC and *Parabacteroides* phages exponentially increasing at late-stage CRC (Nakatsu et al. 2018; Kannen, Parry, and Martin 2019).

9.7 IMMUNE RESPONSE MODULATION DUE TO PHAGEOME-BASED INTERVENTIONS

The impact of the gut microbiota in influencing immune response is immense. It maintains a communication that is bidirectional with the central nervous system, thus causing inflammation resulting in major depressive disorder (Du et al. 2020). A skilled, capable immune system develops, stimulating numerous cells of the immune system and maintaining immune tolerance in the gut (Marrero et al. 2018). It is during neonatal life that the gut microbiota remains unstable and makes space for the development and strengthening of the host immune system (Gensollen et al. 2016). Studies have shown that during the perinatal period, the depletion and gradual absence of microbiota due to treatment with antibiotics have resulted in lower levels of bone marrow neutrophils and circulation (Deshmukh et al. 2014). In turn, we see it rather necessary to have the presence of microbes for the growth and maturation of the systemic immune system. Both the innate and adaptive immunities have explanatory roles in the maintenance of host-pathogen homeostasis (Wang,

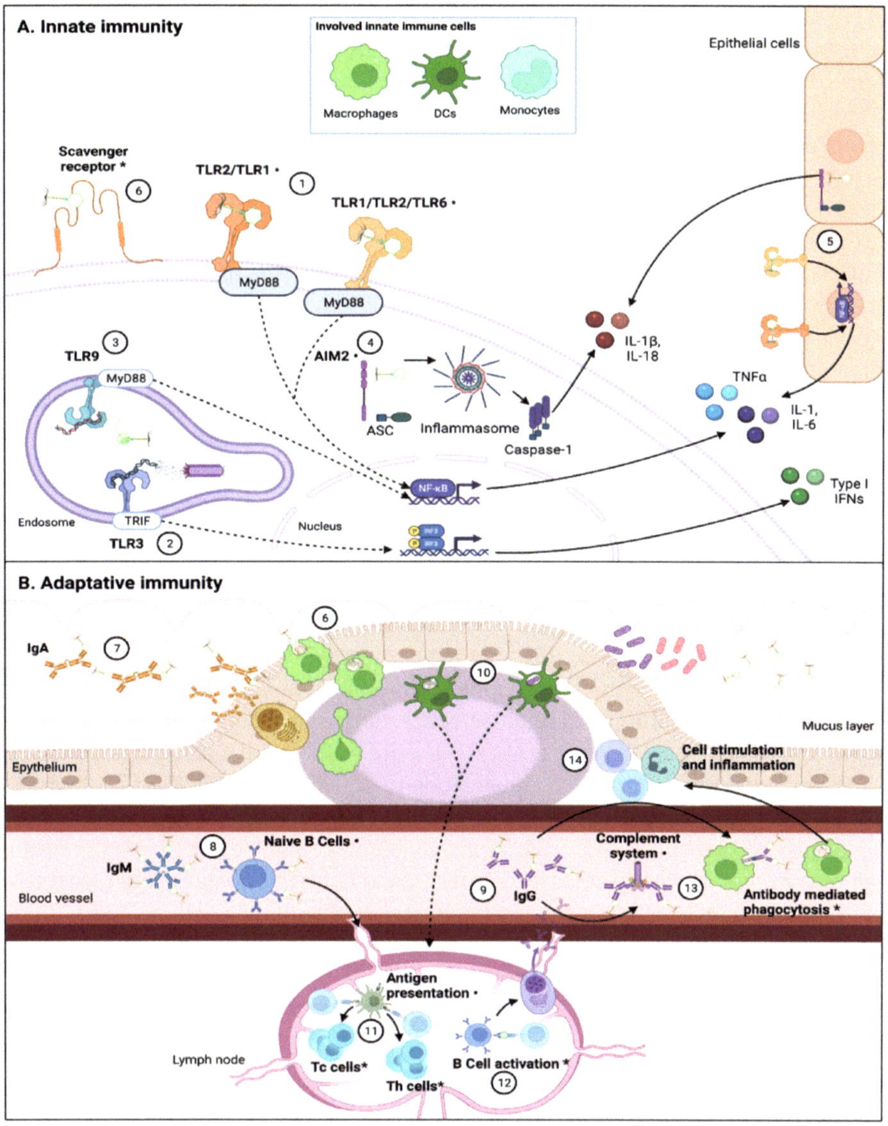

FIGURE 9.2 Immune response modulation due to phageome-based interventions (not at scale) show confirmed, indirectly confirmed (*), and possible interactions (·) between various phages and immune system components. Continuous lines represent immune pathways, and dashed lines represent pathways with intermediary steps. Phages can be recognized by TLR-2/TLR-1 and TLR-1/TLR-2/TLR-6 receptors, resulting in cytokine production through MyD88 coupling and NF-κB activation. Endocytosis of phages by DCs can also lead to cytokine production through TLR-9 sensing of phage genomes and NF-kB activation. Phages can interact with phagocytes, leading to the production of cytokines such as IL-1b and IL-18 through cytosolic phage genome detection by AIM2 receptor recognition and Caspase-1 activation. Epithelial cells can produce cytokines in response to phage detection through TLR and AIM2 receptors. Phages can indirectly impact adaptive immunity by initiating phagocytosis by DCs and macrophages, leading to antigen presentation to T helper (Th) and T cytotoxic (Tc) T cells. Antibody recognition of phages by immunoglobulins such as mucus-excreted IgA, circulating acute-phase IgM, and circulating IgG can play a role in adaptive immunity. Anti-phage IgG can indirectly confirm B cell activation, resulting in memory and secretory B cell maturation. Antibodies binding to phages can result in complement system activation and antibody-mediated phagocytosis, leading to an increase in leukocytes derived from phage interaction with the immune system and inflammation. (Adapted with permission from ref. Farfán, Gonzalez, and Vives 2022. Copyright licensed under a Creative Commons Attribution License.)

Zhu, and Qin 2019). It is after prolonged periods of study that we came to know that normal development of microglia has considerable dependency on the gut microbiota (Erny et al. 2015; Thion and Garel 2018). Endotoxins are found to trigger peripheral inflammation, which in turn triggers various biological pathways leading to the release of different reactive nitrogen and oxygen species, which again contribute to neural toxicity, compromising the blood-brain barrier and causing serious damage to the epithelial cells of the brain (Nguyen et al. 2017). 5-HT, one of the main monoamine neurotransmitters, has several of its receptors in monocytes, lymphocytes, dendritic cells, and macrophages and hence is considered a modulator of our immune system (Gao et al. 2018). Considering metabolites, the response of the innate immunity to bacteria and the regulation of intestinal permeability is often hampered by low propionate levels in the body, thus contributing to inflammation and dysbiosis, having their effect on major depressive disorder pathogenesis. More often than not, direct interaction of phages with the host immune system triggers several immune responses. They can be considered as anti-inflammatory individuals in addition to individuals responsible for inflammation of the immune system (Brown-Jaque, Muniesa, and Navarro 2016). They enter the intestinal epithelial cells, having been recognized by their receptors, by transcytosis, endocytosis, or macropinocytosis, and the dendritic cells by sampling. Once inside, their nucleic acids set off numerous TLR pathways stimulating adaptive immune responses. Thus, they call forth climacteric dysbiotic modifications in ulcerative colitis and Crohn's disease (Łusiak-Szelachowska et al. 2017; Norman et al. 2015). B cell activation ensures phage-specific antibody secretion in the systemic compartment along with the gut. Conversely, activation of T cells advances the production of IFN-γ and cytokines (Sausset et al. 2020). In conclusion, this interaction of phages with the immune system results in downregulation of oxidative stress, performing both immune-modulating and protective effects on gut-lymphoid tissue, interacting with immune cells, and lastly, reducing the production of reactive oxygen species (Łusiak-Szelachowska et al. 2017). Some of the current studies prove bacteriophages to be effective in the reduction of the colonization of *E. coli* in the intestine (Vahedi et al. 2018; Galtier et al. 2017). In contrast to this, bacteriocins not only act upon the immune system but also curb the growth of competitive strains, bringing about niche competition (Kommineni et al. 2015).

Phage therapy was originally developed to target and eliminate bacteria, but recent studies have shown that phages may have other potential applications due to their interactions with eukaryotic cells. Research suggests that phages can interact with the mammalian immune system both directly and indirectly. While phages work to eliminate bacterial hosts, they can also indirectly impact immunity by helping to improve immune responses to bacterial pathogens (Figure 9.2) (Barr 2017).

9.8 FUTURE OF MICROBIOME ENGINEERING

There has been extensive research on synthetic biology for the past few years now, and it has a major impact on microbiome engineering, where microbes with novel functionalities can be engineered (Hwang et al. 2016). This allows strains to be chosen according to the specific requirement of the relevant microbiome engineering technique, such as the construction of *E. coli* to kill the pathogenic microbe *P. aeruginosa* (Hwang et al. 2014; Saeidi et al. 2011) and the production of vaccines against *Vibrio cholerae*, which causes gut infection (Thompson, Oliveira, and Xavier 2014; Duan and March 2010; Hsiao et al. 2014). Microbiome engineering is a potential method to modulate the community of bacteria in the animal or human gut. The understanding of the impact of complex invasions of the entire microbiome on the functioning of the community and its functions needs to be furthered. Moreover, in the current studies of microbiome engineering involving the measurement of functional changes, monitoring inoculant establishment is necessary to check if failure is likely to occur (Albright et al. 2022). Phages, normally used for their antimicrobial effect in humans and agriculture, can also be used to cause changes in host specificity (Manrique et al. 2016). With advancements in this research area, there is an urgent need to utilize it in the study of therapeutic applications and the construction of therapeutic microbes (Sun, Grimm, and Riedel 2015). This, in turn, may regulate host immune responses, which can aid vaccine development, i.e., considering the effects of the microbiome beyond its locale (Valdez, Brown, and Finlay 2014). Thus, the concept of personalized medicine or next-generation precision farming has been given attention and has led to knowledge about several environmental and metabolic diseases (Zmora, Soffer, and Elinav 2019; Jobin 2018). To summarize, overlapping and converging theories, varying analytical tools, and multi-dimensional datasets are urgently required to proceed further (Gilbert and Lynch 2019). On the other hand, considering recent breakthroughs in the plant microbiome world, methods such as predictive and targeted agricultural management, plant breeding, and precision farming can provide revolutionary solutions to improve crop production in a sustainable way by aiming at the particular conditions of fields, like those mentioned in the concept of the "Phytobiomes Roadmap". The seed or rhizosphere microbiome is analogous to the human gut microbiome, and the proposition of "back to the roots" puts forward an intriguing insight into the microbiome of the soil (Sessitsch et al. 2018; Lewin-Epstein, Aharonov, and Hadany 2017; Berg et al. 2014; Pérez-Jaramillo, Mendes, and Raaijmakers 2016). After thorough investigations, it can be concluded that we have limited knowledge of microbiota responses to ecosystems and the functions of the phageome. Even though the progress and development of HMG and omics tools have been immensely helpful to gain information about the microbiome and microbial composition, there is quite a lot yet to discover. There is still a lot of investigation and examination to be done on the applications of the microbiome based on the data obtained from food ecosystems (Cocolin and Ercolini 2015). A lot more attention needs to be given to microbiomes of other parts of the body and the ecosystem.

9.9 CONCLUSION

Manipulating phageomes to modulate the bacterial community in the gut of animals and humans is a promising avenue for improving human health, agricultural productivity, and climate management. The microbiome is known to play a significant role in maintaining the health of animals and humans, and targeting specific communities of bacteria using phage-based interventions has emerged as a simple and effective approach.

With the ability to specifically target pathogenic, commensal, and pathobiont microbiome members, phageomes provide a powerful tool for microbiome engineering. By studying the molecular nature of phage interactions with microbiota, researchers can design new strategies for improving the function of the microbiome.

In this chapter, we have discussed the various strategies involved in modulating the bacterial community using phageomes and their potential lasting effects on animal and human health management. By providing solutions to improve the microbiome's function, we can address some of the significant challenges in biotechnology.

Overall, the use of phageomes for microbiome engineering offers exciting possibilities for enhancing the microbiome's beneficial functions while also providing a better understanding of the intricate relationship between microbiota and their hosts. As research in this field continues to evolve, we can expect to see new and innovative approaches that will revolutionize our ability to improve animal and human health.

REFERENCES

Abedon, Stephen T., Sarah J. Kuhl, Bob G. Blasdel, and Elizabeth Martin Kutter. 2011. "Phage Treatment of Human Infections." *Bacteriophage* 1 (2). Bacteriophage: 66–85. doi:10.4161/BACT.1.2.15845.

Adriaenssens, Evelien M. 2021. "Phage Diversity in the Human Gut Microbiome: A Taxonomist's Perspective." *MSystems* 6 (4). American Society for Microbiology (ASM). doi:10.1128/MSYSTEMS.00799-21.

Alam, Muhammad S. 2018. "Proximity Ligation Assay (PLA)." *Current Protocols in Immunology* 123 (1). Curr Protoc Immunol. doi:10.1002/CPIM.58.

Albright, Michaeline B.N., Stilianos Louca, Daniel E. Winkler, Kelli L. Feeser, Sarah Jane Haig, Katrine L. Whiteson, Joanne B. Emerson, and John Dunbar. 2022. "Solutions in Microbiome Engineering: Prioritizing Barriers to Organism Establishment." *The ISME Journal* 16 (2). ISME J: 331–38. doi:10.1038/S41396-021-01088-5.

Arumugam, Manimozhiyan, Jeroen Raes, Eric Pelletier, Denis Le Paslier, Takuji Yamada, Daniel R. Mende, Gabriel R. Fernandes, et al. 2011. "Enterotypes of the Human Gut Microbiome." *Nature* 473 (7346). Nature: 174–80. doi:10.1038/NATURE09944.

Axelrad, Jordan E., Simon Lichtiger, and Vijay Yajnik. 2016. "Inflammatory Bowel Disease and Cancer: The Role of Inflammation, Immunosuppression, and Cancer Treatment." *World Journal of Gastroenterology* 22 (20). World J Gastroenterol: 4794–801. doi:10.3748/WJG.V22.I20.4794.

Bao, Hong duo, Mao da Pang, Ademola Olaniran, Xu hui Zhang, Hui Zhang, Yan Zhou, Li chang Sun, Stefan Schmidt, and Ran Wang. 2018. "Alterations in the Diversity and Composition of Mice Gut Microbiota by Lytic or Temperate Gut Phage Treatment." *Applied Microbiology and Biotechnology* 102 (23). Springer Verlag: 10219–30. doi:10.1007/s00253-018-9378-6.

Barr, Jeremy J. 2017. "A Bacteriophages Journey through the Human Body." *Immunological Reviews* 279 (1). Immunol Rev: 106–22. doi:10.1111/IMR.12565.

Barrero-Canosa, Jimena, and Cristina Moraru. 2019. "PhageFISH for Monitoring Phage Infections at Single Cell Level." *Methods in Molecular Biology*: 1–26. doi:10.1007/978-1-4939-8940-9_1.

Belleghem, Jonas D. Van, Krystyna Dąbrowska, Mario Vaneechoutte, Jeremy J. Barr, and Paul L. Bollyky. 2018. "Interactions between Bacteriophage, Bacteria, and the Mammalian Immune System." *Viruses* 11 (1). Viruses. doi:10.3390/V11010010.

Berg, Gabriele, Martin Grube, Michael Schloter, and Kornelia Smalla. 2014. "Unraveling the Plant Microbiome: Looking Back and Future Perspectives." *Frontiers in Microbiology* 5 (Jun). Front Microbiol. doi:10.3389/FMICB.2014.00148.

Bikard, David, Chad W. Euler, Wenyan Jiang, Philip M. Nussenzweig, Gregory W. Goldberg, Xavier Duportet, Vincent A. Fischetti, and Luciano A. Marraffini. 2014. "Exploiting CRISPR-Cas Nucleases to Produce Sequence-Specific Antimicrobials." *Nature Biotechnology* 32 (11). Nat Biotechnol: 1146–50. doi:10.1038/NBT.3043.

Brown-Jaque, Maryury, Maite Muniesa, and Ferran Navarro. 2016. "Bacteriophages in Clinical Samples Can Interfere with Microbiological Diagnostic Tools." *Scientific Reports* 6 (1). Nature Publishing Group: 1–8. doi:10.1038/srep33000.

Brunse, Anders, Ling Deng, Xiaoyu Pan, Yan Hui, Josué L. Castro-Mejía, Witold Kot, Duc Ninh Nguyen, Jan Bojsen Møller Secher, Dennis Sandris Nielsen, and Thomas Thymann. 2022. "Fecal Filtrate Transplantation Protects against Necrotizing Enterocolitis." *The ISME Journal* 16 (3). ISME J: 686–94. doi:10.1038/S41396-021-01107-5.

Cho, Ilseung, and Martin J. Blaser. 2012. "The Human Microbiome: At the Interface of Health and Disease." *Nature Reviews Genetics* 13 (4). Nature Publishing Group: 260–70. doi:10.1038/nrg3182.

Clokie, Martha R.J., Andrew D. Millard, Andrey V. Letarov, and Shaun Heaphy. 2011. "Phages in Nature." *Bacteriophage* 1 (1). Taylor & Francis: 31. doi:10.4161/BACT.1.1.14942.

Clooney, Adam G., Thomas D.S. Sutton, Andrey N. Shkoporov, Ross K. Holohan, Karen M. Daly, Orla O'Regan, Feargal J. Ryan, et al. 2019. "Whole-Virome Analysis Sheds Light on Viral Dark Matter in Inflammatory Bowel Disease." *Cell Host & Microbe* 26 (6). Cell Host Microbe: 764–78.e5. doi:10.1016/J.CHOM.2019.10.009.

Cocolin, Luca, and Danilo Ercolini. 2015. "Zooming into Food-Associated Microbial Consortia: A 'Cultural' Evolution." *Current Opinion in Food Science* 2 (April). Elsevier: 43–50. doi:10.1016/J.COFS.2015.01.003.

David, Lawrence A., Arne C. Materna, Jonathan Friedman, Maria I. Campos-Baptista, Matthew C. Blackburn, Allison Perrotta, Susan E. Erdman, and Eric J. Alm. 2014. "Host Lifestyle Affects Human Microbiota on Daily Timescales." *Genome Biology* 15 (7). BioMed Central Ltd.: 1–15. doi:10.1186/GB-2014-15-7-R89/TABLES/1.

DeFilipp, Zachariah, Patricia P. Bloom, Mariam Torres Soto, Michael K. Mansour, Mohamad R.A. Sater, Miriam H. Huntley, Sarah Turbett, Raymond T. Chung, Yi-Bin Chen, and Elizabeth L. Hohmann. 2019. "Drug-Resistant E. Coli Bacteremia Transmitted by Fecal Microbiota Transplant." *The New England Journal of Medicine* 381 (21). N Engl J Med: 2043–50. doi:10.1056/NEJMOA1910437.

Deshmukh, Hitesh S., Yuhong Liu, Ogechukwu R. Menkiti, Junjie Mei, Ning Dai, Claire E. O'Leary, Paula M. Oliver, Jay K. Kolls, Jeffrey N. Weiser, and G. Scott Worthen. 2014. "The Microbiota Regulates Neutrophil Homeostasis and Host Resistance to Escherichia Coli K1 Sepsis in Neonatal Mice." *Nature Medicine* 20 (5). Nat Med: 524–30. doi:10.1038/NM.3542.

Draper, Lorraine A., Feargal J. Ryan, Marion Dalmasso, Pat G. Casey, Angela McCann, Vimalkumar Velayudhan, R. Paul Ross, and Colin Hill. 2020. "Autochthonous Faecal Viral Transfer (FVT) Impacts the Murine Microbiome after Antibiotic Perturbation." *BMC Biology* 18 (1). BioMed Central Ltd: 1–14. doi:10.1186/S12915-020-00906-0/FIGURES/5.

Du, Yu, Xin Ran Gao, Lei Peng, and Jin Fang Ge. 2020. "Crosstalk between the Microbiota-Gut-Brain Axis and Depression." *Heliyon* 6 (6). Heliyon. doi:10.1016/J.HELIYON.2020.E04097.

Duan, Faping, and John C. March. 2010. "Engineered Bacterial Communication Prevents Vibrio Cholerae Virulence in an Infant Mouse Model." *Proceedings of the National Academy of Sciences of the United States of America* 107 (25). National Academy of Sciences: 11260–64. doi:10.1073/PNAS.1001294107/SUPPL_FILE/PNAS.1001294107_SI.PDF.

Duerkop, Breck A., Charmaine V. Clements, Darcy Rollins, Jorge L.M. Rodrigues, and Lora V. Hooper. 2012. "A Composite Bacteriophage Alters Colonization by an Intestinal Commensal Bacterium." *Proceedings of the National Academy of Sciences of the United States of America* 109 (43). Proc Natl Acad Sci U S A: 17621–26. doi:10.1073/PNAS.1206136109.

Džunková, Mária, Soo Jen Low, Joshua N. Daly, Li Deng, Christian Rinke, and Philip Hugenholtz. 2019. "Defining the Human Gut Host-Phage Network through Single-Cell Viral Tagging." *Nature Microbiology* 4 (12). Nat Microbiol: 2192–2203. doi:10.1038/S41564-019-0526-2.

Erny, Daniel, Anna Lena Hrabe De Angelis, Diego Jaitin, Peter Wieghofer, Ori Staszewski, Eyal David, Hadas Keren-Shaul, et al. 2015. "Host Microbiota Constantly Control Maturation and Function of Microglia in the CNS." *Nature Neuroscience* 18 (7). Nature Publishing Group: 965–77. doi:10.1038/nn.4030.

Farfán, Juan, John M. Gonzalez, and Martha Vives. 2022. "The Immunomodulatory Potential of Phage Therapy to Treat Acne: A Review on Bacterial Lysis and Immunomodulation." *PeerJ* 10 (July). PeerJ Inc.: e13553. doi:10.7717/PEERJ.13553/SUPP-2.

Federici, Sara, Sharon Kredo-Russo, Rafael Valdés-Mas, Denise Kviatcovsky, Eyal Weinstock, Yulia Matiuhin, Yael Silberberg, et al. 2022. "Targeted Suppression of Human IBD-Associated Gut Microbiota Commensals by Phage Consortia for Treatment of Intestinal Inflammation." *Cell* 185 (16). Cell: 2879–98.e24. doi:10.1016/J.CELL.2022.07.003.

Galtier, Matthieu, Luisa De Sordi, Adeline Sivignon, Amélie de Vallée, Damien Maura, Christel Neut, Oumaira Rahmouni, et al. 2017. "Bacteriophages Targeting Adherent Invasive Escherichia Coli Strains as a Promising New Treatment for Crohn's Disease." *Journal of Crohn's & Colitis* 11 (7). J Crohns Colitis: 840–47. doi:10.1093/ECCO-JCC/JJW224.

Ganeshan, Sharita Divya, and Zeinab Hosseinidoust. 2019. "Phage Therapy with a Focus on the Human Microbiota." *Antibiotics* 8 (3). Multidisciplinary Digital Publishing Institute (MDPI). doi:10.3390/ANTIBIOTICS8030131.

Gao, Jing, Kang Xu, Hongnan Liu, Gang Liu, Miaomiao Bai, Can Peng, Tiejun Li, and Yulong Yin. 2018. "Impact of the Gut Microbiota on Intestinal Immunity Mediated by Tryptophan Metabolism." *Frontiers in Cellular and Infection Microbiology* 8 (FEB). Frontiers Media S.A.: 13. doi:10.3389/FCIMB.2018.00013/BIBTEX.

Gensollen, Thomas, Shankar S. Iyer, Dennis L. Kasper, and Richard S. Blumberg. 2016. "How Colonization by Microbiota in Early Life Shapes the Immune System." *Science (New York, N.Y.)* 352 (6285). Science: 539–44. doi:10.1126/SCIENCE.AAD9378.

Gilbert, Jack A., and Susan V. Lynch. 2019. "Community Ecology as a Framework for Human Microbiome Research." *Nature Medicine* 25 (6). Nat Med: 884–89. doi:10.1038/S41591-019-0464-9.

Gogokhia, Lasha, Kate Buhrke, Rickesha Bell, Brenden Hoffman, D. Garrett Brown, Christin Hanke-Gogokhia, Nadim J. Ajami, et al. 2019. "Expansion of Bacteriophages Is Linked to Aggravated Intestinal Inflammation and Colitis." *Cell Host & Microbe* 25 (2). NIH Public Access: 285. doi:10.1016/J.CHOM.2019.01.008.

Gordillo Altamirano, Fernando L., and Jeremy J. Barr. 2019. "Phage Therapy in the Postantibiotic Era." *Clinical Microbiology Reviews* 32 (2). American Society for Microbiology. doi:10.1128/CMR.00066-18/ASSET/59E47274-79BC-44BE-9606-AAA673B114B3/ASSETS/GRAPHIC/CMR.00066-18-F0002.JPEG.

Górski, Andrzej, and Beata Weber-Dabrowska. 2005. "The Potential Role of Endogenous Bacteriophages in Controlling Invading Pathogens." *Cellular and Molecular Life Sciences : CMLS* 62 (5). Cell Mol Life Sci: 511–19. doi:10.1007/S00018-004-4403-6.

Górski, Andrzej, Ewa Jończyk-Matysiak, Ryszard Międzybrodzki, Beata Weber-Dabrowska, and Jan Borysowski. 2018. "'Phage Transplantation in Allotransplantation': Possible Treatment in Graft-versus-Host Disease?" *Frontiers in Immunology* 9 (APR). Frontiers Media S.A.: 941. doi:10.3389/FIMMU.2018.00941/BIBTEX.

Grice, Elizabeth A., and Julia A. Segre. 2012. "The Human Microbiome: Our Second Genome." *Annual Review of Genomics and Human Genetics* 13 (September). Annu Rev Genomics Hum Genet: 151–70. doi:10.1146/ANNUREV-GENOM-090711-163814.

Grubb, Diana S., Scott D. Wrigley, Kimberley E. Freedman, Yuren Wei, Allegra R. Vazquez, Roxanne E. Trotter, Taylor C. Wallace, Sarah A. Johnson, and Tiffany L. Weir. 2020. "PHAGE-2 Study: Supplemental Bacteriophages Extend Bifidobacterium Animalis Subsp. Lactis BL04 Benefits on Gut Health and Microbiota in Healthy Adults." *Nutrients* 12 (8). Multidisciplinary Digital Publishing Institute (MDPI): 1–15. doi:10.3390/NU12082474.

Gutiérrez, Beatriz, and Pilar Domingo-Calap. 2020. "Phage Therapy in Gastrointestinal Diseases." *Microorganisms* 8 (9). Multidisciplinary Digital Publishing Institute (MDPI): 1–11. doi:10.3390/MICROORGANISMS8091420.

Handley, Scott A., and Suzanne Devkota. 2019. "Going Viral: A Novel Role for Bacteriophage in Colorectal Cancer." *MBio* 10 (1). American Society for Microbiology (ASM). doi:10.1128/MBIO.02626-18.

Hannigan, Geoffrey D., Melissa B. Duhaime, Mack T. Ruffin, Charlie C. Koumpouras, and Patrick D. Schloss. 2018. "Diagnostic Potential and Interactive Dynamics of the Colorectal Cancer Virome." *MBio* 9 (6). mBio. doi:10.1128/MBIO.02248-18.

"Healthy Human Gut Phageome Manrique, Pilar." 2016. doi:10.1073/pnas.1601060113.

Hillman, Ethan T., Hang Lu, Tianming Yao, and Cindy H. Nakatsu. 2017. "Microbial Ecology along the Gastrointestinal Tract." *Microbes and Environments* 32 (4). Microbes Environ: 300–313. doi:10.1264/JSME2.ME17017.

Holtmann, Gerald J., Alexander C. Ford, and Nicholas J. Talley. 2016. "Pathophysiology of Irritable Bowel Syndrome." *The Lancet. Gastroenterology & Hepatology* 1 (2). Lancet Gastroenterol Hepatol: 133–46. doi:10.1016/S2468-1253(16)30023-1.

Hooper, Lora V., Dan R. Littman, and Andrew J. Macpherson. 2012. "Interactions between the Microbiota and the Immune System." *Science (New York, N.Y.)* 336 (6086). Science: 1268–73. doi:10.1126/SCIENCE.1223490.

Hrncir, Tomas. 2022. "Gut Microbiota Dysbiosis: Triggers, Consequences, Diagnostic and Therapeutic Options." *Microorganisms* 10 (3). Multidisciplinary Digital Publishing Institute: 578. doi:10.3390/MICROORGANISMS10030578.

Hsiao, Ansel, A. M.Shamsir Ahmed, Sathish Subramanian, Nicholas W. Griffin, Lisa L. Drewry, William A. Petri, Rashidul Haque, Tahmeed Ahmed, and Jeffrey I. Gordon. 2014. "Members of the Human Gut Microbiota Involved in Recovery from Vibrio Cholerae Infection." *Nature* 515 (7527). Nature: 423–26. doi:10.1038/NATURE13738.

Hsu, Bryan B., Travis E. Gibson, Vladimir Yeliseyev, Qing Liu, Lorena Lyon, Lynn Bry, Pamela A. Silver, and Georg K. Gerber. 2019. "Dynamic Modulation of the Gut Microbiota and Metabolome by Bacteriophages in a Mouse Model." *Cell Host & Microbe* 25 (6). Cell Host Microbe: 803–14.e5. doi:10.1016/J.CHOM.2019.05.001.

Hwang, In Young, Elvin Koh, Hye Rim Kim, Wen Shan Yew, and Matthew Wook Chang. 2016. "Reprogrammable Microbial Cell-Based Therapeutics against Antibiotic-Resistant Bacteria." *Drug Resistance Updates : Reviews and Commentaries in Antimicrobial and Anticancer Chemotherapy* 27 (July). Drug Resist Updat: 59–71. doi:10.1016/J.DRUP.2016.06.002.

Hwang, In Young, Mui Hua Tan, Elvin Koh, Chun Loong Ho, Chueh Loo Poh, and Matthew Wook Chang. 2014. "Reprogramming Microbes to Be Pathogen-Seeking Killers." *ACS Synthetic Biology* 3 (4). ACS Synth Biol: 228–37. doi:10.1021/SB400077J.

Jobin, Christian. 2018. "Precision Medicine Using Microbiota." *Science (New York, N.Y.)* 359 (6371). Science: 32–34. doi:10.1126/SCIENCE.AAR2946.

Johnson, Constance M., Caimiao Wei, Joe E. Ensor, Derek J. Smolenski, Christopher I. Amos, Bernard Levin, and Donald A. Berry. 2013. "Meta-Analyses of Colorectal Cancer Risk Factors." *Cancer Causes & Control : CCC* 24 (6). Cancer Causes Control: 1207–22. doi:10.1007/S10552-013-0201-5.

Kamada, Nobuhiko, Grace Y. Chen, Naohiro Inohara, and Gabriel Núñez. 2013. "Control of Pathogens and Pathobionts by the Gut Microbiota." *Nature Immunology* 14 (7). Nat Immunol: 685–90. doi:10.1038/NI.2608.

Kannen, Vinicius, Lee Parry, and Francis L. Martin. 2019. "Phages Enter the Fight against Colorectal Cancer." *Trends in Cancer* 5 (10). Trends Cancer: 577–79. doi:10.1016/J.TRECAN.2019.08.002.

Khan Mirzaci, Mohammadali, and Li Deng. 2022. "New Technologies for Developing Phage-Based Tools to Manipulate the Human Microbiome." *Trends in Microbiology* 30 (2). Trends Microbiol: 131–42. doi:10.1016/J.TIM.2021.04.007.

Kim, Min Soo, and Jin Woo Bae. 2016. "Spatial Disturbances in Altered Mucosal and Luminal Gut Viromes of Diet-Induced Obese Mice." *Environmental Microbiology* 18 (5). Environ Microbiol: 1498–510. doi:10.1111/1462-2920.13182.

Kommineni, Sushma, Daniel J. Bretl, Vy Lam, Rajrupa Chakraborty, Michael Hayward, Pippa Simpson, Yumei Cao, Pavlos Bousounis, Christopher J. Kristich, and Nita H. Salzman. 2015. "Bacteriocin Production Augments Niche Competition by Enterococci in the Mammalian Gastrointestinal Tract." *Nature* 526 (7575). Nature: 719–22. doi:10.1038/NATURE15524.

Koonin, Eugene V., Valerian V. Dolja, Mart Krupovic, Arvind Varsani, Yuri I. Wolf, Natalya Yutin, F. Murilo Zerbini, and Jens H. Kuhn. 2020. "Global Organization and Proposed Megataxonomy of the Virus World." *Microbiology and Molecular Biology Reviews : MMBR* 84 (2). Microbiol Mol Biol Rev. doi:10.1128/MMBR.00061-19.

Lee, Hui Ling, Haosheng Shen, In Young Hwang, Hua Ling, Wen Shan Yew, Yung Seng Lee, and Matthew Wook Chang. 2018. "Targeted Approaches for in Situ Gut Microbiome Manipulation." *Genes* 9 (7). Multidisciplinary Digital Publishing Institute (MDPI). doi:10.3390/GENES9070351.

Legoff, Jérôme, Matthieu Resche-Rigon, Jerome Bouquet, Marie Robin, Samia N. Naccache, Séverine Mercier-Delarue, Scot Federman, et al. 2017. "The Eukaryotic Gut Virome in Hematopoietic Stem Cell Transplantation: New Clues in Enteric Graft-versus-Host Disease." *Nature Medicine* 23 (9). Nat Med: 1080–85. doi:10.1038/NM.4380.

Lewin-Epstein, Ohad, Ranit Aharonov, and Lilach Hadany. 2017. "Microbes Can Help Explain the Evolution of Host Altruism." *Nature Communications* 8 (1). Nature Publishing Group: 1–7. doi:10.1038/ncomms14040.

Lin, Derek M., and Henry C. Lin. 2019. "A Theoretical Model of Temperate Phages as Mediators of Gut Microbiome Dysbiosis." *F1000Research* 8. Faculty of 1000 Ltd. doi:10.12688/F1000RESEARCH.18480.1.

Łusiak-Szelachowska, Marzanna, Beata Weber-Dąbrowska, Ewa Jończyk-Matysiak, Renata Wojciechowska, and Andrzej Górski. 2017. "Bacteriophages in the Gastrointestinal Tract and Their Implications." *Gut Pathogens* 9 (1). BioMed Central. doi:10.1186/S13099-017-0196-7.

Ma, Yingfei, Xiaoyan You, Guoqin Mai, Taku Tokuyasu, and Chenli Liu. 2018. "A Human Gut Phage Catalog Correlates the Gut Phageome with Type 2 Diabetes." *Microbiome* 6 (1). BioMed Central Ltd.: 1–12. doi:10.1186/S40168-018-0410-Y/FIGURES/6.

Magro, Fernando, Gonçalo Cordeiro, Andreia Martins Dias, and Maria Manuela Estevinho. 2020. "Inflammatory Bowel Disease - Non-Biological Treatment." *Pharmacological Research* 160 (October). Pharmacol Res. doi:10.1016/J.PHRS.2020.105075.

Maldonado-Contreras, Ana, Sabrina E. Noel, Doyle V. Ward, Martha Velez, and Kelsey M. Mangano. 2020. "Associations between Diet, the Gut Microbiome, and Short-Chain Fatty Acid Production among Older Caribbean Latino Adults." *Journal of the Academy of Nutrition and Dietetics* 120 (12). J Acad Nutr Diet: 2047-2060.e6. doi:10.1016/J.JAND.2020.04.018.

Manrique, Pilar, Benjamin Bolduc, Seth T. Walk, John Der Van Oost, Willem M. De Vos, and Mark J. Young. 2016. "Healthy Human Gut Phageome." *Proceedings of the National Academy of Sciences of the United States of America* 113 (37). National Academy of Sciences: 10400–405. doi:10.1073/PNAS.1601060113/SUPPL_FILE/PNAS.201601060SI.PDF.

Marrero, Idania, Igor Maricic, Ariel E. Feldstein, Rohit Loomba, Bernd Schnabl, Jesus Rivera-Nieves, Lars Eckmann, and Vipin Kumar. 2018. "Complex Network of NKT Cell Subsets Controls Immune Homeostasis in Liver and Gut." *Frontiers in Immunology* 9 (Sep). Frontiers Media S.A.: 2082. doi:10.3389/FIMMU.2018.02082/BIBTEX.

Martín, R., S. Miquel, J. Ulmer, P. Langella, and L. G. Bermúdez-Humarán. 2014. "Gut Ecosystem: How Microbes Help Us." *Beneficial Microbes* 5 (3). Benef Microbes: 219–33. doi:10.3920/BM2013.0057.

Mihindukulasuriya, Kathie A., Ruben A.T. Mars, Abigail J. Johnson, Tonya Ward, Sambhawa Priya, Heather R. Lekatz, Krishna R. Kalari, et al. 2021. "Multi-Omics Analyses Show Disease, Diet, and Transcriptome Interactions With the Virome." *Gastroenterology* 161 (4). W.B. Saunders: 1194-1207.e8. doi:10.1053/J.GASTRO.2021.06.077.

Minot, Samuel, Rohini Sinha, Jun Chen, Hongzhe Li, Sue A. Keilbaugh, Gary D. Wu, James D. Lewis, and Frederic D. Bushman. 2011. "The Human Gut Virome: Inter-Individual Variation and Dynamic Response to Diet." *Genome Research* 21 (10). Genome Res: 1616–25. doi:10.1101/GR.122705.111.

Mirzaei, Mohammadali Khan, and Corinne F. Maurice. 2017. "Ménage à Trois in the Human Gut: Interactions between Host, Bacteria and Phages." *Nature Reviews. Microbiology* 15 (7). Nat Rev Microbiol: 397–408. doi:10.1038/NRMICRO.2017.30.

Moreno-Gallego, J. Leonardo, Shao Pei Chou, Sara C. Di Rienzi, Julia K. Goodrich, Timothy D. Spector, Jordana T. Bell, Nicholas D. Youngblut, Ian Hewson, Alejandro Reyes, and Ruth E. Ley. 2019. "Virome Diversity Correlates with Intestinal Microbiome Diversity in Adult Monozygotic Twins." *Cell Host & Microbe* 25 (2). Cell Host Microbe: 261–72.e5. doi:10.1016/J.CHOM.2019.01.019.

Mueller, U. G., and J. L. Sachs. 2015. "Engineering Microbiomes to Improve Plant and Animal Health." *Trends in Microbiology* 23 (10). Trends Microbiol: 606–17. doi:10.1016/J.TIM.2015.07.009.

Mukhopadhya, Indrani, Jonathan P. Segal, Simon R. Carding, Ailsa L. Hart, and Georgina L. Hold. 2019. "The Gut Virome: The 'missing Link' between Gut Bacteria and Host Immunity?" *Therapeutic Advances in Gastroenterology* 12 (March). Therap Adv Gastroenterol. doi:10.1177/1756284819836620.

Nakatsu, Geicho, Haokui Zhou, William Ka Kei Wu, Sunny Hei Wong, Olabisi Oluwabukola Coker, Zhenwei Dai, Xiangchun Li, et al. 2018. "Alterations in Enteric Virome Are Associated With Colorectal Cancer and Survival Outcomes." *Gastroenterology* 155 (2). Gastroenterology: 529–41.e5. doi:10.1053/J.GASTRO.2018.04.018.

Navarrete, Acacio A., Siu M. Tsai, Lucas W. Mendes, Karoline Faust, Mattias De Hollander, Noriko A. Cassman, Jeroen Raes, Johannes A. Van Veen, and Eiko E. Kuramae. 2015. "Soil Microbiome Responses to the Short-Term Effects of Amazonian Deforestation." *Molecular Ecology* 24 (10). Mol Ecol: 2433–48. doi:10.1111/MEC.13172.

Nguyen, Sophie, Kristi Baker, Benjamin S. Padman, Ruzeen Patwa, Rhys A. Dunstan, Thomas A. Weston, Kyle Schlosser, et al. 2017. "Bacteriophage Transcytosis Provides a Mechanism To Cross Epithelial Cell Layers." *MBio* 8 (6). American Society for Microbiology (ASM). doi:10.1128/MBIO.01874-17.

Norman, Jason M., Scott A. Handley, Megan T. Baldridge, Lindsay Droit, Catherine Y. Liu, Brian C. Keller, Amal Kambal, et al. 2015. "Disease-Specific Alterations in the Enteric Virome in Inflammatory Bowel Disease." *Cell* 160 (3). Cell: 447–60. doi:10.1016/J.CELL.2015.01.002.

Ott, Stephan J., Georg H. Waetzig, Ateequr Rehman, Jacqueline Moltzau-Anderson, Richa Bharti, Juris A. Grasis, Liam Cassidy, et al. 2017. "Efficacy of Sterile Fecal Filtrate Transfer for Treating Patients With Clostridium Difficile Infection." *Gastroenterology* 152 (4). Gastroenterology: 799–811.e7. doi:10.1053/J.GASTRO.2016.11.010.

Penadés, José R., John Chen, Nuria Quiles-Puchalt, Nuria Carpena, and Richard P. Novick. 2015. "Bacteriophage-Mediated Spread of Bacterial Virulence Genes." *Current Opinion in Microbiology* 23 (February). Curr Opin Microbiol: 171–78. doi:10.1016/J.MIB.2014.11.019.

Pérez-Jaramillo, Juan E., Rodrigo Mendes, and Jos M. Raaijmakers. 2016. "Impact of Plant Domestication on Rhizosphere Microbiome Assembly and Functions." *Plant Molecular Biology* 90 (6). Plant Mol Biol: 635–44. doi:10.1007/S11103-015-0337-7.

Peterson, Jane, Susan Garges, Maria Giovanni, Pamela McInnes, Lu Wang, Jeffery A. Schloss, Vivien Bonazzi, et al. 2009. "The NIH Human Microbiome Project." *Genome Research* 19 (12). Genome Res: 2317–23. doi:10.1101/GR.096651.109.

Pfeiffer, Julie K., and Herbert W. Virgin. 2016. "Transkingdom Control of Viral Infection and Immunity in the Mammalian Intestine." *Science (New York, N.Y.)* 351 (6270). NIH Public Access. doi:10.1126/SCIENCE.AAD5872.

Qin, Junjie, Ruiqiang Li, Jeroen Raes, Manimozhiyan Arumugam, Kristoffer Solvsten Burgdorf, Chaysavanh Manichanh, Trine Nielsen, et al. 2010. "A Human Gut Microbial Gene Catalogue Established by Metagenomic Sequencing." *Nature* 464 (7285). Nature Publishing Group: 59–65. doi:10.1038/nature08821.

Reyes, Alejandro, Laura V. Blanton, Song Cao, Guoyan Zhao, Mark Manary, Indi Trehan, Michelle I. Smith, et al. 2015. "Gut DNA Viromes of Malawian Twins Discordant for Severe Acute Malnutrition." *Proceedings of the National Academy of Sciences of the United States of America* 112 (38). National Academy of Sciences: 11941–46. doi:10.1073/PNAS.1514285112/SUPPL_FILE/PNAS.1514285112.ST06.XLSX.

Reyes, Alejandro, Matthew Haynes, Nicole Hanson, Florent E. Angly, Andrew C. Heath, Forest Rohwer, and Jeffrey I. Gordon. 2010. "Viruses in the Faecal Microbiota of Monozygotic Twins and Their Mothers." *Nature* 466 (7304). Nature: 334–38. doi:10.1038/NATURE09199.

Round, June L., and Sarkis K. Mazmanian. 2009. "The Gut Microbiota Shapes Intestinal Immune Responses during Health and Disease." *Nature Reviews. Immunology* 9 (5). Nat Rev Immunol: 313–23. doi:10.1038/NRI2515.

Saeidi, Nazanin, Choon Kit Wong, Tat Ming Lo, Hung Xuan Nguyen, Hua Ling, Susanna Su Jan Leong, Chueh Loo Poh, and Matthew Wook Chang. 2011. "Engineering Microbes to Sense and Eradicate Pseudomonas Aeruginosa, a Human Pathogen." *Molecular Systems Biology* 7. Mol Syst Biol. doi:10.1038/MSB.2011.55.

Sarker, Shafiqul Alam, Shamima Sultana, Gloria Reuteler, Deborah Moine, Patrick Descombes, Florence Charton, Gilles Bourdin, et al. 2016. "Oral Phage Therapy of Acute Bacterial Diarrhea With Two Coliphage Preparations: A Randomized Trial in Children From Bangladesh." *EBioMedicine* 4 (February). EBioMedicine: 124–37. doi:10.1016/J.EBIOM.2015.12.023.

Sausset, R., M. A. Petit, V. Gaboriau-Routhiau, and M. De Paepe. 2020. "New Insights into Intestinal Phages." *Mucosal Immunology* 13 (2). Mucosal Immunol: 205–15. doi:10.1038/S41385-019-0250-5.

Sender, Ron, Shai Fuchs, and Ron Milo. 2016. "Revised Estimates for the Number of Human and Bacteria Cells in the Body." *PLoS Biology* 14 (8). Public Library of Science: e1002533. doi:10.1371/JOURNAL.PBIO.1002533.

Sessitsch, Angela, Günter Brader, Nikolaus Pfaffenbichler, Doris Gusenbauer, and Birgit Mitter. 2018. "The Contribution of Plant Microbiota to Economy Growth." *Microbial Biotechnology* 11 (5). Microb Biotechnol: 801–5. doi:10.1111/1751-7915.13290.

Shkoporov, Andrey N., Adam G. Clooney, Thomas D.S. Sutton, Feargal J. Ryan, Karen M. Daly, James A. Nolan, Siobhan A. McDonnell, et al. 2019. "The Human Gut Virome Is Highly Diverse, Stable, and Individual Specific." *Cell Host & Microbe* 26 (4). Cell Host Microbe: 527–41.e5. doi:10.1016/J.CHOM.2019.09.009.

Singh, Prashant, Eric J. Alm, John M. Kelley, Vivian Cheng, Mark Smith, Zain Kassam, Judy Nee, Johanna Iturrino, and Anthony Lembo. 2022. "Effect of Antibiotic Pretreatment on Bacterial Engraftment after Fecal Microbiota Transplant (FMT) in IBS-D." *Gut Microbes* 14 (1). Gut Microbes. doi:10.1080/19490976.2021.2020067.

Singh, Prashant, Sarah Ballou, Jesse Katon, Eve Takazawa, Vikram Rangan, Rupa Mukherjee, Johanna Iturrino, Judy Nee, and Anthony Lembo. 2020. "Symptom Severity, Mood, and Healthcare Use Are Associated With Satisfaction in Patients With Irritable Bowel Syndrome." *Clinical Gastroenterology and Hepatology : The Official Clinical Practice Journal of the American Gastroenterological Association* 18 (13). NIH Public Access: 2945. doi:10.1016/J.CGH.2020.01.045.

Stidham, Ryan W., and Peter D.R. Higgins. 2018. "Colorectal Cancer in Inflammatory Bowel Disease." *Clinics in Colon and Rectal Surgery* 31 (3). Clin Colon Rectal Surg: 168–78. doi:10.1055/S-0037-1602237.

Sun, Zhongke, Verena Grimm, and Christian U. Riedel. 2015. "AI-2 to the Rescue against Antibiotic-Induced Intestinal Dysbiosis?" *Trends in Microbiology* 23 (6). Trends Microbiol: 327–28. doi:10.1016/J.TIM.2015.04.002.

Thion, Morgane Sonia, and Sonia Garel. 2018. "[Microbiome and Microglia: Prenatal and Postnatal Interactions Diverge According to Sex]." *Medecine Sciences : M/S* 34 (6–7). Med Sci (Paris): 527–29. doi:10.1051/MEDSCI/20183406011.

Thompson, Jessica A., Rita Almeida Oliveira, and Karina B. Xavier. 2014. "Can Chatter between Microbes Prevent Cholera?" *Trends in Microbiology* 22 (12). Trends Microbiol: 660–62. doi:10.1016/J.TIM.2014.10.006.

Thursby, Elizabeth, and Nathalie Juge. 2017. "Introduction to the Human Gut Microbiota." *Biochemical Journal* 474 (11). Portland Press Ltd: 1823. doi:10.1042/BCJ20160510.

Townsend, Eleanor M., Lucy Kelly, George Muscatt, Joshua D. Box, Nicole Hargraves, Daniel Lilley, and Eleanor Jameson. 2021. "The Human Gut Phageome: Origins and Roles in the Human Gut Microbiome." *Frontiers in Cellular and Infection Microbiology* 11 (June). Frontiers Media S.A.: 498. doi:10.3389/FCIMB.2021.643214/BIBTEX.

Turnbaugh, Peter J., Ruth E. Ley, Micah Hamady, Claire M. Fraser-Liggett, Rob Knight, and Jeffrey I. Gordon. 2007. "The Human Microbiome Project." *Nature* 449 (7164). Nature Publishing Group: 804–10. doi:10.1038/nature06244.

Turner, Dann, Andrew M. Kropinski, and Evelien M. Adriaenssens. 2021. "A Roadmap for Genome-Based Phage Taxonomy." *Viruses* 13 (3). Multidisciplinary Digital Publishing Institute (MDPI). doi:10.3390/V13030506.

Ursell, Luke K., Henry J. Haiser, Will Van Treuren, Neha Garg, Lavanya Reddivari, Jairam Vanamala, Pieter C. Dorrestein, Peter J. Turnbaugh, and Rob Knight. 2014. "The Intestinal Metabolome: An Intersection between Microbiota and Host." *Gastroenterology* 146 (6). Gastroenterology: 1470–76. doi:10.1053/J.GASTRO.2014.03.001.

Vahedi, Azadeh, Mohammad Mehdi Soltan Dallal, Masoumeh Douraghi, Farhad Nikkhahi, Zahra Rajabi, Maryam Yousefi, and Maryam Mousavi. 2018. "Isolation and Identification of Specific Bacteriophage against Enteropathogenic Escherichia Coli (EPEC) and in Vitro and in Vivo Characterization of Bacteriophage." *FEMS Microbiology Letters* 365 (16). FEMS Microbiol Lett. doi:10.1093/FEMSLE/FNY136.

Valdez, Yanet, Eric M. Brown, and B. Brett Finlay. 2014. "Influence of the Microbiota on Vaccine Effectiveness." *Trends in Immunology* 35 (11). Trends Immunol: 526–37. doi:10.1016/J.IT.2014.07.003.

Vijay, Amrita, and Ana M. Valdes. 2021. "Role of the Gut Microbiome in Chronic Diseases: A Narrative Review." *European Journal of Clinical Nutrition* 76 (4). Nature Publishing Group: 489–501. doi:10.1038/s41430-021-00991-6.

Virgin, Herbert W. 2014. "The Virome in Mammalian Physiology and Disease." *Cell* 157 (1). Cell: 142–50. doi:10.1016/J.CELL.2014.02.032.

Waldor, Matthew K., Gene Tyson, Elhanan Borenstein, Howard Ochman, Andrew Moeller, B. Brett Finlay, Heidi H. Kong, et al. 2015. "Where Next for Microbiome Research?" *PLoS Biology* 13 (1). Public Library of Science: e1002050. doi:10.1371/JOURNAL.PBIO.1002050.

Waller, Karen M.J., Rupert W. Leong, and Sudarshan Paramsothy. 2022. "An Update on Fecal Microbiota Transplantation for the Treatment of Gastrointestinal Diseases." *Journal of Gastroenterology and Hepatology* 37 (2). J Gastroenterol Hepatol: 246–55. doi:10.1111/JGH.15731.

Wang, Li, Limeng Zhu, and Song Qin. 2019. "Gut Microbiota Modulation on Intestinal Mucosal Adaptive Immunity." *Journal of Immunology Research* 2019. J Immunol Res. doi:10.1155/2019/4735040.

Wróblewska, Barbara. 2006. "Food Allergens." In *Chemical and Functional Properties of Food Components*, Third Edition, January. CRC Press, 275–94. doi:10.4049/jimmunol.20.4.313.

Yang, Ye, and Christian Jobin. 2017. "Novel Insights into Microbiome in Colitis and Colorectal Cancer." *Current Opinion in Gastroenterology* 33 (6). Curr Opin Gastroenterol: 422–27. doi:10.1097/MOG.0000000000000399.

Yarza, Pablo, Pelin Yilmaz, Elmar Pruesse, Frank Oliver Glöckner, Wolfgang Ludwig, Karl Heinz Schleifer, William B. Whitman, Jean Euzéby, Rudolf Amann, and Ramon Rosselló-Móra. 2014. "Uniting the Classification of Cultured and Uncultured Bacteria and Archaea Using 16S RRNA Gene Sequences." *Nature Reviews. Microbiology* 12 (9). Nat Rev Microbiol: 635–45. doi:10.1038/NRMICRO3330.

Yehl, Kevin, Sébastien Lemire, Andrew C. Yang, Hiroki Ando, Mark Mimee, Marcelo Der Torossian Torres, Cesar de la Fuente-Nunez, and Timothy K. Lu. 2019. "Engineering Phage Host-Range and Suppressing Bacterial Resistance through Phage Tail Fiber Mutagenesis." *Cell* 179 (2). Cell: 459–69.e9. doi:10.1016/J.CELL.2019.09.015.

Yosef, Ido, Moran G. Goren, Rea Globus, Shahar Molshanski-Mor, and Udi Qimron. 2017. "Extending the Host Range of Bacteriophage Particles for DNA Transduction." *Molecular Cell* 66 (5). Mol Cell: 721–28.e3. doi:10.1016/J.MOLCEL.2017.04.025.

Yosef, Ido, Miriam Manor, Ruth Kiro, and Udi Qimron. 2015. "Temperate and Lytic Bacteriophages Programmed to Sensitize and Kill Antibiotic-Resistant Bacteria." *Proceedings of the National Academy of Sciences of the United States of America* 112 (23). Proc Natl Acad Sci U S A: 7267–72. doi:10.1073/PNAS.1500107112.

Zheng, Di Wei, Xue Dong, Pei Pan, Ke Wei Chen, Jin Xuan Fan, Si Xue Cheng, and Xian Zheng Zhang. 2019. "Phage-Guided Modulation of the Gut Microbiota of Mouse Models of Colorectal Cancer Augments Their Responses to Chemotherapy." *Nature Biomedical Engineering* 3 (9). Nat Biomed Eng: 717–28. doi:10.1038/S41551-019-0423-2.

Zmora, Niv, Eliran Soffer, and Eran Elinav. 2019. "Transforming Medicine with the Microbiome." *Science Translational Medicine* 11 (477). Sci Transl Med. doi:10.1126/SCITRANSLMED.AAW1815.

Zuppi, Michele, Heather L. Hendrickson, Justin M. O'Sullivan, and Tommi Vatanen. 2022. "Phages in the Gut Ecosystem." *Frontiers in Cellular and Infection Microbiology* 11 (January). Frontiers Media S.A.: 1348. doi:10.3389/FCIMB.2021.822562/BIBTEX.

10 Microbiome Engineering to Improve Biocontrol and Plant Growth-Promoting Mechanisms

Eskalin Shiny, Jissin Mathew, Anu Jacob and HS Rajashree
Karunya Institute of Technology and Sciences
Coimbatore, Tamil Nadu, India

10.1 INTRODUCTION

There are two distinct ways to define the term "microbiome" (1); for instance, some define the microbiota as the set of microorganisms of a specific environment, while they refer to the microbiome as the set of genomes of the microorganisms in a certain habitat. As a result, they use the term "microbiota" to refer to microbial populations. Others use the term "microbiome" to refer to any group of microorganisms within a community, and specifically to the microbes associated with a plant that can survive, grow, and interact with various plant tissues, including roots, shoots, leaves, flowers, and seeds (2–7). Based on this final definition, we are referring to the plant microbiome here. Numerous diverse types of organisms, such as viruses, bacteria, fungi, protozoa, and archaea, make up the microbiome (5). Plant microbiomes can help defend plants from possible infections while also enhancing growth, health, and productivity and giving plants an adaptive edge (3, 8). Specifically, the beneficial interactions that the bacterial microbiome conducts are crucial for sustainable agriculture and have garnered greater attention than other groups of organisms. The development and manipulation of the bacterial microbiome are important. According to Adesemoye and Kloepper (9), bioinoculants allow scientists to influence plant-beneficial activities such as growth limitations, phytopathogen suppression, and boosting plant growth and health, potentially eliminating the need for chemical fertilizers. Plant growth-promoting bacteria (PGPB) are rhizosphere or endophytic bacteria that stimulate plant growth (9, 10).

The microbiome is made up of several different types of organisms, including bacteria, fungi, protozoa, archaea, and viruses. The plant microbiome can play a beneficial role in protecting plants from potential pathogens while improving growth, health, and yield, and conferring adaptive advantages to plants. Bacterial microbiomes that engage in beneficial interactions are important for sustainable agriculture and have attracted more attention than other biota. Manipulation of the bacterial microbiome and production of biological inoculants allow scientists to influence beneficial plant activities, such as growth restriction, action of plant pathogens, and promotion of plant growth and health, potentially reducing the use of chemical fertilizers. PGPB are rhizosphere or endophyte bacteria that promote plant growth. PGPB can promote plant growth by direct or indirect mechanisms. PGPB-facilitated nutrient acquisition typically includes elements such as nitrogen, phosphorus, and iron that are essential for plant development. The regulation of hormone levels may involve the involvement of PGPB in the synthesis of one or more plant hormones, including auxins, cytokinins and gibberellins. Mechanisms include antibiotics (e.g., 2,4-diacetylphloroglucinol), proteases, chitinases, bacteriocins, siderophores, lipopeptides (e.g., iturin A, bacitracin D, and mycocins), and the synthesis volatile organic compounds (11). Although the plant microbiome consists of many

DOI: 10.1201/9781003394662-10

organisms with related roles, this review focuses on bacterial species, especially those with biological control and plant growth promotion activities.

10.2 PLANT MICROBIOME

The "plant microbiome" consists of diverse microbial communities on the external surfaces and internal tissues of the host plant. The rhizosphere and phyllosphere constitute the most important parts of the plant microbiome. The soil microbiome is an important source of the plant microbiome. Plants harbor distinct microbial communities that are specific to each plant organ, such as the phyllosphere, rhizosphere, and endosphere. The rhizosphere is the most studied habitat due to its enormous potential for plant nutrition and health. It has been known for many years that the rhizosphere is enriched with specific microbial species/genotypes relative to the soil and internal tissues, but modern techniques provide deeper insights and expand our understanding of plant-microbe interactions (8). Current models show the emergence of dispersed microbes in seeds and the attraction of microbes to nutrients such as carbohydrates and amino acids, as well as plant-specific secondary metabolites. Plant root exudates play an important role as chemoattractants and repellents. In addition, plant defense signals also play a role in this process (12).

All plant organs are colonized by microorganisms. The composition of the plant microbiome is influenced by a variety of factors, including plant age or stage of development, plant species or cultivar, and plant health. In addition, many abiotic factors regulate the structural and functional diversity of plant-associated microbiomes, including soil properties, nutritional status, and climatic conditions. Microbial colonization of plants is not random; it is a deliberate process, underlined by specific co-occurrence patterns and the presence of microbial networks. These networks are associated with patterns of colonization resistance, which determine the likelihood of alien microbes invading native communities. Also, structural diversity is extremely important to prevent pathogen invasion/epidemic (13).

10.3 MICROBIOME AS RHIZOSPHERE, ENDOSPHERE, OR PHYLLOSPHERE

The plant microbiome comprises the regions or areas in which microorganisms can interact with plants, whether they live outside (rhizosphere), inside (endosphere), or on plants (phyllosphere). The rhizosphere is the part of the soil that is affected by plant roots. This microecosystem is the most important area where chemical communication and the exchange of compounds and nutrients take place between plants and soil microbes. Strategies to improve plant health and development include the selection and engineering of the rhizosphere microbiome. The endophyte microbiome may be a subset of the rhizosphere microbiome. The endophyte is the internal environment of the plant where endophytes, including bacteria, archaea, fungi, and viruses inhabit, colonize, and survive without causing harm to the plant.

The rhizosphere is an important zone for the penetration of micro-organisms into the plant because it is rich in nutrients, brought by root exudates, and therefore has a high concentration of micro-organisms. There are other accessible areas such as lenticels, stomata, wounds, fractures, and nodules. Endophytes can also be inherited vertically through seeds. A plant's internal microbiome has an advantage over external bacteria (phyllosphere or rhizosphere) because it is unaffected by soil conditions, including the presence of bacterial predators (10). Additionally, it has been suggested that chemical communication is more efficient in plant tissues since metabolites secreted by bacterial endophytes at low concentrations have a greater effect on plant secretions than metabolites in open environments such as the rhizosphere, where biotic and abiotic factors can counteract and minimize their effects.

The composition of the rhizosphere is more diverse than that of the endosphere. Rhizosphere and endogenous 16S rRNA genes of the Mexican husk tomato plant (*Physalis ixocarpa*), more operational taxonomic units (OTUs) were found in the rhizosphere clone library compared to the clones from the endosphere of the root. It should be noted that the number of clones randomly obtained from the rhizosphere versus the endosphere was statistically similar. Most of the genera found in the rhizosphere, such as *Cyclophaga*, *Burkholderia*, *Bacillus*, and *Pseudomonas*, were also detected in the internal tissues of the plant. This is consistent with the diversity of OTUs found when culturable bacterial endophytes were isolated from different fields of cape gooseberries and characterized at the molecular level. These results suggest evidence for a core microbiome that has beneficial functions in Physalis ixocarpa plants and is passed down from generation to generation (10).

The phyllosphere consists of the surface of the leaf tissue and the apoplast. The importance of the phyllosphere microbiome for biocontrol and plant growth promotion has been recognized for many years. Indeed, some authors suggest that the leaf microbiota exerts beneficial activities comparable to those observed for the rhizosphere and endophyte microbes. For example, many leaf microbiota can fix nitrogen, which is the main mechanism for supplying nitrogen to plants growing in humid tropical ecosystems (11). Other beneficial effects of the microbiota living on the surface of plants include indirect protection against pathogens and the production of phytohormones. Therefore, leaf microbiota have the potential to be used to reduce the use of agrochemicals to control leaf pathogens and promote plant growth.

10.4 FUNCTIONS AND ECOLOGY OF PLANT MICROBIOME

Many functions of the plant microbiome are essential to the host. To start the germination process, the first stage in the plant's life cycle, many plants cannot start life without microbes. Mosses, one of the Earth's oldest land plants, and orchids, whose very small seeds require specific fungi (usually Rhizoctonia), illustrate this necessity. The germ-promoting fungus Rhizoctonia consists of both beneficial organisms and pathogens. To avoid pathogenic interaction after germination, host plants fully consume their helper fungi (14). In these cases of germination support, microbes are essential, and this may be one of the reasons for the vertical transfer of these key microbes. Positive effects on the germination of cosmopolitan plant-associated bacteria such as *Stenotrophomonas* have also been found.

It is well known that plant-associated microorganisms stimulate plant growth and cultivate plants, such as rhizobia and mycorrhizal fungi. The mechanisms by which these microbes support plant growth include the production of phytohormones, the binding of nitrogen, and the mobilization of minerals such as phosphorus and iron. The plant microbiome is also involved in suppressing pathogens; in particular the root microbiome acts as a protective shield against soil pathogens. These mechanisms are well studied and include various direct interactions with plant pathogens as well as indirect interactions through plants via the stimulation of the immune system. The microbiome is not only involved in the response to biotic stress but also in the defense against abiotic stress. For example, the plant microbiota has been shown to be involved in protection against high salinity and drought (3).

The plant microbiota also influences the composition of plant secondary metabolites and the consequent development of different metabolites. It has been shown to affect strawberry flavor and the production of bioactive compounds in herbal medicines. In *Arabidopsis*, the rhizosphere microbiome may be associated with insect feeding behavior, which is likely the result of microbiome-induced changes in the leaf metabolome. Additionally, the removal of the floral microbiome has been shown to result in a reduced release of floral terpenes, which play a key role in pollination and subsequently in fruit and seed production. Co-evolution of plants and associated microbial communities has been hypothesized based on culture-dependent results of rhizosphere acquisition of ancient and modern wheat cultivars (3, 15). By applying deep sequencing technology, co-evolution has recently been shown to be common in other plants, such as corn, sugar beet, and lettuce. Crop breeding has been identified as a powerful driver of natural evolution. In some cases, breeding strategies are directed against pathogens, but historically this was mostly a process of random selection of plant phenotypes.

10.5 MODIFICATION OF PLANT MICROBIOME

PGPB have the potential to limit the growth of plant pathogens and promote plant growth. All plants are known to contain endophytes in their internal microbiome. The plant microbiome is made up of living organisms that alter plant physiology and development, induce pathogen resistance systems, and tolerance mechanisms to various stresses such as drought, salinity, or soil pollution. However, these functions are not performed by the "whole microbiome" but by one or a few microbial species due to additive or synergistic effects between two or more strains. Thus, these few bacterial species can be used to boost or activate the defense systems of other plant species sensitive to certain pathogens or abiotic factors such as salinity, among other uses such as soil bioremediation or the increase of fruit production. Different plants may have different tolerances to biotic or abiotic stresses, grow at different latitudes or altitudes, or in nutrient-poor soils, and in many cases, their microbiomes allow them to survive such conditions. Therefore, there is a general interest in transferring these capacities to plants that lack them, including for the biological control of potential pathogens. By transferring one or more microbial species, primarily bacteria, larger amounts of genetic material or numbers of genes can be transferred to simultaneously improve multiple functions compared to single gene transfer. For example, some PGPB, such as *Pseudomonas fluorescens* UM270 or *Arthrobacter agilis* UMCV2, exert various direct and indirect growth-promoting activities. These PGPB can exert various beneficial activities from the rhizosphere or endosphere (16).

10.6 TRADITIONAL MICROBIOME ENGINEERING

Soil amendments involve the addition of organic and inorganic additives to soil or agricultural practices that farmers use to optimize productivity and influence and manipulate plant microbiome interactions. Inorganic amendments include lime, vermiculite, perlite, and sand, which have been used to reduce the effects of soil acidity and salinity on crops. Even in nitrogen-poor soils, they lead to increased plant biomass due to increased root exudates and a larger, more active microbiome. Inorganic amendments combined with organic amendments increased yields by 30%, compared to only 8% when organic fertilizers were used alone. Organic improvement involves the application of various additives to the soil as an energy source for plant-beneficial bacteria, fungi, and heterotrophic invertebrates. They have the dual benefit of replenishing the fertility of depleted soils and reducing agricultural waste. Long-term application of conventional organic fertilizers leads to an increase in biomass and microbial diversity, the composition of which depends on the type of amendment. For example, manure has been used for centuries because it is rich in N, P, and organic matter, which can alter the microbiome and the physical and chemical properties of the soil, thereby improving agricultural yields. The addition of sheep manure has reduced lead, cadmium, and zinc levels in alfalfa plants, which improved microbial growth and reduced metal toxicity, resulting in significantly increased crop yields. Bones have been used as a renewable source of phosphorus since the mid-19th century. Soil microorganisms such as Bacillus megaterium var., are well known. Phosphacticum and *Bacillus mucilaginous* can dissolve P present in bone, making it available for uptake by plants by releasing organic acids (17).

The large amount of organic carbon and other nutrient-rich waste in crop residues can be composted or vermicomposted to produce bio-stimulants that improve soil aggregation. Vermicompost end products are environmentally friendly and nutrient-rich and can be used as soil conditioners to promote plant growth. These bio-stimulants are expected to increase nutrient mineralization based on the change in soil microbiome composition and biomass. Composted soil amendments are a popular alternative to chemical fertilizers and also reduce agricultural waste. Biocomposting is a microbial oxidation process that uses a variety of recycled organic materials and is often used in organic farming to protect plants and promote growth. Composts and agricultural wastes have been mentioned in many studies as rich sources of soil microbial communities for biological control, and green composts are also management tools for the control of soil-borne plant pathogens.

Mechanisms and processes similar to those that naturally inhibit soil have also been exploited in greenhouse horticultural soil-less cultivation systems. In the absence of soil fumigants and fungicides, composting is considered a reliable option by farmers who do not have many options for disease control. Anaerobic soil pollution (ASD) with wheat bran or ethanol soil amendments as a carbon source not only reduced the severity of tomato root rot caused by *Pyrenochaeta lycopersici* but also resulted in alterations in the soil microbiome, increasing the presence of abundance of Firmicutes (17–20).

Agricultural practices such as crop rotation, tillage, and intercropping also have a strong impact on the regulation and variation of the soil microbiome. For example, when the root and soil microbial communities of wheat were examined in cropping systems using conventional and organic soil amendments, a change of almost 10% in the root and soil bacterial communities of wheat was observed, and the diversity of microbial populations was strongly influenced by cultivation.

10.7 LESS TRADITIONAL ORGANIC ADDITIONS

The scaling law describes that the top layers of soil harbor more than 1 trillion species of microorganisms, including fungi, lichens, bacteria, actinomycetes, microalgae, and cyanobacteria, all of which are closely related to each other. This biological soil crust (BSC) is believed to play a major role in improving agricultural yields and soil fertility. Many microalgae, including nitrogen-fixing cyanobacteria, contribute to the breakdown of organic waste, the biological control of plant pathogens, the detoxification of harmful chemicals, the recycling of nutrients, and the production of metabolites such as enzymes, hormones, and vitamins that affect the nutrient composition of soil and plants. The partial or complete pyrolysis of organic materials like manure, bones, agricultural leftovers, and abattoir wastes produces biochar, a C-rich substance that typically improves soil nutrient availability and water holding capacity. In addition to fostering soil bacteria, biochar increases crop output. *Phytophthora cactorum*, *Fusarium oxysporum*, *Podosphaera aphanis*, *Pythium aphanidermatum*, *Rhizoctonia solani*, the nematode *Pratylenchus penetrans*, and the parasitic weed *Phelipanche aegyptiaca* are just a few of the pathogens that can cause disease but can be suppressed by biochar (18). Microbial biofertilizers and biopesticides, such as *Rhizobium spp.*, are examples of microbial soil inoculants. These microbes have advantageous properties for increasing plant productivity and defending plants from infections.

As N is the most important nutrient, products containing N-fixing bacteria dominate the global biofertilizer industry. *Actinorhizobium* spp., *Azospirillum* spp., *Azotobacter* spp., and *Rhizobium* spp. are currently the most widely employed N-fixing biofertilizer microorganisms. The majority are utilized for legumes, but some are also useful for other crops, including sugarcane and rice, as free-living N-fixing bacteria. Rhizobia trigger a physiological shift in root hairs, resulting in curling and bulging to produce nodules that rhizobacteria can inhabit once activated by certain flavonoids found in legume root exudates. Inorganic P can also be solubilized by a variety of soil bacteria, making it available for plant uptake (5, 17). The composition and function of microbial communities have become better understood thanks to developments in genome sequencing, metagenomics, 16S or 18S rDNA, internal transcribed spacers (ITS) amplicon community sequencing, qPCR, and non-PCR-based techniques including T-RFLP, FISH, and DNA arrays.

10.8 TARGETED PLANT MICROBIOME ENGINEERING

Through aiding in biotic stress resistance, nutrient uptake, and abiotic stress tolerance, the root microbiome serves as the key determinant for the growth and development of a plant. Manipulating this microbiome has the potential to further improve these characteristics. In the rhizosphere, many microorganisms can collaborate with one another to perform certain tasks. For instance, in a model grassland, PGPR and arbuscular mycorrhizal (AM) fungus worked well together in terms of several limiting nutrients and ecosystem services. By modifying the soil, including adding conventional soil

TABLE 10.1
Examples of Benefits Gained from Microbial Inoculations of Plants

Plant	Beneficial microbe	Benefits to plants
Tomato seeds	*Azospirillum brasilense*	The bacteria increased both shoot and root length
Cucumber and tomato	Bacillus strain QST713	Increased P content by 40% and thereby overall plant growth
Common bean	*Pseudomonas putida* UW4 + Rhizobium PGPR	Significantly enhanced nodule formation in plants under P stress and supplemented growth promotion by Rhizobium
Tomato	*Burkholderia unamae*	Plants inoculated with wild *B. unamae* strains showed better growth than *B. unamae* mutant strains without ACC deaminase activity
Rice	*Azospririllum amazonense*	Significantly improved grain yield in rice by increasing panicle number and N content at maturity

amendments, or by adding or activating signal molecules or substrates (such as root exudates), it is possible to induce or draw the appropriate microbiota into the rhizosphere. Important microbial strains can also be injected into the soil to alter the composition of microbial communities. For instance, a consortium inoculation of the plant-beneficial bacterium *Pseudomonas libanensis* TR1 and the AM fungus *Claroideoglomus claroideum* BEG210 into the rhizosphere of sunflowers improved plant growth under metal and salinity stress alone or in combination. The goal was to harness the potential of plant growth-promoting microbiota to elevate stress tolerance. *Bacillus amyloliquefaciens*, *Bacillus velezensis*, and *Acinetobacter* sp. UQ202 were applied to the roots of growing chili plants to improve plant growth and, consequently, the plants' resistance to soil-borne *Phytophthora capsici* infection. Flavonoids, antioxidants, and P were all raised in the harvested pods when *Agrobacterium* sp. 10C2 was used on the bean plant. After 15 days of inoculation, numerous PGPR species, including *Actinomycetes* spp. and *Brevibacterium* spp., began to colonize the soil near the plant (6).

The seed microbiome can be passed down across generations and has been demonstrated to enhance germination, plant survival, and performance. The growth of tomato seedlings was accelerated by coinoculating the rhizosphere with *Pseudomonas stutzeri* and *Stenotrophomonas maltophilia*, which also released diffusible chemicals (such as dimethyl disulfide) that are effective against the leaf disease Botrytis cinerea. The bacteria that were included as endophytes may multiply and populate the following generation. Roots serve as crucial gatekeepers to choose beneficial bacteria that may enter the plant as endophytes, much like how root exudates enrich for them in the rhizosphere. Although bacteria are also released by plants into the rhizosphere, further research is needed to determine how microorganisms (whether pathogenic or not) enter and leave the plant holobiont (6, 19). Moreover, roots can directly absorb related microorganisms and use them as a source of nitrogen. Crops with greater resilience and ability to withstand pest/pathogen attacks are the goal of plant breeding. Yet, because the corresponding pathways are frequently antagonistic, plants are typically unable to defend themselves against both biotic and abiotic stress at the same time.

REFERENCES

1. Bulgarelli, Davide, Klaus Schlaeppi, Stijn Spaepen, Emiel Ver Loren Van Themaat, and Paul Schulze-Lefert. "Structure and functions of the bacterial microbiota of plants." *Annual Review of Plant Biology* 64 (2013): 807–838.
2. Haney, Cara H., and Frederick M. Ausubel. "Plant microbiome blueprints." *Science* 349, no. 6250 (2015): 788–789.

3. Haney, Cara H., Buck S. Samuel, Jenifer Bush, and Frederick M. Ausubel. "Associations with rhizosphere bacteria can confer an adaptive advantage to plants." *Nature Plants* 1, no. 6 (2015): 1–9.
4. Mendes, Rodrigo, Paolina Garbeva, and Jos M. Raaijmakers. "The rhizosphere microbiome: significance of plant beneficial, plant pathogenic, and human pathogenic microorganisms." *FEMS Microbiology Reviews* 37, no. 5 (2013): 634–663.
5. Mueller, Ulrich G., and Joel L. Sachs. "Engineering microbiomes to improve plant and animal health." *Trends in Microbiology* 23, no. 10 (2015): 606–617.
6. Nelson, Eric B. "The seed microbiome: origins, interactions, and impacts." *Plant and Soil* 422 (2018): 7–34.
7. Turner, Thomas R., Euan K. James, and Philip S. Poole. "The plant microbiome." *Genome Biology* 14, no. 6 (2013): 1–10.
8. Berg, Gabriele, Daria Rybakova, Martin Grube, and Martina Köberl. "The plant microbiome explored: implications for experimental botany." *Journal of Experimental Botany* 67, no. 4 (2016): 995–1002.
9. Adesemoye, Anthony O., and Joseph W. Kloepper. "Plant–microbes interactions in enhanced fertilizer-use efficiency." *Applied Microbiology and Biotechnology* 85 (2009): 1–12.
10. Santoyo, Gustavo, Gabriel Moreno-Hagelsieb, Ma del Carmen Orozco-Mosqueda, and Bernard R. Glick. "Plant growth-promoting bacterial endophytes." *Microbiological Research* 183 (2016): 92–99.
11. del Carmen Orozco-Mosqueda, Ma, Ma del Carmen Rocha-Granados, Bernard R. Glick, and Gustavo Santoyo. "Microbiome engineering to improve biocontrol and plant growth-promoting mechanisms." *Microbiological Research* 208 (2018): 25–31.
12. Badri, Dayakar V., and Jorge M. Vivanco. "Regulation and function of root exudates." *Plant, Cell & Environment* 32, no. 6 (2009): 666–681.
13. Eiler, Alexander, Friederike Heinrich, and Stefan Bertilsson. "Coherent dynamics and association networks among lake bacterioplankton taxa." *The ISME Journal* 6, no. 2 (2012): 330–342.
14. Pierson III, Leland S., and Elizabeth A. Pierson. "Roles of diffusible signals in communication among plant-associated bacteria." *Phytopathology* 97, no. 2 (2007): 227–232.
15. Gaiero, Jonathan R., Crystal A. McCall, Karen A. Thompson, Nicola J. Day, Anna S. Best, and Kari E. Dunfield. "Inside the root microbiome: bacterial root endophytes and plant growth promotion." *American Journal of Botany* 100, no. 9 (2013): 1738–1750.
16. Tan, Chengjia, Mohammad Talib Kalhoro, Yahya Faqir, Jiahua Ma, Matthew Duah Osei, and Ghulam Khaliq. "Climate-resilient microbial biotechnology: A perspective on sustainable agriculture." *Sustainability* 14, no. 9 (2022): 5574.
17. Berendsen, Roeland L., Corné MJ Pieterse, and Peter AHM Bakker. "The rhizosphere microbiome and plant health." *Trends in Plant Science* 17, no. 8 (2012): 478–486.
18. Compant, Stéphane, Abdul Samad, Hanna Faist, and Angela Sessitsch. "A review on the plant microbiome: Ecology, functions, and emerging trends in microbial application." *Journal of Advanced Research* 19 (2019): 29–37.
19. Arif, Inessa, Maria Batool, and Peer M. Schenk. "Plant microbiome engineering: expected benefits for improved crop growth and resilience." *Trends in Biotechnology* 38, no. 12 (2020): 1385–1396.
20. Rasool, Mujahid, Adnan Akhter, Gerhard Soja, and Muhammad Saleem Haider. "Role of biochar, compost and plant growth promoting rhizobacteria in the management of tomato early blight disease." *Scientific Reports* 11, no. 1 (2021): 6092.

11 Plant Microbiome Engineering
A Promising Approach to Safeguard Plant Health

Sudipta Sankar Bora
Assam Agricultural University
Assam, India

Rahul Gogoi
Assam Agricultural University
Assam, India

Madhurjya Ranjan Sharma
Assam Agricultural University
Assam, India

Anshu
All India Institute of Medical Sciences
New Delhi, India
anshu08021998@gmail.com

Romen Singh Naorem
Jorhat Medical College and Hospital
Assam, India

Jugabrata Das
Assam Agricultural University
Assam, India

11.1 INTRODUCTION

In agroecosystems, soil microbiomes play a major role in controlling soil fertility and crop productivity. Agricultural management practices have the potential to cause structural changes in the soil, which in turn can modify native microbial processes. Broader ramifications of these alterations include increased greenhouse gas emissions, diminished soil fertility, and crop health [1]. Sustainable agricultural practices aim to increase microbial biodiversity and improve soil structure. However, how agricultural practices affect these soil microbial and biological processes in diverse environments is yet to be fully understood [2]. To develop climate-smart, stress-resilient, and sustainable agroecosystems, such farm-level decisions must consider research findings that bridge the

 DOI: 10.1201/9781003394662-11

microscale and field scale. It is also necessary to use natural solutions for sustainable agriculture, such as leveraging the soil microbiome [3].

The application of plant-associated microorganisms has immense potential for economically and sustainably solving today's agricultural problems. Plant-associated microbiota can be epiphytic, endophytic, and rhizospheric. Among these, rhizosphere microorganisms exhibiting PGP traits and with root colonization ability remain suitable candidates for biofertilizer development [4]. Plants secrete a portion of their photosynthates to the immediate root-soil environment, i.e., rhizosphere. Rhizo-exudates typically contain primary (simple sugars, amino acids, nucleotides, etc.) and secondary (organic acids, fatty acids, vitamins, sterols, etc.) metabolites that act as carbon sources for various metabolically active rhizobacterial populations [5]. Plants can regulate the rhizo-exudation process as well as modulate the exudate composition to encourage the growth of specific beneficial microorganisms for their advantage. Plant genetic diversity and age, root architecture (types: seminal/nodal; position: apex/base), soil types and edaphic factors, and climatic conditions (season, temperature, precipitation, etc.) also determine rhizosphere microbiome structure [6].

Most of the rhizosphere microbial population supports plant growth and development through nutrient provisioning. These diverse microbial communities execute various functions such as nitrogen fixation (asymbiotic and symbiotic), mineral solubilization (Zn, P, K), and production and regulation of phytohormones (indole acetic acid, auxin, gibberellins, abscisic acid, aminocyclopropane-1-carboxylate deaminase). [7]. Plant growth-promoting rhizobacteria (PGPR) candidates not only improve plant nutrient uptake but also protect the host plants from a variety of biotic and abiotic stress conditions. By diminishing competition among the pathogen populations, the PGPR population lowers the risk of pathogen spread and colonization of plant tissues [8]. However, the success rate of introducing PGP microorganisms to industries for commercial usage has been low. The more resilient current microbial communities, whose makeup has been formed over time by intricate multilateral interactions with the environment, are probably the reason behind the exclusion of the new bacteria [9].

One possible strategy for protecting and promoting plant growth is the careful modification of the soil microbiome. There is a growing interest in using microbiome engineering to address issues related to agriculture, health, and climate change [10]. Microbiome engineering entails the application of traditional (soil amendments and cultivation practices) and less traditional techniques (introducing PGPR) to the latest cutting-edge engineering technologies, including synthetic rhizosphere microbiome (inoculating the host plant with genetically modified microbes) and *in situ* microbiome modifications (manipulating the native microbial community) to improve soil health and crop productivity [11]. Sustained investigation into this topic is required to elucidate the intricate plant-microbe and microbe-microbe interactions, allowing for the manipulation of these relationships for sustainable agriculture. This chapter seeks to illustrate the many methods of microbiome engineering and the advantageous functions of the plant-associated microbiome that may be altered and maximized to improve agricultural productivity.

11.2 TRADITIONAL MICROBIOME ENGINEERING: SOIL STRUCTURE MODIFICATION

It is possible to increase plant health and productivity by modifying the soil structure in a way that fosters the growth of beneficial microbes. Adding organic inputs, such as compost, manure, biochar, or plant leftovers, can have positive effects on soil structure by boosting soil porosity, water-holding ability, and nutrient availability. Additionally, organic matter gives soil bacteria a food source, which can increase the diversity and activity of these organisms. The effects of organic amendments on the makeup, distribution, and organization of the microbial community have been extensively studied in several studies [12]. When compared to conventional mineral fertilizer-based management approaches, the frequent addition of organic amendments, such as alfalfa hay and charcoal, greatly enhanced soil fertility, microbiome structure, and crop yield in Arugula (*Eruca sativa* Mill.) [13].

Various farming techniques such as crop rotation, cover crops, tillage, intercropping, and multi-cropping may also promote endophytic and rhizospheric microbial populations. Crop rotation can disrupt pest and disease cycles by suppressing phytopathogens and pests in the soil [14]. Growing cover crops between cash crops is an additional strategy that improves soil moisture retention and soil organic matter [15]. Compacted soil is broken up through tillage, which increases soil aeration and water infiltration. However, excessive tillage can also damage native microbial populations that are vital to soil health. Tillage should therefore be used sparingly and in conjunction with other methods, such as the addition of organic matter [16].

11.3 LESS-TRADITIONAL MICROBIOME ENGINEERING: APPLICATION OF PGPR

PGPR strains can promote plant growth and development either directly or indirectly. Direct mechanisms are defined as employing those bacterial traits that result in the direct promotion of plant growth. They include biological nitrogen fixation, production, and regulation of phytohormones (auxins, ACC deaminase, cytokinins, and gibberellins), mineral solubilization (phosphate, zinc), siderophore and lytic enzyme production, as well as induction of systemic resistance [17].

Indirect mechanisms include antibiotics, cell wall-degrading enzymes, competition, hydrogen cyanide, induced systemic resistance, quorum quenching, and siderophores [18]. PGPR strains, especially *Pseudomonas fluorescens* and *Bacillus subtilis*, are best recorded as the most promising candidates for indirect stimulation [19]. Apart from nitrogen transformation, increasing the bioavailability of phosphate, iron acquisition, the exhibition of specific enzymatic activity, and plant protection from harmful pathogens with the production of antibiotics can also successfully improve the quality of crops in agriculture [17]. In addition to the aforementioned methods used by PGPR to control phytopathogens, biocontrol of some bacterial phytopathogens may be obtained through the selective use of bacteriophages [18]. Bacteriophages have been used to control phytopathogens such as *Erwinia amylovora* [20] and *Ralstonia solanacearum* [21]. Commercially available Agriphage is used for the control of *Pseudomonas syringae* pv. tomato and *Xanthomonas campestris* pv. Vesicatoria [22].

11.3.1 Direct Mechanisms

11.3.1.1 Biological Nitrogen Fixation

The application of biofertilizer produced from *Bradyrhizobium* and *Streptomyces griseoflavus* improved the growth, nutrient acquisition, and yield in crops like mung bean, soybean, and cowpea [23]. The function of PGPR also extends to enhance the uptake of nitrate (NO_3^-) either through lateral root development or NO_3^- transport systems [24]. A consortium of *Bacillus*, *Paenibacillus*, and *Lysinibacillus* strains was reported to upregulate five nitrate and four ammonium genes and also contributed to improving nutrient uptake, chlorophyll content, and plant biomass in *Arabidopsis thaliana* [25]. Co-inoculation of N_2-fixing *Rhizobium*, PGPR *Bacillus aryabhattai*, and *Azotobacter vinelandii* significantly increased the growth, nutrient uptake, nodulation, and nitrogen fixation in white clover (*Trifolium repens*) under P-deficient conditions [26]. An endophytic N_2-fixing bacterium, *Pseudomonas stutzeri* was reported to improve the growth and nitrogen accumulation of maize, thereby potentially reducing the application of synthetic fertilizers in practice [27].

11.3.1.2 Mineral Solubilization

Bacillus, *Pseudomonas*, and *Gluconoacetobacter* have been reported as zinc-solubilizing bacteria (ZSB) [28]. There are two types of plant growth-related mechanisms that ZSB inoculants can induce: "direct" and "indirect". By secreting organic acids and enzymes, ZSB can directly influence the phytohormone level in the plant and improve the uptake of critical micronutrients. The production of secondary metabolites (antifungal and antibacterial substances) that can inhibit phytopathogen infections is included in the indirect mechanism profile. ZSB can effectively dissolve zinc in soil and facilitate its transition from the soil environment into various plant tissues [29]. The ability of phosphate-solubilizing bacteria (PSB) to convert insoluble P forms into soluble forms increases the amount of P that is subsequently available to plants. Microbial isolates belonging to the genera *Agrobacterium*, *Bacillus*, *Pseudomonas*, etc. have been known as efficient P-solubilizers [30]. The inoculation of rock phosphate-solubilizing *Pseudomonas* sp. FA1 significantly increased soil P availability as well as root and shoot biomass of wheat (*Triticum aestivum* L.) [31]. Similarly, the inoculation of PSBs such as *Pantoea cypripedii* and *P. plecoglossicida* resulted in improved biomass P content and overall yield of maize (20%) and wheat (16%) [32].

11.3.1.3 Phytohormone Production

Indole-3-acetic acid (IAA), which has particularly potent impacts on root architecture and growth, is produced by several PGPRs. Evidence suggests that IAA-producing PGPR increase root length, root biomass, and surface area, decrease stomata size and density, and activate auxin-responsive genes, providing the plant with greater access to soil nutrients. These transcriptional alterations have been linked to hormone, defense, and cell wall-related genes [33]. The biochemistry of gibberellins in bacteria is similar to that of plants, with some small differences. Of 136 identified gibberellin structures (GA_1-GA_{136}), only four GAs, *viz.*, GA_1, GA_3, GA_4, and GA_{20}, have been identified in bacteria [34]. Gibberellins are known for growth stimulation and activation of important growth processes including stem elongation, seed germination, flowering, fruit ripening, photosynthesis rate, and chlorophyll content. Microbial production of GAs has been reported from various genera, *viz.*, *Achromobacter*, *Acinetobacter*, *Azospirillum*, *Azotobacter*, *Bacillus*, *Gluconobacter*, *Herbaspirillum*, and *Rhizobia* [18].

11.3.1.4 Phytohormone Regulation (ACC Deaminase Activity)

ACC deaminase-producing rhizobacteria can modulate plant growth and physiology by regulating phytohormone levels and promoting plant-microbe interactions. They also enhance stress tolerance in various crops, including wheat and rice, through improved photosynthesis, antioxidant capacity, and reduced oxidative stress under drought and salinity stress conditions [35]. Rhizobacteria containing ACC deaminase, such as *Variovorax paradoxus* 5C-2, promote nutrient uptake and growth in maize by reducing ethylene production in the root zone under localized nutrient supply in field conditions [36].

11.3.1.5 Siderophore Activity

Siderophores have been reported as a safer alternative to harmful pesticides [37]. Certain PGPR belonging to the genera Bradyrhizobium, Rhizobium, and Sinorhizobium with plant growth-promoting properties have also been shown to produce siderophores as a crucial mechanism for rhizosphere colonization [38]. A range of iron-chelating compounds produced by siderophore-producing microorganisms (SPMs) can lessen plant stress in soil that is iron-stressed. As a result, SPMs are a trademark for sustainable agriculture. These bacteria produce siderophores while simultaneously functioning as biofertilizers and biocontrol agents for the plants [39].

11.3.2 Indirect Mechanisms: Extracellular Polysaccharides (EPS) and Biofilm Formation

Biofilms are communities of microorganisms that adhere to a surface and are embedded in a matrix of extracellular polysaccharides (EPS) known as the extracellular matrix (ECM). Biofilms can help beneficial microorganisms colonize and compete with phytopathogens for limited resources. This can lead to a reduction in disease incidence and severity, as well as an increase in plant growth and yield. Arabidopsis plants inoculated with biofilm-forming *Xanthomonas*, *Stenotrophomonas*, and *Microbacterium* isolates had a significant increase in plant fresh weight along with inducing systemic resistance to downy mildew [40]. The inoculation of biofilm-forming and tea root inhabiting *Strenotrophomonas* sp. K96 significantly increased the vegetative parameters of *Camellia sinensis* [41]. In addition, biofilms can enhance plant growth-promoting mechanisms by improving nutrient uptake, inducing systemic resistance, and stimulating root growth [42]. By forming a protective layer around the roots, biofilms help to maintain a stable environment and protect the plant from environmental stresses such as drought, salinity, temperature fluctuations, and UV radiation. The EPS matrix surrounding the biofilms helps retain water and nutrients in the soil, creating a microenvironment that is conducive to plant growth.

11.4 TARGETED MICROBIOME ENGINEERING

Targeted plant microbiome engineering employs both top-down and bottom-up methods [10]. In the bottom-up approach, host plant-associated microbes are isolated from various environmental matrices. Following sophisticated genetic engineering manipulation of the necessary features, a microbial core is constituted as synthetic microbial communities. The synthetic microbial core is subsequently inoculated into the rhizosphere or injected into the plant endosphere. The top-down strategy uses horizontal gene transfer (HGT) to *in situ* introduce desired traits into the target host [43]. Mobile genetic elements (MGEs) are another top-down technique that allows the integration of foreign genes into a microbiome subpopulation [44]. Creating bacteriophage systems to engineer or eliminate specific species within populations presents another top-down tactic [45].

11.4.1 Top-Down Approach

In situ microbiome engineering is a novel synthetic biology technique. It allows the insertion and/or removal of specific functionalities from native microbial populations [46].

11.4.1.1 *In Situ* Microbiome Engineering

Brophy and colleagues developed a system of miniaturized integrative and conjugative elements (ICEs) based on the ICE from *B. subtilis*. It enabled heterologous DNA delivery into the genetic makeup of a variety of non-model Firmicutes [47]. A 10 kb-size nitrogen fixation biosynthetic gene cluster could be transferred to the soil microbial community using a *B. subtilis* donor system [47].

11.4.2 Bottom-Up Approach

Broad host-range bottom-up ME includes integrative and conjugative elements (ICEs), chassis-independent recombinase-assisted genome engineering (CRAGE), phage integrases, and others [48].

11.4.2.1 ICE System

ICEs are a broad class of MGEs that spread via host replication after being integrated into a host chromosome by an integrase [49]. In order to avoid disruption, some ICEs have developed a smooth method of integration into the 5′ end of the target gene [50]. It is possible to identify and add genes encoding conjugative machinery, transposases, and integrases linked to MGEs to domesticated mating partners like *Bacillus subtilis* and *E. coli* [51].

11.4.2.2 CRAGE System

CRAGE allows complex biological systems to be accurately and efficiently integrated into the chromosomes of non-model bacteria [52]. CRAGE has enabled genetic engineering in 25 distinct γ-Proteobacteria species by incorporating a uniform landing pad (LP) that facilitates the effective integration of massive BGCs. CRAGE was further expanded to include 40 microbial species.

11.4.2.3 Phage Integrase System

For *Pseudomonas putida* [53] and human commensal Bacteroides species [54], respectively, a phage Bxb1 system and intN2 integrases were employed as dependable integration tools. These systems need native attachment sites that are unique to each integrase. Another effective phage transposon-based genome engineering tool is mini-Tn7. Nevertheless, the mini-Tn7 system is limited to integrating relatively modest constructions [55].

11.5 FUTURE CHALLENGES AND OPPORTUNITIES

Enhancing plant growth and producing high-yield crops through plant microbiome engineering has become a viable alternative that has not yet been leveraged. Nevertheless, plant immunity still faces challenges with plant-microbe interactions, making it unable to distinguish between allies and adversaries in the soil's subsurface ecology [11]. Further research will be required to thoroughly describe the sorts of helpful bacteria for a wide variety of crop plants before this technology is ready for field implementation. In the case of targeted microbiome engineering, while bottom-up genome engineering techniques are effective and resilient, their throughput is restricted to a few dozen rhizobacteria at a time. Since a stable microbial population naturally connects with plants, these robust pre-existing microbiomes may quickly dilute the introduced engineered microbiomes [43]. More knowledge of interactions among engineered microbes, pre-existing microbiomes, and host plants will be necessary for developing a suitable strategy to deliver engineered PGP traits to host plants. In addition, both top-down and bottom-up strategies may use designed features that burden the host's metabolism and lessen its capacity to compete with other microorganisms. Consequently, it's important to carefully weigh the benefits of PGP against any potential drawbacks for plant colonization [43]. Technology selection should be based on prior research for the specific crop and environment, and how these approaches can be implemented in the conventional cropping system. Sustainable agriculture may employ a mix of soil amendments, plant breeding/genetic manipulation, and targeted microbiome engineering.

11.6 CONCLUSION

Plant microbiome engineering is a fundamental strategy to improve the growth and development of crop plants under hostile environments. Research attempting to comprehend this interaction at the community level can improve knowledge of the variables governing the microbiome assemblage and the pertinent input it provides to a host plant. Toward this end, further studies can help understand how the molecular and genetic makeup of microorganisms helps them to mutually associate with the host plant under environmental constraints. This understanding can lead to new developments in biological and ecological practices integrated into mainstream agriculture. Upcoming research endeavors may delve into the highlighted challenges with special reference to various biotic and abiotic elements that conscientiously impact the variety, capabilities, and bonding of microbial communities with their hosts in hostile environments.

REFERENCES

1. Hartmann M, Six J (2023) Soil structure and microbiome functions in agroecosystems. *Nat Rev Earth Environ* 4:4–18. https://doi.org/10.1038/s43017-022-00366-w
2. Bertola M, Ferrarini A, Visioli G (2021) Improvement of soil microbial diversity through sustainable agricultural practices and its evaluation by -omics approaches: A perspective for the environment, food quality and human safety. *Microorganisms* 9. https://doi.org/10.3390/microorganisms9071400

3. Jat SL, Suby SB, Parihar CM, et al (2021) Microbiome for sustainable agriculture: A review with special reference to the corn production system. *Arch Microbiol* 203:2771–2793. https://doi.org/10.1007/s00203-021-02320-8
4. Hakim S, Naqqash T, Nawaz MS, et al (2021) Rhizosphere engineering with plant growth-promoting microorganisms for agriculture and ecological sustainability. *Front Sustain Food Syst* 5. https://doi.org/10.3389/fsufs.2021.617157
5. Molefe RR, Amoo AE, Babalola OO (2023) Communication between plant roots and the soil microbiome; Involvement in plant growth and development. *Symbiosis* 90:231–239. https://doi.org/10.1007/s13199-023-00941-9
6. Brown SP, Grillo MA, Podowski JC, Heath KD (2020) Soil origin and plant genotype structure distinct microbiome compartments in the model legume Medicago truncatula. *Microbiome* 8:139. https://doi.org/10.1186/s40168-020-00915-9
7. Saeed Q, Xiukang W, Haider FU, et al (2021) Rhizosphere bacteria in plant growth promotion, biocontrol, and bioremediation of contaminated sites: A comprehensive review of effects and mechanisms. *Int J Mol Sci* 22:10529. https://doi.org/10.3390/ijms221910529
8. Suman J, Rakshit A, Ogireddy SD, et al (2022) Microbiome as a key player in sustainable agriculture and human health. *Front Soil Sci* 2. https://doi.org/10.3389/fsoil.2022.821589
9. Amaya-Gómez C V, Porcel M, Mesa-Garriga L, Gómez-Álvarez MI (2020) A framework for the selection of plant growth-promoting rhizobacteria based on bacterial competence mechanisms. *Appl Environ Microbiol* 86. https://doi.org/10.1128/AEM.00760-20
10. Orukotan KE, Elughi GN, Abimbola BS, et al (2023) Plant microbiome engineering: Principles, methods, and current trends. In: Isibor PO, Akinduti P, Oranusi SU, Popoola JO (eds) *Biotechnological Approaches to Sustainable Development Goals.* Springer Nature Switzerland, Cham, pp 251–267
11. Afridi MS, Javed MA, Ali S, et al (2022) New opportunities in plant microbiome engineering for increasing agricultural sustainability under stressful conditions. *Front Plant Sci* 13. https://doi.org/10.3389/fpls.2022.899464
12. Cesarano G, De Filippis F, La Storia A, et al (2017) Organic amendment type and application frequency affect crop yields, soil fertility and microbiome composition. *Appl Soil Ecol* 120:254–264. https://doi.org/10.1016/j.apsoil.2017.08.017
13. Bonanomi G, De Filippis F, Zotti M, et al (2020) Repeated applications of organic amendments promote beneficial microbiota, improve soil fertility and increase crop yield. *Appl Soil Ecol* 156:103714. https://doi.org/10.1016/j.apsoil.2020.103714
14. Zohry A, Ouda S (2018) Crop rotation defeats pests and weeds. In: Ouda S, Zohry AE-H, Noreldin T (eds) *Crop Rotation: An Approach to Secure Future Food.* Springer International Publishing, Cham, pp 77–88
15. Scavo A, Fontanazza S, Restuccia A, et al (2022) The role of cover crops in improving soil fertility and plant nutritional status in temperate climates. A review. *Agron Sustain Dev* 42:93. https://doi.org/10.1007/s13593-022-00825-0
16. Kraut-Cohen J, Zolti A, Shaltiel-Harpaz L, et al (2020) Effects of tillage practices on soil microbiome and agricultural parameters. *Sci Total Environ* 705:135791. https://doi.org/https://doi.org/10.1016/j.scitotenv.2019.135791
17. Bhattacharyya PN, Jha DK (2012) Plant growth-promoting rhizobacteria (PGPR): Emergence in agriculture. *World J Microbiol Biotechnol* 28:1327–1350. https://doi.org/10.1007/s11274-011-0979-9
18. Olanrewaju OS, Glick BR, Babalola OO (2017) Mechanisms of action of plant growth promoting bacteria. *World J Microbiol Biotechnol* 33:197. https://doi.org/10.1007/s11274-017-2364-9
19. Xiao-ying G, Chun-e H, Tao L, Zhu O (2015) Effect of bacillus subtilis and pseudomonas fluorescens on growth of greenhouse tomato and rhizosphere microbial community. *J Northeast Agric Univ (English Ed)* 22:32–42. https://doi.org/https://doi.org/10.1016/S1006-8104(16)30004-6
20. Boulé J, Sholberg PL, Lehman SM, et al (2011) Isolation and characterization of eight bacteriophages infecting Erwinia amylovora and their potential as biological control agents in British Columbia, Canada. *Can J Plant Pathol* 33:308–317. https://doi.org/10.1080/07060661.2011.588250
21. Fujiwara A, Fujisawa M, Hamasaki R, et al (2011) Biocontrol of Ralstonia solanacearum by Treatment with Lytic Bacteriophages. *Appl Environ Microbiol* 77:4155–4162. https://doi.org/10.1128/AEM.02847-10
22. Iriarte FB, Balogh B, Momol MT, et al (2007) Factors affecting survival of bacteriophage on tomato leaf surfaces. *Appl Environ Microbiol* 73:1704–1711. https://doi.org/10.1128/AEM.02118-06
23. Htwe AZ, Moh SM, Soe KM, et al (2019) Effects of biofertilizer produced from bradyrhizobium and streptomyces griseoflavus on plant growth, nodulation, nitrogen fixation, nutrient uptake, and seed yield of mung bean, cowpea, and soybean. *Agronomy* 9:77. https://doi.org/10.3390/agronomy9020077

24. Beeckman F, Motte H, Beeckman T (2018) Nitrification in agricultural soils: Impact, actors and mitigation. *Curr Opin Biotechnol* 50:166–173. https://doi.org/10.1016/j.copbio.2018.01.014
25. Calvo P, Zebelo S, McNear D, et al (2019) Plant growth-promoting rhizobacteria induce changes in Arabidopsis thaliana gene expression of nitrate and ammonium uptake genes. *J Plant Interact* 14:224–231. https://doi.org/10.1080/17429145.2019.1602887
26. Matse DT, Huang C-H, Huang Y-M, Yen M-Y (2020) Effects of coinoculation of Rhizobium with plant growth promoting rhizobacteria on the nitrogen fixation and nutrient uptake of Trifolium repens in low phosphorus soil. *J Plant Nutr* 43:739–752. https://doi.org/10.1080/01904167.2019.1702205
27. Ke X, Feng S, Wang J, et al (2019) Effect of inoculation with nitrogen-fixing bacterium Pseudomonas stutzeri A1501 on maize plant growth and the microbiome indigenous to the rhizosphere. *Syst Appl Microbiol* 42:248–260. https://doi.org/https://doi.org/10.1016/j.syapm.2018.10.010
28. Fasim F, Ahmed N, Parsons R, Gadd GM (2002) Solubilization of zinc salts by a bacterium isolated from the air environment of a tannery. *FEMS Microbiol Lett* 213:1–6. https://doi.org/10.1111/j.1574-6968.2002.tb11277.x
29. Shakeel M, Rais A, Hassan MN, Hafeez FY (2015) Root associated bacillus sp. improves growth, yield and zinc translocation for basmati rice (Oryza sativa) varieties. *Front Microbiol* 6. https://doi.org/10.3389/fmicb.2015.01286
30. Elhaissoufi W, Ghoulam C, Barakat A, et al (2022) Phosphate bacterial solubilization: A key rhizosphere driving force enabling higher P use efficiency and crop productivity. *J Adv Res* 38:13–28. https://doi.org/https://doi.org/10.1016/j.jare.2021.08.014
31. Rezakhani L, Motesharezadeh B, Tehrani MM, et al (2019) Phosphate–solubilizing bacteria and silicon synergistically augment phosphorus (P) uptake by wheat (Triticum aestivum L.) plant fertilized with soluble or insoluble P source. *Ecotoxicol Environ Saf* 173:504–513. https://doi.org/https://doi.org/10.1016/j.ecoenv.2019.02.060
32. KAUR G, REDDY MS (2015) Effects of phosphate-solubilizing bacteria, rock phosphate and chemical fertilizers on maize-wheat cropping cycle and economics. *Pedosphere* 25:428–437. https://doi.org/https://doi.org/10.1016/S1002-0160(15)30010-2
33. Backer R, Rokem JS, Ilangumaran G, et al (2018) Plant growth-promoting rhizobacteria: Context, mechanisms of action, and roadmap to commercialization of biostimulants for sustainable agriculture. *Front Plant Sci* 9. https://doi.org/10.3389/fpls.2018.01473
34. Wong WS, Tan SN, Ge L, et al (2015) The importance of phytohormones and microbes in biofertilizers. In: Maheshwari DK (ed) *Bacterial Metabolites in Sustainable Agroecosystem.* Springer International Publishing, Cham, pp 105–158
35. Gupta A, Rai S, Bano A, et al (2022) ACC deaminase produced by PGPR mitigates the adverse effect of osmotic and salinity stresses in pisum sativum through modulating the antioxidants activities. *Plants* 11:3419. https://doi.org/10.3390/plants11243419
36. Chen L, Dodd IC, Theobald JC, et al (2013) The rhizobacterium Variovorax paradoxus 5C-2, containing ACC deaminase, promotes growth and development of Arabidopsis thaliana via an ethylene-dependent pathway. *J Exp Bot* 64:1565–1573. https://doi.org/10.1093/jxb/ert031
37. Schenk PM, Carvalhais LC, Kazan K (2012) Unraveling plant–microbe interactions: Can multi-species transcriptomics help? *Trends Biotechnol* 30:177–184. https://doi.org/10.1016/j.tibtech.2011.11.002
38. El-Tarabily KA, Sivasithamparam K (2006) Non-streptomycete actinomycetes as biocontrol agents of soil-borne fungal plant pathogens and as plant growth promoters. *Soil Biol Biochem* 38:1505–1520. https://doi.org/10.1016/j.soilbio.2005.12.017
39. Alam A (2014) Soil degradation: A challenge to sustainable agriculture. *Int J Sci Res Agric Sci* 50–55. https://doi.org/10.12983/ijsras-2014-p0050-0055
40. Berendsen RL, Vismans G, Yu K, et al (2018) Disease-induced assemblage of a plant-beneficial bacterial consortium. *ISME J* 12:1496–1507. https://doi.org/10.1038/s41396-018-0093-1
41. Hazarika SN, Saikia K, Borah A, Thakur D (2021) Prospecting endophytic bacteria endowed with plant growth promoting potential isolated from camellia sinensis. *Front Microbiol* 12:738058. https://doi.org/10.3389/fmicb.2021.738058
42. Stoll A, Salvatierra-Martínez R, González M, Araya M (2021) The role of surfactin production by bacillus velezensis on colonization, biofilm formation on tomato root and leaf surfaces and subsequent protection (ISR) against Botrytis cinerea. *Microorganisms* 9(11): 2251.
43. Ke J, Wang B, Yoshikuni Y (2021) Microbiome engineering: Synthetic biology of plant-associated microbiomes in sustainable agriculture. *Trends Biotechnol* 39:244–261. https://doi.org/https://doi.org/10.1016/j.tibtech.2020.07.008

44. Frost LS, Leplae R, Summers AO, Toussaint A (2005) Mobile genetic elements: The agents of open source evolution. *Nat Rev Microbiol* 3:722–732. https://doi.org/10.1038/nrmicro1235
45. Gibb B, Hyman P, Schneider CL (2021) The many applications of engineered bacteriophages-an overview. *Pharmaceuticals (Basel)* 14. https://doi.org/10.3390/ph14070634
46. Sheth RU, Cabral V, Chen SP, Wang HH (2016) Manipulating bacterial communities by in situ microbiome engineering. *Trends Genet* 32:189–200. https://doi.org/https://doi.org/10.1016/j.tig.2016.01.005
47. Brophy JAN, Triassi AJ, Adams BL, et al (2018) Engineered integrative and conjugative elements for efficient and inducible DNA transfer to undomesticated bacteria. *Nat Microbiol* 3:1043–1053. https://doi.org/10.1038/s41564-018-0216-5
48. Ke J, Yoshikuni Y (2020) Multi-chassis engineering for heterologous production of microbial natural products. *Curr Opin Biotechnol* 62:88–97. https://doi.org/https://doi.org/10.1016/j.copbio.2019.09.005
49. Johnson CM, Grossman AD (2015) Integrative and conjugative elements (ICEs): What they do and how they work. *Annu Rev Genet* 49:577–601. https://doi.org/10.1146/annurev-genet-112414-055018
50. Touchon M, Bobay L-M, Rocha EPC (2014) The chromosomal accommodation and domestication of mobile genetic elements. *Curr Opin Microbiol* 22:22–29. https://doi.org/https://doi.org/10.1016/j.mib.2014.09.010
51. Inda ME, Broset E, Lu TK, de la Fuente-Nunez C (2019) Emerging frontiers in microbiome engineering. *Trends Immunol* 40:952–973. https://doi.org/https://doi.org/10.1016/j.it.2019.08.007
52. Wang G, Zhao Z, Ke J, et al (2019) CRAGE enables rapid activation of biosynthetic gene clusters in undomesticated bacteria. *Nat Microbiol* 4:2498–2510. https://doi.org/10.1038/s41564-019-0573-8
53. Elmore JR, Furches A, Wolff GN, et al (2017) Development of a high efficiency integration system and promoter library for rapid modification of Pseudomonas putida KT2440. *Metab Eng Commun* 5:1–8. https://doi.org/https://doi.org/10.1016/j.meteno.2017.04.001
54. Mimee M, Tucker AC, Voigt CA, Lu TK (2015) Programming a human commensal bacterium, bacteroides thetaiotaomicron, to sense and respond to stimuli in the murine gut microbiota. *Cell Syst* 1:62–71. https://doi.org/https://doi.org/10.1016/j.cels.2015.06.001
55. Choi K-H, Gaynor JB, White KG, et al (2005) A Tn7-based broad-range bacterial cloning and expression system. *Nat Methods* 2:443–448. https://doi.org/10.1038/nmeth765

12 Root-Associated Microbiota Transplantation

D Nivethini and A Vijaya Chitra
Sri Ramakrishna College of Arts and Science for Women
Coimbatore

A microbiome can be defined as a collection of highly interactive microbial species. The environment in which it resides, including host, exogenous environment, and human forces, shapes the microbiome. As microbiomes exist in all ecosystems and are made up of various microbial communities, a disturbance to them results in undesirable phenotypes in the hosts, which may lead to several diseases and disorders and also cause an imbalance in the associated ecosystems. Interaction between the microorganisms and their host leads to a synergistic relationship. These interactions influence the characteristics of ecosystems as well as affect the physiology of the host and/or condition of the locality. Just as human gut microbiota influences the physiological and psychological health of the host, soil microbiota can affect the productivity of crops. Engineering of microbiomes can lead to an alteration in microbial compositions, which helps in the actions of host phenotypes and benefits ecosystems. Several different strategies are used for the engineering of microbiomes associated with humans, animals, plants, and soil. Engineering methods can be employed for feed enzymes, prebiotics, probiotics, fecal microbiota transplants in humans, root-associated microbiota transplants in plants, administration of antibiotics, etc. Root microbiome or rhizosphere microbiome is a community of microbes correlated with the roots of plants, which may include mutualists, pathogens, and commensals. As the root microbiome is rich in a variety of carbon compounds, the plant roots provide a unique environment for a distinct assembly of soil organisms such as bacteria, fungi, and archaea. Plant microbiomes participate in a variety of plant functions such as nutrient absorption, growth promotion, and resistance towards biotic and abiotic stresses. Mutualistic structures such as arbuscular mycorrhizae help in phosphate absorption, and rhizobia colonize legumes to form nitrogen-fixing root nodules. Soil microbiota, especially the rhizosphere microbiome, provides protection and resistance against soil-borne diseases. Several researchers suggest that optimization of the root microbiome may act as a solution for food scarcity. Root-associated microbiome transplantation has the potential to reduce the risk of plant diseases and increase agricultural production. It also helps in the reduction of greenhouse gas emissions and the need for chemical inputs. In addition, manipulating the plant microbiome by transplantation could increase the plant-mediated removal of pollutants. Root-associated microbiota transplantation, with all its advantages, could provide long-lasting effects on plant growth characteristics, productivity, and other applications such as bioremediation.

12.1 INTRODUCTION

The microbiome is a collection of microbes such as bacteria, fungi, viruses, and their genes that reside on and inside our body. Root microbiome, or rhizosphere microbiome, is a community of microbes associated with plant roots (Hartmann et al., 2015). The microbiome may be mutualistic or parasitic. When the soil microbiome is healthy, it influences plant health and often provides protection against drought, pests, etc. The soil microbiome also confers protection against pathogens and protect plants by producing toxins.

DOI: 10.1201/9781003394662-12

12.1.1 Plants

Plants are immobile organisms fixed to the soil by their roots. As plants are exposed to various stresses from pathogens, pests, unfavorable environmental conditions, heavy metals, etc., they have evolved a range of sophisticated immune responses to protect themselves. For several decades, individual plant-microbe relationships have been studied (Mendes *et al.*, 2011). Plant roots are colonized by complex communities of bacterial and archaeal microbiota from the soil, which have the potential to affect plant nutrition and fitness. Only in recent years have studies on plant-associated microorganisms and their beneficial activities in increasing the immune response of plants become evident (Berendsen *et al.*, 2012). Additionally, it was found that plants produce certain compounds that attract soil microbes beneficial for their growth.

12.1.2 Plant Roots and Microbiome

Plants anchor a diverse variety of microbes above and below tissues, collectively called plant microbiota. Genomes of microbiota that live in close association with plants are referred to as the plant microbiome (Bulgarelli *et al.*, 2013). The interaction of these microbiomes with the plant roots exhibits several functions in providing durability and fitness to the plant (De Coninck *et al.*, 2015). The main functions include fixing the plant in position, uptake and storage of nutrients and water, and intervention of interaction with soil-inhabiting microbes. Roots and the soil surrounding them constitute one of the richest and distinct ecosystems on Earth. Microbial concentration in the rhizosphere releases carbon-rich products of photosynthesis, which act as nutrients for the microbes (Bais *et al.*, 2006). Various rhizo deposits such as organic acids, amino acids, sugars, and products of secondary metabolism are found (Dakora and Phillips, 2002). The microbes that multiply in the rhizosphere get exposed to plant-derived compounds and signaling molecules and represent a subset of the highly complex microbial communities of the bulk soil.

12.2 PLANT INTERACTIONS WITH BENEFICIAL ORGANISMS

Researchers unearthed several beneficial interactions between plants and microbes several years ago. Studies revealed that land plants were colonized by ancestral filamentous fungi which facilitated the absorption of water and nutrient acquisition, while simultaneously, the fungi received photosynthetically fixed carbon (Field *et al.*, 2015; Martin *et al.*, 2017).

This symbiosis evolved in a successful manner, i.e., more than 90% of plant species form symbiosis with mycorrhizal fungi, and 80% of these are classified as arbuscular mycorrhizal fungi (AMF) (Bonfante and Genre, 2010). A continuous exchange of nutrients occurs between the host plant and microbes when the symbiosis is established. Thus, arbuscular mycorrhizal fungi (AMF) uptake and transport low water-soluble inorganic orthophosphate (Pi) and other major nutrients such as nitrogen, in the form of nitrate, ammonium, and amino acids, using specialized transporters (Parniske, 2008). Even in the absence of symbiosis with AMF or rhizobia, plants uptake nutrients by a wide array of microbes. An example of this nutrient uptake is the non-host plant *Arabidopsis*, which acquires Pi with the help of the natural root endophytic symbiont *Colletotrichum tofieldiae* (Hiruma *et al.*, 2016). Some endophytic fungi such as *Serendipita indica* promote plant growth by Pi acquisition. In a similar manner, *Trichoderma* sp. produces chelating metabolites that solubilize phosphate and increase plant growth. In non-leguminous plants, organisms not belonging to the nitrogen-fixing bacterial group are involved in nitrogen acquisition. Studies also show that some beneficial organisms colonize the roots and cause the induction of gene expression for the uptake of iron. These genes are commonly used commonly by plants for the mobilization and uptake of iron when iron is present in unavailable forms in the soil (Yadav *et al.*, 2010).

Beneficial microbes have the ability to promote plant growth by influencing the hormonal balance of plants. This effect is induced by the secretion of microbial small secondary metabolites (SM), which act as hormone-like plant growth regulators (Verbon and Liberman, 2016). Several microbial species of plant-associated bacteria and fungi produce indole-3-acetic acid (IAA) or auxin-like molecules that play a direct role in the development and growth of plants. Other microbial phytohormones or phytohormone-like molecules such as cytokinins and gibberellins are produced primarily to facilitate the colonization of microbes through modulating the immunity of plants (Stringlis *et al.*, 2018a).

12.3 INTERACTIONS BETWEEN PLANT AND PATHOGENS

Plant roots support beneficial associations with soil organisms while needing to cope up with infections caused by pathogenic organisms. Soil pathogens can cause disease in various economically important crops and lead to decreased crop yield (Pimental *et al.*, 1991). Soil-borne pathogens reside in the soil for shorter or longer periods and survive as saprophytes or as resting structures like chlamydospores and oospores. For example, the root exudates of watermelon plants, including phenolic acids, sugars, and free amino acids, increase the germination of spores of *F.oxysporum*. In a similar manner. exudates of tomato roots cause the germination of pathogens like *F. oxysporum f. sp. lycopersici* and *F. oxysporum f. sp. radicis* (Bruehl, 1987). Soil pathogens, such as fungal pathogens, detect root exudates. During favorable environmental conditions, pathogens invade plants through the root system. Plants affected by soil-borne pathogens suffer from root rot, stunted growth, seedling damping-off, stem and collar rots, wilting, or even death of plants (Katan, 2017).

It is difficult to control soil-borne pathogens for several reasons such as soil borne pathogens producing resting structures that can reside in soil for long periods, even in the absence of susceptible hosts. Controlling these structures by using chemicals is unsuitable due to public health concerns and environmental issues. A negative effect was also observed in pathogens and other organisms present in soil, such as antibiosis, and competence for food and nutrients. Different soil-borne pathogens like species of fungi, oomycetes, bacteria, viruses, and nematodes are observed. Examples of soil-borne fungal pathogens include *F. oxysporum*, *F. solani*, *Rhizoctonia solani*, *Verticillum*sp., and *Sclerotinia sclerotiorum*; oomycetes include *Phytophthora sp.* and *Pythium sp.*; and bacterial pathogens include *Ralstonia solanacearum* (Peeters *et al.*, 2013) and *A. Tumefaciens*, which requires an opening or wound to penetrate and cause infection. Similarly, a few viruses cause infections that require an opening for penetration. Viruses are generally transmitted in the soil by zoosporic fungi or nematodes (Campbell, 1996; Brown *et al.*, 1995).

12.4 ROOT IMMUNITY AND MICROBIOME

The soil microbial population consists of both beneficial and harmful organisms, and plants need to differentiate between them for proper growth and yield. Thus, plants possess several defense strategies in order to allow or defend the colonization of microbes (Zamioudis and Pieterse, 2012). For this purpose, plants possess a multi-layered detection system that is effective and timely in responding to microbial signals (Dodds and Rathjen, 2010). Surface localized pattern recognition receptors (PRRs) are the first line of this defense system, recognizing microbe-derived molecules termed microbe-associated molecular patterns (MAMPs) (Couto and Zipfel, 2016). For example, in Arabidopsis, a few MAMP/PRR pairs are observed. Bacterial flagellin and the immunogenic epitope of flagellin flg22 are recognized by receptor kinase FLAGELLIN-SENSING 2 (FLS2), and similarly, bacterial elongation factor Tu and the immunogenic peptide elf18 are recognized by Elongation Factor-TU Receptor (EFR). Thus, detection MAMP results in the induction of immune responses in host plants, referred as MAMP-triggered immunity (MTI) (Vos *et al.*, 2013).

Depending on the time taken, the activated responses range from rapid, early, and latc. During rapid response alkalization, oxidative burst (ROS) and protein phosphorylation take place. Ethylene biosynthesis and defense gene activation occur in the early response, and callose deposition and growth inhibition occur in the late response. These responses are directed towards arresting further growth and colonization of microbes (Hacquard *et al.*, 2017). This activation of PRRs in microbes and elicitor-enriched environments like roots and the surrounding rhizosphere may lead to the MTI response, which simultaneously may result in retardation or inhibition of growth and yield. Several researchers conducted studies to find the involvement of different organs and the contribution of root tissue in the induction of MTI. It was found that inner tissues show higher expression than epidermal tissues. It is also evidenced that not only plants but also microbes adapt to the presence of MAMPs and PRRs, respectively. A few PRRs can also identify "self" molecules known as host-derived damage-associated molecular patterns (DAMPs) (Poncini *et al.*, 2017; Marhavy *et al.*, 2019). During any fungal or nematode attack, DAMPs are released in response to cellular rupture, and the effectiveness of DAMP defense responses is comparatively higher than MAMP.

12.5 BENEFICIAL MICROBES AND SUPPRESSION OF ROOT DEFENSES

Several beneficial root microbiomes possess various strategies in order to suppress MTI and/or maneuver the defense hormones to colonize and benefit the host. In addition, different strategies are employed by beneficial microbes to help in colonization (Carvalhais *et al.*, 2015). The type III secretion system (T3SS) plays a major role in facilitating symbiotic relationships between rhizobia and their legume partners. T3SS is a multicomponent apparatus used to secrete effector molecules into host cells by pathogenic, gram-negative bacteria, with the aim to restrict defense responses caused by their recognition and succeed in colonization of the host. On the other hand, pathogenic bacteria such as *P. aeruginosa* and *P. syringae* releases extracellular enzyme alkaline protease AprA, which have the ability to degrade flagellin monomers, and this facilitates microbes to have their MAMPs undetected by the host immune system. Recent studies show that beneficial rhizobacteria, *Pseudomonas* sp., secrete organic acids that lower environmental pH and suppress root immune responses (Bardoel *et al.*, 2011; Pel *et al.*, 2014).

12.6 DISEASE SUPPRESSIVENESS

Microbial communities in soil provide several benefits to the ecosystem. Among them, disease suppression is a well-known phenomenon where the microbiome mediates the first line defense against infections caused by soil-borne pathogens. Soil with disease suppressiveness can be defined as "soil where pathogen doesn't prevail; pathogen prevails but causes minimum or no damage; or causes disease which is less important as the disease may disappear even though the pathogen exists in soil". On the other hand, disease occurs readily in conductive soil (Baker and Cook, 1974). Generally, two types of soil suppressiveness are observed: "general" and "specific" suppression. In the case of general suppression, multiplication and activity of the pathogen are suppressed to an extent, and it depends on the microbial community associated with competition for available resources (Weller *et al.*, 2002). In contrast, specific suppression occurs when a single species or specific subsets of soil organisms interfere with the infection cycle of the pathogen. This type of specific suppression can be transferred to conductive soil by inoculating a small amount of suppressive soil (Schlatter et al., 2017). Specific suppression is more effective than general suppression, as the suppression retains for a longer period of time even when the soil is left bare (Raaijmakers and Mazzola, 2016).

Activation of pathogen-suppressive microorganisms occurs due to the secretion of pathogen- or plant-derived metabolites that are produced before any disease outbreak. It is found that the suppression activity may be due to the production of antibiotics that inhibit the microbial population. More interestingly, it was found that this ability can be carried over to next-generation plants by transplantation of soil, which paves way for exploiting microbiome for disease resistance (Chapelle *et al.*, 2016).

12.7 MODULATION OF MICROBIOME

Microbiome can be modulated by coumarins, benzoxazinoids, and other exudates produced by roots. Coumarins are phenolic compounds that have been studied to possess resistance towards disease and also in iron deficiency. Whenever iron is unavailable in soil, the plant roots produce coumarins, which makes more iron available before it is imported inside the roots (Tsai and Schmidt, 2017a, 2017b). Recent studies showed that coumarins have an important role in shaping the microbiome around the roots. Benzoxazinoids are compounds found to attract beneficial microbes in the rhizosphere and also possess defense against pathogens. In maize, it was observed that the release of benzoxazinoids provides protection against herbivore insects (Ahmad *et al.*, 2011). Thus, studies have shown that a microbiome can be selected through exudation, allowing us to breed plants that can alter their microbiome composition in order to increase growth as well as health benefits (Cotton *et al.*, 2019).

Triterpenes and Camaloxin are also found to be involved in the shaping of microbiome. These are produced during plant metabolism, which provides disease resistance and antimicrobial effects (Thimmappa *et al.*, 2014; Huang *et al.*, 2019). Whenever plants experience any infection by phytopathogens or insects, they recruit beneficial microbes from the rhizosphere microbiota, which may help them overcome biotic stresses. This phenomenon is called a "cry for help" (Bakker *et al.*, 2018). This build-up of beneficial microbes is due to change in gene expression and variations in the exudates produced by roots in response to pathogenic attacks (Yuan *et al.*, 2018). These pieces of evidence show that the microbiome provides protection against both biotic and abiotic stresses. The presence of rhizosphere bacteria in plants elicits induced systemic tolerance toward extreme salinity, drought, and nutrient deficiency or excess, etc.

12.8 RHIZOSPHERE MICROBIOME AGAINST BIOTIC AND ABIOTIC STRESSES

Studies showed that exudates produced by roots attract several organisms that are beneficial for plants as well as organisms that have pathogen-suppressing abilities. Thus, this reshapes the microbiome assembly which in turn suppresses the disease symptoms (Kwak *et al.*, 2018). Growth of pathogens is suppressed by different mechanisms such as competition between microbes, secretion of antimicrobial compounds, and increased rate of parasitism. Production of phytohormones by the rhizosphere microbiome alters plant growth (Duran *et al.*, 2018). Apart from resistance against biotic stresses, the rhizosphere microbiome confers resistance against abiotic stresses (Rolli *et al.*, 2015). These rhizosphere bacteria elicit induced systemic tolerance against high salt concentration, drought, and nutrient deficiency. Recent studies revealed that root-associated bacteria of soybean and wheat such as *Pseudomonas* sp., *Pantoea* sp., and *Paraburkholderia* sp. possess different mechanisms that are involved in efficient nutrient uptake, plant growth, stress tolerance, etc. (Rascovan *et al.*, 2016). Accumulation of heavy metals, hydrocarbons, and pesticides can cause damage to soil properties, thus affecting growth of the plant or making it unsuitable for consumption. An interesting factor was found that the enhancement of microbial functions may help them in degrading aromatic compounds and thus, in turn, helping in bioremediation (Perez-Garcia *et al.*, 2016; Eng and Borenstein, 2019). Another study has shown that modifying the assembly of microbiota by transplanting the natural microbiome showed effective degradation of contaminants in comparison to the introduction of artificial selection (Bell *et al.*, 2016).

12.9 ENGINEERING OF PLANT MICROBIOME

The plant microbiome can be represented as a root-associated microbiome. Root-associated microbiota can be defined as microbial communities present in roots as well as the rhizosphere, which comprises thousands of bacteria and fungi. As the microbiome acts by influencing the plant physiology and also plays an important role in protecting against pathogens, engineering these organisms helps in altering microbiome composition and also in improving plant growth, fitness, and health (Agler *et al.*, 2016).

Different methods such as microbiome transfer, synthetic microbiome, and host-mediated artificial selection can be employed for engineering root-associated microbiomes. In the case of microbiome transfer for plant disease management, an approach where disease-suppressive soil was mixed with disease-conductive soil was found to be helpful. This was found to inhibit sugar beet infection, potato common scab, and tobacco black root rot (Gopal *et al.*, 2013). It was also found that in the *Arabidopsis thaliana* plant model, microbiome transfer from sympatric soil showed the capability to mitigate drought stress and also boost biomass under drought conditions (Zolla *et al.*, 2013). In the case of production of synthetic microbiomes, microbial strains with relevant functions can be selected and combined. For example, five root-associated bacteria were found to protect *Nicotiana attenuate* from sudden wilt disease (Santhanam *et al.*, 2015). Synthetic microbiomes have the main advantage that we can control their microbial composition; however, the development of synthetic microbiomes is limited, due to our understanding of core microbiomes, as only culturable microbes can be used in synthetic microbiomes. In the case of host-mediated artificial selection, the ecology of the plant microbiome can be altered (Glick *et al.*, 2012; Lebeis *et al.*, 2015).

12.10 ENGINEERING OF SOIL MICROBIOME

Soil microbiome refers to the microbial population in soil beyond the rhizosphere. Through engineering of these soil microbiomes, varied and balanced soil microbial communities can be established, which may improve the health of soil and fertility of plants. Engineering of soil microbiomes can be achieved by employing agricultural management techniques. One of these is the implementation of organic farming (Chaparro *et al.*, 2012). This was found to promote and enrich soil microbial diversity, biological cycles, and biological activity, as well as minimize the usage of chemical fertilizers and herbicides. Implementation of organic farming was found to increase richness, reduce evenness and dispersion, and also alter the structure of the soil microbiome when compared to the usage of chemical fertilizers. Another way of modifying the soil microbiome is to change land utilization. During the change of vegetation, a significant shift occurs in the structure and function of the soil microbiome. Navarrette *et al.* (2015) has reported that there were both taxonomic and functional adaptations in the soil microbiome due to deforestation in the Amazon region, which resulted in increased soil fertility. In addition to this, agricultural practices such as logging, tillage, and cropping systems have been shown to play a vital role in influencing structural and functional changes in the soil microbiome, which leads to changes in soil functions (Xiong *et al.*, 2015).

12.11 TECHNIQUES TO ENGINEER BENEFICIAL MICROBIOMES

Engineering of microbiomes can promote fitness of plants as well as their health. This approach holds great promise in realizing sustainable future agriculture. However, the implementation of this technique has several limitations. In order to unweave the mechanisms behind host and microbiome interaction, we have to integrate multiple omics techniques (Jansson and Baker, 2016). Some approaches, like metabolomics, metagenomics, plant transcriptomics, metatranscriptomics, and plant genetics, can be combined to disentangle the interactions between holobiont members. A recent study combined metabolomics and amplicon-based metagenomics analysis on two maize genotypes. It revealed the formation of defense-related benzoxazinoid metabolite-related structures of bacterial and fungal community in the maize rhizosphere.

12.12 GENETIC ENGINEERING OF MICROBIOME: LAB TO FIELD

Successful application of microbial consortia as inoculum in order to protect plants from stresses and improve their productivity depends on the effectiveness of microbes that are stable enough to overcome the hurdles and retain their characteristics when applied in the field (Sessitsch *et al.*, 2019). There are two main principles behind this strategy: (i) selection and combination of microorganisms that are not closely related, i.e., they have different or complementing characteristics which enhance plant growth, suppress pathogens, or withstand unfavorable conditions (Compant *et al.*,

2019); and (ii) selection and combination of closely related microorganisms (Wei *et al.*, 2015; Hu *et al.*, 2016). It was found that when *Pseudomonas* sp. is introduced, it reduced *R. solanacearum* density in the rhizosphere soil of tomato plants and, thus, the incidence of disease (Hu *et al.*, 2016). In addition, plant biomass was enhanced, and there was effective assimilation of nutrients. Plant growth promotion was observed which was associated with increased production of plant hormones, siderophores, and solubilization of phosphorus in vitro (Hu *et al.*, 2017). Recent studies revealed that microbial synthetic communities can be used as inoculants to produce plant growth substances with desired characteristics, such as biocontrol of specific pathogens and plant growth promotion (Tsolakidou *et al.*, 2019a). The composition of the synthetic community determines the plant growth factor and pathogen inhibition. Studies also show that complex inocula can result in increased disease resistance and growth promotion compared to single strains (Rolli *et al.*, 2015; Santhanam *et al.*, 2015; Niu *et al.*, 2017).

12.13 CONCLUSION

Research in the field of root microbiota transplantation is in progress, and the transplantation process varies depending on the target plant species and desired outcome. Specific techniques and protocols can bring successful impacts to root-associated microbiota transplantation. However, risks associated with it should also to be considered, as introducing foreign microorganisms into new environments can have unintended ecological consequences. Overall, root-associated microbiota transplantation is a promising tool to enhance plant health, improve crop productivity, and aid in ecological restoration efforts. However, further research is needed to fully understand the mechanisms involved and optimize the techniques for practical applications.

REFERENCES

Agler MT, Ruhe J, Kroll S, Morhenn C, Kim ST, Weigel D, Kemen EM. 2016. Microbial hub taxa link host and abiotic factors to plant microbiome variation. *PLoS Biol.* Jan 20;14(1), e1002352.

Ahmad S, Veyrat N, Gordon-Weeks R, Zhang Y, Martin J, Smart L, et al. 2011. Benzoxazinoid metabolites regulate innate immunity against aphids and fungi in maize. *Plant Physiol.* 157, 317–327. doi:10.1104/pp.111.180224

Bais HP, Weir TL, Perry LG, Gilroy S, Vivanco JM. 2006. Therole of root exudates in rhizosphere interactions with plants and other organisms. *Annu Rev Plant Biol.* 57, 233–266.

Baker KF, Cook RJ. 1974. *Biological control of plant pathogens* (San Francisco: WH Freeman and Company).

Bakker PAHM, Pieterse CMJ, de Jonge R, Berendsen RL. 2018. The soil-borne legacy. *Cell.* Mar 8;172(6), 1178–1180.

Bardoel BW, van der Ent S, Pel MJ, Tommassen J, Pieterse CM, van Kessel KP, van Strijp JA. 2011. Pseudomonas evades immune recognition of flagellin in both mammals and plants. *PLoS Pathog*, 7(8), e1002206.

Bell TH, Stefani FO, Abram K, Champagne J, Yergeau E, Hijri M, St-Arnaud M. 2016. A diverse soil microbiome degrades more crude oil than specialized bacterial assemblages obtained in culture. *Appl Environ Microbiol.* Aug 30;82(18):5530–5541.

Berendsen RL, Pieterse CMJ, Bakker PAHM. 2012. The rhizosphere microbiome and plant health. *Trends Plant Sci.* 17, 478–486.

Bonfante P, Genre A. 2010. Mechanisms underlying beneficial plant- fungus interactions in mycorrhizal symbiosis. *Nat. Commun.* 1, 48.

Bruehl, G. W.,1987. Soilborne plant pathogens (New York, USA: Macmillan Publishing Company). doi: 10.1086/415660

Brown DJ, Robertson WM, Trudgill DL. 1995. Transmissionofviruses by plant nematodes. *Annu Rev Phytopathol.* 33, 223–249.

Bulgarelli D, Schlaeppi K, Spaepen S, Ver Loren Van Themaat E, Schulze- Lefert P. 2013. Structure and functions of the bacterial microbiota of plants. *Annu Rev Plant Biol.* 64, 807–838.

Campbell R. N., 1996. Fungal transmission of plant viruses. Annual review of phytopathology, 34, 87–108. https://doi.org/10.1146/annurev.phyto.34.1.87

Carvalhais, L. C., Dennis, P. G., Badri, D. V., Kidd, B. N., Vivanco, J. M., Schenk, P. M., 2015. Linking Jasmonic Acid Signaling, Root Exudates, and Rhizosphere Microbiomes. Molecular plant-microbe interactions : MPMI, 28(9), 1049–1058. https://doi.org/10.1094/MPMI-01-15-0016-R
Chaparro J, Sheflin A, Manter D, Vivanco J. 2012 Manipulating the soil microbiome to increase soil health and plant fertility. *Biol Fertil Soil.* 48, 489–499.
Chapelle E, Mendes R, Bakker PA, Raaijmakers JM. 2016. Fungal invasion of the rhizosphere microbiome. *ISME J.* 10(1), 265–268.
Compant S, Samad A, Faist H, Sessitsch A. 2019. A review on the plant microbiome: Ecology, functions, and emerging trends in microbial application. *J Adv Res.* 20;19:29–37.
Cotton TEA, Pétriacq P, Cameron DD, Meselmani MA, Schwarzenbacher R, Rolfe SA, Ton J. 2019. Metabolic regulation of the maize rhizobiome by benzoxazinoids. *ISME J.* 13(7), 1647–1658.
Couto, D., and Zipfel, C., 2016. Regulation of pattern recognition receptor signaling in plants. Nature reviews. Immunology, 16(9), 537–552.
Dakora FD, Phillips DA. 2002. Root exudates as mediators of mineral acquisition in low-nutrient environments. *Plant Soil.* 245, 35–47.
DeConinck B, Timmermans P, Vos C, Cammue BP, Kazan K. 2015. What lies beneath: belowground defense strategies in plants. *Trends In Plant Sci.* 20, 91–101.
Dodds PN, Rathjen JP. 2010. Plant immunity: Towards an integrated view of plant-pathogen interactions. *Nat Rev Genet.* 11, 539–548.
Durán P, Thiergart T, Garrido-Oter R, Agler M, Kemen E, Schulze-Lefert P, Hacquard S. 2018. Microbial interkingdom interactions in roots promote arabidopsis survival. *Cell.* 1;175(4), 973–983.
Eng A, Borenstein E. 2019. Microbial community design: Methods, applications, and opportunities. *Curr Opin Biotechnol.* 58, 117–128.
Field KJ, Pressel S, Duckett JG, Rimington WR, Bidartondo MI. 2015. Symbiotic options for the conquest of land. *Trends Ecol Evol.* 30 (8), 477–486.
Glick B. R., 2012. Plant growth-promoting bacteria: mechanisms and applications. Scientifica, 2012, 963401. https://doi.org/10.6064/2012/963401
Gopal M, Gupta A, Thomas GV, 2013. Bespoke microbiome therapy to manage plant diseases. *Front Microbiol.* 4, 355.
Hacquard S, Spaepen S, Garrido-Oter R, Schulze-Lefert P. 2017. Interplay between innate immunity and the plant microbiota. *Annu Rev Phytopathol.* 55, 565–589.
Hartmann M, Frey B, Mayer J, Mader P, Widmer F. 2015. Distinct soil microbial diversity under long-term organic and conventional farm- ing. *ISME J.* 9, 1177–1194.
Hiruma K, Gerlach N, Sacristán S, Nakano RT, Hacquard S, Kracher B, Neumann U, Ramírez D, Bucher M, O'Connell RJ, Schulze-Lefert P. 2016. Root endophyte colletotrichum tofieldiae confers plant fitness benefits that are phosphate status dependent. *Cell.* Apr 7;165(2), 464–474.
Hu, J, Wei Z, Friman VP, Gu, SH, Wang XF, Eisenhauer N, et al. 2016. Probiotic diversity enhances rhizosphere microbiome functionand plant disease suppression. *MBio.* 7, e01790–e01716.
Hu, J, Wei Z, Weidner S, Friman VP, Xu, YC, Shen QR, et al. 2017. Probiotic *Pseudomonas* communities enhance plant growth and nutrient assimilation *via* diversity-mediated ecosystem functioning. *Soil Biol Biochem.* 113, 122–129.
Huang AC, Jiang T, Liu YX, Bai YC, Reed J, Qu B, Goossens A, Nützmann HW, Bai Y, Osbourn A. 2019. A specialized metabolic network selectively modulates Arabidopsis root microbiota. *Science.* May 10;364(6440), eaau6389.
Jansson JK, Baker ES. 2016. A multi-omic future for microbiome studies. *Nat Microbiol.* Apr 26;1, 16049.
Katan J. 2017. Diseases caused by soil borne pathogens: Biology, management and challenges. *J. Plant Pathol.* 99, 305–315.
Kwak MJ, Kong HG, Choi K, Kwon SK, Song JY, Lee J, Lee PA, Choi SY, Seo M, Lee HJ, Jung EJ, Park H, Roy N, Kim H, Lee MM, Rubin EM, Lee SW, Kim JF. 2018. Rhizosphere microbiome structure alters to enable wilt resistance in tomato. *Nat Biotechnol.* Oct 8. PMID: 30295674.
Lebeis SL, Paredes SH, Lundberg DS, Breakfield N, Gehring J, McDonald M, Malfatti S, Glavina del Rio T, Jones CD, Tringe SG, Dangl JL. 2015. PLANT MICROBIOME. Salicylic acid modulates colonization of the root microbiome by specific bacterial taxa. *Science.* Aug 21;349(6250), 860–864.
Marhavý P, Kurenda A, Siddique S, Dénervaud Tendon V, Zhou F, Holbein J, Hasan MS, Grundler FM, Farmer EE, Geldner N. 2019. Single-cell damage elicits regional, nematode-restricting ethylene responses in roots. *EMBO J.* May 15;38(10), e100972.
Martin FM, Uroz S, Barker DG. 2017. Ancestral alliances: Plant mutualistic symbioses with fungi and bacteria. *Science.* May 26;356(6340), eaad4501.

Mendes R, Kruijt M, de Bruijn I, Dekkers E, van der Voort M, Schneider JH, Piceno YM, DeSantis TZ, Andersen GL, Bakker PA, Raaijmakers JM. 2011. Deciphering the rhizosphere microbiome for disease-suppressive bacteria. *Science*. May 27; 332(6033), 1097–1100.

Navarrete AA, Tsai SM, Mendes LW, Faust K, et al. 2015. Soil microbiome responses to the short-term effects of Amazonian deforestation. *Mol Ecol*. 24, 2433–2448.

Niu B, Paulson JN, Zheng X, Kolter R. 2017. Simplified and representative bacterial community of maize roots. *Proc Natl Acad Sci USA*. 114, 2450–2459.

Parniske M. 2008. Arbuscular mycorrhiza: The mother of plant root endosymbioses. *Nat Rev Microbiol*. 6, 763–775. doi:10.1038/nrmicro1987

Peeters N, Guidot A, Vailleau F, Valls M. 2013. *Ralstonia solanacearum*, a widespread bacterial plant pathogen in the post-genomic era. *Mol Plant Pathol*. 14, 651–662. doi:10.1111/mpp.12038

Pel, M. J., van Dijken, A. J., Bardoel, B. W., Seidl, M. F., van der Ent, S., van Strijp, J. A., Pieterse, C. M., 2014. Pseudomonas syringae evades host immunity by degrading flagellin monomers with alkaline protease AprA. Molecular plant-microbe interactions : MPMI, 27(7),603–610. https://doi.org/10.1094/MPMI-02-14-0032-R

Perez-Garcia O, Lear G, Singhal N. 2016. Metabolic network modelling of microbial interactions in natural and engineered environmental systems. *Front Microbiol*. 7, 673. doi:10.3389/fmicb.2016.00673

Pimentel DL, McLaughlin A, Zepp B, Lakitan T, Kraus P, Kleinman F, et al. 1991. Environmental and economic effects of reducing pesticide use. *Bioscience*. 41, 402–409.

Poncini L, Wyrsch I, Denervaud Tendon V, Vorley T, Boller T, Geldner N, et al. 2017. In roots of *Arabidopsis thaliana*, the damage-associated molecular pattern AtPepl is a stronger elicitor of immune signalling than flg22 or the chitin heptamer. *PloS One*. 12, e0185808.

Raaijmakers JM, Mazzola M. 2016. Soil immune responses. *Science*. 352, 1392–1393.

Rascovan N, Carbonetto B, Perrig D, Diaz M, Canciani W, Abalo M, et al. 2016. Integrated analysis of root microbiomes of soybean and wheat from agricultural fields. *Sci Rep*. 6, 28084. doi:10.1038/srep28084

Rolli E, Marasco R, Vigani G, Ettoumi B, Mapelli F, Deangelis ML, et al. 2015. Improved plant resistance to drought is promoted by the root- associated microbiome as a water stress-dependent trait. *Environ Microbiol*. 17, 316–331. doi:10.1111/1462-2920.12439

Santhanam R, Luu V.T, Weinhold A, Goldberg J, Oh Y, Baldwin IT. 2015. Native root-associated bacteria rescue a plant from a sudden-wilt disease that emerged during continuous cropping. *Proc Natl Acad Sci USA*. 112, 5013–5020.

Schlatter D, Kinkel L, Thomashow L, Weller D, Paulitz T. 2017. Disease suppressive soils: New insights from the soil microbiome. *Phytopathology*. 107(11):1284–1297.

Sessitsch A, Pfaffenbichler N, Mitter B. 2019. Microbiome applications from lab to field: Facing complexity. *Trends Plant Sci*. 24, 194–198. doi:10.1016/j.tplants.2018.12.004

Stringlis IA, Proietti S, Hickman R, Van Verk MC, Zamioudis C, Pieterse CMJ. 2018a. Root transcriptional dynamics induced by beneficial rhizobacteria and microbial immuneelicitors reveal signatures of adaptation to mutualists. *Plant J*. 93, 166–180. doi:10.1111/tpj.13741

Thimmappa R, Geisler K, Louveau T, O'maille P, Osbourn A. 2014. Triterpene biosynthesis in plants. *Annu Rev Plant Biol*. 65, 225–257.

Tsai HH, Schmidt W. 2017a. Mobilization of iron by plant-borne coumarins. *Trends Plant Sci*. 22, 538–548.

Tsai HH, Schmidt W. 2017b. One way. Or another? Iron uptake in plants. *New Phytol*. 214, 500–505. doi:10.1111/nph.14477

Tsolakidou MD, Stringlis IA, Fanega-Sleziak N, Papageorgiou S, Tsalakou A, Pantelides IS. 2019a. Rhizosphere-enriched microbes as a pool to design synthetic communities for reproducible beneficial outputs. *FEMS Microbiol Ecol*. 95, fiz138.

Verbon EH, Liberman LM. 2016. Beneficial microbes affect endogenous mechanisms controlling root development. *Trends Plant Sci*. 21, 218–229.

Vos IA, Pieterse CMJ, VanWees SCM. 2013. Costs and benefits of hormone-regulated plant defences. *Plant Pathol*. 62, 43–55.

Wei Z, Yang T, Friman VP, Xu Y, Shen Q, Jousset A. 2015. Trophic network architecture of root-associated bacterial communities determines pathogen invasion and plant health. *Nat Commun*. 6, 9413.

Weller DM, Raaijmakers JM, Gardener BB, Thomashow LS. 2002. Microbial populations responsible for specific soil suppressiveness to plant pathogens. *Annu Rev Phytopathol*. 40, 309–348. doi:10.1146/annurev.phyto.40.030402.110010

Xiong W, Zhao Q, Zhao J, Xun W, Li R, Zhang R, Wu H, Shen Q, 2015. Different continuous cropping spans significantly affect microbial community member- ship and structure in a vanilla-grown soil as revealed by deep pyrosequencing. *Microb Ecol*. 70, 209–218.

Yadav V, Kumar M, Deep DK, Kumar H, Sharma R, Tripathi T, et al. 2010. A phosphate transporter from the root endophytic fungus Piriformospora indicaplays a role in phosphate transport to the host plant. *J Biol Chem.* 285, 26532–26544. doi:10.1074/jbc.M110.111021

Yuan J, Zhao J, Wen T, Zhao M, Li, R, Goossens P, et al. 2018. Root exudates drive the soil-borne legacy of aboveground pathogen infection. *Microbiome.* 6, 156.

Zamioudis C, Pieterse CMJ. 2012. Modulation of host immunity by beneficial microbes. *Mol Plant-Microbe Interact.* 25, 139–150.

Zolla G, Badri DV, Bakker MG, Manter DK, Vivanco JM. 2013. Soil microbiomes vary in their ability to confer drought tolerance to Arabidopsis. *Appl Soil Ecol.* 68, 1–9.

13 Manipulating the Rhizosphere Microbiome for Plant Health

Amol D. Gholap and Sagar R. Pardeshi*
Department of Pharmaceutics, St. John
Institute of Pharmacy and Research
Palghar, Maharashtra, India

Pankaj R. Khuspe
Department of Pharmaceutics, Shriram Shikshan
Sanstha's, College of Pharmacy
Solapur, Maharashtra, India

Sadikali F. Sayyad and Machindra Chavan
Amrutvahini College of Pharmacy
Sangamner, Maharashtra, India

Navnath T. Hatvate
Institute of Chemical Technology Mumbai,
Marathwada Campus
Jalna, Maharashtra, India

Md Faiyazuddin
School of Pharmacy, Al - Karim University
Katihar, Bihar, India

Md Jasim Uddin
Department of Pharmaceutical Technology,
Faculty of Pharmacy, Universiti Malaya
Kuala Lumpur, Malaysia

13.1 INTRODUCTION

The phytomicrobiome, also known as the plant microbiome, has lately emerged as an essential element of plant scientists' research. This is mainly duc to abundant evidence that the crop microbiome has an essential function in crop and plant health output (Purahong et al. 2018). A greater understanding of plant microbiome and plant microbial interactions is likely to boost crop yield greatly in the future. This chapter provides an overview of current developments in this domain, indicating that it is critical to investigate in order to boost crop yield in the future (Berg et al. 2020; Rowland et al. 2018).

* Corresponding author. Email: amolgholap16@gmail.com

DOI: 10.1201/9781003394662-13

The plant microbiome is made up of a large number of commensal, helpful, and pathogenic bacteria that have an essential function in plant growth and health maintenance (Song, Jin, and Raaijmakers 2021; J. Zhang et al. 2021; Noman et al. 2021). Recent research has demonstrated that soil microorganisms, in collaboration with plant microbiomes, particularly in the rhizosphere, facilitate a variety of plant functions. They accomplish this by producing hormones such as auxins, which aid in root development modifications. Some bacteria absorb iron from the soil, while others solubilize phosphorus and provide it to plants. The plant microbiome protects plants against diseases by indirectly stimulating immune responses. Nitrogen fixers also participate in the transfer of atmospheric nitrogen into ammonia. Plant microbiomes are exceedingly specialized, and their significance in highly integrated microbial networks critical for ecosystem functioning and plant health has recently been investigated (Santoyo 2022; Dlamini, Akanmu, and Babalola 2022; R. Jacoby et al. 2017; Pantigoso, Newberger, and Vivanco 2022). It is important to note that plant microbiology is particular to plant species, plant genotype, and other environmental variables, although the responsibilities of this component are challenging to separate out in natural situations. These microbial residents in the soil and surrounding the plants also produce fungal cell wall and antibiotic-destroying enzymes. It has also been shown that modifying the genetic structure of keystone microbial communities can improve plant performance. A variety of biological and ecological variables influence soil microbiome activity. The crop microbiome is separated into four parts: the rhizosphere, the endosphere, the phyllosphere, and the seed microbiome (Singh et al. 2023; F. Ramos et al. 2023; Ali, Tyagi, and Bae 2023; Sohrabi et al. 2023).

13.1.1 Rhizosphere Microbiome

The microbiome of the rhizosphere is critical for plant health. For example, rhizosphere microbiome associations have been shown to benefit plants by enhancing stress tolerance and offering disease resistance, as presented in Figure 13.1. The availability of naphthalene, limonene, and pinene in the soybean rhizosphere led to greater detoxification and decreased carbon metabolic pathways such as mannose and starch pathways, and the available chemicals were utilized as a source of carbon (Li et al. 2021). Rhizosphere bacteria such as *Pseudomonas fluorescens* used pinene as a carbon source, whereas naphthalene was degraded by numerous other bacterial species such as *Pseudomonas*, *Ralstonia*, *Mycobacterium*, *Streptomyces*, and others. In another study, it was discovered that *Lactococcus* sp., which generates nisin, a lantibiotic, gave resistance to disease in Cassava towards *Xanthomonas axonopodis pv. manihotis* (*Xam*) (Liu et al. 2019; Mendes, Garbeva, and Raaijmakers 2013). It was also shown that *Enterobacter cloacae* is responsible for the amylose content of cassava, while *Pseudomonas* and *Stenotrophomonas* sp. are responsible for the carbohydrate content and viscosity.

Secondary metabolites are produced by genera such as *Lysobacter* and *Aeromicrobium*, which are prevalent in higher pH agricultural soils. These compounds are responsible for the antibacterial activity of the common bean (*Phaseolus vulgaris*) plant against numerous soilborne diseases. When compared to the decade-old microbiome, recent rhizosphere microbiome research of maize plants demonstrated that the oldest and most current germplasms have developed mechanisms for nitrogen fixation conversion and boosted amino acid synthesis (Lata et al. 2022; Berendsen, Pieterse, and Bakker 2012).

It was also proposed that the maize plant had a copiotrophic microbiome, which resulted in lower output in modern agro-systems. Such findings necessitate interbreeding techniques to improve the microbiome and may result in various soil-management methods. The rhizosphere microbiome can potentially be used to predict soil contamination and thus the health of plants. It can also serve as an indicator of soil contamination and thus crop viability (Dlamini et al. 2023; Ayangbenro et al. 2022). Bamboo soil affected with heavy metals such as chromium contained lower levels of the phylum Proteobacteria. Certain Actinobacteria like *Arthrobacter* and *Iamia*, and Acidobacteria like *Granulicella* and *Bryobacter* species have the ability to decrease heavy metals

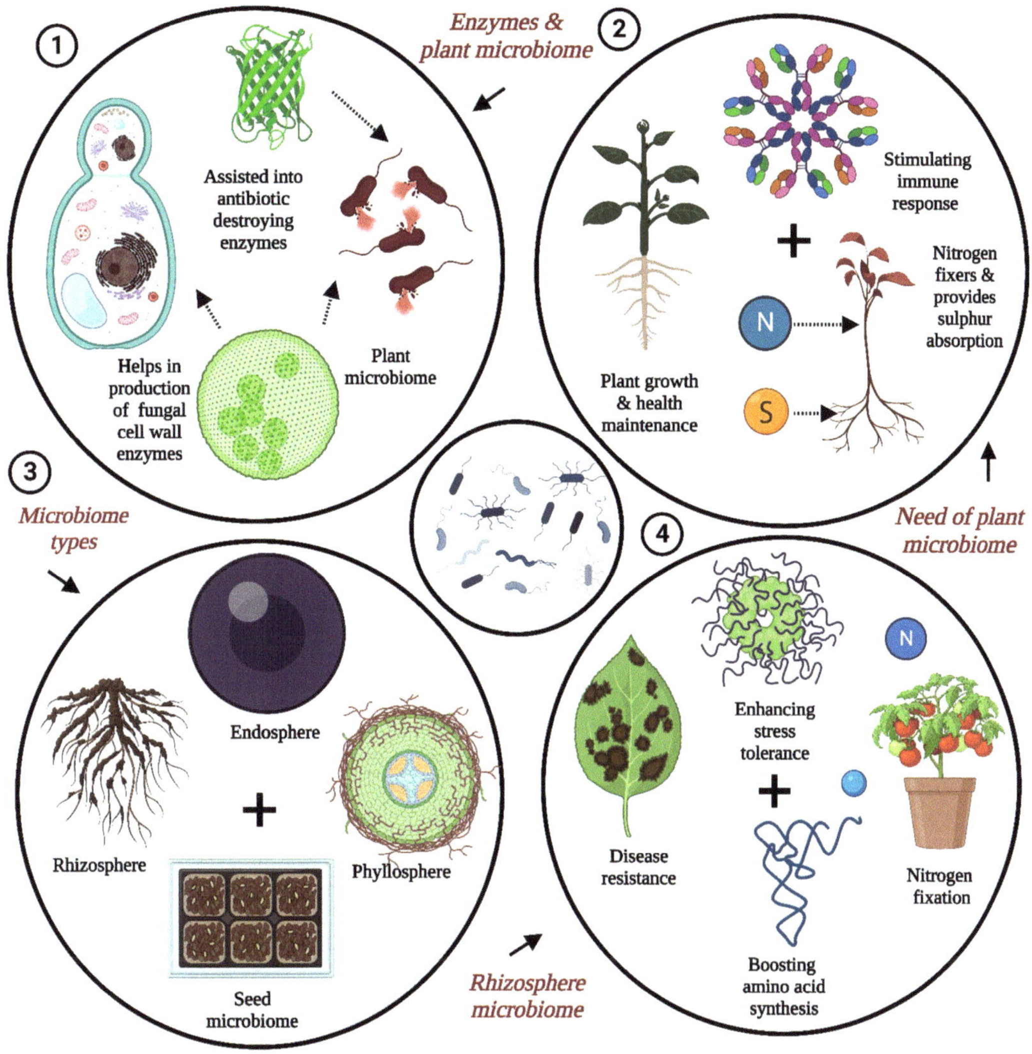

FIGURE 13.1 The summary of the plant microbiome is illustrated as 1. interactions of enzymes of plant microbiome, 2. need for the plant microbiome, 3. types of microbiomes, 4. rhizosphere microbiome.

such as chromium within soils (Chebotar et al. 2023; Kavamura et al. 2021). Anoxygenic photosynthesis is known to occur in the phylum Chloroflexi, with multiple inorganic elements serving as electron donors. Drought resistance can also be achieved through altering plants' rhizospheric microbiota. Extracellular polysaccharides produced by the microbiota of the phyla Actinobacteria and Betaproteobacteria resulted in improved plant development, increased water retention, and a high respiratory rate in wheat plants (Delitte et al. 2021; Nelson 2018; Breitkreuz et al. 2021).

13.1.2 Phyllosphere Microbiome

Despite being subjected to a variety of abiotic challenges such as nutritional stress, UV radiation, and desiccation, the plant leaf surface, called the phyllosphere, harbors a distinct and diversified microbiome. Fungi, cyanobacteria, bacteria, and viruses are the most common phyllosphere invaders (Sivakumar et al. 2020; Sohrabi et al. 2023). Proteobacteria like gammaproteobacteria and beta-, Firmicutes, and Actinobacteria are the main phyla among bacteria, with Pseudomonas and Bacillus being the most prevalent genera. Erwinia, Pantoea, Sphingomonas, Acinetobacter, Gluconobacter, and Xanthomonas are also frequent. *Aureobasidium pullulans*, a yeast-like fungus, predominates in the phyllosphere and on the exterior of fruits. *Alternaria*, *Cladosporium*, *Acremonium*, Penicillium, Aspergillus, and Mucor are common fungal genera found in the leaves, while *Cryptococcus*, *Rhodotorula*, and *Sporobolomyces* are prevalent yeasts (Thapa and Prasanna 2018). Cyanobacteria diversity is regulated by the plant species it occupies as well as the habitat. Nostocales and Oscillatoriales are the major orders. These bacteria react with the plant host in a variety of ways, promoting plant development directly by regulating phytohormones, enhancing nutrient absorption by enhancing bioavailability, and suppressing disease indirectly. The investigation of the phyllosphere microbiome and population is critical in developing sustainable agricultural policy and designing consortiums for their usage as inoculants for phytostimulation, rhizoremediation, biocontrol, and general plant growth improvement. Emerging research reveals that nearby plants, through aerial propagation and genetic variables, also influence the host phyllosphere bacteria (Mendes, Garbeva, and Raaijmakers 2013; Vincent et al. 2022; Fortin Faubert, Labrecque, and Hijri 2022).

13.1.3 Endosphere Microbiome

Sophisticated microorganisms inhabit the inside of plant endospheres and play an important role in plant health and growth by, for example, rendering plants immune to many diseases or through other cooperative associations. Hence, studying the endosphere microbiome is an important aspect of understanding and modulating plant characteristics, ecosystems, and physiological functioning. The organization of the endosphere microbiome in crops was discovered to be constantly changing during withering processes and growth. For example, bacterial and fungal variety improved with the growth of banana plantlets, with bacterial species from the Enterobacteriaceae family observed in varying relative abundance in all specimens. The endospore microbiota was modified to promote resistance to Fusarium wilt in banana plants. Inoculation with modified *Enterobacter* sp. and *Kosakonia* sp. that express 1-Aminocyclopropane-1-Carboxylate (ACC) deaminase in bacterial cell walls enhanced resistance to Fusarium wilt disease in comparison to controls who did not get inoculations. Pseudomonas strain (LTGT-11-2Z) was also discovered to be common in the root endosphere microbiome of the desert plant *Alhagi sparsifolia* . LTGT-11-2Z was discovered to enhance drought tolerance in wheat plants, indicating its potential as a biotechnological product capable of making crop plants more drought-resistant, which would be very significant for farming (Vandenkoornhuyse et al. 2015; Compant et al. 2021; Köberl et al. 2013). To determine the effect of soil on the variety of endosphere microbiota of leaves and roots, sterilized dirt was mixed using 10% grass-grown sand, and Chrysanthemum crops were planted in the mixed soils (Shimasaki et al. 2021).

The control plants were solely grown on sterilized soil. All of the plants were thrips-infested, and the leaves and roots were harvested from both the control plants and Chrysanthemum.

The analysis of the endosphere microbiome demonstrated a rise in bacterial diversity only in the roots and not in the leaf tissues (Shimasaki et al. 2021; French, Kaplan, and Enders 2021). The Pseudomonadaceae family dominated the endosphere, including both leaves and roots. Pseudomonadaceae was more abundant in infected plants than in control plants in the form of leaves. In the case of roots, control plants had a greater quantity. Rare bacterial families were predominantly detected in leaves, indicating that microbial communities in both leaves and roots differed significantly and had different effects on plant growth. Endosphere microbiota investigation of ginseng (Panax sp.) has recently demonstrated that a vast number and species of fungal, archaeal, bacterial, and viral endophytes are found in its endosphere. Bacterial and fungal endophytes are plentiful in its roots, and certain types of endophytes are specific to tissues above and below ground. Several endophytes have been discovered to metabolize ginsenosides, the main active pharmacological constituent of ginseng. Endosphere microorganisms' capacity to metabolize and generate ginsenosides indicates that they have acquired unique or changed genes that allow them to flourish in the ginseng apoplast. Endosphere microbiome investigations clearly contribute data to enhance our knowledge of endophytes and plant evolution, adaptations, and plant interactions. This can help in the design and development of more efficient microbe-based agriculture systems (Hubbard et al. 2019; Köberl et al. 2013; Shimasaki et al. 2021).

13.1.4 Seed Microbiome

Microbes linked with both the root (rhizosphere) and the shoot (phyllosphere) have been widely investigated in plants. So far, little attention has been paid to the function of the microbiome that resides as a component of the seed (Glick and Gamalero 2021; Nelson 2018; De Mandal and Jeon 2023). Recent research has demonstrated that microbiomes influence plant fertilization, health, and productivity in both natural and agricultural settings. Epiphytic and endophytic microbial associations have been documented in all plant tissues, and seeds have both forms of microbial interactions known as the spermosphere and endosperm sphere. Plant seeds are connected with a variety of microbial communities that vary depending on the genome of the plant, their environment, and, to a certain extent, anthropogenic actions (Glick and Gamalero 2021; Berendsen, Pieterse, and Bakker 2012). Fungal and bacterial populations are thought to be the most prevalent occupants of the seed microbiome; nevertheless, archaea have been discovered as recently found seed endophytes in the embryos of native alpine plants. Understanding the good microbiome can help boost crop productivity and crop resistance to abiotic and biotic stress. It can be targeted by reintroducing missing helpful seed microorganisms into wild agricultural cultivar seed microbial populations. The importance of soil has also been demonstrated, as microbial transmission to seed is primarily through the soil. It has recently been demonstrated that the endophytic seedling microbiome can be influenced by both abiotic and climatic factors (Johnston-Monje, Gutiérrez, and Lopez-Lavalle 2021; Ding et al. 2019; Shao et al. 2021).

13.2 FACTORS DRIVING THE ASSEMBLY OF THE RHIZOSPHERE MICROBIOME

The majority of the rhizosphere microbial community's prominent members are often rapidly r-strategists that are drawn to and feed on the copious carbon substrates supplied by the plant root and Actinobacteria (Shao et al. 2021; Lata et al. 2022). According to independent research, the majority of bacterial species found in the rhizosphere belong to the family Proteobacteria, followed by Firmicutes and Actinobacteria. Several studies have revealed that the structure and relative frequency of rhizosphere bacterial communities are plant species- or location-specific (land-dominated) (Gao et al. 2019; Bulgarelli et al. 2013). The abiotic parameters of soil characteristics and climate, as well as the biological characteristics of plant species, plant immunology, and biotic interactions, regulate the formation of the microbiome. The summary of factors associated with the rhizosphere microbiome is illustrated in Figure 13.2. Initially, the soil microbiome is regarded as the rhizosphere microbiome's microbial seed bank (Yang Yang et al. 2023).

The organization of the soil microbe community is shaped by physicochemical parameters such as nutrient availability, temperature, soil pH, and soil moisture, which vary between types of soil along biogeographical distance variations (Yang Yang et al. 2023; Xue et al. 2018).

Anthropogenic activities, in particular agricultural management methods, have a significant impact on the physical properties of the soil, impacting the soil and rhizosphere-associated microorganisms (Hermans et al. 2020; Naylor, McClure, and Jansson 2022). Amplicon sequencing-based research on the rhizosphere bacterial diversity of several type A soils is the primary variable impacting the rhizosphere microbial strain in thaliana populations, although temperature is more significant for the rhizosphere fungal community (Naylor, McClure, and Jansson 2022; Guajardo-Leiva et al. 2022). The host plant, in addition to soil type, is a key predictor of the rhizosphere microbiota

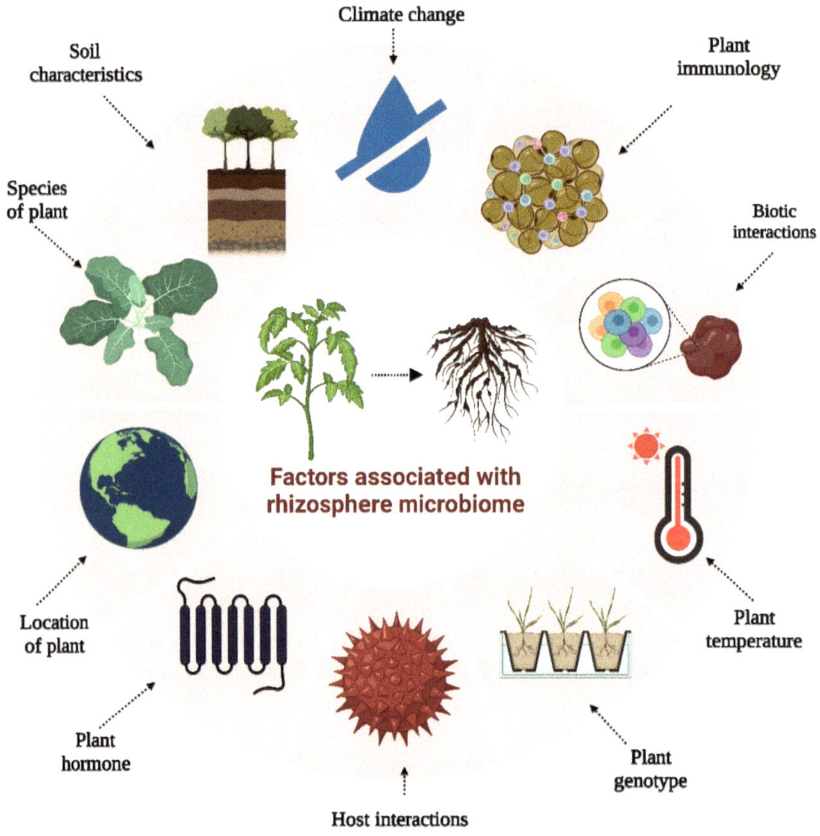

FIGURE 13.2 The factors associated with the rhizosphere microbiome.

makeup. Variations in plant genotype-induced unique rhizodeposits, both inter- and intraspecies, can have a significant impact on rhizosphere microbial colonization (Xun et al. 2021). Many identified genetic variants within the sorghum genome's chromosome region are expected to govern the high heritability of Verrucomicrobia and Planctomycetes in the rhizosphere microbiome, according to a current genome-wide association analysis. Similarly, a group of core bacterial taxa including Proteobacteria, Acidobacteria, Actinobacteria, and Verrucomicrobia were consistently discovered in the rhizosphere, as well as archaeal communities of common bean, regardless of plant development, growing region, or divergent genotypes (A. Kumar and Dubey 2020; C. Yang et al. 2022).

Plant genotypic variances are anticipated to attract diverse bacteria due to changes in metabolite production, which may induce varied reactions in various soil microbes (Gholap et al., 2023; Gholap and Dattatray, 2012b; Gholap and Santhosh, 2012a; Gholap et al., 2023; Gholap and Deepak, 2011; Gholap et al., 2023). Three distinct routes for the manufacture of root triterpenes in *A. thaliana* roots were reconstructed, and their crucial function as carbon sources in the establishment of Arabidopsis-specific root bacterial communities was validated. Furthermore, the manufacture of the plant defense hormone salicylic acid can change the composition of the *A. thaliana* rhizosphere microbial community by selecting tolerant microorganisms (C. Yang et al. 2022; R. P. Jacoby et al. 2020). These heritable or particular metabolite-attractive/resistant rhizosphere microbiome members are consistent with host choice. This host preference is most likely linked to particular transcription reprogramming within roots in order to generate host-specific rhizosphere habitats; therefore, commensal bacteria offer a comparative benefit in their cognate host (Dhar et al. 2015; Hiruma 2019). These host species-specific interactions, however, are not constant. For example, the advancement of high-yielding variants and modern agricultural agronomy inadvertently altered root microbe interactions in maize, recruiting fewer microorganisms responsible for nitrogen (N) acquisition but more members contributing to N losses within the rhizosphere of newly invented cultivars.

Adaptive evolution, on the other hand, may support a non-plant-associated microbe's established reciprocal symbiotic connection to the host plant by increasing its competition for root exudates or increasing tolerance to plant-generated antimicrobial compounds (Chaparro et al. 2013; Carvalhais et al. 2013).

In addition to soil conditions and particular metabolites, discovering the factors that influence the abundance and distribution of plant rhizosphere microorganisms has become a priority.

Many commensal species have been shown to express immunostimulatory bacterial molecular patterns (MAMPs) that can be observed in both phytopathogens and beneficial members, implying that these microorganisms have the capacity to induce immunological responses in their host plants, a phenomenon known as MAMP-triggered immunity (MTI) (Li et al. 2021; Saeed et al. 2021). These rhizosphere microorganisms have evolved methods to decrease host innate immunity in response. For example, the deadly fungus *Magnaporthe oryzae* can create linear actuators that attack the host jasmonic acid (JA) signaling system, but commensal and beneficial microorganisms can avoid detection by programming various flagellin (flg22) peptide variants. In addition to MAMP detection, the reactive oxygen species (ROS) outburst is an important stage in plant innate immune activation (Xun et al. 2021; Jamil et al. 2022). The experimental pathogenic Pseudomonas bacterium syringae pv. tomato DC3000 can produce phetamine A, a newly discovered small chemical that can reduce the stimulatory impacts of L-arginine and spermidine on the ROS burst produced upon bacterial flagellin recognition. Besides that, FERONIA-mediated ROS generation in *A. thaliana*, which is essentially independent of its immunological scaffold function, has been shown to control beneficial pseudomonads exclusively in the rhizosphere (Alawiye and Babalola 2019).

13.3 PLANT-BENEFICIAL FUNCTIONS OF THE RHIZOSPHERE MICROBIOME

Plants recruit a variety of rhizosphere microbial communities to perform key plant development and health tasks. Several research studies are currently being conducted to describe the taxonomic makeup of the microbiome. Understanding the roles of the microbiome is critical for enhancing plant fitness. Yet, because of the wide range of the rhizosphere microbiome, dissecting plant-microbiome interactions is extremely difficult (Mendes, Garbeva, and Raaijmakers 2013; Park, Seo, and Mannaa 2023). A reductionist method to the synthetic population (SynCom) has been established in recent years to examine the fundamental processes that influence community construction and interactions among distinct components in a gnotobiotic environment. For example, inoculation with a basic community of seven strains of bacteria revealed the continuous differences between maize characteristics of breeding lines and hybrids, as well as the makeup of rhizosphere fungal and bacterial communities (Ali, Tyagi, and Bae 2023). This reductionist method would enable global research to readily reproduce studies using two standardized transmission lines: higher microbial cultivation and characterization and plant growth system establishment. The rhizosphere microbiome provides plant hosts with fitness benefits such as growth promotion, nutrient acquisition, disease control, and stress resistance (Figure 13.1b). Several rhizosphere microbiome members have been demonstrated to regulate plant development. Recent research has shown that a wide variety of rhizosphere strains of bacteria can restrict root development and switch the response of plants to smaller concentrations of glyphosate from improving plant growth to growth inhibition. Even so, the readjustment of root growth inhibition of growth can be explained by one bacterial genus, Variovorax, that also contains a preserved auxin-degradation subunit (Berg et al. 2020; Berendsen, Pieterse, and Bakker 2012).

13.3.1 Nutrient Acquisition

The relevant objectives of such a rhizosphere microbiome in enhancing plant nutrition are triggered by interactions between microbes and plants. By increasing the availability of soil quality or by altering the plant root structure of host plants, PGPRs can boost plant nutrient uptake (Hakim et al. 2021). For instance, a larger percentage of N, cycle-related microorganisms are recruited into the rhizosphere of Oryza indica variants, which is regulated by the host's NRT1.1B gene, leading to better nitrogen usage efficiency (NUE) of O. indica varieties than O. japonica cultivars. Particularly noteworthy is that inoculation of a 16-member indica-enriched SynCom verified its contribution to greater NUE in O. indica variants. A recent study on maize types found that origin flavones mostly encourage Oxalobacteraceae concentration inside the rhizosphere, which in turn enhances LRT1-mediated root system formation and fosters maize growth and N uptake. Among several maize germplasm groupings, sweetcorn exhibits a distinctive root exudate composition and attracts N-fixing bacterium taxa of Rhizobium, Burkholderia, and Sphingomonas in the rhizosphere (Pantigoso, Newberger, and Vivanco 2022; Mendes, Garbeva, and Raaijmakers 2013). Some studies demonstrated synthetic paths for the production of rhizopine scyllo-inosamine in both barley and *Medicago truncatula*, which was identified by rhizobia connected with both the cereal tissues and legume in order to take advantage of the role of metabolic pathways in initiating microbe-plant interactions. In phosphate hunger and iron deprivation situations, the plant rhizosphere microbiota can also enhance iron nutrition and phosphorus (P) (Saeed et al. 2021; Mahmud et al. 2021). The phosphate starvation response (PSR) in *A. thaliana*, which is mediated by the master transcriptional regulator PHR1, contributes to the formation of the rhizosphere microbiome under phosphate-limiting environments. Subsequent inoculations of SynComs with various levels of complexity show that the rhizosphere microbiome can focus on promoting the uptake of inorganic phosphate and improve the transcription of plant PSR genes (Ahkami et al. 2017). Opportunistic invaders have been shown to make the host plant more susceptible to phosphate deprivation, suggesting that this PSR-induced rhizosphere microbial assembly may include opportunistic as well as plant-adaptive strategies (Lu et al. 2023; Hartman and Tringe 2019). Moreover, it has been shown that non-mycorrhizal plants' rhizospheres can enhance the hitherto underappreciated significance of fungus-plant interactions in P cycling, which in turn may enhance P uptake in highly P-limited soil and plant growth (Xun et al. 2021; Bongard 2012).

It should be highlighted that the study mentioned above implies that the rhizosphere microbiota regulates the immunological state in conjunction with P acquisition (Stringlis et al. 2018). Similar to an iron chelator with specific antimicrobial activity, such as coumarin scopoletin, the root-specific MYB72-regulated biosynthetic pathway and BGLU42-dependent secretion altered the structure of the rhizosphere microbiome with the help of variable antibacterial activities on particular microbial taxa. It also played significant roles in both immune modulation under iron-limiting conditions and rhizobacteria-mediated iron acquisition (Lurthy et al. 2021; Stringlis, de Jonge, and Pieterse 2019).

13.3.2 Stress Tolerance

Under biotic or abiotic stress circumstances, the host plant is capable of creating a rhizosphere microbiome, which alters plant reactions to external challenges. Among the most frequent environmental challenges is drought. Metabolism functions and bacterial iron transport inside the sorghum rhizosphere are significantly linked to their abundance under drought stress, according to a combined research of comparative genomics and genome-resolved metagenomics. In particular, drought stress led to the depletion of a phytosiderophore iron transporter, which in turn enriched Actinobacteria. This enrichment, however, could be reversed by adding exogenous iron (Abdul Rahman, Abdul Hamid, and Nadarajah 2021; Meena et al. 2017). Rice and wheat may be negatively impacted by the long-term drought-induced abundance of endospheric Actinobacteria, particularly *Streptomyces*. Under salt stress, several microbial species like *Rhizobium*, *Thiomonas*, and *Dyella*may be recruited in order to control the salt-tolerant gene in rice (Saeed et al. 2021). The bacterial consortium in the rhizosphere, rather than a single member, provides long-lasting tolerance to salt stress (Gholap et al., 2023). Additionally, it has been demonstrated that the rhizosphere microbiome affects the timing of plant flowering. It has been shown that root microbial inhibition to the abscisic acid reactions in the root mediates the establishment of root diffusing barriers in the endodermis, specifically suberin deposition, which is essential for the mineral nutrient content of plants. The potential to create crops that are more suited to harsh environmental circumstances has been made possible by these microbe-mediated controls on the flowering time of plants and developing diffusion barriers for the root (Fadiji et al. 2023; Enagbonma et al. 2023; Bogati and Walczak 2022).

13.3.3 Pathogen Suppression

Plant pathogens release effectors that are chemicals aiding host colonization and enhancing disease progression during host colonization in order to establish infection (Tushar V Magar and Amol D Gholap 2011; Dattatray Tupe and Amol D Gholap* 2012; Ajay Khedkar and Dattatray Tupe 2012a; Gholap 2012; Vikas G Rajurkar and Amol D Gholap 2012a, 2012b; Gholap, Bhowmik et al. 2023; Amol D Gholap 2012; Gholap, Gupta et al. 2023; Bothiraja et al. 2014; S. B. D. Amol D Gholap 2012c; Ajay Khedkar and Dattatray Tupe 2012b; Gholap, Said et al. 2023; S. B. D. Amol D Gholap 2012b; Vikas G Rajurkar and Amol D Gholap 2012b; D. T. Amol D Gholap 2012a; Vikas G Rajurkar and Amol D Gholap n.d.; Vikas G Rajurkar 2012; Gholap et al. 2011). For instance, the fungal pathogen *Verticillium dahliae* can control and repress hostile bacteria by secreting the virulence effector VdAve1 in the rhizosphere microbiome. This may make it easier for it to colonize cotton and tomato (Snelders et al. 2021). The dysbiosis of the gram-positive microbial community favors the invasion of pathogens and disease progression. Rhizosphere microorganisms have the ability to alter plant evolution responses in a variety of ways in relation to invasion of pathogens (de Jonge et al. 2012). First, the microbiome of the rhizosphere offers a direct defense against soil-borne diseases, including the release of antimicrobial chemicals to reduce pathogen virulence and growth, as well as competition for resources like nitrogen, carbon, and iron. Then, rhizosphere microorganisms may trigger plant immune responses to improve resistance to disease. To modify host disease susceptibility, individual rhizosphere microorganisms can either trigger or suppress MTI responses (Yuankun Yang et al. 2018).

It's interesting to note that soil management and organic crop cultivation results in *Micrococcaceae* enrichment in the rhizosphere, which increases plant salicylic acid levels and, as a result, increases resistance in plants to insect herbivores. The host plant releases b-caryophyllene, a volatile organic molecule of plants, in response to the PGPR strain of Bacillus amyloliquefaciens GB03, which encourages salicylic acid production in the roots of nearby seedlings (Bano and Muqarab 2017). Finally, the host plant's growth-defense trade-off can be regulated by the rhizosphere microbiota. For instance, under insufficient light, the transcriptional regulator MYC2 in A. thaliana controls a microbiome-root-shoot circuit that favors development over defense (Piechulla, Lemfack, and Kai 2017). Although cereal roots' release of protective secondary metabolites of benzoxazinoids boosts jasmonate signaling pathways for plant resistance, plant growth may be inhibited, and the microbial composition of the rhizosphere may alter to enhance insect herbivore tolerance, even in the following plant generation. The idea of a "soil-borne legacy" denotes an inhibitory memory for plants that grow on the same soil in the future. It is believed that the change of plant root exudates is what causes this suppressing memory to improve plant health (Xun et al. 2021).

13.4 UNRAVELLING THE RHIZOBIOME COMPLEX: PAST AND FUTURE

When the search term "rhizobiome" was used to look up research on the plant microbiome on PubMed and Google Scholar, the results were 1,230 and 18,591 reports, respectively. These papers included both research and review studies; however, for the purpose of this review, researchers were only concerned with research studies that looked at how the field of plant rhizosphere research had developed. Also, we eliminated any studies that didn't exclusively address plant health and the rhizosphere. The majority of the investigations concentrated on how plant age, plant genotype, nutrient availability, water availability, and soil type affected the microbiome's variability and structure.

Biocontrol, bioremediation, and biofertilization investigations were less prevalent (Compant et al. 2019). Due to recent technological advancements and a decline in the price of genome sequencing, scientific attention has recently switched to an in-depth examination of individual organisms (Gholap, Hatvate et al. 2023; Gholap et al. 2024; Gholap, Sayyad et al. 2023; Vilas Bhagat 2012; Rajurkar 2012; Anand M. Mahajan and Amol D. Gholap* 2011; D. S. M. Amol D. Gholap 2011; Anand M. Mahajan and Amol D. Gholap* n.d.; Amol D. Gholap and Mahajan 2011). To completely understand the traits of rhizosphere bacteria and how they interact with one another, interaction between the host plants and microbes is required. To untangle the complex interactions in the rhizobiome, methods including transcriptomics, genomics, proteomics, metagenomics, and metabolic engineering have been at the forefront of microbiota study (Olanrewaju and Babalola 2022). Co-colonization is a vital aspect in the rhizobiome's success in boosting plant growth, and some research has shown this feature as a sign of a sustained rhizobiome. Fewer studies have gone beyond explaining the mechanisms needed for successful co-colonization, which serves as the foundation of the complicated interplay in the rhizosphere (Backer et al. 2018).

According to Molina-Santiago et al. (2019), the outer layer of *Bacillus subtilis* prevents Pseudomonas invasion and improves/modifies the founder of plant development bacteria genera on melon (Molina-Santiago et al. 2019). The rhizome is the soil's microbiologically contested area, and various groups of bacteria, fungi, and other helpful creatures, as well as non-beneficial organisms, are drawn to it. Conflict and the synthesis of antibiotics, poisons, and other chemicals are all a component of one microbe's survival mechanism. As shown in Marchaim et al.'s clinical experiment, co-colonization can provide extra insights into the mechanisms of resistance used by some microorganisms. They discovered a rise in resistance to antibiotics and risk of mortality in individuals co-colonized with *Acinetobacter baumannii*, carbapenem-resistant *Enterobacteriaceae*, or *Pseudomonas aeruginosa* (Olanrewaju and Babalola 2019; Olanrewaju and Babalola 2022; Compant et al. 2019). Although this research focused on people, the findings can provide insight into what could occur in the rhizosphere. Certain genus co-colonization can lead to higher resistance to antibiotics inside the rhizosphere. Only a subset of a microbial community in the process and soil can be cultivated in the lab. Because the vast majority of these microbes are unculturable, research into their characteristics and functions has been difficult until recently (Benmrid et al. 2023; Fasusi, Cruz, and Babalola 2021; Wei et al. 2023).

13.4.1 Plant-Microbe Rhizosphere Interactions: Evolution of the Holobiont

Plants gave rise to a microscopic ecology. As the initial plants pushed their roots into primaeval soil, they discovered a thriving ecosystem of fungal and bacterial species (Gholap et al., 2023). Plants most likely started having an impact on the rhizosphere microbiome from the beginning. The physical aspect of soils was transformed by their roots (Bell et al. 2015). They competed with bacteria to take nutrients from the soil. They drew soil water, causing bacteria to experience a different moisture regime. Their debris accumulated organic carbon that was subsequently digested by heterotrophic microorganisms to produce organic matter in the soil (SOM). When plants started to bleed carbon-rich substrates via the root, such fragile substrates likely favored microorganisms that could digest them quickly (Yue et al. 2023; Li et al. 2021).

Plants influenced the viability of diverse bacterial groups, relationships among microorganisms, and the emergence of new microbes better suited to live in the rhizosphere through changing both the chemical and physical environment inside the rhizobia. Plants design the rhizosphere microbiota, according to compelling evidence (Saeed et al. 2021; Goyal and Habtewold 2023). The most primitive plant lineages have a great potential to change the relative abundance of microbial groups in the soil around the rhizosphere. Plant species sustain various microbial populations. These disparities in microbiomes have been related to variances in root fluid composition and plant nutrient uptake rates (Dinnage et al. 2019; Fasusi, Cruz, and Babalola 2021; Wei et al. 2023). Knowing that plants can influence the rhizosphere microbiome via a variety of mechanisms and that the microbiome can influence plant growth, a plausible hypothesis emerges: plant characteristics that promote microbiomes, improve plant nutrient supply, avoid infections, or generally improve plant health, growth, and performance gain a fitness advantage. As a result, this capacity of a crop to sustain a positive microbiome is indeed a plant feature that is being bred. The microbiome's fitness advantage must compensate for the expense to the plant in misdirected C and energy. Plant-microbe symbiosis can be considered an integrated biological entity called a holobiont. The holobiont could have served as the unit under selection for hundreds of millions of years, directing evolution towards plant characteristics that allowed favorable microbiomes (Wang et al. 2022).

13.4.2 Plant Domestication and Crop Breeding: Decoupling the Holobiont

Manufacturers started to artificially choose crops based on attributes apart from reproductive fitness as humans began to civilize plants and construct cropping systems, departing from biological evolution. Humans unwittingly changed plant features that regulate the microbiome by selecting for increased seed size, decreased bitterness (a critical defensive mechanism), and other traits. Domestication appears to have affected particular microbial populations attracted to crops, albeit the functional implications are uncertain. Several of the plant features being bred might well be related to the microbiome. Plant phenology, for example, has been found to be impacted by soils microbiota (Wang et al. 2022; Bell et al. 2015; Li et al. 2021). Crops that maintained favorable microbial populations may have been chosen in early agriculture because they demonstrated desirable features such as faster growth and productivity and improved disease resistance. The greatest departure from biological evolution is considered to have occurred with the introduction of Haber-Bosch fertilizer. We then partially eliminated the necessity for another of the most important microbial processes, nutrient cycling (Chu and Karr 2017; Rodríguez-Frías et al. 2021). Crop productivity was no longer directly related to a plant's capacity to sustain microbial nutrient cycling. N fertilization, in fact, decreases microbial population as well as variety, favoring copiotrophs over oligotrophs. Plant selection under high fertilization regimes most likely hastened the dissociation of soil microbiota from plant performance. Humans might have been selecting for genotypes that facilitate microbial N metabolism when we started developing plants under high fertilization regimes, as there must be major trade-offs. As a consequence, current cultivars may no longer be able to maintain microbial populations that break down biological nitrogen as well as solubilize necessary nutrients such as phosphorus (Rodríguez-Frías et al. 2021; Ignatieva et al. 2023).

13.4.3 Rhizosphere Engineering: A Systems Perspective

Consider having the capacity to create the optimal rhizosphere. The significance of bacteria in organic matter in soil creation, mineral metabolism and solubility, and disease assault, rhizosphere engineering ideas, and instruments that enable us to modify the microbiota would be critical. Our capacity to manage as well as control the rhizosphere microbiota is currently limited (Hakim et al. 2021). Inoculation represents the most direct technique to influence the microbiota. For decades, foods that contain two or even more types of microbes or fungi have indeed been available commercially. Unfortunately, the majority of these species were obtained using typical culturing methods that do not mimic the microenvironment of soil. Inoculants frequently produce attractive outcomes in controlled laboratory as well as greenhouse settings that are not reliably replicable in wild field settings, yet there is minimal proof that these injected microorganisms establish, compete, and operate in agricultural soils (Saeed et al. 2021; Mishra et al. 2021). This is owing, in part, to the wide variety of environments encountered by bacteria in the field. In addition, essential properties like acidity, mineral composition, and surface vary between soils, but so do the climate regimes encountered by microorganisms in the field. For the inoculated bacteria to survive, reproduce, and function, the circumstances should coincide with their multifunctional niche. They must, however, integrate not just with the native microbiome but also within the food chain (K. Zhang et al. 2023). Most inoculants fail in the field because they are swiftly devoured by predators or are driven out of business for resources by natural bacteria. Healthy plants in nature recruit microorganisms from extremely diverse but weakly connected bulk soils and prefer rhizobial microbial populations, which are not as varied yet strictly regimented into modules of highly interacting bacteria with soil animals. In order to be effective, inoculants must create relationships with the rest of the microbiome, mimicking the highly organized networks found in native roots and shoots (S. Kumar et al. 2022). Other options include adding organic materials containing rich but undefined and unknown microbial mixes to soils, hoping that increasing diversity will result in greater functioning as well as disease tolerance. If this is true, diversity may decrease the capacity of helpful inoculants to flourish. However, evidence for a substantial positive link between microbial diversity and function is limited to the most arid lands (Tahri et al. 2013; Timmis and Ramos 2021).

Additionally, resource pulses might temporarily reduce competition, potentially resulting in disease infestation. This strategy is more likely to succeed in extremely poor soil with little variety and poor food web linkages. While nature has provided an immense diversity of microbes with varying features and purposes, modern genome tinkering and synthetic biology methods provide another avenue for engineering bacteria with specific capabilities (Timmis and Ramos 2021; Hibbing et al. 2010). While these procedures are effective, they pose issues about the introduction of novel genes and encounter negative consumer attitudes regarding gene changes. According to a systems perspective, they are unlikely to function in field settings provided their connections with the ecosystem as well as the soil nutrient chain are taken into account in their design—a challenging task (Gupta, Gupta, and Singh 2017). Plant breeding or engineering is another method of engineering the rhizosphere. We will be able to build certain extracts, rhizome architecture, or other plant features that promote good microbiomes into crops using CRISPR as well as other genome engineering technologies when we identify them. The technique shows potential because it mimics the interactions that sustain beneficial microorganisms in natural systems and had been chosen via holobiont development (Havel et al. 2015). Because SOM quantity and quality have a considerable influence on the microbiome, management strategies that attempt to improve these soil properties may benefit the microbiome. In addition to management measures such as no-till and cover crops, organic material inputs may significantly enhance the organic matter in the soil. Little input of compost, for example, can have long-term effects. Despite this, millions of tonnes of food-related waste are disposed of in landfills every year (Chen, Modi, and Picot 2023).

This organic waste has a huge chance to be processed and returned to the farm. If those same energy bursts overlap alongside beneficial inoculants, the transient drop in competition by native bacteria may improve their likelihood of thriving. There will be good potential to purposely engineer the rhizosphere with greater sophistication in the next decades (Wen et al. 2023; Baudoin, Benizri, and Guckert 2003). The rhizosphere is a complex adaptive system with ecological, pharmacological, as well as mechanical parts that communicate with one another. A systems approach will be required for successful rhizosphere engineering. There is great scope to improve agricultural production efficiency and sustainability by imitating the symbiotic relationships among crops, sediments, and microorganisms, which developed through thousands of years in nature, as we discover how to build the interacting rhizosphere (Rana et al. 2020; Wallenstein 2017).

13.5 CONCLUSION AND FUTURE PERSPECTIVES

The current state of knowledge on the complex plant-microbe relationships that end up taking position inside the rhizobia remains in its adolescence. The role of a root microbiota in plant development is supported by experimental research, and it's increasingly obvious that the plant may alter the makeup of its microbiome. Plants that regulate the microbiota in a manner that enhances their reproductive success are likely to be rewarded through genetic adaptation. That selection pressure appears to have resulted in numerous unique relationships among microbes and plants, and there is increasing proof that crops rely on bacterial assistance in times of need. Several deep insights into selection pressures that create the root microbiota as well as how it impacts the plant are expected in the near future, so next-generation sequencing tools will surely provide new opportunities to examine bacterial interactions between the crop and its associated microbiota. To date, metagenomic investigations of the root microbiome have concentrated on phylogenetic composition, leaving little data regarding the existence of particular functional taxonomic groups. The advancement in operational genomic studies as well as phenotyping will provide insight into the microbiome's functions and responsibilities. Uncovering the processes by which plants manage their microbiome and how the microbiome influences plant health will eventually offer new paths for increasing agricultural quality and efficiency.

REFERENCES

Abdul Rahman, Nur Sabrina Natasha, Nur Wahida Abdul Hamid, and Kalaivani Nadarajah. 2021. "Effects of Abiotic Stress on Soil Microbiome." *International Journal of Molecular Sciences* 22 (16): 9036. doi:10.3390/ijms22169036.

Ahkami, Amir H., Richard Allen White, Pubudu P. Handakumbura, and Christer Jansson. 2017. "Rhizosphere Engineering: Enhancing Sustainable Plant Ecosystem Productivity." *Rhizosphere* 3 (June): 233–43. doi:10.1016/j.rhisph.2017.04.012.

Alawiye, Temitayo, and Olubukola Babalola. 2019. "Bacterial Diversity and Community Structure in Typical Plant Rhizosphere." *Diversity* 11 (10): 179. doi:10.3390/d11100179.

Ali, Sajad, Anshika Tyagi, and Hanhong Bae. 2023. "Plant Microbiome: An Ocean of Possibilities for Improving Disease Resistance in Plants." *Microorganisms* 11 (2): 392. doi:10.3390/microorganisms11020392.

Ayangbenro, Ayansina Segun, Chinenyenwa Fortune Chukwuneme, Modupe Stella Ayilara, Funso Raphael Kutu, Motlagomang Khantsi, Bartholomew Saanu Adeleke, Bernard R. Glick, and Olubukola Oluranti Babalola. 2022. "Harnessing the Rhizosphere Soil Microbiome of Organically Amended Soil for Plant Productivity." *Agronomy* 12 (12): 3179. doi:10.3390/agronomy12123179.

Backer, Rachel, J. Stefan Rokem, Gayathri Ilangumaran, John Lamont, Dana Praslickova, Emily Ricci, Sowmyalakshmi Subramanian, and Donald L. Smith. 2018. "Plant Growth-Promoting Rhizobacteria: Context, Mechanisms of Action, and Roadmap to Commercialization of Biostimulants for Sustainable Agriculture." *Frontiers in Plant Science* 9 (October). doi:10.3389/fpls.2018.01473.

Bano, A., and R. Muqarab. 2017. "Plant Defence Induced by <scp>PGPR</Scp> against *Spodoptera Litura* in Tomato (*Solanum Lycopersicum* L.)." Edited by H. Papen. *Plant Biology* 19 (3): 406–12. doi:10.1111/plb.12535.

Baudoin, Ezékiel, Emile Benizri, and Armand Guckert. 2003. "Impact of Artificial Root Exudates on the Bacterial Community Structure in Bulk Soil and Maize Rhizosphere." *Soil Biology and Biochemistry* 35 (9): 1183–92. doi:10.1016/S0038-0717(03)00179-2.

Bell, Terrence H., Benoît Cloutier-Hurteau, Fahad Al-Otaibi, Marie-Claude Turmel, Etienne Yergeau, François Courchesne, and Marc St-Arnaud. 2015. "Early Rhizosphere Microbiome Composition Is Related to the Growth and <scp> <scp>Zn</Scp> </Scp> Uptake of Willows Introduced to a Former Landfill." *Environmental Microbiology* 17 (8): 3025–38. doi:10.1111/1462-2920.12900.

Benmrid, Bouchra, Cherki Ghoulam, Youssef Zeroual, Lamfeddal Kouisni, and Adnane Bargaz. 2023. "Bioinoculants as a Means of Increasing Crop Tolerance to Drought and Phosphorus Deficiency in Legume-Cereal Intercropping Systems." *Communications Biology* 6 (1): 1016. doi:10.1038/s42003-023-05399-5.

Berendsen, Roeland L., Corné M.J. Pieterse, and Peter A.H.M. Bakker. 2012. "The Rhizosphere Microbiome and Plant Health." *Trends in Plant Science* 17 (8): 478–86. doi:10.1016/j.tplants.2012.04.001.

Berg, Gabriele, Daria Rybakova, Doreen Fischer, Tomislav Cernava, Marie-Christine Champomier Vergès, Trevor Charles, Xiaoyulong Chen, et al. 2020. "Microbiome Definition Re-Visited: Old Concepts and New Challenges." *Microbiome* 8 (1): 103. doi:10.1186/s40168-020-00875-0.

Bogati, Kalisa, and Maciej Walczak. 2022. "The Impact of Drought Stress on Soil Microbial Community, Enzyme Activities and Plants." *Agronomy* 12 (1): 189. doi:10.3390/agronomy12010189.

Bongard, Cindy. 2012. "A Review of the Influence of Root-Associating Fungi and Root Exudates on the Success of Invasive Plants." *NeoBiota* 14 (August): 21–45. doi:10.3897/neobiota.14.2927.

Bothiraja, Chellampillai, Amol D Gholap, Karimunnisa S Shaikh, and Atmaram P Pawar. 2014. "Investigation of Ethyl Cellulose Microsponge Gel for Topical Delivery of Eberconazole Nitrate for Fungal Therapy." *Therapeutic Delivery* 5 (7): 781–94. doi:10.4155/tde.14.43.

Breitkreuz, Claudia, Laura Herzig, François Buscot, Thomas Reitz, and Mika Tarkka. 2021. "Interactions between Soil Properties, Agricultural Management and Cultivar Type Drive Structural and Functional Adaptations of the Wheat Rhizosphere Microbiome to Drought." *Environmental Microbiology* 23 (10): 5866–82. doi:10.1111/1462-2920.15607.

Bulgarelli, Davide, Klaus Schlaeppi, Stijn Spaepen, Emiel Ver Loren van Themaat, and Paul Schulze-Lefert. 2013. "Structure and Functions of the Bacterial Microbiota of Plants." *Annual Review of Plant Biology* 64 (1): 807–38. doi:10.1146/annurev-arplant-050312-120106.

Carvalhais, Lilia C., Frederico Muzzi, Chin-Hong Tan, Jin Hsien-Choo, and Peer M. Schenk. 2013. "Plant Growth in Arabidopsis Is Assisted by Compost Soil-Derived Microbial Communities." *Frontiers in Plant Science* 4. doi:10.3389/fpls.2013.00235.

Chaparro, Jacqueline M., Dayakar V. Badri, Matthew G. Bakker, Akifumi Sugiyama, Daniel K. Manter, and Jorge M. Vivanco. 2013. "Root Exudation of Phytochemicals in Arabidopsis Follows Specific Patterns That Are Developmentally Programmed and Correlate with Soil Microbial Functions." Edited by Keqiang Wu. *PLoS One* 8 (2): e55731. doi:10.1371/journal.pone.0055731.

Chebotar, Vladimir K., Elena P. Chizhevskaya, Evgeny E. Andronov, Nikolai I. Vorobyov, Oksana V. Keleinikova, Maria E. Baganova, Sergey N. Konovalov, Polina S. Filippova, and Veronika N. Pishchik. 2023. "Assessment of the Rhizosphere Bacterial Community under Maize Growth Using Various Agricultural Technologies with Biomodified Mineral Fertilizers." *Agronomy* 13 (7): 1855. doi:10.3390/agronomy13071855.

Chen, Wen, Dixi Modi, and Adeline Picot. 2023. "Soil and Phytomicrobiome for Plant Disease Suppression and Management under Climate Change: A Review." *Plants* 12 (14): 2736. doi:10.3390/plants12142736.

Chu, E.W., and J.R. Karr. 2017. "Environmental Impact: Concept, Consequences, Measurement ☆." In *Reference Module in Life Sciences*. Elsevier. doi:10.1016/B978-0-12-809633-8.02380-3.

Compant, Stéphane, Marine C. Cambon, Corinne Vacher, Birgit Mitter, Abdul Samad, and Angela Sessitsch. 2021. "The Plant Endosphere World – Bacterial Life within Plants." *Environmental Microbiology* 23 (4): 1812–29. doi:10.1111/1462-2920.15240.

Compant, Stéphane, Abdul Samad, Hanna Faist, and Angela Sessitsch. 2019. "A Review on the Plant Microbiome: Ecology, Functions, and Emerging Trends in Microbial Application." *Journal of Advanced Research* 19 (September): 29–37. doi:10.1016/j.jare.2019.03.004.

Dattatray Tupe Amol D Gholap*. 2012. "The Threat of Photoaging: Clinical Conditions, Evaluation and Treatment." *Inventi Rapid: Cosmeceuticals [ISSN 0976-3864]* 4: 1–3.

Delitte, Mathieu, Simon Caulier, Claude Bragard, and Nicolas Desoignies. 2021. "Plant Microbiota Beyond Farming Practices: A Review." *Frontiers in Sustainable Food Systems* 5 (March). doi:10.3389/fsufs.2021.624203.

Dhar, Niha, Sumeer Razdan, Satiander Rana, Wajid W. Bhat, Ram Vishwakarma, and Surrinder K. Lattoo. 2015. "A Decade of Molecular Understanding of Withanolide Biosynthesis and In Vitro Studies in Withania Somnifera (L.) Dunal: Prospects and Perspectives for Pathway Engineering." *Frontiers in Plant Science* 6 (November). doi:10.3389/fpls.2015.01031.

Ding, Long-Jun, Hui-Ling Cui, San-An Nie, Xi-En Long, Gui-Lan Duan, and Yong-Guan Zhu. 2019. "Microbiomes Inhabiting Rice Roots and Rhizosphere." *FEMS Microbiology Ecology* 95 (5). doi:10.1093/femsec/fiz040.

Dinnage, Russell, Anna K. Simonsen, Luke G. Barrett, Marcel Cardillo, Nat Raisbeck-Brown, Peter H. Thrall, and Suzanne M. Prober. 2019. "Larger Plants Promote a Greater Diversity of Symbiotic Nitrogen-fixing Soil Bacteria Associated with an Australian Endemic Legume." Edited by Richard Shefferson. *Journal of Ecology* 107 (2): 977–91. doi:10.1111/1365-2745.13083.

Dlamini, Siphiwe Prudence, Akinlolu Olalekan Akanmu, and Olubukola Oluranti Babalola. 2022. "Rhizospheric Microorganisms: The Gateway to a Sustainable Plant Health." *Frontiers in Sustainable Food Systems* 6 (August). doi:10.3389/fsufs.2022.925802.

Dlamini, Siphiwe Prudence, Akinlolu Olalekan Akanmu, Ayomide Emmanuel Fadiji, and Olubukola Oluranti Babalola. 2023. "Maize Rhizosphere Modulates the Microbiome Diversity and Community Structure to Enhance Plant Health." *Saudi Journal of Biological Sciences* 30 (1): 103499. doi:10.1016/j.sjbs.2022.103499.

Enagbonma, Ben Jesuorsemwen, Ayomide Emmanuel Fadiji, Ayansina Segun Ayangbenro, and Olubukola Oluranti Babalola. 2023. "Communication between Plants and Rhizosphere Microbiome: Exploring the Root Microbiome for Sustainable Agriculture." *Microorganisms* 11 (8): 2003. doi:10.3390/microorganisms11082003.

F. Ramos, Rodrigo, Lisiane Sobucki, Estéfany Pawlowski, Janaina S. Sarzi, Jessica E. Rabuske, Lucas G. Savian, Tiago E. Kaspary, and Cristiano Bellé. 2023. "Perspective Chapter: Microorganisms and Their Relationship with Tree Health." In *Current and Emerging Challenges in the Diseases of Trees*. IntechOpen. doi:10.5772/intechopen.110461.

Fadiji, Ayomide Emmanuel, Ajar Nath Yadav, Gustavo Santoyo, and Olubukola Oluranti Babalola. 2023. "Understanding the Plant-Microbe Interactions in Environments Exposed to Abiotic Stresses: An Overview." *Microbiological Research* 271 (June): 127368. doi:10.1016/j.micres.2023.127368.

Fasusi, Oluwaseun Adeyinka, Cristina Cruz, and Olubukola Oluranti Babalola. 2021. "Agricultural Sustainability: Microbial Biofertilizers in Rhizosphere Management." *Agriculture* 11 (2): 163. doi:10.3390/agriculture11020163.
Fortin Faubert, Maxime, Michel Labrecque, and Mohamed Hijri. 2022. "Ectomycorrhizal Fungi Dominated the Root and Rhizosphere Microbial Communities of Two Willow Cultivars Grown for Six-Years in a Mixed-Contaminated Environment." *Journal of Fungi* 8 (2): 145. doi:10.3390/jof8020145.
French, Elizabeth, Ian Kaplan, and Laramy Enders. 2021. "Foliar Aphid Herbivory Alters the Tomato Rhizosphere Microbiome, but Initial Soil Community Determines the Legacy Effects." *Frontiers in Sustainable Food Systems* 5 (April). doi:10.3389/fsufs.2021.629684.
Gao, Xiaoning, Zilin Wu, Rui Liu, Jiayun Wu, Qiaoying Zeng, and Yongwen Qi. 2019. "Rhizosphere Bacterial Community Characteristics over Different Years of Sugarcane Ratooning in Consecutive Monoculture." *BioMed Research International* 2019 (November): 1–10. doi:10.1155/2019/4943150.
Gholap, Amol D. 2012. "Translational Nanomedicine: Status Assessment and Opportunities." *International Journal of Research in Pharmaceutical and Biomedical Sciences* 3 (4): 1454–63.
Gholap, Amol D., Deblina D. Bhowmik, Aditya Y. Deshmukh, and Navnath T. Hatvate. 2023. "Quintessential Impact of Dendrimer Bioconjugates in Targeted Drug Delivery." *Bioconjugated Materials Part 2 Applications in Drug Delivery, Vaccine Formulations and Important Conjugates for Cancer Therapy.* doi:10.1016/bs.coac.2023.02.009.
Gholap, Amol D., Santosh S. Borude, and Anand M. Mahajan. 2011. "SMART CRYSTALS TECHNOLOGY: A REVIEW." *Pharmacologyonline* 3: 238–43.
Gholap, Amol D, Santosh S Borude, Anand M Mahajan, and Mr Amol Dilip Gholap. 2011. "Aquasomes: A Potential Drug Delivery Carrier." *Pharmacologyonline* 3: 230–37.
Gholap, Amol D., and Dattatray Tupe. 2012a. "Potential Applications of Botanical Extract." *Inventi Rapid: Ethnopharmacology [ISSN 0976-3805]* 4: 1–4.
Gholap, Amol D., and Dattatray Tupe. 2012b. "Artificial Viral Envelopes: Emerging Vehicle For Gene Delivery." *PHARMATUTOR-ART-1439*: 1–3.
Gholap, Amol D., and Santosh B. Dighe. 2012a. "Dandruff: Pathogenesis And Treatment." *Inventi Rapid: Cosmeceuticals [ISSN 0976-3864]* 4: 1–4.
Gholap, Amol D., and Santosh B. Dighe. 2012b. "Role of Sebum in Human Health & Acne." *Inventi Rapid: Cosmeceuticals [ISSN 0976-3864]* 4: 1–3.
Gholap, Amol D., Juhi S. Gupta, Pallavi A. Kamandar, Gauri V. Banchhod, and Navnath T. Hatvate. 2023. "Antibody-Drug Conjugates for Cancer Therapy: An up-to-Date Review on the Chemistry and Pharmacology." In *Bioconjugated Materials Part 2 - Applications in Drug Delivery, Vaccine Formulations and Important Conjugates for Cancer Therapy.* doi:10.1016/bs.coac.2023.02.007.
Gholap, Amol D., Navnath T. Hatvate, Pankaj R. Khuspe, Trushali A. Mandhare, Pooja Kashid, and Vinay D. Gaikwad. 2023. "Proteomics in Oncology: Retrospect and Prospects." In *Novel Technologies in Biosystems, Biomedical & Drug Delivery*, 243–69. Singapore: Springer Nature Singapore. doi:10.1007/978-981-99-5281-6_10.
Gholap, Amol D., and Deepak S. Musmade. 2011. "Advances in the Field of Nanooncology: An Overview." *Inventi Impact: NDDS [ISSN 0976-3791]* 4: 1–3.
Gholap, Amol D., and Vikas G. Rajurkar. 2012a. "Lipid Nanocapsules: A Novel Promising Tool as Drug Delivery System." *Inventi Rapid: NDDS [ISSN 0976-3791]* 4: 1–4.
Gholap, Amol D., and Vikas G. Rajurkar. 2012b. "New Threat of Clinical Toxicities Evoked from Nanocarrier Systems." *Inventi Rapid: NDDS [ISSN 0976-3791]* 4: 1–4.
Gholap, Amol D., and Vikas G. Rajurkar. 2012c. "Potential Applications of Superparmagnetic Nanoparticles Technology." *Inventi Rapid: NDDS [ISSN 0976-3791]* 4: 1–3.
Gholap, Amol D., and Vikas G. Rajurkar. n.d. "Potential Applications of Capillary Electromigration Methods For The Improvement Of Separation Efficiency." *Inventi Rapid: Pharm Analysis & Quality Assurance [ISSN 0976-3813]* 2012 (4): 1–3.
Gholap, Amol D., Rushikesh P. Said, Rutuja D. Pawar, Gauri S. Ambore, and Navnath T. Hatvate. 2023. "Importance of Carbohydrate-Drug Conjugates in Vaccine Development: A Detailed Review." In *Bioconjugated Materials Part 2 Applications in Drug Delivery, Vaccine Formulations and Important Conjugates for Cancer Therapy.* doi:10.1016/bs.coac.2023.03.002.
Gholap, Amol D., Sadikali F. Sayyad, Navnath T. Hatvate, Vilas V. Dhumal, Sagar R. Pardeshi, Vivek P. Chavda, and Lalitkumar K. Vora. 2023. "Drug Delivery Strategies for Avobenzone: A Case Study of Photostabilization." *Pharmaceutics* 15 (3): 1008. doi:10.3390/pharmaceutics15031008.
Gholap, Amol D., Satish Rojekar, Harshad S. Kapare, Nikhar Vishwakarma, Sarjana Raikwar, Atul Garkal, Tejal A. Mehta, Harsh Jadhav, Mahendra Kumar Prajapati, and Uday Annapure. 2024. "Chitosan

Scaffolds: Expanding Horizons in Biomedical Applications." *Carbohydrate Polymers* 323 (January): 121394. doi:10.1016/j.carbpol.2023.121394.

Glick, Bernard R., and Elisa Gamalero. 2021. "Recent Developments in the Study of Plant Microbiomes." *Microorganisms* 9 (7): 1533. doi:10.3390/microorganisms9071533.

Goyal, Ravinder K., and Jemaneh Z. Habtewold. 2023. "Evaluation of Legume–Rhizobial Symbiotic Interactions Beyond Nitrogen Fixation That Help the Host Survival and Diversification in Hostile Environments." *Microorganisms* 11 (6): 1454. doi:10.3390/microorganisms11061454.

Guajardo-Leiva, Sergio, Jaime Alarcón, Florence Gutzwiller, Jorge Gallardo-Cerda, Ian S. Acuña-Rodríguez, Marco Molina-Montenegro, Keith A. Crandall, Marcos Pérez-Losada, and Eduardo Castro-Nallar. 2022. "Source and Acquisition of Rhizosphere Microbes in Antarctic Vascular Plants." *Frontiers in Microbiology* 13 (September). doi:10.3389/fmicb.2022.916210.

Gupta, Ankit, Rasna Gupta, and Ram Lakhan Singh. 2017. "Microbes and Environment." In *Principles and Applications of Environmental Biotechnology for a Sustainable Future*, 43–84. Springer Singapore. doi:10.1007/978-981-10-1866-4_3.

Hakim, Sughra, Tahir Naqqash, Muhammad Shoib Nawaz, Iqra Laraib, Muhammad Jawad Siddique, Rabisa Zia, Muhammad Sajjad Mirza, and Asma Imran. 2021. "Rhizosphere Engineering With Plant Growth-Promoting Microorganisms for Agriculture and Ecological Sustainability." *Frontiers in Sustainable Food Systems* 5 (February). doi:10.3389/fsufs.2021.617157.

Hartman, Kyle, and Susannah G. Tringe. 2019. "Interactions between Plants and Soil Shaping the Root Microbiome under Abiotic Stress." *Biochemical Journal* 476 (19): 2705–24. doi:10.1042/BCJ20180615.

Havel, John E., Katya E. Kovalenko, Sidinei Magela Thomaz, Stefano Amalfitano, and Lee B. Kats. 2015. "Aquatic Invasive Species: Challenges for the Future." *Hydrobiologia* 750 (1): 147–70. doi:10.1007/s10750-014-2166-0.

Hermans, Syrie M., Hannah L. Buckley, Bradley S. Case, Fiona Curran-Cournane, Matthew Taylor, and Gavin Lear. 2020. "Using Soil Bacterial Communities to Predict Physico-Chemical Variables and Soil Quality." *Microbiome* 8 (1): 79. doi:10.1186/s40168-020-00858-1.

Hibbing, Michael E., Clay Fuqua, Matthew R. Parsek, and S. Brook Peterson. 2010. "Bacterial Competition: Surviving and Thriving in the Microbial Jungle." *Nature Reviews Microbiology* 8 (1): 15–25. doi:10.1038/nrmicro2259.

Hiruma, Kei. 2019. "Roles of Plant-Derived Secondary Metabolites during Interactions with Pathogenic and Beneficial Microbes under Conditions of Environmental Stress." *Microorganisms* 7 (9): 362. doi:10.3390/microorganisms7090362.

Hubbard, Charley J., Baohua Li, Robby McMinn, Marcus T. Brock, Lois Maignien, Brent E. Ewers, Daniel Kliebenstein, and Cynthia Weinig. 2019. "The Effect of Rhizosphere Microbes Outweighs Host Plant Genetics in Reducing Insect Herbivory." *Molecular Ecology* 28 (7): 1801–11. doi:10.1111/mec.14989.

Ignatieva, Maria, Diana Dushkova, Daniel Jan Martin, Fahimeh Mofrad, Katherine Stewart, and Michael Hughes. 2023. "From One to Many Natures: Integrating Divergent Urban Nature Visions to Support Nature-Based Solutions in Australia and Europe." *Sustainability* 15 (5): 4640. doi:10.3390/su15054640.

Jacoby, Richard P., Li Chen, Melina Schwier, Anna Koprivova, and Stanislav Kopriva. 2020. "Recent Advances in the Role of Plant Metabolites in Shaping the Root Microbiome." *F1000Research* 9 (February): 151. doi:10.12688/f1000research.21796.1.

Jacoby, Richard P., Manuela Peukert, Antonella Succurro, Anna Koprivova, and Stanislav Kopriva. 2017. "The Role of Soil Microorganisms in Plant Mineral Nutrition—Current Knowledge and Future Directions." *Frontiers in Plant Science* 8 (September). doi:10.3389/fpls.2017.01617.

Jamil, Fatima, Hamid Mukhtar, Mireille Fouillaud, and Laurent Dufossé. 2022. "Rhizosphere Signaling: Insights into Plant–Rhizomicrobiome Interactions for Sustainable Agronomy." *Microorganisms* 10 (5): 899. doi:10.3390/microorganisms10050899.

Johnston-Monje, David, Janneth P. Gutiérrez, and Luis Augusto Becerra Lopez-Lavalle. 2021. "Seed-Transmitted Bacteria and Fungi Dominate Juvenile Plant Microbiomes." *Frontiers in Microbiology* 12 (October). doi:10.3389/fmicb.2021.737616.

Jonge, Ronnie de, H. Peter van Esse, Karunakaran Maruthachalam, Melvin D. Bolton, Parthasarathy Santhanam, Mojtaba Keykha Saber, Zhao Zhang, et al. 2012. "Tomato Immune Receptor Ve1 Recognizes Effector of Multiple Fungal Pathogens Uncovered by Genome and RNA Sequencing." *Proceedings of the National Academy of Sciences* 109 (13): 5110–15. doi:10.1073/pnas.1119623109.

Kavamura, Vanessa N., Rodrigo Mendes, Adnane Bargaz, and Tim H. Mauchline. 2021. "Defining the Wheat Microbiome: Towards Microbiome-Facilitated Crop Production." *Computational and Structural Biotechnology Journal* 19: 1200–1213. doi:10.1016/j.csbj.2021.01.045.

Khedkar, Ajay, Dattatray Tupe, and Amol Gholap. 2012a. "Cubosomes: Emerging Drug Delivery System." *Inventi Rapid: NDDS [ISSN 0976-3791]* 4: 1–3.

Khedkar, Ajay, Dattatray Tupe, and Amol Gholap. 2012b. "Pharmacosomes: An Emerging Vesicular System." *Inventi Rapid: NDDS [ISSN 0976-3791]* 4: 1–3.
Köberl, Martina, Ruth Schmidt, Elshahat M. Ramadan, Rudolf Bauer, and Gabriele Berg. 2013. "The Microbiome of Medicinal Plants: Diversity and Importance for Plant Growth, Quality and Health." *Frontiers in Microbiology* 4. doi:10.3389/fmicb.2013.00400.
Kumar, Ashwani, and Anamika Dubey. 2020. "Rhizosphere Microbiome: Engineering Bacterial Competitiveness for Enhancing Crop Production." *Journal of Advanced Research* 24 (July): 337–52. doi:10.1016/j.jare.2020.04.014.
Kumar, Satish, Diksha, Satyavir S. Sindhu, and Rakesh Kumar. 2022. "Biofertilizers: An Ecofriendly Technology for Nutrient Recycling and Environmental Sustainability." *Current Research in Microbial Sciences* 3: 100094. doi:10.1016/j.crmicr.2021.100094.
Lata, Pushp, Kiran Bala, Piyush Goel, Utkarsh Sood, Princy Hira, Vatsala Dwivedi, Abhilash Kumar, Charu Dogra Rawat, and Rup Lal. 2022. "Microbiome and Plant Health." *Indian Journal of Plant Genetic Resources* 35 (3): 355–59. doi:10.5958/0976-1926.2022.00099.7.
Li, Jingtao, Chenyang Wang, Wenxing Liang, and Sihui Liu. 2021. "Rhizosphere Microbiome: The Emerging Barrier in Plant-Pathogen Interactions." *Frontiers in Microbiology* 12 (October). doi:10.3389/fmicb.2021.772420.
Liu, Fang, Tarek Hewezi, Sarah L. Lebeis, Vince Pantalone, Parwinder S. Grewal, and Margaret E. Staton. 2019. "Soil Indigenous Microbiome and Plant Genotypes Cooperatively Modify Soybean Rhizosphere Microbiome Assembly." *BMC Microbiology* 19 (1): 201. doi:10.1186/s12866-019-1572-x.
Lu, Yufan, Yixiu Yan, Jie Qin, Luyan Ou, Xinyu Yang, Fang Liu, and Yunjian Xu. 2023. "Arbuscular Mycorrhizal Fungi Enhance Phosphate Uptake and Alter Bacterial Communities in Maize Rhizosphere Soil." *Frontiers in Plant Science* 14 (June). doi:10.3389/fpls.2023.1206870.
Lurthy, Tristan, Barbara Pivato, Philippe Lemanceau, and Sylvie Mazurier. 2021. "Importance of the Rhizosphere Microbiota in Iron Biofortification of Plants." *Frontiers in Plant Science* 12 (December). doi:10.3389/fpls.2021.744445.
Mahajan, Anand M., and Amol D. Gholap*, Santosh S. Borude. 2011. "POEMS Syndrome: A Review." *Pharmacologyonline* 3: 265–74.
Mahajan, Anand M., and Amol D. Gholap*, Santosh S. Borude. n.d. "Recent Updates on Q Fever: Diagnosis And Treatment." *Pharmacologyonline* 3: 254–64.
Mahmud, Kishan, Ali Missaoui, Kendall Lee, Bhawana Ghimire, Holly W. Presley, and Shiva Makaju. 2021. "Rhizosphere Microbiome Manipulation for Sustainable Crop Production." *Current Plant Biology* 27 (September): 100210. doi:10.1016/j.cpb.2021.100210.
Mandal, Surajit De, and Junhyun Jeon. 2023. "Phyllosphere Microbiome in Plant Health and Disease." *Plants* 12 (19): 3481. doi:10.3390/plants12193481.
Meena, Kamlesh K., Ajay M. Sorty, Utkarsh M. Bitla, Khushboo Choudhary, Priyanka Gupta, Ashwani Pareek, Dhananjaya P. Singh, et al. 2017. "Abiotic Stress Responses and Microbe-Mediated Mitigation in Plants: The Omics Strategies." *Frontiers in Plant Science* 8 (February). doi:10.3389/fpls.2017.00172.
Mendes, Rodrigo, Paolina Garbeva, and Jos M. Raaijmakers. 2013. "The Rhizosphere Microbiome: Significance of Plant Beneficial, Plant Pathogenic, and Human Pathogenic Microorganisms." *FEMS Microbiology Reviews* 37 (5): 634–63. doi:10.1111/1574-6976.12028.
Mishra, Reema, Tripti Grover, Pooja Gulati, and Aparajita Mohanty. 2021. "Rhizosphere Engineering." In *Phytomicrobiome Interactions and Sustainable Agriculture*, 91–117. Wiley. doi:10.1002/9781119644798.ch6.
Molina-Santiago, Carlos, John R. Pearson, Yurena Navarro, María Victoria Berlanga-Clavero, Andrés Mauricio Caraballo-Rodriguez, Daniel Petras, María Luisa García-Martín, et al. 2019. "The Extracellular Matrix Protects Bacillus Subtilis Colonies from Pseudomonas Invasion and Modulates Plant Co-Colonization." *Nature Communications* 10 (1): 1919. doi:10.1038/s41467-019-09944-x.
Naylor, Dan, Ryan McClure, and Janet Jansson. 2022. "Trends in Microbial Community Composition and Function by Soil Depth." *Microorganisms* 10 (3): 540. doi:10.3390/microorganisms10030540.
Nelson, Eric B. 2018. "The Seed Microbiome: Origins, Interactions, and Impacts." *Plant and Soil* 422 (1–2): 7–34. doi:10.1007/s11104-017-3289-7.
Noman, Muhammad, Temoor Ahmed, Usman Ijaz, Muhammad Shahid, Azizullah, Dayong Li, Irfan Manzoor, and Fengming Song. 2021. "Plant–Microbiome Crosstalk: Dawning from Composition and Assembly of Microbial Community to Improvement of Disease Resilience in Plants." *International Journal of Molecular Sciences* 22 (13): 6852. doi:10.3390/ijms22136852.
Olanrewaju, Oluwaseyi Samuel, and Olubukola Oluranti Babalola. 2019. "Bacterial Consortium for Improved Maize (Zea Mays L.) Production." *Microorganisms* 7 (11): 519. doi:10.3390/microorganisms7110519.

Olanrewaju, Oluwaseyi Samuel, and Olubukola Oluranti Babalola. 2022. "The Rhizosphere Microbial Complex in Plant Health: A Review of Interaction Dynamics." *Journal of Integrative Agriculture* 21 (8): 2168–82. doi:10.1016/S2095-3119(21)63817-0.

Pantigoso, Hugo A., Derek Newberger, and Jorge M. Vivanco. 2022. "The Rhizosphere Microbiome: Plant–Microbial Interactions for Resource Acquisition." *Journal of Applied Microbiology* 133 (5): 2864–76. doi:10.1111/jam.15686.

Park, Inmyoung, Young-Su Seo, and Mohamed Mannaa. 2023. "Recruitment of the Rhizo-Microbiome Army: Assembly Determinants and Engineering of the Rhizosphere Microbiome as a Key to Unlocking Plant Potential." *Frontiers in Microbiology* 14 (May). doi:10.3389/fmicb.2023.1163832.

Piechulla, Birgit, Marie Chantal Lemfack, and Marco Kai. 2017. "Effects of Discrete Bioactive Microbial Volatiles on Plants and Fungi." *Plant, Cell & Environment* 40 (10): 2042–67. doi:10.1111/pce.13011.

Purahong, Witoon, Luigi Orrù, Irene Donati, Giorgia Perpetuini, Antonio Cellini, Antonella Lamontanara, Vania Michelotti, Gianni Tacconi, and Francesco Spinelli. 2018. "Plant Microbiome and Its Link to Plant Health: Host Species, Organs and Pseudomonas Syringae Pv. Actinidiae Infection Shaping Bacterial Phyllosphere Communities of Kiwifruit Plants." *Frontiers in Plant Science* 9 (November). doi:10.3389/fpls.2018.01563.

Rajurkar, Vikas G., and Amol D. Gholap. 2012. "Recent Technologies in Nanomedicine for Treatment of Cancer." *Inventi Rapid: Pharm Tech [ISSN 0976-3783]* 4: 1–3.

Rana, Kusam Lata, Divjot Kour, Tanvir Kaur, Imran Sheikh, Ajar Nath Yadav, Vinod Kumar, Archna Suman, and Harcharan Singh Dhaliwal. 2020. "Endophytic Microbes from Diverse Wheat Genotypes and Their Potential Biotechnological Applications in Plant Growth Promotion and Nutrient Uptake." *Proceedings of the National Academy of Sciences, India Section B: Biological Sciences* 90 (5): 969–79. doi:10.1007/s40011-020-01168-0.

Rodríguez-Frías, Francisco, Josep Quer, David Tabernero, Maria Francesca Cortese, Selene Garcia-Garcia, Ariadna Rando-Segura, and Tomas Pumarola. 2021. "Microorganisms as Shapers of Human Civilization, from Pandemics to Even Our Genomes: Villains or Friends? A Historical Approach." *Microorganisms* 9 (12): 2518. doi:10.3390/microorganisms9122518.

Rowland, Ian, Glenn Gibson, Almut Heinken, Karen Scott, Jonathan Swann, Ines Thiele, and Kieran Tuohy. 2018. "Gut Microbiota Functions: Metabolism of Nutrients and Other Food Components." *European Journal of Nutrition* 57 (1): 1–24. doi:10.1007/s00394-017-1445-8.

Saeed, Qudsia, Wang Xiukang, Fasih Ullah Haider, Jiří Kučerik, Muhammad Zahid Mumtaz, Jiri Holatko, Munaza Naseem, et al. 2021. "Rhizosphere Bacteria in Plant Growth Promotion, Biocontrol, and Bioremediation of Contaminated Sites: A Comprehensive Review of Effects and Mechanisms." *International Journal of Molecular Sciences* 22 (19): 10529. doi:10.3390/ijms221910529.

Santoyo, Gustavo. 2022. "How Plants Recruit Their Microbiome? New Insights into Beneficial Interactions." *Journal of Advanced Research* 40 (September): 45–58. doi:10.1016/j.jare.2021.11.020.

Shao, Jiahui, Youzhi Miao, Kaiming Liu, Yi Ren, Zhihui Xu, Nan Zhang, Haichao Feng, Qirong Shen, Ruifu Zhang, and Weibing Xun. 2021. "Rhizosphere Microbiome Assembly Involves Seed-Borne Bacteria in Compensatory Phosphate Solubilization." *Soil Biology and Biochemistry* 159 (August): 108273. doi:10.1016/j.soilbio.2021.108273.

Shimasaki, Tomohisa, Sachiko Masuda, Ruben Garrido-Oter, Takashi Kawasaki, Yuichi Aoki, Arisa Shibata, Wataru Suda, et al. 2021. "Tobacco Root Endophytic Arthrobacter Harbors Genomic Features Enabling the Catabolism of Host-Specific Plant Specialized Metabolites." Edited by David S. Guttman. *MBio* 12 (3). doi:10.1128/mBio.00846-21.

Singh, Brajesh K., Manuel Delgado-Baquerizo, Eleonora Egidi, Emilio Guirado, Jan E. Leach, Hongwei Liu, and Pankaj Trivedi. 2023. "Climate Change Impacts on Plant Pathogens, Food Security and Paths Forward." *Nature Reviews Microbiology* 21 (10): 640–56. doi:10.1038/s41579-023-00900-7.

Sivakumar, Natesan, Ramamoorthy Sathishkumar, Gopal Selvakumar, Rajaram Shyamkumar, and Kalimuthu Arjunekumar. 2020. "Phyllospheric Microbiomes: Diversity, Ecological Significance, and Biotechnological Applications." *Plant Microbiomes for Sustainable Agriculture* 25: 113–72. doi:10.1007/978-3-030-38453-1_5.

Snelders, Nick C., Gabriella C. Petti, Grardy C. M. van den Berg, Michael F. Seidl, and Bart P. H. J. Thomma. 2021. "An Ancient Antimicrobial Protein Co-Opted by a Fungal Plant Pathogen for in Planta Mycobiome Manipulation." *Proceedings of the National Academy of Sciences* 118 (49). doi:10.1073/pnas.2110968118.

Sohrabi, Reza, Bradley C. Paasch, Julian A. Liber, and Sheng Yang He. 2023. "Phyllosphere Microbiome." *Annual Review of Plant Biology* 74 (1): 539–68. doi:10.1146/annurev-arplant-102820-032704.

Song, Chunxu, Kemo Jin, and Jos M Raaijmakers. 2021. "Designing a Home for Beneficial Plant Microbiomes." *Current Opinion in Plant Biology* 62 (August): 102025. doi:10.1016/j.pbi.2021.102025.

Stringlis, Ioannis A., Ronnie de Jonge, and Corn M J Pieterse. 2019. "The Age of Coumarins in Plant–Microbe Interactions." *Plant and Cell Physiology* 60 (7): 1405–19. doi:10.1093/pcp/pcz076.

Stringlis, Ioannis A., Ke Yu, Kirstin Feussner, Ronnie de Jonge, Sietske Van Bentum, Marcel C. Van Verk, Roeland L. Berendsen, Peter A. H. M. Bakker, Ivo Feussner, and Corné M. J. Pieterse. 2018. "MYB72-Dependent Coumarin Exudation Shapes Root Microbiome Assembly to Promote Plant Health." *Proceedings of the National Academy of Sciences* 115 (22). doi:10.1073/pnas.1722335115.

Tahri, Nezha, Wifak Bahafid, Hanane Sayel, and Nama El Ghachtouli. 2013. "Biodegradation: Involved Microorganisms and Genetically Engineered Microorganisms." In *Biodegradation - Life of Science*. InTech. doi:10.5772/56194.

Thapa, Shobit, and Radha Prasanna. 2018. "Prospecting the Characteristics and Significance of the Phyllosphere Microbiome." *Annals of Microbiology* 68 (5): 229–45. doi:10.1007/s13213-018-1331-5.

Timmis, Kenneth, and Juan Luis Ramos. 2021. "The Soil Crisis: The Need to Treat as a Global Health Problem and the Pivotal Role of Microbes in Prophylaxis and Therapy." *Microbial Biotechnology* 14 (3): 769–97. doi:10.1111/1751-7915.13771.

Tushar V Magar, Amol D Gholap, Deepak S Musmade. 2011. "Elastic Vesicles as Drugs Carriers Through The Skin." *Inventi Rapid: NDDS [ISSN 0976-3791]* 4: 1–2.

Vandenkoornhuyse, Philippe, Achim Quaiser, Marie Duhamel, Amandine Le Van, and Alexis Dufresne. 2015. "The Importance of the Microbiome of the Plant Holobiont." *New Phytologist* 206 (4): 1196–1206. doi:10.1111/nph.13312.

Vilas Bhagat, Amol Gholap. 2012. "Gold Nanorods: A Promising Drug Delivery Vehicle." *Inventi Rapid: NDDS [ISSN 0976-3864]* 4: 1–4.

Vincent, Stacey A., Andreas Ebertz, Pietro D. Spanu, and Paul F. Devlin. 2022. "Salicylic Acid-Mediated Disturbance Increases Bacterial Diversity in the Phyllosphere but Is Overcome by a Dominant Core Community." *Frontiers in Microbiology* 13 (February). doi:10.3389/fmicb.2022.809940.

Wallenstein, Matthew D. 2017. "Managing and Manipulating the Rhizosphere Microbiome for Plant Health: A Systems Approach." *Rhizosphere* 3 (June): 230–32. doi:10.1016/j.rhisph.2017.04.004.

Wang, Liyang, Zed Rengel, Kai Zhang, Kemo Jin, Yang Lyu, Lin Zhang, Lingyun Cheng, Fusuo Zhang, and Jianbo Shen. 2022. "Ensuring Future Food Security and Resource Sustainability: Insights into the Rhizosphere." *IScience* 25 (4): 104168. doi:10.1016/j.isci.2022.104168.

Wei, Zhao, Zhu Sixi, Gu Baojing, Yang Xiuqing, Xia Guodong, and Wang Baichun. 2023. "Effects of Cr Stress on Bacterial Community Structure and Composition in Rhizosphere Soil of Iris Tectorum under Different Cultivation Modes." *Microbiology Research* 14 (1): 243–61. doi:10.3390/microbiolres14010020.

Wen, Tao, Penghao Xie, Hongwei Liu, Ting Liu, Mengli Zhao, Shengdie Yang, Guoqing Niu, et al. 2023. "Tapping the Rhizosphere Metabolites for the Prebiotic Control of Soil-Borne Bacterial Wilt Disease." *Nature Communications* 14 (1): 4497. doi:10.1038/s41467-023-40184-2.

Xue, Pei-Pei, Yolima Carrillo, Vanessa Pino, Budiman Minasny, and Alex. B. McBratney. 2018. "Soil Properties Drive Microbial Community Structure in a Large Scale Transect in South Eastern Australia." *Scientific Reports* 8 (1): 11725. doi:10.1038/s41598-018-30005-8.

Xun, Weibing, Jiahui Shao, Qirong Shen, and Ruifu Zhang. 2021. "Rhizosphere Microbiome: Functional Compensatory Assembly for Plant Fitness." *Computational and Structural Biotechnology Journal* 19: 5487–93. doi:10.1016/j.csbj.2021.09.035.

Yang, Chuanzhen, Hongchen Yue, Zheng Ma, Zili Feng, Hongjie Feng, Lihong Zhao, Yalin Zhang, et al. 2022. "Influence of Plant Genotype and Soil on the Cotton Rhizosphere Microbiome." *Frontiers in Microbiology* 13 (September). doi:10.3389/fmicb.2022.1021064.

Yang, Yang, Weiguo Liu, Tingwen Huang, and Yaru Yang. 2023. "Similarities and Differences in the Rhizosphere Biota among Different Ephemeral Desert Plants in Gurbantünggüt Desert." *Environmental Sciences Europe* 35 (1): 18. doi:10.1186/s12302-023-00724-5.

Yang, Yuankun, Yi Zhang, Beibei Li, Xiufen Yang, Yijie Dong, and Dewen Qiu. 2018. "A Verticillium Dahliae Pectate Lyase Induces Plant Immune Responses and Contributes to Virulence." *Frontiers in Plant Science* 9 (September). doi:10.3389/fpls.2018.01271.

Yue, Hong, Wenjie Yue, Shuo Jiao, Hyun Kim, Yong-Hwan Lee, Gehong Wei, Weining Song, and Duntao Shu. 2023. "Plant Domestication Shapes Rhizosphere Microbiome Assembly and Metabolic Functions." *Microbiome* 11 (1): 70. doi:10.1186/s40168-023-01513-1.

Zhang, Janie, Jamie Cook, Jacob T. Nearing, Junzeng Zhang, Renee Raudonis, Bernard R. Glick, Morgan G.I. Langille, and Zhenyu Cheng. 2021. "Harnessing the Plant Microbiome to Promote the Growth of Agricultural Crops." *Microbiological Research* 245 (April): 126690. doi:10.1016/j.micres.2020.126690.

Zhang, Kai, Zed Rengel, Fusuo Zhang, Philip J. White, and Jianbo Shen. 2023. "Rhizosphere Engineering for Sustainable Crop Production: Entropy-Based Insights." *Trends in Plant Science* 28 (4): 390–98. doi:10.1016/j.tplants.2022.11.008.

14 Organic Fertilizer for Engineering the Soil Microbiome

N Hemashenpagam
Department of Microbiology,
Hindusthan College of Arts & Science
Coimbatore, Tamilnadu, India

S Selvajeyanthi
Department of Microbiology, Theeran Chinnamalai
College of Arts and Science for Women
Tirupur, Tamil Nadu, India

14.1 INTRODUCTION

To meet the needs of a population that is always growing, innovative approaches for sustainable food production are essential in the modern world. Agriculture was practiced for thousands of years by ancient people without the use of synthetic chemicals. In order to meet the demand for food, increased agricultural productivity is required. In the middle of the 19th century, work on artificial fertilizers began, and although artificial fertilizer had some short-term benefits, it had negative long-term side effects, such as soil compaction, erosion, a reduction in the overall fertility of the land, and safety concerns regarding dangerous substances contaminating the food supply (Sisay Assefa, 2019). Chemical fertilizers significantly boost crop yields but create a negative impact on natural environments, where nutrient over-enrichment results in a loss of habitats, species richness and functionality, and altering ecosystems. The majority of the soil utilized for agriculture is infertile due to excessive usage of chemical fertilizers. To increase crop yield or biomass and restore the microbiome in soil, standard agricultural practices to improve soil fertility are needed (Robertson and Vitousek, 2009; Sabir *et al.*, 2021). A promising approach to addressing the issue of environmental pollution, which leads to imbalanced biomass and soil infertility due to excessive usage of synthetic chemical fertilizers, is required.

To balance the biomass in the soil by biological approaches, several microbial species are known to contribute to these ecological inputs through processes that are beneficial to plant growth. Microorganisms have adapted to establish ecological communities on earth during millions of years of co-evolution. These microbiomes, or microbial communities, may endure for a time in or on living hosts, including people, animals, and plants, in addition to soil, water, and air. Microbiomes develop synergistic connections through dynamic interactions with their hosts and among themselves. These interactions shape the microbiomes and have a significant impact on the ecosystems and physiology of the hosts and the environment. For instance, the soil microbiome can affect agricultural productivity (Chaparro *et al.*, 2012). Organic fertilizers (OFs) such as earthworm manure and mushroom residues were used to improve the biomass of *Leymus chinensis*, increase the number of *Actinobacteria*, *Proteobacteria* sp., and increase the diversity of the soil bacterial community in degraded grasslands (Shang *et al.*, 2020). Usage of OFs in tea orchards shows that the relative

DOI: 10.1201/9781003394662-14

abundance of organisms belonging to the Burkholderiales genus is significantly increased (Lin *et al.*, 2019). There has been a rise in interest in recent years to study the microbiome and acquire insights into variables that contribute to differences in microbiomes as a result of the widespread impact of the microbiome on ecosystems. There is growing interest in engineering microbiomes to shape microbiota to change desired ecosystems. The application of microbiome engineering to enhance agricultural productivity is also being explored. OFs improve the efficacy of plant-growth-promoting microorganisms present in the soil.

14.2 DIVERSITY OF MICROBIAL COMMUNITIES IN SOIL

The earth is about 4.5 billion years old, and the diversity of microorganisms has a much longer evolutionary history. The soil is enriched with a diversified microbiome, which includes bacteria, fungi, and algae. The microclimate, the availability of soil resources, soil structure, and soil chemistry all have an impact on the size, distribution, and composition of the soil microbiome community (Giri *et al.*, 2001). The dynamic biological systems work in soil to improve fertility through the nutrient cycling process. Microbial diversity in soil is classified into three categories: common microbial flora, microbial diversity based on biological spheres, and chemical transformation, as shown in Table 14.1.

14.3 ORGANIC FERTILIZER

Organic fertilizers are substances with a known chemical composition and high analytical values that provide plant nutrients in a form the plants can use (Gupta PK, 2004). The term "organic fertilizers" refers to fertilizers made from plant, human, or animal remains (such as compost or manure). Natural ingredients are used to make organic fertilizers, which are typically related to our biodegradable wet suit. Compost is often created through the breakdown of biodegradable garbage such as paper, leaves, fruit peelings, and leftover meals. The soil benefits from the addition of organic fertilizers and improves the soil's ability to support plants to grow healthily (Sisay Assefa, 2019). Animal byproducts from the meat industry, peat, and slurry are instances of naturally occurring organic substances as raw material. Organic fertilizers are substances with a carbon basis that improve the productivity and quality of plant development. Organic fertilizers are usually prepared from waste of agricultural and agriculture-based industries. There are six types of organic fertilizers: cattle manure, organic compost, sedimentary rock phosphate, chicken litter, amino acid rich bone meal, and heterogeneous food waste, which are used to prepare vermicompost. They are rich in nutrients and easily biodegradable materials. Slaughterhouse waste, agricultural waste, and sewage sludge are the main sources for OFs. Bio-organic fertilizers are harmless and considered conventional inorganic green fertilizers and are used to improve soil fertility and promote plant growth. The soil microbial community plays an essential role in the regulation of plant growth and development, which is associated with the plant growth process and also regulates soil fertility and quality resistance cycling of carbon, nitrogen, and other nutrients. For example, earthworm manure and mushroom residue enhance the diversity of the bacterial community in the soil and increase bacterial species biomass. They also contribute to the colonization of microorganisms, including bacteria and fungi. OFs are substances with a carbon structure that improves plant growth and productivity. OFs are complex substances that add a variety of secondary and micronutrients, far from being refined and simplified chemicals. Important micronutrients are found in organic materials, including manures, powdered rocks (such as lime, rock phosphate, and greensand), blood meal, bone meal, wood ash, and compost. Additionally, the texture of these materials enhances soil quality rather than degrade it. Emphasis is placed on using only organic fertilizers by organic farmers to maintain fertility. In many ways, organic farming is a way of life rather than just a technique. Organic farming greatly aids a farmer in reducing costs and being self-sufficient in terms of needs

TABLE 14.1
Classification of Diversity of Microbial Communities in Soil

Classification		Microorganisms	Action	Reference
1) Common soil microbial flora		*Arthrobacteria* and *Nocardia*	Nutrition is derived from native soil organic or mineral matter	Herman *et al.*, 1993
		Nitrosomonas	Cellulose decomposition	NA
		Nitrobacter	Nitrogen-utilizing bacteria	NA
		Schizomycetes (*Pseudomonas*, *Eubacteria*, and *Actinomycetes*)	Predominantly reported in the soil	Gaskins *et al.*, 1984; Benson 1988; Paul and Clark, 1989; Liesack and Stackebrandt, 1992; Benizri *et al.*, 2001
		Myxococcus, *Chondrococcus*, *Archangium*, *Polyangium*, *Cytophaga*, and *Sporocytophaga*	Cellulolytic activity	Slater, 1988; Benizri *et al.*, 2001
		Cyanobacteria	Atmospheric nitrogen fixation	Prescott *et al.*, 1996
		Actinomycetes	Organic matter decomposing	Tate, 1995
		Methanogens	Methanogenesis process	Prescott *et al.*, 1996
		Fungi	Degrade cellulose and lignin	Lynch and Hobbie, 1988
		Alternaria, *Aspergillus*, *Cladosporium*, *Dematium*, *Gliocladium*, *Helminthosporium*, *Humicola*, and *Metarhizium*	Maintenance of soil organic matter	Hawksworth, 1991
2) Diversity based on biological spheres	Detritusphere	Fungi	Decomposition	Wicklow *et al.*, 1974
	Drilosphere	Earthworm, bacteria, and fungi	Homogenize the surface soil horizons and aid N2-fixing bacteria	Wolters, 1991
	Porosphere	Bacteria, protozoa, nematodes, and fungi	Phosphorus (increased plant growth)	NA
	Aggregatusphere	Fungi and bacteria	Aggregated stabilization	Beare *et al.*, 1994
	Rhizosphere	Bacteria and Actinomycetes	Physical, chemical, and biological properties of soils fertility	Fitter, 1985
		Mycorrhizal symbionts	Increases in fine root proliferation	Meyer and Linderman, 1986
3) Diversity based chemical transformation	Nitrogen rransformation	Bacteria	Processes of N2 fixation, nitrification, and denitrification	Postgate, 1987
		Eubacteria and archaea	Ammonification	Kaplan, 1983
		Obligate aerobic soil bacteria (ammonium oxidizers and nitrite oxidizers)	Chemoautotrophic nitrification	
		Nitrosomonas, *Nitrococcus*	Oxidizes NH3 to NO2	
		Arthrobacter, Actinomycetes	Heterotrophic nitrification	Payne, 1981
		Mycobacterium, *Clostridium*	Dissimilatory NO3 and NO2 reduction	
		Pseudomonas, *Bacillus*, *Thiobacillus*	Denitrification	

(*Continued*)

TABLE 14.1 (CONTINUED)
Classification of Diversity of Microbial Communities in Soil

Classification	Microorganisms	Action	Reference
	Ectotrophic mycorrhizal species	Ammonification	Wainwright, 1992
	Aspergillus flavus	Nitrification	Killham, 1987
	Bacteria and fungi	Nitrification in acid forest soils	Schimel *et al.* , 1984
Phosphorus transformation	*Pseudomonas* and *Bacillus*	Solubilization of inorganic phosphorus	Barea, 2000
	Fungi	Better agents in the dissolution of phosphates	Chalot *et al.*, 2002
	Soil bacteria	Immobilization and mineralization of organic phosphorus	Subba Rao, 1997
	Thiobacilli (Biosuper)	Enhances the phosphorus nutrition of plants	Widdel and Hansen, 1991
Sulfur transformation	Bacteria, *Desulfovibrio desulfuricans*	Sulfur transformation	Hurst, 2002
Iron transformation	*Bacillus*	Metal reduction	Mehta *et al.*, 1979
	Chladobacteriaceae	Precipitation of iron	
	Thiobacillus ferroxidans	Chemolithotrophic oxidation of Fe2+	
	Aspergillus and *Penicillium*	Fe siderophores	Sollins *et al.*, 1981
	Gallionella, *Siderophacus*, *Siderocapsa*, *Siderophaera*, *Ferribacterium*, *Naumanniela*, *Ochrobium*, *Sideromanas*, *Sideronema*, *Ferrobacillus*, *Siderobacter*, *Siderococcus* Algae *Leptothrix*, *Sphaerotilus*, *Toxothrix*, *Crenothrix*, and *Colnothrix*	Iron transformation	Quastel, 1995

for agro-inputs. Organic farming strives to achieve a balance of ethical, social, environmental, and organic goals.

14.4 SOURCES OF OFS

The following materials are frequently used as raw materials and retained for composting: alfalfa meal, cottonseed meal, corn gluten meal, soybean meal, seaweed, grass clippings, palm kernel, cow manure, poultry manure, earthworm castings, blood meal, bone meal, feather meal, seabird guano, fish meal, fish emulsion, and shell meal. Organic fertilizer is a term for finely formed ingredients. Organic fertilizer is a by-product of waste from animals, vegetables, and minerals. It contains three important nutrients: nitrogen, phosphorus, and potassium, which are released slowly into the soil and increase the productivity of the plant. Organic matter breaks down and converts to a water-soluble form that plants can absorb easily, so it helps to improve soil quality by breaking up heavy clay and improving air circulation in the soil. The organic compost is called organic fertilizer, which is made from commercial sources like garden waste, household waste, and food industry waste. These fertilizers are a good source of essential nutrients needed for plant growth. The impurities and pathogens are killed in the decomposition process by different thermal cycles. In addition, animal manure is the best kind of organic fertilizer and contains a good source of nitrogenous compounds. Organic fertilizers from plant by-products include alfalfa meal, which contains 3% nitrogen, and cottonseed meal, which contains 7% nitrogen and can be used as an insecticide on cotton in the US. Soybean meal contains 7% nitrogen but is expensive to use as a fertilizer. The trace minerals present in seaweed are also used as biofertilizers, as are ashes from wood and other plant residues. These sources are good source for calcium and potash and balance the nutrient level in soil, which is usually alkaline and can have an advanced effect on soil pH. Rocks such as limestone, potassium sulfate, magnesium sulfate, and phosphate gypsum are the ancient methods of improving soil fertility. Mined minerals are added to organic fertilizers and are considered composted manure-based waste materials. Organic fertilizer is made from the waste material of paper leaves, fruit pulps, food waste, industrial waste, garbage, and animal manure. With proper composting, these become biofertilizers, which are organic fertilizers that boost soil health and fertility and provide sufficient nutrition for soil microbes and plants. Different sources of raw material used for organic fertilizer preparation include animal waste sewage, sludge waste, agricultural waste, and numerous secondary components such as powdered rocks, blood meal, bone meal, food industry waste, and wood ash, which are combined to make compost. This compost contains all the nutrients that help maintain soil quality and promote the growth of plants. The positive effects of organic fertilizer on agricultural inputs reduce costs. Animal waste, including slaughterhouse waste such as animal blood meal and bone meal etc., are typically precursors for organic fertilizer. Chicken litter, fish waste, and other animal waste are also used for organic fertilizer. Sewage containing human excreta is also used as well as plant waste from leguminous plants. These are supplements in adding organic matter to the soil (Sisay Assefa, 2019).

14.5 MICROBIOME

Microbiome engineering is evolving to improve the ecosystem, especially in the agricultural field. We have to design for improved function and the establishment of functional design by using organic fertilizer. Soil microbiome establishment has received lot of attention to improve soil fertility and enhance plant growth microbial engineering. The establishment of microbial communities in soil has two general strategies: manipulating the endogenous microbes or introducing new members through organic fertilizer. Organic fertilizers have different microbial populations, such as composted waste. Rhizosphere soil is mainly influenced by agricultural management practices, resulting in a diverse and balanced microbial community that supports soil health and fertility. Good agricultural practices using organic fertilizers play a crucial role in implementing and promoting microbial diversity, biological cycles, and biological activity without harmful chemical fertilizers. This has been achieved by OFs (Albright *et al.*, 2022; Inda *et al.*, 2019; Qiu *et al.*, 2019).

14.6 MECHANISMS USED BY MICROBES IN SOIL IN PROMOTING PLANT HEALTH

Composed organic fertilizer containing consortia of different microbiomes plays a crucial role in soil fertility, and physical and biodiversity of microbial populations in agricultural land. Maintaining soil fertility, plant health and production, stress resilience, and the ecological recycling of carbon, nitrogen, and other nutrients depends upon the microorganisms in the soil (Zuo *et al., 2021*; Li *et al.*, 2023). To sustain soil fertility, organic farming primarily relies on the cycling of organic materials. Organic fertilizers are free or almost free of hazardous chemicals or metals. To increase productivity, organic fields are treated with compost, plant waste, animal dung, and other biological substances. Other organic fertilizers include blood, fish emulsion, bone meal, cottonseed meal, wood ash, and residues of organic waste or plant and animal byproducts from industry. Together with the major and minor organic nutrients, organic fertilizer raises the soil's organic content. Furthermore, natural components including minerals that have been mined are permitted. Organic fertilizers can be either natural or manmade. The carbon, more specifically the carbon-hydrogen bond in organic fertilizers, which inhibits the release of nutrient ions, distinguishes the two. In comparison to inorganic fertilizers, the slower nutrient release enhances the soil's overall physical properties, improving the nutrients' long-term availability, soil aeration, and the prevention of root burn and leaching losses. Organic fertilizers improve soil structure and plant growth by providing microorganisms in the soil with an energy source. Moreover, they release nitrogen, phosphate, and potash in a way that is simple for plants to absorb. Agricultural land for many years has been protected by organic fertilizers from the damaging effects of inorganic fertilizers, which result in soil deterioration and pollution.

There are different microorganisms that play a key role in interactions between plants and the environment. Microbial components of the plant are called the plant microbiome or phyto microbiome. Microorganisms present inside the plants are endophytic, those above the plant parts are epiphytic, and the organisms associated with the roots are rhizospheric (Lakshmanan *et al.*, 2014).

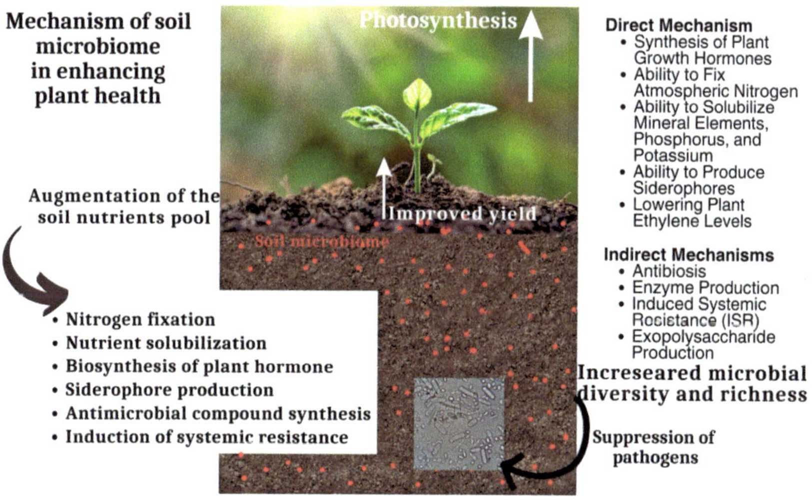

FIGURE 14.1 Mechanism of soil microbiome in enhancing plant health.

Organic fertilizer containing microbes plays a vital role in soil chemistry and physiology, helping to derive substantial ecological processes that provide primary services to plant health and productivity. These microbes also interplay with plants to absorb nutrients and contribute to the microbial community's role in the environment. Rhizobacteria and mycorrhizae organisms can help increase stress tolerance by promoting plant growth, stimulating the production of phytohormones and siderophores, and solubilizing phosphates, lowering ethylene levels, and upregulating the expression of dehydration response and antioxidant genes. The soil remains an unpredictable environment due to numerous factors and a dynamic population with various interactions. Soil food webs are traditionally considered to have distinct energy channels through which resources flow below ground (E. Morriën, 2016).

Rhizosphere engineering or soil microbiome engineering would benefit greatly from our ability to manage the microbiome, as well as the important functions that bacteria play in the creation of soil organic matter, nutrient mineralization and solubilization, and pathogen pressure. Plant genotypes that supported microbiomes that improved their own health during the course of plant and microbial co-evolution probably had a fitness advantage under natural selection. The rhizosphere microbiome and plant selection have been largely separated as a result of plant domestication and crop breeding under fertilization (Matthew, 2017). Plants have developed with a variety of microorganisms that play critical functions in plant growth and health (Stéphane Compant *et al.*, 2019).

Directly, microbes influence plant growth hormone synthesis and its attributes for promoting plant growth, health, and stability. For example, rhizobacteria produce indole acetic acid (IAA) and ethylene, the biomolecules produced by plants, which are always associated with microbes. Rhizobacteria in the root zone producing IAA contribute to rooting development and plant growth. IAA helps in plant root number and development, facilitating nutrient absorption by plants from soil. The dominant bacterial genera are *Pseudomonas* sp. and *Bacillus* sp., as documented. Cytokinin and gibberellins are important plant growth hormones synthesized by the soil microbiome, which contribute to plant growth promotion and ethylene for leaf abscission and fruiting (Glick, 2020; Chandra *et al.*, 2018). Atmospheric air contains 78% nitrogen, which is fixed by nitrogen-fixing microorganisms and is crucial to crop production. This mechanism is carried out by rhizobacteria and plant roots, and fosters soil microbial interaction that leads to the activation of the biological nitrogen fixation process (Xiubin Ke *et al.*, 2019). The lack of potassium in the soil results in poor plant growth and root development. This is managed by plant growth-promoting potassium-stabilizing bacterial communities, such as *Enterobacter* sp., which enable the stabilization of potassium in soil (Liu *et al.*, 2017). For photosynthesis respiration and signal transduction, and energy transfer, phosphorus is needed. Ninety-eight percent of the phosphorus found in soil is insoluble, which is actively immobilized. Phosphorus that is transformed into soluble mono and di forms by biotransformation of microbes occurs due to phosphate-solubilizing bacterial genera, which includes *Bacillus* sp. (Gomez-Ramirez, 2021).

Organic fertilizers typically enhance the microbial proliferation in rhizosphere soil. They contribute to root development for increased nutrient absorption and resistance to plant pathogens. Iron plays a role in energy metabolism as a component of the electron transport chain. *Pseudomonas* sp. which is a plant pathogen, can develop. In iron-deficient soil, the presence of siderophore-producing bacteria facilitates nitrogen fixation, with the help of iron and molybdenum acting as cofactors for the synthesis of the enzyme nitrogenase. Organically fertilized soil contains siderophore-producing microbiomes, including *Enterobacter*, *Azospirillum*, *Rhizobium*, and *Pseudomonas* (Ferreira *et al.*, 2019; Złoch *et al.*, 2016). Ethylene plays a major role in plant growth by promoting the ability of microbes to synthesize the enzyme 1-aminocyclopropane-1-carboxylic (ACC) acid, a precursor of ethylene production. Ethylene supports overall root development, ripening of fruit, and flowering. This aids in the plant growth and biological processes such as xylem formation and flowering. Exogenous ethylene is crucial role in ripening and reducing ethylene levels, while auxin signal transduction enables bacterial action to enhance plant growth. Organic fertilized soil exhibiting

ACC deaminase activity has been identified. It includes *Pseudomonas*, *Alcaligenes*, *Enterobacter*, and *Bacillus* (Khan *et al.*, 2015; Wang *et al.*, 2009; Glick *et al.*, 2007)

14.7 ROLE OF ORGANIC FERTILIZER FOR ENGINEERING SOIL MICROBIOME

The application of organic fertilizers utilizing as many locally or farm-produced materials is feasible. These organic fertilizers are affordable, effective, and eco friendly. The advantage of this kind of fertilizer is that it is cheap, improves soil structure, texture, and airing, increases soil water preservation abilities, and stimulates healthy root development (Twarg, 2006). Incorporating organic fertilizers to increase soil microbial diversity, Bulluck *et al.* (2002) found that plant diseases Phytophthora and Pythium reproduced at much lower rates. The bacterial community structure in healthy soil was similar to that of soil infected by *Phytophthora cinnamani*, but the soil infected by *Phytophthora cinnamomi* was significantly different. The bacterial diversity index of the infested soil also decreased significantly, according to Yang *et al.* (2001) in their comparison of the bacterial community structure in the avocado rhizosphere soil.

Minerals such as limestone, rock phosphate, and other salt organic matter contains nitrogen, phosphorus, sulfate, and potassium, which increase crop growth. From agricultural waste known as green manure, sufficient amounts of phosphorus can be provided from the vegetable parts of stems and leaves, which helps to produce good flowers and fruits. Potassium is another component needed for plant health. Nutrients such as calcium and magnesium are also provided by organic fertilizer in addition to the carbon shortage benefits of adding compost to agricultural soil. This leads to improved soil quality, productivity, and cost savings. The various sources of organic fertilizers have a major impact on nutrient availability and composition, moisture content, microbial ecology, and changes in biochemical and biophysical properties. On the other hand, research has shown that higher levels of native microorganisms aid in the inactivation of *E. coli* O157:H7 and that higher levels of manure in the soil contribute to an increase in the number of background microflora in the

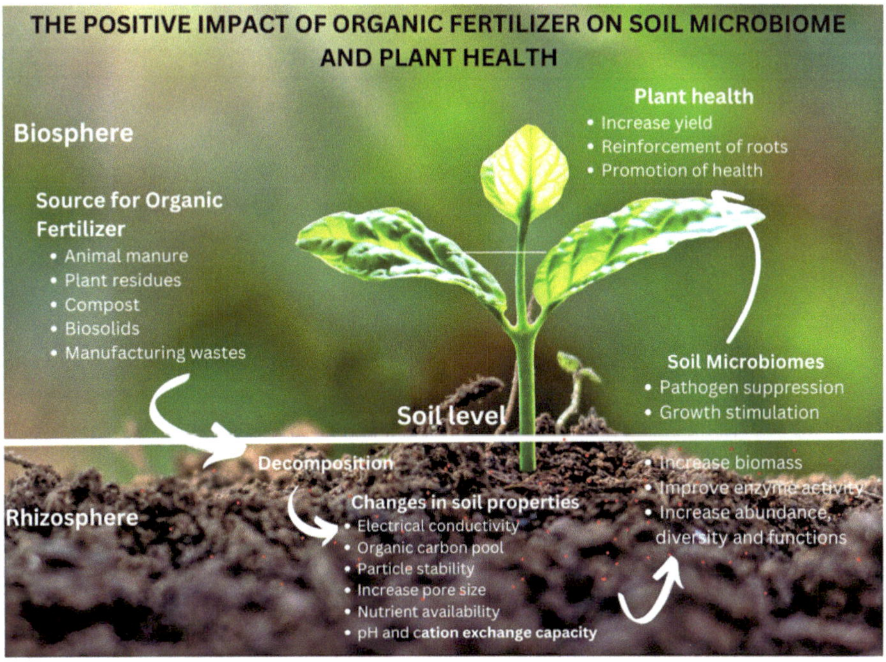

FIGURE 14.2 The positive impact of OFs on soil microbiome and plant health.

soil (Jiang *et al.*, 2002). The activity of soil peroxidase and sucrase can be increased by applying organic fertilizer after fumigation (Li *et al.*, 2022).

Organic fertilizer helps with soil drainage and air circulation. Synthetic fertilizers usually contain chemicals that mix inton the water system, negatively affecting birds and wildlife. Organic fertilizer increases biodiversity and enhances the structure of soil and infiltration of water. The nutrient availability greatly reduce the risk of groundwater pollution, greenhouse effect, and global warming through the carbon in the soil (Sisay Assefa, 2019).

Organic fertilizer helps with soil pH maintenance and variability, depending on the usage of organic fertilizer, which is made from different sources depending on the manure source and soil characteristics. The pH of the soil depends on the microbes that act as a buffering system. Organic fertilizer has three important characteristics: physio-chemical properties, buffering abilities, and biodiversity of biological activity, which have a positive influence on organic matter on soil functionality, and maintaining improved soil quality. It enhances soil structure, increases the porosity and water holding capacity, and decreases the evaporation rate, especially with cattle feed manure or farmyard organic fertilizer. Physical soil quality improvements are generally indicated by increases in water infiltration, macroporosity, aggregate size stability, and soil organic matter (Gupta PK, 2004; Sisay Assefa, 2019).

14.8 CONCLUSION

For plants to access recalcitrant soil-borne nutrients, they are dependent upon the metabolic activities of soil microbiota. Considering the environmental damage associated with current fertilization practices, a current research priority is to optimize plant–microbe nutritional interactions for more sustainable agricultural systems. However, the specific mechanisms governing the assembly of the plant microbiome and its modulation according to plant nutritional status are extremely complex and difficult to predict (Jacoby Richard *et al.*, 2017). The study of the microbiome has advanced greatly in recent years. As a result, while being a young field, microbiome engineering has been successfully used to change microbiota and shows excellent possibilities for agricultural applications to raise agricultural productivity. It is essential to manage agricultural soils for sustainable productivity by encouraging beneficial microorganisms in the rhizosphere and creating a foundation for reducing the application of chemical fertilizers, especially given the significance of the environmental implications, such as soil degradation and the accumulation of pollutants as a result of the effect of chemical fertilizer application. For agricultural production to be sustainable, healthy soils are crucial. The quantity and nutritional worth of the crops are significantly influenced by the soils in which the crops are grown. Subsistence farmers largely rely on soil biota and the ecosystem services it offers to sustain production because they often do not have access to agricultural inputs. Similarly to the way it is in low-input agricultural systems, soil biota is significant. The transformation of nutrients into forms that are more or less accessible to plants is one of the crucial roles that soil organisms play in the process of the nutrient cycle. Moreover, cutting-edge techniques and technologies are being created, such as through synthetic biology, to advance microbiome engineering projects and offer deeper insights into the microbiome. With continued research into the microbiome and its manipulation, microbiome engineering will become a common approach to enhancing both human health and agriculture.

REFERENCES

Albright MBN, Louca S, Winkler DE, et al (2022) Solutions in microbiome engineering: prioritizing barriers to organism establishment. *ISME J* 16(2):331–338. doi:10.1038/s41396-021-01088-5

Barea JM (2000) Rhizosphere and mycorrhiza of field crops. In: Touant A (ed) *Biological resource management: Connecting science and policy.* OECD, INRA Editions and Springer, Berlin Heidelberg, New York, pp. 110–125

Beare MH, Cabrera ML, Hendrix PF, Coleman CD (1994) Aggregate-protected and unprotected pools of organic matter in conventional and no-tillage soils. *Soil Sci Soc Am J* 57:392–399
Benizri E, Baudin E, Guckert A (2001) Root colonization by plant growth promoting Rhizobacteria. *Biocont Sci Technol* 5(11):557–574
Benson DR (1988) The genus Frankia: actinimycetes symbionts of plants. *Microb Sci* 5:9–12
Bulluck, L., Brosius, M., Evanylo, G., & Ristaino, J. (2002) Organic and synthetic fertility amendments influence soil microbial, physical and chemical properties on organic and conventional farms. *Applied Soil Ecology,* 19(2), 147–160 https://doi.org/10.1016/s0929-1393(01)00187-1
Chalot M, Javelle A, Blaudez D, Lambilliote R, Cooke R, Sentenac H, Wipf D, Botton B (2002) An uptake on nutrient transport processes in ectomycorrhizas. *Plant Soil* 244:165–175
Chandra S, Askari K, Kumari M (2018) Optimization of indole acetic acid production by isolated bacteria from Stevia rebaudiana rhizosphere and its effects on plant growth. *J Genet Eng Biotechnol* 16:581–586
Chaparro JM, Sheflin AM, Manter DK, Vivanco JM (2012) Manip- ulating the soil microbiome to increase soil health and plant fertility. *Biol Fertil Soils* 48:489–499
Ferreira MJ, Silva H, Cunha A (2019) Siderophore-producing rhizobacteria as a promising tool for empowering plants to cope with iron limitation in saline soils: A review. *Pedosphere* 29:409–420
Fitter AH (1985) Functional significance of root morphology and root system architecture. In: Fitter AH, Atkinson D, Read DJ, Usher MB (eds) *Ecological interactions in soil.* Blackwell, Oxford, pp. 87–106
Gaskins MH, Albrecht SL, Hubell DH (1984) Rhizosphere bacteria and their use to increase plant productivity: A review. *Agric Ecosyst Environ* 12:99–116
Giri B, Giang P, Kumari R, Prasad R, Varma, A (2005) Microbial diversity in soils. doi:10.1007/3-540-26609-7_2
Giri, A., Dhingra, V., Giri, C., Singh, A., Ward, O. P., & Narasu, M. (2001). Biotransformations using plant cells, organ cultures and enzyme systems: current trends and future prospects. *Biotechnology Advances, 19*(3), 175–199. https://doi.org/10.1016/s0734-9750(01)00054-4
Glick BR (2020) *Beneficial plant-bacterial interactions*, 2nd edn. Springer: Cham, Switzerland, p. 383
Glick BR, Cheng Z, Czarny J, Duan J (2007) Promotion of plant growth by ACC deaminase-producing soil bacteria. *New Perspect. Approaches Plant Growth-Promot. Rhizobact Res* 26(5-6):329–339
Gomez-Ramirez LF, Uribe-Velez D (2021) Phosphorus solubilizing and mineralizing Bacillus spp. contribute to rice growth promotion using soil amended with rice straw. *Curr Microbiol* 78:932–943
Gupta PK (2004) *A handbook of soil, fertilizer and manure*, 2nd edn by Publisher Agrobios (india)
Hawksworth DL (1991) The fungal dimension of diversity: Magnitude, significance, and conservation. *Mycol Res* 95:641–655
Herman RP, Provencio KR, Torrez RJ, Seager GM (1993) Effect of water and nitrogen additions on free-living nitrogen fixer populations in desert grass root zones. *Appl Environ Microbiol* 59:3021–3026
Hurst CJ (2002) An introduction to viral taxonomy and the proposal of Akamara, a potential domain for the genomic acellular agents. In: Hurst CJ (ed) *Viral ecology.* Academic Press, San Diego, pp. 41–62
Inda ME, Broset E, Lu TK, de la Fuente-Nunez C (2019) Emerging frontiers in microbiome engineering. *Trends Immunol* 40:952–973
Jacoby, R., Peukert, M., Succurro, A., Koprivova, A., & Kopriva, S. (2017) The Role of Soil Microorganisms in Plant Mineral Nutrition-Current Knowledge and Future Directions. *Frontiers in plant science, 8, 1617. https://doi.org/10.3389/fpls.2017.01617*
Jiang, X., Morgan, J., & Doyle, M. P. (2002) Fate of Escherichia coli O157:H7 in manure-amended soil. *Applied and environmental microbiology, 68(5), 2605–2609 https://doi.org/10.1128/AEM.68.5.2605-2609.2002*
Kaplan WA (1983) Nitrification. In: Carpenter EJ, Capone DG (eds) *Nitrogen in the marine environment.* Academic Press, New York, pp. 139–190
Ke X, Feng S, Wang J, Lu W, Zhang W, Chen M, Lin M (2019) Effect of inoculation with nitrogen-fixing bacterium Pseudomonas stutzeri A1501 on maize plant growth and the microbiome indigenous to the rhizosphere. *Syst Appl Microbiol* 42:248–260
Khan M, Trivellini A, Fatma M, Masood A, Francini A, Iqbal N, Ferrante A, Khan NA (2015) Role of ethylene in responses of plants to nitrogen availability. *Front. Plant Sci* 6, 927
Killham K (1987) Heterotrophic nitrification. In: Prosser JI (ed) *Nitrification. Society of general microbiology.* Spec Public IRL Press, Oxford, pp. 117–126
Lakshmanan V, Selvaraj G, Bais HP (2014) Functional soil microbiome: belowground solutions to an aboveground problem. *Plant Physiol* Oct;166(2):689–700. doi: 10.1104/pp.114.245811. Epub 2014 Jul 24. PMID: 25059708; PMCID: PMC4213098

Li, Q., Zhang, D., Song, Z., Ren, L., Jin, X., Fang, W., Yan, D., Li, Y., Wang, Q., & Cao, A. (2022) Organic fertilizer activates soil beneficial microorganisms to promote strawberry growth and soil health after fumigation. *Environmental pollution (Barking, Essex : 1987), 295, 118653. https://doi.org/10.1016/j.envpol.2021.118653*
Li S, Fan W, Xu G, Cao Y, Zhao X, Hao S, Deng B, Ren S, Hu S (2023) Bio-organic fertilizers improve Dendrocalamus farinosus growth by remolding the soil microbiome and metabolome. *Front Microbiol*
Liesack W, Stackebrandt E (1992) Occurrence of novel groups of the domain bacteria as revealed by analysis of genetic material isolated from an Australian terrestrial environment. *J Bacteriol* 174:5072–5078
Lin W, Lin M, Zhou H, Wu H, Li Z, Lin W (2019) The effects of chemical and organic fertilizer usage on rhizosphere soil in tea orchards. *PLoS One* 14:e0217018. doi: 10.1371/journal.pone.0217018
Liu S, Tang W, Yang F, Meng J, Chen W, Li X (2017) Influence of biochar application on potassium-solubilizing Bacillus mucilaginosus as potential biofertilizer. *Prep Biochem Biotechnol* 47:32–37
Lynch JM, Hobbie JB (1988) *Microorganismsin action: Concepts and applicationin microbial ecology.* Blackwell, Oxford, 363 pp
Matthew D (2017) Wallenstein, Managing and manipulating the rhizosphere microbiome for plant health: A systems approach. *Rhizosphere* 3, Part 2:230–232, ISSN 2452-2198
Mehta AP, Torma AE, Murr LE (1979) Effect of environmental parameters on the efficiency of biodegradation of basalt rock by fungi. *Biotechnol Bioeng* 21:875–885
Meyer JR, Linderman RG (1986) Selective influences on population of rhizosphere or rhizoplane bacteria and actinomycetes by mycorrhizas formed by Glomus fasciculatum. *Soil Biol Biochem* 18:191–196
Morriën E (2016) Understanding soil food web dynamics, how close do we get? *Soil Biology and Biochemistry* 102:10–13, ISSN 0038-0717
Paul EA, Clark FE (1989) *Soil microbiology and biochemistry.* Academic Press, San Diego
Payne JW (1981) *Denitrification.* Wiley, New York
Postgate JR (1987) *Nitrogen fixation*, 2nd edn. Arnold, London
Prescott LM, Harley JP, Klein DA (1996) The diversity of the microbial world. In: Prescott LM, Harley JP, Klein DA (eds) *Microbiology.* WCB Publishers, Dubuque, Iowa
Qiu ZG, Egidi E, Liu HW, Kaur S, Singh BK (2019) New frontiers in agriculture productivity: optimised microbial inoculants and in situ microbiome engineering. *Biotechnol Adv* 37:107371
Quastel JH (1995) Soil metabolism. *Proc R Soc* 143:159–179
Robertson GP, Vitousek PM (2009) Nitrogen in agriculture: balancing the cost of an essential resource. *Annu Rev Environ Resour* 34:97–125. doi: 10.1146/annurev. Environ.032108.105046
Sabir MS, Shahzadi F, Ali F, Shakeela Q, Niaz Z, Ahmed S (2021) Comparative effect of fertilization practices on soil microbial diversity and activity: an overview. *Curr Microbiol* 78, 3644–3655. doi: 10.1007/s00284-021-02634-2
Schimel JP, Firestone MK, Killham K (1984) Identification of heterotrophic nitrification in a Sieerrs forest soil. *Appl Environ Microbiol* 48:802–806
Shang L, Wan L, Zhou X, Li S, Li X (2020) Effects of organic fertilizer on soil nutrient status, enzyme activity, and bacterial community diversity in Leymus chinensis steppe in Inner Mongolia. China. *PLoS One* 15:e0240559. doi: 10.1371/journal. Pone.0240559
Sisay Assefa (2019) The principal role of organic fertilizer on soil properties and agricultural productivity -a review. *Agri Res & Tech: Open Access J* 22(2):556192. doi: 10.19080/ARTOAJ.2019.22.556192
Slater JH (1988) Microbial population and community dynamics. In: Lunch JM, Hobbie JB (eds) *Microorganisms in action: Concepts and application in microbial ecology.* Blackwell, Oxford, pp. 51–74
Sollins P, Cromack K Jr, Li CY, Fogel R (1981) Role of low-molecular weight organic acids in the inorganic nutrition of fungi and higher plants. In: Carroll GC, Wicklow DT (eds) *The fungal community, its organization and role in ecosystem.* Dekker, New York
Stéphane Compant, Abdul Samad, Hanna Faist, Angela Sessitsch (2019) A review on the plant microbiome: Ecology, functions, and emerging trends in microbial application, Journal of Advanced Research, Volume 19, 2019, Pages 29-37, ISSN 2090-1232
Subba Roa NS (1997) *Soil microbiology.* IBH Publ, Oxford
Tate RL III (1995) *Soil microbiology.* Wiley, New York
Twarg, Sophia, 2006. 'Organic Agriculture: A Trade And Sustainable Development Opportunity For Developing Countries.' In Trade and Environment Review 2006. New York and Geneva: United Nations
Wainwright M (1992) The impact of fungi on environmental biogeochemistry. In: Carroll GC, Wicklow DT (eds) *The fungal community,its organization and role in the ecosystem.* Dekker, New York, pp. 601–618

Wang B, Wang J, Feng X, Lin L, Zhao Y, Jiang W (2009) Effects of 1-MCP and exogenous ethylene on fruit ripening and antioxidants in stored mango. *Plant Growth Regul.* 2009, 57, 185–192

Wicklow MC, Billen WB, Denison WC (1974) Comparison of soil microfungi in 40 year-old stands of pure alder, pure conifer, and alder-conifer mixtures. *Soil Biol Biochem* 6:73–78

Widdel F, Hansen TA (1991) The dissimilatory sulfate and sulfur-reducing bacteria. In: Balows A, Truper HG, Dworkin M, Harder W, Schleifer KH (eds) *The prokaryotes*, 2nd edn. Springer, Berlin Heidelberg New York, pp. 583–634

Wolters V (1991) Soil invertebrates – effects on nutrient turnover and soil structure: A review. Z *Pflanzenern Bodenkd* 154: 389–402

Yang, C. H., Crowley, D. E., Borneman, J., & Keen, N. T. (2001) Microbial phyllosphere populations are more complex than previously realized. *Proceedings of the National Academy of Sciences,* 98(7), 3889–3894 https://doi.org/10.1073/pnas.051633898

Złoch M, Thiem D, Gadzała-Kopciuch R, Hrynkiewicz K (2016) Synthesis of siderophores by plant-associated metallotolerant bacteria under exposure to Cd2+. *Chemosphere* 156:312–325

Zuo J, Zu M, Liu L. et al (2021) Composition and diversity of bacterial communities in the rhizosphere of the Chinese medicinal herb Dendrobium. *BMC Plant Biol* 21:127

15 Land Utilization Pattern and Effects on Soil Microbiome

Abhra Abir Chowdhury
Department of Botany, Darjeeling Government College
Darjeeling, West Bengal, India

Anmol Kumar Sharma
Department of Biosciences, Jamia Millia Islamia
New Delhi, India

Chandan Naskar
Department of Botany, Darjeeling Government College
Darjeeling, West Bengal, India

Sharmistha Ganguly
Department of Dravyaguna, Institute of Post
Graduate Ayurvedic Education and Research
Kolkata, India

Pratyusha De
Department of Botany, University of Burdwan
Burdwan, West Bengal

Arun Kumar Maurya
Department of Botany, Multanimal Modi College, Modinagar
Ghaziabad, India

Sanchita Seal
Department of Botany, Polba Mahavidyalaya
Polba, India

Swastika Banerjee
Department of Botany, Sanjivani Arts, Commerce and Science College
Kopargaon, Maharashtra

 DOI: 10.1201/9781003394662-15

Veena Kumari
Department of Botany, Buddha P.G. College
Kushinagar, UP, India

Dwaipayan Sinha
Department of Botany, Government General Degree College
Mohanpur, Paschim Medinipur, India

15.1 INTRODUCTION

Human activity is causing significant changes to the Earth's surface, with land use patterns changing quickly in response to urbanization, overgrowth or abundance of industries, infrastructure development, agriculture, and urbanization (Rahman, Huq, and Mukul 2023). The intricate colonies of microbes that live beneath the surface, as well as the ecosystems they construct, are profoundly affected by these alterations. The rapid growth of cities worldwide in recent decades presents several significant environmental issues for urban areas, including major land and soil impacts (Xueqin Li, Stringer, and Dallimer 2022). Fertile soils are surrounded and deteriorated by urban sprawl and infrastructures, which also encapsulate or compound them, modify them through excavation or combining with other substances, or exploit them to carry and disperse pollution from transportation, industry, incineration, and waste disposal (Rahman, Huq, and Mukul 2023; Ayilara et al. 2020). The collection of microorganisms that includes bacteria, viruses, and fungi is referred to as the microbiome. The microbiome inhabits a specific ecological niche. Within the context of human biology, the term "microbiota" is frequently employed to denote the collective assemblage of microorganisms that inhabit or colonize specific anatomical regions, including but not limited to the skin and gastrointestinal tract ("Microbiome" 2023). The wide array of microbes known as the soil microbiome is essential for the provisioning of essential ecosystem services like disease management, nutrient cycling, and soil structure preservation (M. Wang et al. 2021). The breakdown of organic material, nitrogen cycling, and ecological stability are all facilitated by the soil microbiome (Benbow et al. 2019; Ranheim Sveen, Hannula, and Bahram 2023). Therefore, it is crucial to comprehend how changes in land use impact the makeup, variety, and performance of soil-microbial communities for the purpose of sustainable land management and ecosystem preservation.

The physical, chemical, and biological makeup of the soil, the deposition of litter, and environmental conditions—all of which depend on daily and seasonal changes—have a very subtle effect on soil bacteria (Jansson and Hofmockel 2019). The structure and makeup of the soil microbial population can be delicately changed at a short depth in topsoil. This might regulate a variety of ecological soil processes, including nutrient breakdown and mineralization in topsoil (Hao, et al., 2021). Understanding the connection between land use patterns and the soil microbiome is crucial for understanding ecosystem sustainability and health. Soil microbiomes are necessary for nutrient cycling, decomposition of organic matter (OM), and soil fertility. Low microbial productivity and biodiversity are frequently expected when taking into account the features of urban soils, such as low fertility, impaired soil health, and a risk for elevated pollution levels (Epp Schmidt et al. 2017). In this context, this chapter aims to investigate how different land utilization patterns affect soil microbe (SM) diversity, abundance, and functionality.

Grazing can influence the diversity of SM. Intense grazing pressure can reduce microbial diversity as certain species become dominant, while rotational grazing or rest periods for pastures can promote greater diversity (Liu et al. 2023). The effects of grazing on plant populations and soil characteristics can affect the composition of the microbial community (Yang et al., 2018). Due to the deposition of OM and nutrients from grazing animals, pastures are rich in microbiological activity and improved nutrient-cycling processes (Garcia et al. 2021). This can stimulate microbial activity and enhance nutrient cycling processes (Mencel, Mocek-Płóciniak, and Kryszak 2022). Hoof action from grazing animals can compact the soil, affecting its physical structure. Soil compaction can reduce aeration and water infiltration, impacting the microbial communities that rely on these conditions (Whalley, Dumitru, and Dexter 1995). The interaction between SM and plant roots in pastures plays a role in carbon sequestration (Ussiri, Lal, and Jacinthe 2006). A healthy soil microbial community (SMC) can be preserved with the use of sustainable grazing management techniques, such as rotational grazing and appropriate stocking rates (Moore, Manter, and Maczko 2023). This balance is essential for soil health and fertility. Grazing and pasture management can influence the soil microbiome in various ways, impacting soil health, nutrient cycling, and plant productivity (Yang et al. 2019). Grazing arid ecosystems occupy about 30% of the land area on Earth (HilleRisLambers et al. 2001). Nonetheless, grazing results in biomass loss, and one of the primary issues with seaweed aquaculture is sharp drops in overall crop productivity (Behera et al. 2022). It also has an impact on the bacterial populations involved, which are crucial for the growth and well-being of the host (Bodenhausen, Horton, and Bergelson 2013).

A multitude of intricate elements, including mineral content, pH, and soil texture, can impact the variety and quantity of soil microorganisms, which in turn impacts the soil microbiome (Abdul Rahman, Abdul Hamid, and Nadarajah 2021; Köster et al. 2021). The diversity and number of soil microorganisms can be influenced by the form, pH, and mineral makeup of the soil (Xia, Rufty, and Shi 2020; Zhalnina et al. 2015). Temperature and precipitation patterns have a significant impact on the microbiomes of soil; extreme conditions can limit microbial diversity (D. Xiao et al. 2021). Over the years, rapid urbanization and infrastructure development, industrial agriculture, mining and extraction, land fragmentation and degradation, etc. that differ from the traditional, sustainable use of land have been put in place. These practices have resulted in land being utilized or abandoned (Pinto-Correia and Mascarenhas 1999). Land use practices that are not sustainable can even lead to severe soil degradation processes (Costa, Freitas, and Sousa 2013). Commercial and domestic construction that impacts the soil microbiome depends on the extent of land disturbance, the use of chemicals, and the post-construction land management techniques (Korneykova et al. 2021; Kosheleva et al. 2018). Sustainable construction practices that prioritize soil health and consider erosion control, native plant restoration, and responsible chemical usage can mitigate adverse effects on soil microbiota and promote long-term soil fertility and ecosystem health. Nitrogen addition also impacts soil microbial diversity (J. Zhang et al. 2017). However, these changes depend on nitrogen sources, soil types, and soil management practices (Zhalnina et al. 2014).

Land management practices, like using fertilizers, pesticides, and tilling, can alter microbial composition and function (Smith et al. 2016; Tripathi et al. 2020). Soil microbes interact with one another and can influence microbial diversity and community structure. Plants can help break down organic pollutants, making it easier for soil microorganisms to eliminate them (Kuiper et al. 2004).

The geographic location of a soil sample can affect its microbiome. Regional climate, vegetation, and historical differences can lead to distinct microbial communities (Jangid et al. 2011; Bardgett and Caruso 2020). The soil microbiome can change over time due to seasonal variations and long-term environmental shifts (X. Zhang et al. 2020). Invasive plant species or invasive microbes can disrupt native soil ecosystems and lead to shifts in microbial composition. The presence and quality of OM in soil influence microbial activity. Microorganisms utilize organic matter as a supply of carbon and energy (Morrissey et al. 2014; Beyer et al. 1995).

Construction activities like excavation and grading can physically disrupt the soil, altering its structure and aeration (Elam and Björdal 2020). This can impact the distribution and diversity of microorganisms. Heavy machinery and construction equipment can compact the soil, reducing pore space and limiting air and water movement (Berli et al. 2003). The alterations in circumstances may result in modifications to the microbial populations in the soil. Construction materials, such as concrete and asphalt, can introduce chemicals into the soil that may harm some microorganisms

FIGURE 15.1 A pictorial representation of various land encroachment activities affecting soil microbiome composition and diversity.

(Azizian et al. 2003). Clearing land for construction often involves the removal of vegetation, which can disrupt the symbiotic relationships amongst plants and SM. This may impact the diversity of soil microorganisms (Porter and Sachs 2020). After construction, the development of urban areas can lead to changes in land use and increased pollution, further altering SM (Sazykina et al. 2022). Figure 15.1 depicts the various land encroachment patterns by humans.

In order to better understand how land alteration caused by humans affects this vital component of the Earth's ecosystem, we examine the intricate interaction among land use trends and the soil microbiome in this chapter. The chapter also aims to elucidate the intricate interactions that drive this critical relationship and their impact on soil health, agricultural efficiency, and ecosystem sustainability.

15.2 CONCEPT OF SOIL MICROBIOME AND ITS DIFFERENT COMPONENTS

A key element in shaping the composition, variety, and ecological functioning of plant communities is the communication between plants and SM (Reynolds et al. 2003). Comprising a wide variety of microorganisms, the soil microbiome is essential to various ecosystem functions and services, including nutrient recycling, disease suppression, and soil structure preservation (M. Wang et al. 2021). Each ecosystem has unique soil properties that harbor various microbes, mainly composed of bacterial communities but also including fungi, viruses, protists, archaea, and other abundant microorganisms. The health of the environment surrounding the soil also depends on these microbes (Omotayo and Babalola 2021). Soil comprises different components, including soil OM, minerals, water, living organisms, and gas (Kalev and Toor 2018). Soil OM refers to the residues of animals, plants, and microorganisms in various phases of decay, which determine the soil quality of agricultural land by supplying essential mineral nutrients, oxygen, water, and root support to crop plants that require them for proper development (Franzluebbers 2002). Soil minerals like clay, silt, and sand determine the soil texture, which affects water retention capacity, air content, and the rate of water movement through the soil (Plante et al. 2006) .

Together, soil and plants can support thousands of microbes that facilitate proper plant growth, effective nutrient utilization, and control over pests and phytopathogens (Ray et al. 2020). Soil microorganisms comprise the greatest diversity and can exist in large numbers as long as they obtain a carbon source for energy. The soil microbiome includes bacteria, viruses (bacteriophages), archaea, fungi, protozoa, algae, microarthropods, actinomycetes, nematodes, and protists (Ray et al. 2020). Bacteria are essential for nutrient cycling, decaying organic litter, fixing nitrogen, and providing it to host plants. Certain bacterial strains can protect plants from phytopathogens (Attiwill and Adams 1993; Dellagi, Quillere, and Hirel 2020). Archaea are well-suited to the harsh environmental conditions required in the nutrient cycle in specific soils and are genetically distinct from bacteria. Fungi are a crucial part of the soil microbiome, acting as decomposers and establishing mycorrhizal associations with plant roots. Single-celled eukaryotes called protozoa typically feed on bacteria to control soil bacterial populations (Rønn et al. 2002). The interaction of these elements, along with above-ground plants and several soil characteristics that influence soil health, plant growth, and nutrient cycling, results in a complex ecosystem. The fungal population is smaller but significantly impacts soil biomass when the soil is left undisturbed (found in untilled soil). In contrast, actinomycetes, protozoa, and bacteria can tolerate more soil disturbance than fungi (found in tilled soil) (Mcgonigle and Miller 1996). By storing soil carbon and regulating nutrient cycling, living soil microbes are crucial to agricultural climate change mitigation (Mondini and Sequi 2008).

15.3 FACTORS CONTROLLING SOIL MICROBIOME

Soil microbiome consists of various organisms, such as bacteria, viruses, fungi, protozoa, and archaea, which interact to form a diversified ecosystem (Jansson and Hofmockel 2019).

Some major significant factors affecting the soil microbiome include: 1. chemical properties of soil such as pH, composition, etc. (Qi et al. 2018); 2. climate, such as temperature and rainfall; 3. land use such as mining and agriculture (Abdul Rahman, Abdul Hamid, and Nadarajah 2021); 4. plant communities such as local flora (Yuan et al. 2015); and 5. fauna such as earthworms and insects (Thakur and Geisen 2019).

Soil pH is one of the most deterministic factors that can be used to evaluate soil characteristics, such as the presence of free mineral acid, the ratio of water, the acidic and essential nature of the soil, and the accessibility of nutrients. Soil pH is one of the best parameters to determine bacterial diversity and composition; soil with a higher pH has been estimated to have more bacterial diversity than soil with a more acidic pH (Fierer and Jackson 2006). Soil pH is related to the parent material of earth, as bedrock, soil weathering, and plant cover are crucial for soil development and pH modulation (Lazzaro, Abegg, and Zeyer 2009). In a study carried out in the Swiss Alps' GLORIA Mountain Summits, an increase in soil pH was associated with an increase in Shannon diversity and bacterial richness. However, the impact on the diversity of fungal species was less pronounced (Adamczyk et al. 2019). It is hypothesized that this relationship results from fungal populations' greater pH tolerance than bacterial communities, which have a more limited pH range for ideal growth (Rousk et al. 2010). Soil as a habitat always changes its environmental conditions. Various soil types result from different nutrient compositions, moisture, and gas content, which leads to multiple microenvironments for soil-dwelling organisms. Additionally, the soil microbiome is essential to many biogeochemical cycles, including those involving carbon, nitrogen, and phosphorus (Tecon and Or 2017).

Both the composition of the microbial population and microbial activity are significantly influenced by soil moisture (Bian et al. 2022). Soil moisture affects the soil microbial population more than nutrient availability (Banerjee et al. 2016). The McMurdo Dry Valleys are a harsh ecosystem where bacterial communities experience increased water availability due to melting ice caps and increased stream flow, which further mobilize nutrients and salts, stimulating microbial activity and eventually leading to copiotroph organisms (Van Horn et al. 2014). The bacterial population has also has been reported to decrease with low precipitation. Still, the bacterial community diversity is not dependent on the precipitation gradient (Bachar et al. 2010). Microbes in the soil react rapidly to environmental alterations such as rainfall, but the composition of the community does not respond similarly to land use or change in soil properties and takes years to develop a change in community composition (Buckley and Schmidt 2003; Kulmatiski and Beard 2011). Soil moisture concentration influences soil microbiota conditions, causing variations in diversity and microbial activity (Kim et al. 2008). Organotrophic bacteria and fungi showed higher diversity in air-dry soil than wet soil (Borowik and Wyszkowska 2016). Dry soils have shown more diversity than irrigated soil (Schjønning et al. 2011). Still, studies have shown that bacterial diversity doesn't increase when the soil is watered (Kim et al. 2008). Both arid and wet soil can lead to decreased microbial biomass (H. Zhang et al. 2018).

Most soil microorganisms cannot survive at temperatures above 50 °C due to the temperature not being optimal for essential enzyme and cell component activity (Bárcenas-Moreno et al. 2009). Higher soil temperatures can also diminish soil moisture content, making it difficult for soil microbiomes to survive, maintain motility, and rhizosphere colonization (Carson et al. 2010). Soil bacterial and fungal communities showed a positive growth response to temperature ranges from –7.5 to 35 °C. Nonetheless, continued exposure to high temperatures resulted in a decline in the microbial community (Nottingham et al. 2019). Due to factors such as climate change leading to warmer temperatures, it is expected that burn severity in boreal forests will increase, which will lead to more fire-related soil microbiome mortality (Köster et al. 2021).

Land use intensification by practices such as agriculture, mining, deforestation, and urbanization can also affect the soil microbiome (Abdul Rahman, Abdul Hamid, and Nadarajah 2021; E. Xiao et al. 2021; Berkelmann et al. 2020; Korneykova et al. 2021).

Agricultural output increases over time with more fertilizers, pesticides, increased irrigation, and cultivation. However, the increased use of fertilizers and pesticides harms the soil microbiota and has complicated and varied effects (Abdul Rahman, Abdul Hamid, and Nadarajah 2021). Organically managed soils have shown more microbial biomass and diversity than conventionally managed soils (Liao, Liang, and Huang 2018; Hartmann et al. 2015). The decline in microbial variety in conventional farming can be linked to excessive use of insecticides, fungicides, and herbicides, which impact soil microbiome and their ecosystem services (Lupatini et al. 2017; Liao et al. 2019; Harkes et al. 2019). Monoculture crops and uncontrolled use of chemicals and fertilizers result in damage to biodiversity, altered functional gene pool, and microbiome homogeneity (Constancias et al. 2013; Figuerola et al. 2014).

Mining activities are a key source of worry since they emit harmful metals into the soil, such as Mn, Ni, Cr, and Fe (Fernandes et al., 2018). In coal mines, the microbiome's composition after mining was significantly different compared to undisturbed sites, as post-mines sites showed less microbial diversity and biomass. Soil microbiome diversity increased as the site recovered, but the composition and microbial abundance remained unchanged (Quadros et al., 2016). Lead contamination has been demonstrated to decrease soil microbial biomass, catalase activity, and carbon and nitrogen levels, while increasing invertase activity (Sun et al., 2022). However, several studies have shown that heavy metal (HM) concentrations in soil have no effect on microbial diversity and that their diversity is more dependent on soil physicochemical parameters such as texture, pH, organic carbon, and so on (Yáñez et al., 2022; Mohamed and Abdelmajid, 2017; Jiang et al., 2019).

Deforestation is a major source of worry. The physicochemical and biological properties of soil change with the transformation of forest area to grazing land using practices such as "slash and burn" (Melo et al. 2017). The alterations in soil characteristics have an impact on the variety and organization of the soil microbiome, which is involved in biological activities including phosphorus transformation (Pedrinho et al. 2023).

The influence of urbanization on soil microbiota is still a little-studied area. A study conducted in the Arctic region comparing natural soil to urban soil showed that urban soil had a pH shift and increased carbon and nutrient content. The total microbial biomass in urban topsoil was lower than that of natural soil. Still, the urban sub-soil showed a 2–2.5-times increase in microbial biomass compared to natural conditions (Korneykova et al. 2021). In other studies, it has been seen that land-use type accounted for most of the β-diversity of soil microbiota, and it also had an effect in shaping the soil microbiome by manipulating the overall richness of the taxa (C. Wang et al. 2021).

Plant root exudates are released in the rhizosphere, acting as nutrition for bacteria and contributing to their growth (Islam et al. 2020). Plants affect soil microbiota, as some plants, such as *Lupinus albus*, release root exudates like protons, which impact bacterial growth around the rhizosphere (Lambers, Clements, and Nelson 2013). Distinct soil microbial taxa can be associated with many plant species, whereas some mycorrhizal fungi, fungal pathogens, and N-fixing bacteria are linked to specific plant species. Therefore, it can be said that specific plant species influence some microbial taxa. In contrast, others are cosmopolitan and are not affected by vegetation but primarily by soil properties (Islam et al. 2020). Additionally, the roots of plants such as *Arabidopsis* can rapidly select their microbiome during distinct developmental stages without being impacted by environmental or soil property factors (Yuan et al. 2015).

In the soil ecosystem, some organisms feed on bacteria and fungi and affect the soil microbiome assembly and role by selective and non-selective feeding. Protists, nematodes, and micro arthropods are examples of direct predators of the soil microbiome. Indirect predators include earthworms and isopods, which consume the soil microbiome while feeding on soil OM (Thakur and Geisen 2019; Bonkowski 2004). These predators can change the composition of microbes in the soil if the presence of predators that feed on fungi is higher in the soil. Thus, these predators can shape the microbiome (Mamilov et al. 2001).

15.4 TYPES OF LAND UTILIZATION PATTERNS AND THE EFFECTS ON SOIL MICROBIOME

15.4.1 Agricultural Land

In the twenty-first century, the major cause of biodiversity loss is agricultural intensification (Sala et al. 2000). Extensive usage of agricultural land destroys the multifunctional aspects of the ecosystem (Allan et al. 2015) which thereby causes imbalances in nature through actions like the elevated greenhouse effect (Burney et al., (2010) and loss of nitrogen through the process of leaching (Cameron, Di, and Moir 2013). Shifting vegetation has been the major cause of land intensification, affecting the soil microbiome and their ecosystem. It is a well-established fact that the organic carbon exudated from the vegetation or soil property modifications is in close vicinity to the roots of the plant (Dedeyn and Vanderputten 2005; Philippot et al. 2013). Furthermore, the soil microbiome is directly impacted by the use of excessive tillage techniques as well as the practical application of fertilizers and pesticides (Lauber et al. 2008; Hartmann et al. 2015; Leff et al. 2015). Land usage in the past has left significant effects on the soil microbiome in the contemporary use of land (Xiaogang Li et al. 2018; Turley et al. 2020). The microbial communities were found to be directly affected before and after the tillage concept was applied on pine savannas (Turley et al. 2020). The biogeochemical cycle is highly influenced and controlled by the microbiome, which is reflected in the function of an ecosystem (van der Heijden, Bardgett, and van Straalen 2008; Wagg et al. 2014).

Though the microbiome of the soil is always a highly diverse and complex system, several investigations have only centered on bacterial and fungal diversity individually (Tardy et al. 2015; Estendorfer et al. 2017; Banerjee et al. 2016; Hartman et al. 2018). While microbes do not influence the ecosystem by their populations alone, they form a complex system that includes not only includes bacteria and fungi but also protists. An assemblage of all these, and their interactions among each other, crucially affects the functioning of the ecosystem (Faust and Raes 2012; de Vos et al. 2017; Friedman, Higgins, and Gore 2017; Russel et al. 2017; Geisen et al. 2018) The protist predates the microbial community and hence affects the agricultural ecosystem (Geisen et al. 2018; Z. Gao et al. 2019; Xiong et al. 2020) and nutrient cycling (Bonkowski 2004). The predation of protists on bacterial communities increases their nitrogen mobilization, contributing to the mineralization of nitrogen in the soil (Clarholm 1985; Barberán et al. 2011). These interactions are exemplified across different domain microbes in the soil and affect its functioning. The methods of co-concurrence networks are often applied to demystify the relationships among microorganisms (Faust et al. 2015). This method does not provide a clear picture of understanding the complexities of microbial interactions but can help in depicting the changes acquired by microbial populations due to changes in land usage of land (Barberán et al. 2011; Faust et al. 2015; Wagg et al. 2019). The availability of water and fluctuations in temperature control the microbial populations of the soil, especially in arid and semi-arid zones (Lacerda-Júnior et al. 2019). Short-term water status is critical in regulating microbial populations in agricultural fields, which vary in pH, texture, and OM (Liao, Liang, and Huang 2018). In forest and grassland ecosystems, a tilled and fertilized system of land use has been immensely studied and is a matter of considerable focus for many researchers (Lacerda-Júnior et al. 2019). Growing crops is the most damaging type of land use in dry-land regions, feeding approximately one-third of the global population while simultaneously affecting soil health and quality (Cowie et al. 2011).

Recent research revealed that types of land use in agriculture can affect the β-diversity of the SMC (Flores-Rentería et al. 2016). Human-made forests, cropland, arid rivers, shallow artificial lakes, and oil wells have different land-use types which drastically and prominently affect the soil biota (C. Wang et al. 2021). The use of various irrigation techniques, fungicides, and herbicides has a significant impact on soil-microbial diversity and ecosystem performance (Ding et al. 2013). Dryland influences the biogeochemical changes of the ecosystem because it covers about 41.5% of the earth's surface ("A Spatial Analysis Approach to the Global Delineation of Dryland Areas of Relevance to the CBD Program of Work on Dry and Subhumid Lands—DocsLib" 2023; Maestre et al. 2016). In semi-arid regions, irrigation is done using different qualities and quantities of water, such as untreated water, freshwater, and untreated wastewater, which widely impact the microbial diversity of the soil. An abundance of genes is noticed in carbon, nitrogen, and phosphorus cycles, affected by different land use and revealed by RNA-based analysis (Lacerda-Júnior et al. 2019).

15.4.2 Grazing and Pastures

Grazing by herbivores imposes negative pressure on plant biomass and creates a decline in crop production (Tan et al. 2020). The process of grazing reduces the carbon (C) and nitrogen (N) reservoirs inside soils, as well as in below-ground plants, microbial communities, and those accumulated through OM decomposition (Zhou et al. 2017). However, occasionally it is employed as a tactic to preserve plant species richness and evenness (Z. Wang et al. 2014). The physico-chemical characteristics of the soil are altered when plant communities are consumed due to grazing by herbivores (Huhe et al. 2017; Mårtensson and Olsson 2012). Overgrazing has been widely established to destroy grassland vegetation by reducing plant growth and diversity and weakening plant health (Y. Bai et al. 2012). Animal excreta, including dung and urine, can alter the soil's carbon and nitrogen contents as well as its air permeability and water potential (Kohler et al. 2005; Z. Wang et al. 2022). All of these factors can influence the makeup of the soil microbiome. Grazing affects microbial composition more than seasonal variations do (Xun et al. 2018). A field study in a desert steppe habitat was done to explore the effects of varying degrees of imposed grazing intensities on soil microorganisms. The study revealed that intensive and excessive grazing practices were associated with an elevated abundance of ascomycete fungi while concurrently leading to a decline in the populations of Glomeromycota and Actinobacteria (Z. Wang et al. 2022). Gentle grazing promotes the growth of Firmicutes and Actinobacteria (Z. Wang et al. 2022). The composition and quantity of the SMC of bacteria and fungi alter when dominant plant species shift due to grazing (Z. Wang et al. 2022). Another study was carried out in the Hulunber Grassland Ecosystem Research Station (HGERS) of CAAS in a typical meadow steppe vegetation, where the grazing intensities are split into G0, G2, G4, and G8 (G stands for grazing) (Xun et al. 2018). Since 2009, these grazing areas have served as summer pastures from June through September. While Chloroflexi, Planctomyces, and Verrucomicrobia predominated in G0 and G2, Glomeromycota, Ascomycota, and Zygomycota increased in relative abundance in G0 and G2 and decreased in the higher intensities. Fast-growing bacteria, including Bacteroidetes, Firmicutes, and Proteobacteria, have become more prevalent in environments with higher grazing intensities, such as G4 and G8. The high bacteria-to-fungal ratio at higher grazing intensities indicates that when grazing pressure rises, bacteria take over as the dominating species (Xun et al. 2018). The Basidiomycetes fungus is slow-growing and susceptible to environmental changes because of its protracted dikaryotic stage (Osono 2007). The arbuscular mycorrhizal fungi (AMF) of the Glomeromycota group of fungi are positively correlated with plant biomass (Guo et al. 2016). Grazing modifies plant biomass, which in turn modifies the relative abundance and mode of fungal colonization (Macdonald et al. 2015). Due to the change in microbial diversity brought about by grazing, dominant microorganisms have an advantage over subordinate microbes (Eldridge et al. 2017). Firmicutes, Bacteroidetes, Nitrospira, Proteobacteria, etc., which are proportional to soil organic carbon concentrations, are favored by intensive grazing (Klumpp et al. 2009; Nelson and Carlson 2012). For grazing fields to resume plant growth, more nutrients are required. To recover the available nutrients, a forced SMC will be established (Strickland and Rousk 2010; Kallenbach, Frey, and Grandy 2016). Animal waste contains readily available nutrients that promote the development of coprophagic, quickly expanding bacteria that can break down labile soil organic carbon (SOC) using their own SOC decomposition enzymes (Gougoulias et al., 2014). . To show the bacterial makeup, an experiment was done on two economically significant rhodophycean algae species, *Gracilaria firma* and

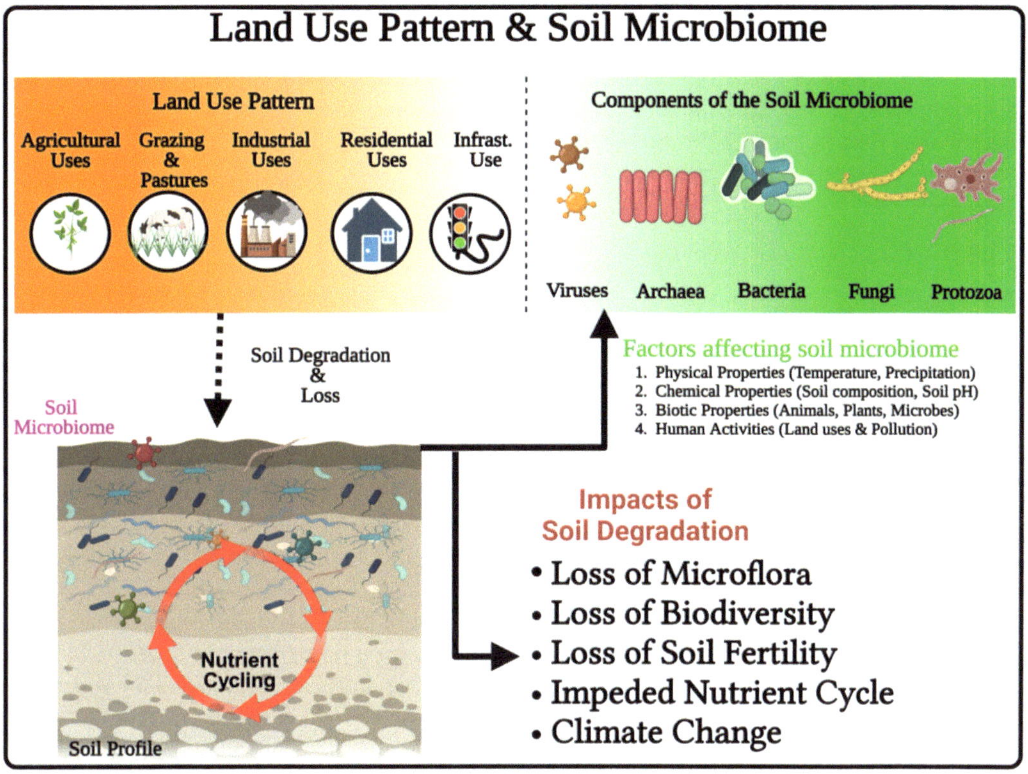

FIGURE 15.2 Land use patterns and the resulting varied impact on the soil microbiome.

Gracilaria salicornia, based on ungrazed and grazed fronds (Tan et al. 2020). Both types of algae support various bacterial populations and exhibit interactions with other microorganisms. For algal growth, bacterial species fix nitrogen and create IAA, which protects against environmental stress and, in turn, receives organic carbon and shelter from algal members. Both mesograzers (fish, amphipods, crabs) and micrograzers (gastropods, crustaceans, etc.) degrade the quality and biomass of seaweeds (Wahl et al. 2012; Cruz-Rivera and Friedlander 2011). Grazed *G. firma* tissues exhibit high amounts of nucleotide and aromatic chemical degradation, whereas the tissues that have not been grazed exhibit high levels of cell wall formation, aerobic respiration, and phospholipid production (Tan et al. 2020). Grazing is harmful to *G. firma* because it causes some agarolytic bacteria, such as *Marinagarivorans* sp. and *Algisphaera* sp., to destroy polysaccharides present in tissues of grazed *G. firma* but not in the ungrazed counterpart. Grazing, in comparison, had no substantial impact on the microbiota of *G. salicornia* (Tan et al. 2020). According to a study, grazing by a coral predator and eutrophication due to nutrient enrichment have a deleterious effect on the staghorn coral Acropora cervicornis and its associated microbiomes. Predation slows coral growth and increases tissue loss and death rates (Shaver et al. 2017). Through plant consumption, excretion, and trampling, herbivores significantly alter soil characteristics, nutrient recycling, and rhizosphere microbial communities (Berman et al. 2018). Mammalian herbivores are now said to impact the microbiome profiles of other animals (Hammer et al. 2016). Research was conducted on the grassland vegetation in Israel's Karei Deshe farm, where *Echinops*, *Trifolium*, and *Medicago* were numerous (Berman et al. 2018). The study looked at how cow grazing affected the microbiome makeup of polyphagous herbivorous spring caterpillars (*Ocnogyna* spp.) in both early and late instars. Gregarious caterpillars spend their formative years in a communal web nest, restricting their exposure to the outside world and resulting in microbiomes that are similar under both grazing and non-grazing settings (Berman et al. 2018). However, isolated caterpillars in grazed and non-grazed paddocks had different microbiota compositions. Cattle grazing alters the vegetation and compacts the soil, forcing solitary caterpillars to adjust their diets, which may impact the microbiome (Henkin et al. 2011; Sternberg et al. 2015). Figure 15.2 illustrates the land use pattern and its impact on the soil microbiome.

15.4.3 Industrial (Factories)

Due to the loss of vegetation, industrialization is the primary cause of soil habitat damage and fragmentation. For the soil microbiome to meet its need for OM, above-ground plants typically partner with it. Thus, the alteration of the soil microbiota's composition is more influenced by decreased plant diversity. Soil microorganisms are the most important factors in recovering contaminated soil. They are able to modify the physicochemical characteristics of the soil (Naz et al. 2022). However, a great deal of research has been conducted on how HMs and soil pH affect a variety of microbial populations in soil (Jiang et al. 2019). The increased level of HM concentrations has a detrimental effect on bacteria and actinomycetes (Y. Gao et al. 2009). In a research study at an e-waste dumpsite in Nigeria, it was shown that low levels of HM concentrations had less of an effect on the composition and variety of the soil microbiome than did high levels of pH, cation exchange capacity, soil organic carbon, and soil textures (Jiang et al. 2019). Lead, cadmium, mercury, and nickel are examples of HMs that can affect soil-dwelling bacteria and fungi (Jiang et al. 2019). Industrial activities can also unintentionally, or through sewage sludge, introduce non-native microbes to the soil, which can upset the usual equilibrium of the local microbiome. One study found that numerous non-native species spread from South Georgia to the Antarctic Peninsula via four cars, posing a survival risk to nearby inhabitants (Hughes et al. 2009). Heavy gear needed in industrial operations leads to compacted soil by lowering pore space, limiting the aerobic soil microbiomes' access to oxygen. Reducing the pore space impacts microbial biomass (Weisskopf et al. 2010). Anaerobes have an advantage over aerobic microbes because of the limited oxygen concentration. Anaerobic bacterial populations such as *Geobacter*, *Dechlorosoma*, *Desulfuromonas*, and *Methanosarcina* considerably rise due to soil compaction (Li, et al., 2004; Hartmann M et al. 2014). However, the population of aerobic bacteria like *Nitrospira*, *Mycobacterium*, and *Nitrosphaera* has decreased (De Neve and Hofman 2000). One soil compaction event has been found to drastically modify several soil metrics, including soil air permeability and gas diffusion, which lowers the pliability of the SMC throughout the following four seasons (Longepierre et al. 2021). Some industrial activities, such as using chemical fertilizers, may enhance crop yields but also lead to an imbalance in the accessibility of nutrients, which may change the pattern of the soil microbiome and favor some organisms over others. In a field experiment, three soil sites in Gongzhuling (black soil), Xuchang (fluvo-aquic soil), and Taoyua (red soil) were examined to see how fertilizer-induced functioning affects phagotrophic protists (Zhao et al. 2020). Utilizing nitrogen fertilizers increases the amount of soil ammonia that, in basic fluvo-aquic soil, mobilizes via cell membranes and toxicity to the cell. Fertilization results in a decrease in ammonia-oxidizing bacteria in acidic black and red soil (Liu et al., 2023). Another research examined how long-term fertilization on wild and cultivated walnut trees in southwest China affected the diversity and makeup of the SM. Natural walnut trees were fertilized with organic fertilizers for just one year, whereas farmed trees were fertilized for the next four years. The outcome indicates that while the microbial makeup has altered, the microbial diversity has not. Long-term fertilization, which raises the ammonium-nitrogen content in the soil, promotes soil acidity, thereby reducing nutrient availability (Zhao et al. 2020). Chemical fertilizers reduce proteobacteria's relative abundance. Long-term fertilization also increases the accessibility of phosphorus in the soil, reducing fungal diversity (Y.-C. Bai et al. 2020). Hazardous polycyclic aromatic hydrocarbons (PAHs), two-fused aromatic rings discharged into the soil environment and harmful to human health, are emitted by industrial gas facilities. Soils that are high in PAHs are susceptible to soil erosion. The development of well-adapted soil microbiomes is hampered by limited nutrient availability. When OM, such as wood sawdust and clay minerals, is purposefully added to soil that has been contaminated with PAHs, it changes the composition of bacteria that are capable of decomposing PAHs (Cébron et al. 2015). Phenanthrene has altered the local PAH-degrading microbiomes' activity and composition (Corgié, Beguiristain, and Leyval 2004). Synthetic fertilizers and insecticides frequently employed in industrial farming methods can affect the nutrient cycle. These usages also have an impact on soil microbiomes' makeup and function. The overwhelming majority of chemical pesticides can disrupt the soil microbiota's metabolism and enzymatic function (Arora and Sahni 2016). To evaluate the negative effects of utilizing these chemicals, a study on the fungicide Azoxystrobin was carried out on the soil microbiota of the annelid worm *Enchytraeus crypticus* (Q. Zhang et al. 2019). There was a noticeable reduction of beneficial bacteria in the intestine, according to (Baćmaga et al., 2015). Proteobacteria grow more when azoxystrobin concentration is increased. Penconazole, fludioxonil, and carbendazim causes dramatic change in soil microbial abundance (Y.-S. Wang et al. 2009; Marinozzi et al. 2013), and pesticides weaken soil animals' immune systems by altering the microbiota in their intestines. Industrial impacts on soil microbiota may differ significantly depending on the local environment, the pollutants involved, the industries engaged, and the mitigating efforts (Syromyatnikov et al. 2020)

15.4.4 Commercial and Domestic Construction

The rapid increase in the human population strains the environment, causing resource overuse and land degradation (Bilsborrow 1992). The Earth's carrying capacity is stressed by the increasing depletion of resources. Unplanned urbanization reduces the amount of land. Humans predominantly live in metropolitan areas, turning formerly used agricultural land into urban real estate (Ghatak and Mookherjee 2014). Unlimited human activity causes deforestation, which in turn causes biodiversity to dwindle (Y. Li, Wang, and Mao 2020). Soil's physico-chemical characteristics significantly impact the makeup and diversity of its microbiome (Pawar et al. 2020). The microbial populations in the soil change as a result of modifications in the soil's properties (Dixit et al., 2024). Excavation and grading during the building of flat and residential structures can have a range of consequences on soil microbial ecosystems. The soil microbiome can be affected both directly or indirectly by the building site's location and scope. According to (Borrelli et al. 2017), converting forests into agricultural fields results in soil loss that has a detrimental effect on soil nitrogen and carbon (Rodrigues et al. 2013). The forest is a rich source of organic carbon and has a higher microbial population than croplands used for agriculture (Fichtner et al. 2014). Heavy machinery used in construction causes soil compaction, which reduces oxygen transport and prevents aerobic bacteria from flourishing. The construction site disturbs the native vegetation, deteriorating the topsoil and releasing organic matter that causes soil erosion. Different levels of soil compaction were tested to see how they affected the soil microbiome over four years in terms of resilience or resistance. Some anaerobic species, such as Firmicutes, which exhibit nearly maximum stability over four years, benefit from soil compaction. When soil is compacted for four years, some bacteria, such as the deltaproteobacteria *Geobacter*, lose their survival ability and exhibit short-term resilience (Hartmann et al. 2014).

In addition to reducing oxygen availability for the aerobic microbiota, soil compaction reduces water infiltration(Hartmann et al., 2014), bulk density (Longepierre et al., 2022), air permeability (Longepierre et al., 2022), and macroporosity (Hartmann et al., 2014). More delicate ectomycorrhizal species see a decline in abundance due to this compacted soil (Querejeta et al., 2021); Hartmann, M. et al., 2014). Sedimentation from construction sites can transport a variety of pollutants, affecting the soil microbiome makeup (Fichtner et al. 2014). The distribution of indoor microbiomes in the universities old and recently constructed buildings is influenced by human occupancy (Cao et al. 2021). Older buildings are usually observed to possess higher species richness of proteobacteria, actinobacteria, and firmicutes than freshly constructed buildings, which, on the other hand, are reported to have higher human activity levels. *Rhizobium*, however, exhibits higher abundance in recently constructed, less populated buildings, demonstrating its outdoor nature (Cao et al. 2021). Due to the presence of calcium oxide (lime) and elevated soil pH, concrete, cement, and mortar used in construction activities are alkaline. This alters the soil microbiome's makeup and diversity by decreasing acid-loving bacteria and enhancing alkaline-loving microbes. These items should be properly disposed of in approved sites to prevent them from leaking into the soil. As a result of habitat fragmentation brought on by building construction, soil-borne microorganisms' ability to adapt and long-term survival may be hampered. Long-term commercial and residential development produces noise and vibrations that may harm the soil-dwelling species. Sub-lethal consequences for the soil microbiota could result from this. According to certain research, these will lessen their microbial biomass by stressing them, changing their physiology, and affecting gene expression, ultimately impeding plant-microbe interactions. However, low vibration and noise levels are advantageous to the symbiotic relationship between plants and microbes. It was evident from a study that noise and vibration impact the microbiota of grapevine leaves (Wassermann, Korsten, and Berg 2021). The microbiome diversity of the sound vibration microbiomes and the non-music exposed microbiomes (control) differs, as shown by 16S rRNA and ITS analyses. The host plant becomes more tolerant due to a change in microbiome diversity brought on by classical music. By inhibiting necrotrophic *Botrytis* spp., beneficial bacteria like *Bacillus, Kocuria*, and the fungus *Hortaea* rescue the host plants. *Candidatus* sp., the organism responsible for grapevine yellows, exhibits antagonistic effects against *Methylobacterium* and *Sphingomonas* (Wassermann, Korsten, and Berg 2021) produced. Waterlogging and excessive drainage patterns will be produced by changing drainage systems during building construction, which influences moisture change and varied soil microbial populations. Petroleum-based products with long-lasting harmful effects on soil microorganisms may be released into the environment. Following construction, landscaping activities like sod installation and planting may cause soil compaction, reducing the soil microbiome's variety. Applications of sustainable building techniques, restoring plants, and soil erosion control measures could reduce the negative effects on soil microbiota.

15.5 CONCLUSION

In this review, we have examined the complex relationship between land use patterns and their significant influence on soil microbial communities. Numerous studies have found that land management patterns have a major effect on the composition, variation, and usefulness of soil microbial populations. Over time, practices that prioritize the preservation and restoration of soil microbiomes can improve soil fertility, reduce erosion, and enhance ecosystem resilience. Agriculture, deforestation, and urbanization, among other human actions, all a counter-impact on the equilibrium of soil microbiomes, frequently result in a decline in microbial diversity and changes to community structures. This is generally harmful to soil health and ecological stability. Nutrient cycling and carbon sequestration, as well as disease suppression, are among the essential ecosystem services that soil microbes provide, and their disruption can have far-reaching consequences. These important biological forces may help in the design of sustainable farming systems in dry-land habitats. This chapter highlights the higher inconsistency in all microbiological variables across the soils, reflecting the typical spatial diversity of urban soils, where soil organic matter plays a greater role in explaining microbial activity and composition than land use. Bacterial communities appear to be more homogeneous and land use-dependent, while fungi seem to be heterogeneous and less affected by group-related characteristics.

Hence, to sum up,

1. Understanding how land use interacts with soil microbiomes is critical to sustainable land management.
2. More interdisciplinary research and policy, as well as land-use practices, are needed to balance human development with the conservation of soil microbiome diversity.

3. The multifunctional components of the ecosystem are destroyed by extensive agricultural use, which leads to an imbalance in nature via acts like the increased greenhouse impact and nitrogen loss through leaching.
4. Pesticides disrupt the microbiota in soil animals' intestines, which weakens their immune system and hence should be judiciously used.
5. Grazing intensity has a major impact on soil microbial communities as well as ecosystem functioning and is known to affect microbial composition more than seasonal variations do.
6. A variety of effects on soil microbial communities may result from excavation and grading during the construction of flat and residential structures.
7. The soil becomes compacted as a result of heavy construction equipment use, which decreases oxygen delivery, reduces water infiltration, and inhibits the growth of aerobic microorganisms.
8. Petroleum-based products that affect soil microorganisms over a lengthy period of time may be discharged into the environment.
9. Since soil OM plays a major role in biological activity in urban soils, measures to protect, increase, and maintain its content are necessary to ensure urban soil function and health.
10. Applications of environmentally friendly building methods, plant restoration, and soil erosion management strategies should lessen the detrimental impact on soil microbiota.

REFERENCES

"A Spatial Analysis Approach to the Global Delineation of Dryland Areas of Relevance to the CBD Programme of Work on Dry and Subhumid Lands - DocsLib." 2023. Accessed September 25. https://docslib.org/doc/2573884/a-spatial-analysis-approach-to-the-global-delineation-of-dryland-areas-of-relevance-to-the-cbd-programme-of-work-on-dry-and-subhumid-lands.

Abdul Rahman, Nur Sabrina Natasha, Nur Wahida Abdul Hamid, and Kalaivani Nadarajah. 2021. "Effects of Abiotic Stress on Soil Microbiome." *International Journal of Molecular Sciences* 22 (16). Switzerland: 9036. doi:10.3390/ijms22169036.

Adamczyk, Magdalene, Frank Hagedorn, Sonja Wipf, Johanna Donhauser, Pascal Vittoz, Christian Rixen, Aline Frossard, Jean-Paul Theurillat, and Beat Frey. 2019. "The Soil Microbiome of GLORIA Mountain Summits in the Swiss Alps." *Frontiers in Microbiology* 10 (May). Switzerland: 1080. doi:10.3389/fmicb.2019.01080.

Allan, Eric, Pete Manning, Fabian Alt, Julia Binkenstein, Stefan Blaser, Nico Blüthgen, Stefan Böhm, et al. 2015. "Land Use Intensification Alters Ecosystem Multifunctionality via Loss of Biodiversity and Changes to Functional Composition." *Ecology Letters* 18 (8). Wiley: 834–43. doi:10.1111/ele.12469.

Arora, Sanjay, and Divya Sahni. 2016. "Pesticides Effect on Soil Microbial Ecology and Enzyme Activity- An Overview." *Journal of Applied and Natural Science* 8 (2). ANSF Publications: 1126–32. doi:10.31018/jans.v8i2.929.

Attiwill, Peter M., and Mark A. Adams. 1993. "Nutrient Cycling in Forests." *New Phytologist* 124 (4): 561–82. doi:10.1111/j.1469-8137.1993.tb03847.x.

Ayilara, Modupe, Oluwaseyi Olanrewaju, Olubukola Babalola, and Olu Odeyemi. 2020. "Waste Management through Composting: Challenges and Potentials." *Sustainability* 12 (11): 4456. doi:10.3390/su12114456.

Azizian, Mohammad F, Peter O. Nelson, Pugazhendhi Thayumanavan, and Kenneth J. Williamson. 2003. "Environmental Impact of Highway Construction and Repair Materials on Surface and Ground Waters." *Waste Management* 23 (8): 719–28. doi:10.1016/S0956-053X(03)00024-2.

Bachar, Ami, Ashraf Al-Ashhab, M. Ines M. Soares, Menachem Y. Sklarz, Roey Angel, Eugene D. Ungar, and Osnat Gillor. 2010. "Soil Microbial Abundance and Diversity Along a Low Precipitation Gradient." *Microbial Ecology* 60 (2). Springer Science and Business Media LLC: 453–61. doi:10.1007/s00248-010-9727-1.

Bai, Yong-Chao, Ying-Ying Chang, Muzammil Hussain, Bin Lu, Jun-Pei Zhang, Xiao-Bo Song, Xia-Shuo Lei, and Dong Pei. 2020. "Soil Chemical and Microbiological Properties Are Changed by Long-Term Chemical Fertilizers That Limit Ecosystem Functioning." *Microorganisms* 8 (5). MDPI AG: 694. doi:10.3390/microorganisms8050694.

Bai, Yongfei, Jianguo Wu, Chris M. Clark, Qingmin Pan, Lixia Zhang, Shiping Chen, Qibing Wang, and Xingguo Han. 2012. “Grazing Alters Ecosystem Functioning and C:N:P Stoichiometry of Grasslands along a Regional Precipitation Gradient.” *Journal of Applied Ecology* 49 (6): 1204–15. doi:10.1111/j.1365-2664.2012.02205.x.

Banerjee, Samiran, Bobbi Helgason, Lianfeng Wang, Tristrom Winsley, Belinda C. Ferrari, and Steven D. Siciliano. 2016. “Legacy Effects of Soil Moisture on Microbial Community Structure and N2O Emissions.” *Soil Biology and Biochemistry* 95. Elsevier BV: 40–50. doi:10.1016/j.soilbio.2015.12.004.

Barberán, Albert, Scott T. Bates, Emilio O. Casamayor, and Noah Fierer. 2011. “Using Network Analysis to Explore Co-Occurrence Patterns in Soil Microbial Communities.” *The ISME Journal* 6 (2). Springer Science and Business Media LLC: 343–51. doi:10.1038/ismej.2011.119.

Bárcenas-Moreno, Gema, Maria Gómez-Brandón, Johannes Rousk, and Erland Bååth. 2009. “Adaptation of Soil Microbial Communities to Temperature: Comparison of Fungi and Bacteria in a Laboratory Experiment.” *Global Change Biology* 15 (12). Wiley: 2950–57. doi:10.1111/j.1365-2486.2009.01882.x.

Bardgett, Richard D., and Tancredi Caruso. 2020. “Soil Microbial Community Responses to Climate Extremes: Resistance, Resilience and Transitions to Alternative States.” *Philosophical Transactions of the Royal Society B: Biological Sciences* 375 (1794): 20190112. doi:10.1098/rstb.2019.0112.

Behera, Durga Prasad, Kapilkumar N. Ingle, Doniya Elze Mathew, Asmita Dhimmar, Harshal Sahastrabudhe, Sunil Kumar Sahu, M. Gobala Krishnan, Pramod. B. Shinde, M. Ganesan, and Vaibhav. A. Mantri. 2022. “Epiphytism, Diseases and Grazing in Seaweed Aquaculture: A Comprehensive Review.” *Reviews in Aquaculture* 14 (3): 1345–70. doi:10.1111/raq.12653.

Benbow, M. Eric, Philip S. Barton, Michael D. Ulyshen, James C. Beasley, Travis L. DeVault, Michael S. Strickland, Jeffery K. Tomberlin, Heather R. Jordan, and Jennifer L. Pechal. 2019. “Necrobiome Framework for Bridging Decomposition Ecology of Autotrophically and Heterotrophically Derived Organic Matter.” *Ecological Monographs* 89 (1). doi:10.1002/ecm.1331.

Berkelmann, Dirk, Dominik Schneider, Anja Meryandini, and Rolf Daniel. 2020. “Unravelling the Effects of Tropical Land Use Conversion on the Soil Microbiome.” *Environmental Microbiome* 15 (1). England: 5. doi:10.1186/s40793-020-0353-3.

Berli, M., J. M. Kirby, S. M. Springman, and R. Schulin. 2003. “Modelling Compaction of Agricultural Subsoils by Tracked Heavy Construction Machinery under Various Moisture Conditions in Switzerland.” *Soil and Tillage Research* 73 (1–2): 57–66. doi:10.1016/S0167-1987(03)00099-0.

Berman, Tali S., Sivan Laviad-Shitrit, Maya Lalzar, Malka Halpern, and Moshe Inbar. 2018. “Cascading Effects on Bacterial Communities: Cattle Grazing Causes a Shift in the Microbiome of a Herbivorous Caterpillar.” *The ISME Journal* 12 (8). Springer Science and Business Media LLC: 1952–63. doi:10.1038/s41396-018-0102-4.

Beyer, Lothar, Hans-Peter Blume, Dirk-Christian Elsner, and Antje Willnow. 1995. “Soil Organic Matter Composition and Microbial Activity in Urban Soils.” *Science of the Total Environment* 168 (3): 267–78. doi:10.1016/0048-9697(95)04704-5.

Bian, Hongfeng, Chao Li, Jianxing Zhu, Li Xu, Mingxu Li, Shuai Zheng, and Nianpeng He. 2022. “Soil Moisture Affects the Rapid Response of Microbes to Labile Organic C Addition.” *Frontiers in Ecology and Evolution* 10. Frontiers Media SA. doi:10.3389/fevo.2022.857185.

Bilsborrow, Richard E. 1992. “Population Growth, Internal Migration, and Environmental Degradation in Rural Areas of Developing Countries.” *European Journal of Population* 8 (2): 125–48. doi:10.1007/BF01797549.

Bodenhausen, Natacha, Matthew W. Horton, and Joy Bergelson. 2013. “Bacterial Communities Associated with the Leaves and the Roots of Arabidopsis Thaliana.” *PLoS One* 8 (2): e56329. doi:10.1371/journal.pone.0056329.

Bonkowski, Michael. 2004. “Protozoa and Plant Growth: The Microbial Loop in Soil Revisited.” *New Phytologist* 162 (3). Wiley: 617–31. doi:10.1111/j.1469-8137.2004.01066.x.

Borowik, A., and J. Wyszkowska. 2016. “Soil Moisture as a Factor Affecting the Microbiological and Biochemical Activity of Soil.” *Plant, Soil and Environment* 62 (6). Czech Academy of Agricultural Sciences: 250–55. doi:10.17221/158/2016-pse.

Borrelli, Pasquale, David A. Robinson, Larissa R. Fleischer, Emanuele Lugato, Cristiano Ballabio, Christine Alewell, Katrin Meusburger, et al. 2017. “An Assessment of the Global Impact of 21st Century Land Use Change on Soil Erosion.” *Nature Communications* 8 (1): 2013. doi:10.1038/s41467-017-02142-7.

Buckley, Daniel H., and Thomas M. Schmidt. 2003. “Diversity and Dynamics of Microbial Communities in Soils from Agro-Ecosystems.” *Environmental Microbiology* 5 (6). Wiley: 441–52. doi:10.1046/j.1462-2920.2003.00404.x.

Cameron, K. C., H. J. Di, and J. L. Moir. 2013. "Nitrogen Losses from the Soil/Plant System: A Review." *Annals of Applied Biology* 162 (2): 145–73. doi:10.1111/aab.12014.

Cao, Liu, Lu Yang, Clifford S. Swanson, Shuai Li, and Qiang He. 2021. "Comparative Analysis of Impact of Human Occupancy on Indoor Microbiomes." *Frontiers of Environmental Science & Engineering* 15 (5): 89. doi:10.1007/s11783-020-1383-1.

Carson, Jennifer K., Vanesa Gonzalez-Quiñones Daniel V. Murphy, Christoph Hinz, Jeremy A. Shaw, and Deirdre B. Gleeson. 2010. "Low Pore Connectivity Increases Bacterial Diversity in Soil." *Applied and Environmental Microbiology* 76 (12). United States: 3936–42. doi:10.1128/AEM.03085-09.

Cébron, Aurélie, Thierry Beguiristain, Jeanne Bongoua-Devisme, Jérémie Denonfoux, Pierre Faure, Catherine Lorgeoux, Stéphanie Ouvrard, Nicolas Parisot, Pierre Peyret, and Corinne Leyval. 2015. "Impact of Clay Mineral, Wood Sawdust or Root Organic Matter on the Bacterial and Fungal Community Structures in Two Aged PAH-Contaminated Soils." *Environmental Science and Pollution Research* 22 (18). Springer Science and Business Media LLC: 13724–38. doi:10.1007/s11356-015-4117-3.

Clarholm, Marianne. 1985. "Interactions of Bacteria, Protozoa and Plants Leading to Mineralization of Soil Nitrogen." *Soil Biology and Biochemistry* 17 (2). Elsevier BV: 181–87. doi:10.1016/0038-0717(85)90113-0.

Constancias, Florentin, Nicolas Chemidlin Prévost-Bouré, Sébastien Terrat, Simon Aussems, Virginie Nowak, Jean-Philippe Guillemin, Aline Bonnotte, et al. 2013. "Microscale Evidence for a High Decrease of Soil Bacterial Density and Diversity by Cropping." *Agronomy for Sustainable Development* 34 (4). Springer Science and Business Media LLC: 831–40. doi:10.1007/s13593-013-0204-3.

Corgié, S. C., T. Beguiristain, and C. Leyval. 2004. "Spatial Distribution of Bacterial Communities and Phenanthrene Degradation in the Rhizosphere of *Lolium Perenne* L." *Applied and Environmental Microbiology* 70 (6). American Society for Microbiology: 3552–57. doi:10.1128/aem.70.6.3552-3557.2004.

Costa, Dalila, Helena Freitas, and José Paulo Sousa. 2013. "Influence of Seasons and Land-Use Practices on Soil Microbial Activity and Metabolic Diversity in the 'Montado Ecosystem.'" *European Journal of Soil Biology* 59. Elsevier BV: 22–30. doi:10.1016/j.ejsobi.2013.08.003.

Cowie, A. L., T. D. Penman, L. Gorissen, M. D. Winslow, J. Lehmann, T. D. Tyrrell, S. Twomlow, et al. 2011. "Towards Sustainable Land Management in the Drylands: Scientific Connections in Monitoring and Assessing Dryland Degradation, Climate Change and Biodiversity." *Land Degradation & Development* 22 (2): 248–60. doi:10.1002/ldr.1086.

Cruz-Rivera, Edwin, and Michael Friedlander. 2011. "Feeding Preferences of Mesograzers on Aquacultured Gracilaria and Sympatric Algae." *Aquaculture* 322–323 (December): 218–22. doi:10.1016/j.aquaculture.2011.09.035.

Dedeyn, G., and W. Vanderputten. 2005. "Linking Aboveground and Belowground Diversity." *Trends in Ecology & Evolution* 20 (11): 625–33. doi:10.1016/j.tree.2005.08.009.

Dellagi, Alia, Isabelle Quillere, and Bertrand Hirel. 2020. "Beneficial Soil-Borne Bacteria and Fungi: A Promising Way to Improve Plant Nitrogen Acquisition." *Journal of Experimental Botany* 71 (15): 4469–79. doi:10.1093/jxb/eraa112.

Ding, Guo-Chun, Yvette M. Piceno, Holger Heuer, Nicole Weinert, Anja B. Dohrmann, Angel Carrillo, Gary L. Andersen, Thelma Castellanos, Christoph C. Tebbe, and Kornelia Smalla. 2013. "Changes of Soil Bacterial Diversity as a Consequence of Agricultural Land Use in a Semi-Arid Ecosystem." *PLoS One* 8 (3). Public Library of Science (PLoS): e59497. doi:10.1371/journal.pone.0059497.

Elam, Johanna, and Charlotte Björdal. 2020. "A Review and Case Studies of Factors Affecting the Stability of Wooden Foundation Piles in Urban Environments Exposed to Construction Work." *International Biodeterioration & Biodegradation* 148 (March): 104913. doi:10.1016/j.ibiod.2020.104913.

Eldridge, David J., Manuel Delgado-Baquerizo, Samantha K. Travers, James Val, Ian Oliver, Kelly Hamonts, and Brajesh K. Singh. 2017. "Competition Drives the Response of Soil Microbial Diversity to Increased Grazing by Vertebrate Herbivores." *Ecology* 98 (7). Wiley: 1922–31. doi:10.1002/ecy.1879.

Epp Schmidt, Dietrich J., Richard Pouyat, Katalin Szlavecz, Heikki Setälä, D. Johan Kotze, Ian Yesilonis, Sarel Cilliers, Erzsébet Hornung, Miklós Dombos, and Stephanie A. Yarwood. 2017. "Urbanization Erodes Ectomycorrhizal Fungal Diversity and May Cause Microbial Communities to Converge." *Nature Ecology & Evolution* 1 (5). Springer Science and Business Media LLC. doi:10.1038/s41559-017-0123.

Estendorfer, Jennifer, Barbara Stempfhuber, Paula Haury, Gisle Vestergaard, Matthias C. Rillig, Jasmin Joshi, Peter Schröder, and Michael Schloter. 2017. "The Influence of Land Use Intensity on the Plant-Associated Microbiome of Dactylis Glomerata L." *Frontiers in Plant Science* 8. Frontiers Media SA. doi:10.3389/fpls.2017.00930.

Faust, Karoline, and Jeroen Raes. 2012. "Microbial Interactions: From Networks to Models." *Nature Reviews Microbiology* 10 (8). Springer Science and Business Media LLC: 538–50. doi:10.1038/nrmicro2832.

Faust, Karoline, Gipsi Lima-Mendez, Jean-Sébastien Lerat, Jarupon F. Sathirapongsasuti, Rob Knight, Curtis Huttenhower, Tom Lenaerts, and Jeroen Raes. 2015. "Cross-Biome Comparison of Microbial Association Networks." *Frontiers in Microbiology* 6. Frontiers Media SA. doi:10.3389/fmicb.2015.01200.

Fernandes, Camila Cesário, Luciano Takeshi Kishi, Erica Mendes Lopes, Wellington Pine Omori, Jackson Antonio Marcondes de Souza, Lucia Maria Carareto Alves, and Eliana Gertrudes de Macedo Lemos. 2018. "Bacterial Communities in Mining Soils and Surrounding Areas under Regeneration Process in a Former Ore Mine." *Brazilian Journal of Microbiology : [Publication of the Brazilian Society for Microbiology]* 49 (3). Brazil: 489–502. doi:10.1016/j.bjm.2017.12.006.

Fichtner, A., G. von Oheimb, W. Härdtle, C. Wilken, and J. L. M. Gutknecht. 2014. "Effects of Anthropogenic Disturbances on Soil Microbial Communities in Oak Forests Persist for More than 100 Years." *Soil Biology and Biochemistry* 70 (March): 79–87. doi:10.1016/j.soilbio.2013.12.015.

Fierer, Noah, and Robert B. Jackson. 2006. "The Diversity and Biogeography of Soil Bacterial Communities." *Proceedings of the National Academy of Sciences of the United States of America* 103 (3). United States: 626–31. doi:10.1073/pnas.0507535103.

Figuerola, Eva L. M., Leandro D. Guerrero, Dominique Türkowsky, Luis G. Wall, and Leonardo Erijman. 2014. "Crop Monoculture Rather than Agriculture Reduces the Spatial Turnover of Soil Bacterial Communities at a Regional Scale." *Environmental Microbiology* 17 (3). Wiley: 678–88. doi:10.1111/1462-2920.12497.

Flores-Rentería, Dulce, Ana Rincón, Fernando Valladares, and Jorge Curiel Yuste. 2016. "Agricultural Matrix Affects Differently the Alpha and Beta Structural and Functional Diversity of Soil Microbial Communities in a Fragmented Mediterranean Holm Oak Forest." *Soil Biology and Biochemistry* 92 (January): 79–90. doi:10.1016/j.soilbio.2015.09.015.

Franzluebbers, A. J. 2002. "Soil Organic Matter Stratification Ratio as an Indicator of Soil Quality." *Soil and Tillage Research* 66 (2): 95–106. doi:10.1016/S0167-1987(02)00018-1.

Friedman, Jonathan, Logan M. Higgins, and Jeff Gore. 2017. "Community Structure Follows Simple Assembly Rules in Microbial Microcosms." *Nature Ecology & Evolution* 1 (5). Springer Science and Business Media LLC. doi:10.1038/s41559-017-0109.

Gao, Yang, Pei Zhou, Liang Mao, Yue-er Zhi, and Wan-jun Shi. 2009. "Assessment of Effects of Heavy Metals Combined Pollution on Soil Enzyme Activities and Microbial Community Structure: Modified Ecological Dose–Response Model and PCR-RAPD." *Environmental Earth Sciences* 60 (3). Springer Science and Business Media LLC: 603–12. doi:10.1007/s12665-009-0200-8.

Gao, Zhilei, Ida Karlsson, Stefan Geisen, George Kowalchuk, and Alexandre Jousset. 2019. "Protists: Puppet Masters of the Rhizosphere Microbiome." *Trends in Plant Science* 24 (2). Elsevier BV: 165–76. doi:10.1016/j.tplants.2018.10.011.

Garcia, Liza, José C. B. Dubeux, Lynn E. Sollenberger, Joao M. B. Vendramini, Nicolas DiLorenzo, Erick R. S. Santos, David M. Jaramillo, and Martin Ruiz-Moreno. 2021. "Nutrient Excretion from Cattle Grazing Nitrogen-fertilized Grass or Grass–Legume Pastures." *Agronomy Journal* 113 (4): 3110–23. doi:10.1002/agj2.20675.

Geisen, Stefan, Edward A. D. Mitchell, Sina Adl, Michael Bonkowski, Micah Dunthorn, Flemming Ekelund, Leonardo D. Fernández, et al. 2018. "Soil Protists: A Fertile Frontier in Soil Biology Research." *FEMS Microbiology Reviews* 42 (3). Oxford University Press (OUP): 293–323. doi:10.1093/femsre/fuy006.

Ghatak, Maitreesh, and Dilip Mookherjee. 2014. "Land Acquisition for Industrialization and Compensation of Displaced Farmers." *Journal of Development Economics* 110 (September): 303–12. doi:10.1016/j.jdeveco.2013.01.001.

Guo, Yanjun, Qingfeng Du, Guangdi Li, Yu Ni, Zhen Zhang, Weibo Ren, and Xiangyang Hou. 2016. "Soil Phosphorus Fractions and Arbuscular Mycorrhizal Fungi Diversity Following Long-Term Grazing Exclusion on Semi-Arid Steppes in Inner Mongolia." *Geoderma* 269. Elsevier BV: 79–90. doi:10.1016/j.geoderma.2016.01.039.

Hammer, Tobin J., Noah Fierer, Bess Hardwick, Asko Simojoki, Eleanor Slade, Juhani Taponen, Heidi Viljanen, and Tomas Roslin. 2016. "Treating Cattle with Antibiotics Affects Greenhouse Gas Emissions, and Microbiota in Dung and Dung Beetles." *Proceedings of the Royal Society B: Biological Sciences* 283 (1831): 20160150. doi:10.1098/rspb.2016.0150.

Harkes, Paula, Afnan K. A. Suleiman, Sven J. J. van den Elsen, Johannes J. de Haan, Martijn Holterman, Eiko E. Kuramae, and Johannes Helder. 2019. "Conventional and Organic Soil Management as Divergent Drivers of Resident and Active Fractions of Major Soil Food Web Constituents." *Scientific Reports* 9 (1). England: 13521. doi:10.1038/s41598-019-49854-y.

Hartman, Kyle, Marcel G. A. van der Heijden, Raphaël A. Wittwer, Samiran Banerjee, Jean-Claude Walser, and Klaus Schlaeppi. 2018. "Correction to: Cropping Practices Manipulate Abundance Patterns of Root and Soil Microbiome Members Paving the Way to Smart Farming." *Microbiome* 6 (1). Springer Science and Business Media LLC. doi:10.1186/s40168-018-0456-x.

Hartmann, Martin, Beat Frey, Jochen Mayer, Paul Mäder, and Franco Widmer. 2014. "Distinct Soil Microbial Diversity under Long-Term Organic and Conventional Farming." *The ISME Journal* 9 (5). Springer Science and Business Media LLC: 1177–94. doi:10.1038/ismej.2014.210.

Hartmann, Martin, Beat Frey, Jochen Mayer, Paul Mäder, and Franco Widmer. 2015. "Distinct Soil Microbial Diversity under Long-Term Organic and Conventional Farming." *The ISME Journal* 9 (5). England: 1177–94. doi:10.1038/ismej.2014.210.

Heijden, Marcel G. A. van der, Richard D. Bardgett, and Nico M. van Straalen. 2008. "The Unseen Majority: Soil Microbes as Drivers of Plant Diversity and Productivity in Terrestrial Ecosystems." *Ecology Letters* 11 (3). Wiley: 296–310. doi:10.1111/j.1461-0248.2007.01139.x.

Henkin, Z., E. D. Ungar, L. Dvash, A. Perevolotsky, Y. Yehuda, M. Sternberg, H. Voet, and S. Y. Landau. 2011. "Effects of Cattle Grazing on Herbage Quality in a Herbaceous Mediterranean Rangeland." *Grass and Forage Science* 66 (4). Wiley: 516–25. doi:10.1111/j.1365-2494.2011.00808.x.

HilleRisLambers, Reinier, Max Rietkerk, Frank van den Bosch, Herbert H. T. Prins, and Hans de Kroon. 2001. "Vegetation Pattern Formation in Semi-Arid Grazing Systems." *Ecology* 82 (1). Wiley: 50–61. doi:10.1890/0012-9658(2001)082[0050:vpfisa]2.0.co;2.

Horn, David J. Van, Jordan G. Okie, Heather N. Buelow, Michael N. Gooseff, John E. Barrett, and Cristina D. Takacs-Vesbach. 2014. "Soil Microbial Responses to Increased Moisture and Organic Resources along a Salinity Gradient in a Polar Desert." *Applied and Environmental Microbiology* 80 (10). United States: 3034–43. doi:10.1128/AEM.03414-13.

Hughes, K. A., P. Convey, N. R. Maslen, and R. I. L. Smith. 2009. "Accidental Transfer of Non-Native Soil Organisms into Antarctica on Construction Vehicles." *Biological Invasions* 12 (4). Springer Science and Business Media LLC: 875–91. doi:10.1007/s10530-009-9508-2.

Huhe, Xianjiang Chen, Fujiang Hou, Yanpei Wu, and Yunxiang Cheng. 2017. "Bacterial and Fungal Community Structures in Loess Plateau Grasslands with Different Grazing Intensities." *Frontiers in Microbiology* 8. Frontiers Media SA. doi:10.3389/fmicb.2017.00606.

Islam, Waqar, Ali Noman, Hassan Naveed, Zhiqun Huang, and Han Y. H. Chen. 2020. "Role of Environmental Factors in Shaping the Soil Microbiome." *Environmental Science and Pollution Research* 27 (33). Springer Science and Business Media LLC: 41225–47. doi:10.1007/s11356-020-10471-2.

Jangid, Kamlesh, Mark A. Williams, Alan J. Franzluebbers, Thomas M. Schmidt, David C. Coleman, and William B. Whitman. 2011. "Land-Use History Has a Stronger Impact on Soil Microbial Community Composition than Aboveground Vegetation and Soil Properties." *Soil Biology and Biochemistry* 43 (10): 2184–93. doi:10.1016/j.soilbio.2011.06.022.

Jansson, Janet K., and Kirsten S. Hofmockel. 2019. "Soil Microbiomes and Climate Change." *Nature Reviews Microbiology* 18 (1). Springer Science and Business Media LLC: 35–46. doi:10.1038/s41579-019-0265-7.

Jiang, Bo, Adedoyin Adebayo, Jianli Jia, Yi Xing, Songqiang Deng, Limin Guo, Yuting Liang, and Dayi Zhang. 2019. "Impacts of Heavy Metals and Soil Properties at a Nigerian E-Waste Site on Soil Microbial Community." *Journal of Hazardous Materials* 362. Elsevier BV: 187–95. doi:10.1016/j.jhazmat.2018.08.060.

Kalev, Stefan D., and Gurpal S. Toor. 2018. "The Composition of Soils and Sediments." In *Green Chemistry*, 339–57. Elsevier. doi:10.1016/B978-0-12-809270-5.00014-5.

Kallenbach, Cynthia M., Serita D. Frey, and A. Stuart Grandy. 2016. "Direct Evidence for Microbial-Derived Soil Organic Matter Formation and Its Ecophysiological Controls." *Nature Communications* 7 (1). Springer Science and Business Media LLC. doi:10.1038/ncomms13630.

Kim, Seon-Young, Seung-Hoon Lee, Chris Freeman, Nathalie Fenner, and Hojeong Kang. 2008. "Comparative Analysis of Soil Microbial Communities and Their Responses to the Short-Term Drought in Bog, Fen, and Riparian Wetlands." *Soil Biology and Biochemistry* 40 (11). Elsevier BV: 2874–80. doi:10.1016/j.soilbio.2008.08.004.

Klumpp, Katja, Sébastien Fontaine, Eléonore Attard, Xavier Le Roux, Gerd Gleixner, and Jean-Francois Soussana. 2009. "Grazing Triggers Soil Carbon Loss by Altering Plant Roots and Their Control on Soil Microbial Community." *Journal of Ecology* 97 (5). Wiley: 876–85. doi:10.1111/j.1365-2745.2009.01549.x.

Kohler, F., J. Hamelin, F. Gillet, J.-M. Gobat, and A. Buttler. 2005. "Soil Microbial Community Changes in Wooded Mountain Pastures Due to Simulated Effects of Cattle Grazing." *Plant and Soil* 278 (1–2). Springer Science and Business Media LLC: 327–40. doi:10.1007/s11104-005-8809-1.

Korneykova, Maria V., Viacheslav I. Vasenev, Dmitry A. Nikitin, Anastasia S. Soshina, Andrey V. Dolgikh, and Yulia L. Sotnikova. 2021. "Urbanization Affects Soil Microbiome Profile Distribution in the Russian Arctic Region." *International Journal of Environmental Research and Public Health* 18 (21). Switzerland: 11665. doi:10.3390/ijerph182111665.

Kosheleva, Natalia E., Dmitry V. Vlasov, Ilya D. Korlyakov, and Nikolay S. Kasimov. 2018. "Contamination of Urban Soils with Heavy Metals in Moscow as Affected by Building Development." *Science of the Total Environment* 636 (September): 854–63. doi:10.1016/j.scitotenv.2018.04.308.

Köster, Kajar, Heidi Aaltonen, Frank Berninger, Jussi Heinonsalo, Egle Köster, Caius Ribeiro-Kumara, Hui Sun, Leho Tedersoo, Xuan Zhou, and Jukka Pumpanen. 2021. "Impacts of Wildfire on Soil Microbiome in Boreal Environments." *Current Opinion in Environmental Science & Health* 22. Elsevier BV: 100258. doi:10.1016/j.coesh.2021.100258.

Kuiper, Irene, Ellen L. Lagendijk, Guido V. Bloemberg, and Ben J. J. Lugtenberg. 2004. "Rhizoremediation: A Beneficial Plant-Microbe Interaction." *Molecular Plant-Microbe Interactions®* 17 (1). Scientific Societies: 6–15. doi:10.1094/mpmi.2004.17.1.6.

Kulmatiski, Andrew, and Karen H. Beard. 2011. "Long-Term Plant Growth Legacies Overwhelm Short-Term Plant Growth Effects on Soil Microbial Community Structure." *Soil Biology and Biochemistry* 43 (4). Elsevier BV: 823–30. doi:10.1016/j.soilbio.2010.12.018.

Lacerda-Júnior, Gileno V., Melline F. Noronha, Lucélia Cabral, Tiago P. Delforno, Sanderson Tarciso Pereira de Sousa, Paulo I. Fernandes-Júnior, Itamar S. Melo, and Valéria M. Oliveira. 2019. "Land Use and Seasonal Effects on the Soil Microbiome of a Brazilian Dry Forest." *Frontiers in Microbiology* 10. Frontiers Media SA. doi:10.3389/fmicb.2019.00648.

Lambers, Hans, Jon C. Clements, and Matthew N. Nelson. 2013. "How a Phosphorus-acquisition Strategy Based on Carboxylate Exudation Powers the Success and Agronomic Potential of Lupines (*Lupinus*, Fabaceae)." *American Journal of Botany* 100 (2). Wiley: 263–88. doi:10.3732/ajb.1200474.

Lauber, Christian L., Michael S. Strickland, Mark A. Bradford, and Noah Fierer. 2008. "The Influence of Soil Properties on the Structure of Bacterial and Fungal Communities across Land-Use Types." *Soil Biology and Biochemistry* 40 (9). Elsevier BV: 2407–15. doi:10.1016/j.soilbio.2008.05.021.

Lazzaro, A., C. Abegg, and J. Zeyer. 2009. "Bacterial Community Structure of Glacier Forefields on Siliceous and Calcareous Bedrock." *European Journal of Soil Science* 60 (6). Wiley: 860–70. doi:10.1111/j.1365-2389.2009.01182.x.

Leff, Jonathan W., Stuart E. Jones, Suzanne M. Prober, Albert Barberán, Elizabeth T. Borer, Jennifer L. Firn, W. Stanley Harpole, et al. 2015. "Consistent Responses of Soil Microbial Communities to Elevated Nutrient Inputs in Grasslands across the Globe." *Proceedings of the National Academy of Sciences* 112 (35). Proceedings of the National Academy of Sciences: 10967–72. doi:10.1073/pnas.1508382112.

Li Xiaogang, Alexandre Jousset, Wietse de Boer, Víctor J. Carrión, Taolin Zhang, Xingxiang Wang, and Eiko E. Kuramae. 2018. "Legacy of Land Use History Determines Reprogramming of Plant Physiology by Soil Microbiome." *The ISME Journal* 13 (3). Springer Science and Business Media LLC: 738–51. doi:10.1038/s41396-018-0300-0.

Li Xueqin, Lindsay C. Stringer, and Martin Dallimer. 2022. "The Impacts of Urbanisation and Climate Change on the Urban Thermal Environment in Africa." *Climate* 10 (11): 164. doi:10.3390/cli10110164.

Li, Yongwei, Xianmin Wang, and Hang Mao. 2020. "Influence of Human Activity on Landslide Susceptibility Development in the Three Gorges Area." *Natural Hazards* 104 (3): 2115–51. doi:10.1007/s11069-020-04264-6.

Liao, Jianli, Yun Liang, and Danfeng Huang. 2018. "Organic Farming Improves Soil Microbial Abundance and Diversity under Greenhouse Condition: A Case Study in Shanghai (Eastern China)." *Sustainability* 10 (10). MDPI AG: 3825. doi:10.3390/su10103825.

Liao, Jianli, Qicong Xu, Huilian Xu, and Danfeng Huang. 2019. "Natural Farming Improves Soil Quality and Alters Microbial Diversity in a Cabbage Field in Japan." *Sustainability* 11 (11). MDPI AG: 3131. doi:10.3390/su11113131.

Liu, Huaiqiang, Frank Yonghong Li, Jiayue Liu, Chunjun Shi, Kuanyan Tang, Qianhui Yang, Yu Liu, et al. 2023. "The Reciprocal Changes in Dominant Species with Complete Metabolic Functions Explain the Decoupling Phenomenon of Microbial Taxonomic and Functional Composition in a Grassland." *Frontiers in Microbiology* 14 (March). doi:10.3389/fmicb.2023.1113157.

Longepierre, Manon, Franco Widmer, Thomas Keller, Peter Weisskopf, Tino Colombi, Johan Six, and Martin Hartmann. 2021. "Limited Resilience of the Soil Microbiome to Mechanical Compaction within Four Growing Seasons of Agricultural Management." *ISME Communications* 1 (1). Springer Science and Business Media LLC. doi:10.1038/s43705-021-00046-8.

Lupatini, Manoeli, Gerard W. Korthals, Mattias de Hollander, Thierry K. S. Janssens, and Eiko E. Kuramae. 2017. "Soil Microbiome Is More Heterogeneous in Organic Than in Conventional Farming System." *Frontiers in Microbiology* 7 (January). Switzerland: 2064. doi:10.3389/fmicb.2016.02064.

Macdonald, Catriona A., Michael J. Crawley, Denis J. Wright, Justin Kuczynski, Lucinda Robinson, Rob Knight, Waleed Abu Al-Soud, et al. 2015. "Identifying Qualitative Effects of Different Grazing Types on Below-ground Communities and Function in a Long-term Field Experiment." *Environmental Microbiology* 17 (3): 841–54. doi:10.1111/1462-2920.12539.

Maestre, Fernando T., David J. Eldridge, Santiago Soliveres, Sonia Kéfi, Manuel Delgado-Baquerizo, Matthew A. Bowker, Pablo García-Palacios, et al. 2016. "Structure and Functioning of Dryland Ecosystems in a Changing World." *Annual Review of Ecology, Evolution, and Systematics* 47 (1). Annual Reviews: 215–37. doi:10.1146/annurev-ecolsys-121415-032311.

Mamilov, Anvar Sh, Boris A. Byzov, Dmitri G. Zvyagintsev, and Oliver M. Dilly. 2001. "Predation on Fungal and Bacterial Biomass in a Soddy-Podzolic Soil Amended with Starch, Wheat Straw and Alfalfa Meal." *Applied Soil Ecology* 16 (2). Elsevier BV: 131–39. doi:10.1016/s0929-1393(00)00109-8.

Marinozzi, Maria, Laura Coppola, Elga Monaci, Dimitrios G. Karpouzas, Evangelia Papadopoulou, Urania Menkissoglu-Spiroudi, and Costantino Vischetti. 2013. "The Dissipation of Three Fungicides in a Biobed Organic Substrate and Their Impact on the Structure and Activity of the Microbial Community." *Environmental Science and Pollution Research* 20 (4): 2546–55. doi:10.1007/s11356-012-1165-9.

Mårtensson, Linda-Maria, and Pål Axel Olsson. 2012. "Reductions in Microbial Biomass along Disturbance Gradients in a Semi-Natural Grassland." *Applied Soil Ecology* 62. Elsevier BV: 8–13. doi:10.1016/j.apsoil.2012.07.003.

Mcgonigle, Terence P., and Murray H. Miller. 1996. "Development of Fungi below Ground in Association with Plants Growing in Disturbed and Undisturbed Soils." *Soil Biology and Biochemistry* 28 (3): 263–69. doi:10.1016/0038-0717(95)00129-8.

Melo, Vander Freitas, Alessandro Góis Orrutéa, Antônio Carlos Vargas Motta, and Samara Alves Testoni. 2017. "Land Use and Changes in Soil Morphology and Physical-Chemical Properties in Southern Amazon." *Revista Brasileira de Ciência Do Solo* 41 (0). FapUNIFESP (SciELO). doi:10.1590/18069657rbcs20170034.

Mencel, Justyna, Agnieszka Mocek-Płóciniak, and Anna Kryszak. 2022. "Soil Microbial Community and Enzymatic Activity of Grasslands under Different Use Practices: A Review." *Agronomy* 12 (5): 1136. doi:10.3390/agronomy12051136.

"Microbiome." 2023. Accessed September 29. https://www.genome.gov/genetics-glossary/Microbiome.

Mohamed, Nouri, and Haddioui Abdelmajid. 2017. "Diversity of Soil Microbial Communities from an Iron Mining Area (Oued Zem, Morocco)." *Materials and Geoenvironment* 64 (1). Walter de Gruyter GmbH: 21–34. doi:10.1515/rmzmag-2017-0002.

Mondini, C., and P. Sequi. 2008. "Implication of Soil C Sequestration on Sustainable Agriculture and Environment." *Waste Management* 28 (4): 678–84. doi:10.1016/j.wasman.2007.09.026.

Moore, Jennifer M., Daniel K. Manter, and Kristie A. Maczko. 2023. "Rotational Grazing Strategies Minimally Impact Soil Microbial Communities and Carbon Dynamics—A Texas Case Study." *Land* 12 (8): 1517. doi:10.3390/land12081517.

Morrissey, Ember M., Jaimie L. Gillespie, Joseph C. Morina, and Rima B. Franklin. 2014. "Salinity Affects Microbial Activity and Soil Organic Matter Content in Tidal Wetlands." *Global Change Biology* 20 (4): 1351–62. doi:10.1111/gcb.12431.

Naz, Misbah, Zhicong Dai, Sajid Hussain, Muhammad Tariq, Subhan Danish, Irfan Ullah Khan, Shanshan Qi, and Daolin Du. 2022. "The Soil PH and Heavy Metals Revealed Their Impact on Soil Microbial Community." *Journal of Environmental Management* 321. Elsevier BV: 115770. doi:10.1016/j.jenvman.2022.115770.

Nelson, Craig E., and Craig A. Carlson. 2012. "Tracking Differential Incorporation of Dissolved Organic Carbon Types among Diverse Lineages of Sargasso Sea Bacterioplankton." *Environmental Microbiology* 14 (6). Wiley: 1500–16. doi:10.1111/j.1462-2920.2012.02738.x.

Neve, S. De, and G. Hofman. 2000. "Influence of Soil Compaction on Carbon and Nitrogen Mineralization of Soil Organic Matter and Crop Residues." *Biology and Fertility of Soils* 30 (5–6). Springer Science and Business Media LLC: 544–49. doi:10.1007/s003740050034.

Nottingham, Andrew T., Erland Bååth, Stephanie Reischke, Norma Salinas, and Patrick Meir. 2019. "Adaptation of Soil Microbial Growth to Temperature: Using a Tropical Elevation Gradient to Predict Future Changes." *Global Change Biology* 25 (3). England: 827–38. doi:10.1111/gcb.14502.

Omotayo, Oluwadara Pelumi, and Olubukola Oluranti Babalola. 2021. "Resident Rhizosphere Microbiome's Ecological Dynamics and Conservation: Towards Achieving the Envisioned Sustainable Development Goals, a Review." *International Soil and Water Conservation Research* 9 (1). Elsevier BV: 127–42. doi:10.1016/j.iswcr.2020.08.002.

Osono, Takashi. 2007. "Ecology of Ligninolytic Fungi Associated with Leaf Litter Decomposition." *Ecological Research* 22 (6). Wiley: 955–74. doi:10.1007/s11284-007-0390-z.

Pawar, Rishabh, Rakesh Sharma, Anil Kumar, and Swapana Sepehya. 2020. "Impact of Land Use Change on Soil Erosion, Sedimentation and Soil Microbiome." *International Journal of Chemical Studies* 8 (2): 881–86. doi:10.22271/chemi.2020.v8.i2m.8877.

Pedrinho, Alexandre, Lucas William Mendes, Felipe Martins do Rêgo Barros, Luis Fernando Merloti, Mayara Martins e Martins, Simone Raposo Cotta, Fernando Dini Andreote, and Siu Mui Tsai. 2023. "Impacts of Deforestation and Forest Regeneration on Soil Bacterial Communities Associated with Phosphorus Transformation Processes in the Brazilian Amazon Region." *Ecological Indicators* 146. Elsevier BV: 109779. doi:10.1016/j.ecolind.2022.109779.

Philippot, Laurent, Jos M. Raaijmakers, Philippe Lemanceau, and Wim H. van der Putten. 2013. "Going Back to the Roots: The Microbial Ecology of the Rhizosphere." *Nature Reviews Microbiology* 11 (11). Springer Science and Business Media LLC: 789–99. doi:10.1038/nrmicro3109.

Pinto-Correia, Teresa, and José Mascarenhas. 1999. "Contribution to the Extensification/Intensification Debate: New Trends in the Portuguese Montado." *Landscape and Urban Planning* 46 (1–3). Elsevier BV: 125–31. doi:10.1016/s0169-2046(99)00036-5.

Plante, Alain F., Richard T. Conant, Catherine E. Stewart, Keith Paustian, and Johan Six. 2006. "Impact of Soil Texture on the Distribution of Soil Organic Matter in Physical and Chemical Fractions." *Soil Science Society of America Journal* 70 (1): 287–96. doi:10.2136/sssaj2004.0363.

Porter, Stephanie S., and Joel L. Sachs. 2020. "Agriculture and the Disruption of Plant–Microbial Symbiosis." *Trends in Ecology & Evolution* 35 (5): 426–39. doi:10.1016/j.tree.2020.01.006.

Qi, Daihua, Xuwen Wieneke, Jianping Tao, Xu Zhou, and Udaya Desilva. 2018. "Soil PH Is the Primary Factor Correlating With Soil Microbiome in Karst Rocky Desertification Regions in the Wushan County, Chongqing, China." *Frontiers in Microbiology* 9. Frontiers Media SA. doi:10.3389/fmicb.2018.01027.

Quadros, Patricia Dorr de, Kateryna Zhalnina, Austin G. Davis-Richardson, Jennifer C. Drew, Fátima B. Menezes, Flávio A. de O Camargo, and Eric W. Triplett. 2016. "Coal Mining Practices Reduce the Microbial Biomass, Richness and Diversity of Soil." *Applied Soil Ecology* 98. Elsevier BV: 195–203. doi:10.1016/j.apsoil.2015.10.016.

Rahman, Md. Mizanur, Hamidul Huq, and Sharif A. Mukul. 2023. "Implications of Changing Urban Land Use on the Livelihoods of Local People in Northwestern Bangladesh." *Sustainability* 15 (15): 11769. doi:10.3390/su151511769.

Ranheim Sveen, T., S.E. Hannula, and M. Bahram. 2023. "Microbial Regulation of Feedbacks to Ecosystem Change." *Trends in Microbiology*, July. doi:10.1016/j.tim.2023.06.006.

Ray, Prasun, Venkatachalam Lakshmanan, Jessy L. Labbé, and Kelly D. Craven. 2020. "Microbe to Microbiome: A Paradigm Shift in the Application of Microorganisms for Sustainable Agriculture." *Frontiers in Microbiology* 11. Frontiers Media SA. doi:10.3389/fmicb.2020.622926.

Reynolds, Heather L., Alissa Packer, James D. Bever, and Keith Clay. 2003. "GRASSROOTS ECOLOGY: PLANT–MICROBE–SOIL INTERACTIONS AS DRIVERS OF PLANT COMMUNITY STRUCTURE AND DYNAMICS." *Ecology* 84 (9). Wiley: 2281–91. doi:10.1890/02-0298.

Rodrigues, Jorge L. M., Vivian H. Pellizari, Rebecca Mueller, Kyunghwa Baek, Ederson da C. Jesus, Fabiana S. Paula, Babur Mirza, et al. 2013. "Conversion of the Amazon Rainforest to Agriculture Results in Biotic Homogenization of Soil Bacterial Communities." *Proceedings of the National Academy of Sciences* 110 (3): 988–93. doi:10.1073/pnas.1220608110.

Rønn, Regin, Allison E. McCaig, Bryan S. Griffiths, and James I. Prosser. 2002. "Impact of Protozoan Grazing on Bacterial Community Structure in Soil Microcosms." *Applied and Environmental Microbiology* 68 (12): 6094–6105. doi:10.1128/AEM.68.12.6094-6105.2002.

Rousk, Johannes, Erland Bååth, Philip C. Brookes, Christian L. Lauber, Catherine Lozupone, J. Gregory Caporaso, Rob Knight, and Noah Fierer. 2010. "Soil Bacterial and Fungal Communities across a PH Gradient in an Arable Soil." *The ISME Journal* 4 (10). Springer Science and Business Media LLC: 1340–51. doi:10.1038/ismej.2010.58.

Russel, Jakob, Henriette L. Røder, Jonas S. Madsen, Mette Burmølle, and Søren J. Sørensen. 2017. "Antagonism Correlates with Metabolic Similarity in Diverse Bacteria." *Proceedings of the National Academy of Sciences* 114 (40). Proceedings of the National Academy of Sciences: 10684–88. doi:10.1073/pnas.1706016114.

Sala, Osvaldo E., F. Stuart Chapin, I. I. I., Juan J. Armesto, Eric Berlow, Janine Bloomfield, Rodolfo Dirzo, et al. 2000. "Global Biodiversity Scenarios for the Year 2100." *Science* 287 (5459). American Association for the Advancement of Science (AAAS): 1770–74. doi:10.1126/science.287.5459.1770.

Sazykina, M. A., T. M. Minkina, E. Yu. Konstantinova, L. E. Khmelevtsova, T. N. Azhogina, E. M. Antonenko, Sh. K. Karchava, et al. 2022. "Pollution Impact on Microbial Communities Composition in Natural and Anthropogenically Modified Soils of Southern Russia." *Microbiological Research* 254 (January): 126913. doi:10.1016/j.micres.2021.126913.

Schjønning, Per, Ingrid K. Thomsen, Søren O. Petersen, Kristian Kristensen, and Bent T. Christensen. 2011. "Relating Soil Microbial Activity to Water Content and Tillage-Induced Differences in Soil Structure." *Geoderma* 163 (3–4). Elsevier BV: 256–64. doi:10.1016/j.geoderma.2011.04.022.

Shaver, Elizabeth C., Andrew A. Shantz, Ryan McMinds, Deron E. Burkepile, Rebecca L. Vega Thurber, and Brian R. Silliman. 2017. "Effects of Predation and Nutrient Enrichment on the Success and Microbiome of a Foundational Coral." *Ecology* 98 (3). Wiley: 830–39. doi:10.1002/ecy.1709.

Smith, Chris R., Peter L. Blair, Charlie Boyd, Brianne Cody, Alexander Hazel, Ashley Hedrick, Hitesh Kathuria, et al. 2016. "Microbial Community Responses to Soil Tillage and Crop Rotation in a Corn/Soybean Agroecosystem." *Ecology and Evolution* 6 (22): 8075–84. doi:10.1002/ece3.2553.

Sternberg, Marcelo, Carly Golodets, Mario Gutman, Avi Perevolotsky, Eugene D. Ungar, Jaime Kigel, and Zalmen Henkin. 2015. "Testing the Limits of Resistance: A 19-year Study of Mediterranean Grassland Response to Grazing Regimes." *Global Change Biology* 21 (5). Wiley: 1939–50. doi:10.1111/gcb.12866.

Strickland, Michael S., and Johannes Rousk. 2010. "Considering Fungal:Bacterial Dominance in Soils – Methods, Controls, and Ecosystem Implications." *Soil Biology and Biochemistry* 42 (9). Elsevier BV: 1385–95. doi:10.1016/j.soilbio.2010.05.007.

Sun, Xin, Mingjie Sun, Ying Chao, Xiaoyang Shang, Hui Wang, Hong Pan, Quangang Yang, Yanhong Lou, and Yuping Zhuge. 2022. "Effects of Lead Pollution on Soil Microbial Community Diversity and Biomass and on Invertase Activity." *Soil Ecology Letters* 5 (1). Springer Science and Business Media LLC: 118–27. doi:10.1007/s42832-022-0134-6.

Syromyatnikov, Mikhail Y., Mariya M. Isuwa, Olga V. Savinkova, Mariya I. Derevshchikova, and Vasily N. Popov. 2020. "The Effect of Pesticides on the Microbiome of Animals." *Agriculture* 10 (3). MDPI AG: 79. doi:10.3390/agriculture10030079.

Tan, Teck-Toh, Sze-Looi Song, Sze-Wan Poong, Georgia M. Ward, Juliet Brodie, and Phaik-Eem Lim. 2020. "The Effect of Grazing on the Microbiome of Two Commercially Important Agarophytes, Gracilaria Firma and G. Salicornia (Gracilariaceae, Rhodophyta)." *Journal of Applied Phycology* 32 (4). Springer Science and Business Media LLC: 2549–59. doi:10.1007/s10811-020-02062-y.

Tardy, Vincent, Aymé Spor, Olivier Mathieu, Jean Lévèque, Sébastien Terrat, Pierre Plassart, Tiffanie Regnier, et al. 2015. "Shifts in Microbial Diversity through Land Use Intensity as Drivers of Carbon Mineralization in Soil." *Soil Biology and Biochemistry* 90. Elsevier BV: 204–13. doi:10.1016/j.soilbio.2015.08.010.

Tecon, Robin, and Dani Or. 2017. "Biophysical Processes Supporting the Diversity of Microbial Life in Soil." *FEMS Microbiology Reviews* 41 (5). England: 599–623. doi:10.1093/femsre/fux039.

Thakur, Madhav P., and Stefan Geisen. 2019. "Trophic Regulations of the Soil Microbiome." *Trends in Microbiology* 27 (9). Elsevier BV: 771–80. doi:10.1016/j.tim.2019.04.008.

Tripathi, Sachchidanand, Pratap Srivastava, Rajkumari S. Devi, and Rahul Bhadouria. 2020. "Influence of Synthetic Fertilizers and Pesticides on Soil Health and Soil Microbiology." In *Agrochemicals Detection, Treatment and Remediation*, 25–54. Elsevier. doi:10.1016/B978-0-08-103017-2.00002-7.

Turley, Nash E., Lukas Bell-Dereske, Sarah E. Evans, and Lars A. Brudvig. 2020. "Agricultural Land-use History and Restoration Impact Soil Microbial Biodiversity." *Journal of Applied Ecology* 57 (5). Wiley: 852–63. doi:10.1111/1365-2664.13591.

Ussiri, D. A. N., R. Lal, and P. A. Jacinthe. 2006. "Soil Properties and Carbon Sequestration of Afforested Pastures in Reclaimed Minesoils of Ohio." *Soil Science Society of America Journal* 70 (5): 1797–1806. doi:10.2136/sssaj2005.0352.

Vos, Marjon G. J. de, Marcin Zagorski, Alan McNally, and Tobias Bollenbach. 2017. "Interaction Networks, Ecological Stability, and Collective Antibiotic Tolerance in Polymicrobial Infections." *Proceedings of the National Academy of Sciences* 114 (40). Proceedings of the National Academy of Sciences: 10666–71. doi:10.1073/pnas.1713372114.

Wagg, Cameron, S. Franz Bender, Franco Widmer, and Marcel G. A. van der Heijden. 2014. "Soil Biodiversity and Soil Community Composition Determine Ecosystem Multifunctionality." *Proceedings of the National Academy of Sciences* 111 (14): 5266–70. doi:10.1073/pnas.1320054111.

Wagg, Cameron, Klaus Schlaeppi, Samiran Banerjee, Eiko E. Kuramae, and Marcel G. A. van der Heijden. 2019. "Fungal-Bacterial Diversity and Microbiome Complexity Predict Ecosystem Functioning." *Nature Communications* 10 (1). Springer Science and Business Media LLC. doi:10.1038/s41467-019-12798-y.

Wahl, Martin, Franz Goecke, Antje Labes, Sergey Dobretsov, and Florian Weinberger. 2012. "The Second Skin: Ecological Role of Epibiotic Biofilms on Marine Organisms." *Frontiers in Microbiology* 3. Frontiers Media SA. doi:10.3389/fmicb.2012.00292.

Wang, Can, Abolfazl Masoudi, Min Wang, Jia Yang, Zhijun Yu, and Jingze Liu. 2021. "Land-Use Types Shape Soil Microbial Compositions under Rapid Urbanization in the Xiong'an New Area, China." *Science of The Total Environment* 777. Elsevier BV: 145976. doi:10.1016/j.scitotenv.2021.145976.

Wang, M., D. Garrido-Sanz, P. Sansegundo-Lobato, M. Redondo-Nieto, R. Conlon, M. Martin, R. Mali, et al. 2021. "Soil Microbiome Structure and Function in Ecopiles Used to Remediate Petroleum-Contaminated Soil." *Frontiers in Environmental Science* 9 (March). Frontiers Media SA. doi:10.3389/fenvs.2021.624070.

Wang, Yei-Shung, Yu-Ju Huang, Wen-Ching Chen, and Jui-Hung Yen. 2009. "Effect of Carbendazim and Pencycuron on Soil Bacterial Community." *Journal of Hazardous Materials* 172 (1). Elsevier BV: 84–91. doi:10.1016/j.jhazmat.2009.06.142.

Wang, Zhen, Xiangyang Hou, Michael P. Schellenberg, Yan Qin, Xiangjun Yun, Zhijun Wei, Chao Jiang, and Yunfeng Wang. 2014. "Different Responses of Plant Species to Deferment of Sheep Grazing in a Desert Steppe of Inner Mongolia, China." *The Rangeland Journal* 36 (6). CSIRO Publishing: 583. doi:10.1071/rj13115.

Wang, Zhen, Shenyi Jiang, Paul C. Struik, Hai Wang, Ke Jin, Riliga Wu, Risu Na, Huabing Mu, and Na Ta. 2022. "Plant and Soil Responses to Grazing Intensity Drive Changes in the Soil Microbiome in a Desert Steppe." *Plant and Soil*, April. doi:10.1007/s11104-022-05409-1.

Wassermann, Birgit, Lise Korsten, and Gabriele Berg. 2021. "Plant Health and Sound Vibration: Analyzing Implications of the Microbiome in Grape Wine Leaves." *Pathogens* 10 (1): 63. doi:10.3390/pathogens10010063.

Weisskopf, P., R. Reiser, J. Rek, and H.-R. Oberholzer. 2010. "Effect of Different Compaction Impacts and Varying Subsequent Management Practices on Soil Structure, Air Regime and Microbiological Parameters." *Soil and Tillage Research* 111 (1). Elsevier BV: 65–74. doi:10.1016/j.still.2010.08.007.

Whalley, W. R., E. Dumitru, and A. R. Dexter. 1995. "Biological Effects of Soil Compaction." *Soil and Tillage Research* 35 (1–2): 53–68. doi:10.1016/0167-1987(95)00473-6.

Xia, Qing, Thomas Rufty, and Wei Shi. 2020. "Soil Microbial Diversity and Composition: Links to Soil Texture and Associated Properties." *Soil Biology and Biochemistry* 149 (October): 107953. doi:10.1016/j.soilbio.2020.107953.

Xiao, Dan, You Chen, Xunyang He, Zhihong Xu, Shahla Hosseini Bai, Wei Zhang, Ming Cheng, Peilei Hu, and Kelin Wang. 2021. "Temperature and Precipitation Significantly Influence the Interactions between Arbuscular Mycorrhizal Fungi and Diazotrophs in Karst Ecosystems." *Forest Ecology and Management* 497 (October): 119464. doi:10.1016/j.foreco.2021.119464.

Xiao, Enzong, Zengping Ning, Tangfu Xiao, Weimin Sun, and Shiming Jiang. 2021. "Soil Bacterial Community Functions and Distribution after Mining Disturbance." *Soil Biology and Biochemistry* 157. Elsevier BV: 108232. doi:10.1016/j.soilbio.2021.108232.

Xiong, Wu, Yuqi Song, Keming Yang, Yian Gu, Zhong Wei, George A. Kowalchuk, Yangchun Xu, Alexandre Jousset, Qirong Shen, and Stefan Geisen. 2020. "Rhizosphere Protists Are Key Determinants of Plant Health." *Microbiome* 8 (1). Springer Science and Business Media LLC. doi:10.1186/s40168-020-00799-9.

Xun, Weibing, Ruirui Yan, Yi Ren, Dongyan Jin, Wu Xiong, Guishan Zhang, Zhongli Cui, Xiaoping Xin, and Ruifu Zhang. 2018. "Grazing-Induced Microbiome Alterations Drive Soil Organic Carbon Turnover and Productivity in Meadow Steppe." *Microbiome* 6 (1). Springer Science and Business Media LLC: 170. doi:10.1186/s40168-018-0544-y.

Yáñez, Carolina, José Verdejo, Héctor Moya, Pamela Donoso, Constanza Rojas, Elvira A. Dovletyarova, Olga A. Shapoval, Yurii A. Krutyakov, and Alexander Neaman. 2022. "Microbial Responses Are Unreliable Indicators of Copper Ecotoxicity in Soils Contaminated by Mining Activities." *Chemosphere* 300. Elsevier BV: 134517. doi:10.1016/j.chemosphere.2022.134517.

Yang, Yichao, Amanda J. Ashworth, Jennifer M. DeBruyn, Cammy Willett, Lisa M. Durso, Kim Cook, Jr., Philip A. Moore, and Phillip R. Owens. 2019. "Soil Bacterial Biodiversity Is Driven by Long-Term Pasture Management, Poultry Litter, and Cattle Manure Inputs." *PeerJ* 7 (October): e7839. doi:10.7717/peerj.7839.

Yuan, Jun, Jacqueline M. Chaparro, Daniel K. Manter, Ruifu Zhang, Jorge M. Vivanco, and Qirong Shen. 2015. "Roots from Distinct Plant Developmental Stages Are Capable of Rapidly Selecting Their Own Microbiome without the Influence of Environmental and Soil Edaphic Factors." *Soil Biology and Biochemistry* 89. Elsevier BV: 206–9. doi:10.1016/j.soilbio.2015.07.009.

Zhalnina, Kateryna, Raquel Dias, Patricia Dörr de Quadros, Austin Davis-Richardson, Flavio A. O. Camargo, Ian M. Clark, Steve P. McGrath, Penny R. Hirsch, and Eric W. Triplett. 2014. "Soil PH Determines Microbial Diversity and Composition in the Park Grass Experiment." *Microbial Ecology* 69 (2). Springer Science and Business Media LLC: 395–406. doi:10.1007/s00248-014-0530-2.

Zhalnina, Kateryna, Raquel Dias, Patricia Dörr de Quadros, Austin Davis-Richardson, Flavio A. O. Camargo, Ian M. Clark, Steve P. McGrath, Penny R. Hirsch, and Eric W. Triplett. 2015. "Soil PH Determines Microbial Diversity and Composition in the Park Grass Experiment." *Microbial Ecology* 69 (2): 395–406. doi:10.1007/s00248-014-0530-2.

Zhang, Hui, Wenjun Liu, Xiaoming Kang, Xiaoyong Cui, Yanfen Wang, Haitao Zhao, Xiaoqing Qian, and Yanbin Hao. 2018. "Changes in Soil Microbial Community Response to Precipitation Events in a Semi-Arid Steppe of the Xilin River Basin, China." *Journal of Arid Land* 11 (1). Springer Science and Business Media LLC: 97–110. doi:10.1007/s40333-018-0071-5.

Zhang, Jiaoyang, Zemin Ai, Chutao Liang, Guoliang Wang, and Sha Xue. 2017. "Response of Soil Microbial Communities and Nitrogen Thresholds of Bothriochloa Ischaemum to Short-Term Nitrogen Addition on the Loess Plateau." *Geoderma* 308 (December): 112–19. doi:10.1016/j.geoderma.2017.08.034.

Zhang, Qi, Dong Zhu, Jing Ding, Fei Zheng, Shuyidan Zhou, Tao Lu, Yong-Guan Zhu, and Haifeng Qian. 2019. "The Fungicide Azoxystrobin Perturbs the Gut Microbiota Community and Enriches Antibiotic Resistance Genes in Enchytraeus Crypticus." *Environment International* 131. Elsevier BV: 104965. doi:10.1016/j.envint.2019.104965.

Zhang, Xuwang, Zhe Ji, Yating Shao, Chaochen Guo, Hao Zhou, Lifen Liu, and Yuanyuan Qu. 2020. "Seasonal Variations of Soil Bacterial Communities in Suaeda Wetland of Shuangtaizi River Estuary, Northeast China." *Journal of Environmental Sciences* 97 (November): 45–53. doi:10.1016/j.jes.2020.04.012.

Zhao, Zhi-Bo, Ji-Zheng He, Zhi Quan, Chuan-Fa Wu, Rong Sheng, Li-Mei Zhang, and Stefan Geisen. 2020. "Fertilization Changes Soil Microbiome Functioning, Especially Phagotrophic Protists." *Soil Biology and Biochemistry* 148 (September): 107863. doi:10.1016/j.soilbio.2020.107863.

Zhou, Guiyao, Xuhui Zhou, Yanghui He, Junjiong Shao, Zhenhong Hu, Ruiqiang Liu, Huimin Zhou, and Shahla Hosseinibai. 2017. "Grazing Intensity Significantly Affects Belowground Carbon and Nitrogen Cycling in Grassland Ecosystems: A Meta-analysis." *Global Change Biology* 23 (3): 1167–79. doi:10.1111/gcb.13431.

Hao, J., Chai, Y. N., Lopes, L. D., Ordóñez, R. A., Wright, E. E., Archontoulis, S., & Schachtman, D. P. (2021). The Effects of Soil Depth on the Structure of Microbial Communities in Agricultural Soils in Iowa, USA. Applied and environmental microbiology, 87(4), e02673-20. https://doi.org/10.1128/AEM.02673-20

Yang, F., Niu, K., Collins, C. G., Yan, X., Ji, Y., Ling, N., Zhou, X., Du, G., Guo, H., & Hu, S. (2018). Grazing practices affect the soil microbial community composition in a Tibetan alpine meadow. Land Degradation & Development, 30(1), 49–59. Portico. https://doi.org/10.1002/ldr.3189

Burney, J. A., Davis, S. J., & Lobell, D. B. (2010). Greenhouse gas mitigation by agricultural intensification. Proceedings of the National Academy of Sciences, 107(26), 12052–12057. https://doi.org/10.1073/pnas.0914216107

Gougoulias, C., Clark, J. M., & Shaw, L. J. (2014). The role of soil microbes in the global carbon cycle: tracking the below-ground microbial processing of plant-derived carbon for manipulating carbon dynamics in agricultural systems. Journal of the science of food and agriculture, 94(12), 2362–2371. https://doi.org/10.1002/jsfa.6577

Li, Q., Lee Allen, H., & Wollum, A. G. (2004). Microbial biomass and bacterial functional diversity in forest soils: effects of organic matter removal, compaction, and vegetation control. Soil Biology and Biochemistry, 36(4), 571–579. https://doi.org/10.1016/j.soilbio.2003.12.001

Liu, H., Qin, S., Li, Y., Zhao, P., Nie, Z., & Liu, H. (2023). Comammox Nitrospira and AOB communities are more sensitive than AOA community to different fertilization strategies in a fluvo-aquic soil. Agriculture, Ecosystems & Environment, 342, 108224. https://doi.org/10.1016/j.agee.2022.108224

Baćmaga, M., Kucharski, J., & Wyszkowska, J. (2015). Microbial and enzymatic activity of soil contaminated with azoxystrobin. Environmental monitoring and assessment, 187(10), 615. https://doi.org/10.1007/s10661-015-4827-5

Dixit, M., Ghoshal, D., Lal Meena, A., Ghasal, P. C., Rai, A. K., Choudhary, J., & Dutta, D. (2024). Changes in soil microbial diversity under present land degradation scenario. Total Environment Advances, 10, 200104. https://doi.org/10.1016/j.teadva.2024.200104

Hartmann, M., Niklaus, P. A., Zimmermann, S., Schmutz, S., Kremer, J., Abarenkov, K., Lüscher, P., Widmer, F., & Frey, B. (2014). Resistance and resilience of the forest soil microbiome to logging-associated compaction. The ISME journal, 8(1), 226–244. https://doi.org/10.1038/ismej.2013.141

Hartmann, M., Niklaus, P. A., Zimmermann, S., Schmutz, S., Kremer, J., Abarenkov, K., Lüscher, P., Widmer, F., & Frey, B. (2014). Resistance and resilience of the forest soil microbiome to logging-associated compaction. The ISME journal, 8(1), 226–244. https://doi.org/10.1038/ismej.2013.141

Longepierre, M., Feola Conz, R., Barthel, M., Bru, D., Philippot, L., Six, J., & Hartmann, M. (2022). Mixed Effects of Soil Compaction on the Nitrogen Cycle Under Pea and Wheat. Frontiers in microbiology, 12, 822487. https://doi.org/10.3389/fmicb.2021.822487

16 Synthetic Biology and Generation of Smart Microbes

Safoora Ahmed, Subhashree Sunderay, Ankan Chakraborty and E Sumitha
Department of Biotechnology and Bioinformatics JSS Academy of Higher Education and Research Mysore, Karnataka, India

Yuvaraj Sivamani
Crescent School Of Pharmacy, B.S. Abdur Rahman Crescent Institute of Science & Technology Tamil Nadu, India

16.1 INTRODUCTION

In the human body, there are at least as many host cells as there are microbial cells. Microbes, also referred to as the microbiota, have a number of important impacts on how the host functions. According to the "hygiene theory", the immune system's development depends on early exposure to various microorganisms. Various recent research has demonstrated the connection between microbiome dysbiosis and human illnesses [1]. The utilization of the microbiome as a therapeutic target is being emphasized more and more, as the significance of the connection between the microbiome and human health is becoming clearer [2]. The development of several microbes to constitutively express therapeutic compounds that effectively treat disease is still ongoing. A number of current sensors may be used in microbial engineering, whose inputs include various chemical and environmental signals [1]

16.2 WHAT IS SYNTHETIC BIOLOGY

Synthetic biology combines the fields of engineering and biology to build novel biomolecular parts, networks, and pathways and use such structures to rewire and reprogram organisms [1]. These re-engineered organisms will revolutionize our way of life in the years to come by making it possible to provide more inexpensive drugs, ecologically friendly fuels for our cars, and specialized therapies for "superbugs" and diseases like cancer [3]. These significant issues are starting to be addressed by the de novo engineering of genetic circuits, biological modules, and synthetic pathways, which are also being employed in relevant real-world applications [1].

16.2.1 Role of Synthetic Biology

In the science of "synthetic biology", organisms are engineered to have new capabilities in order to be redesigned for practical uses as shown in Figure 16.1. To address issues in agriculture, industry, and health, synthetic biology researchers, and businesses all around the world are utilizing the power of nature [2]. Redesigning organisms to produce a chemical, such as a medication or fuel, or

DOI: 10.1201/9781003394662-16

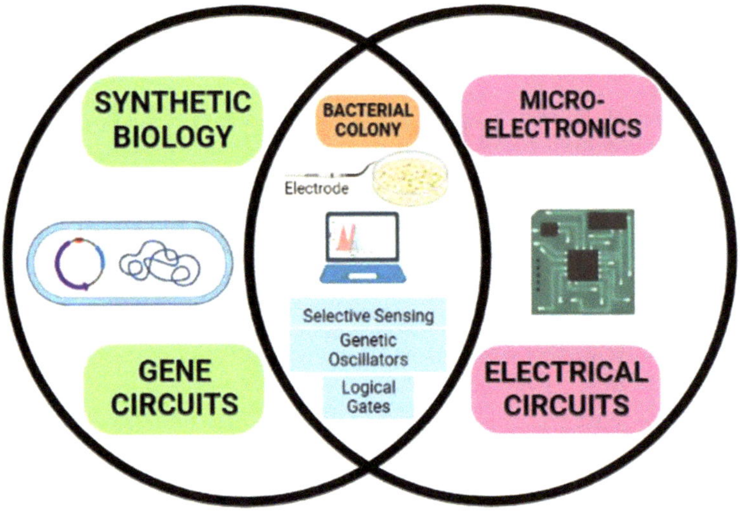

FIGURE 16.1 Synthetic biology overview.

to acquire a new ability, such as the capacity to perceive something in the environment, are common goals of synthetic biology initiatives. Moreover, it makes it possible to develop new enzymes, pathways, and complete cell systems. Human, bacterial, viral, and plant cells have all been manipulated using synthetic biology to create substances or change biological processes that are beneficial to human health. Synthetic biology has had a huge impact on human health and the treatment of disease, from crop fortification to the modification of T cells. Synthetic biology aims to create organisms that produce a substance, such as a drug, food, or fuel, or acquire a new capability, such as the capacity to perceive something in the environment [2]. Additional applications include the creation of disease-resistant crops, tissue engineering, and engineered living materials that can self-assemble or alter a material's function.

16.2.2 Human Healthcare

In 2020, a method was found for producing messenger RNA (mRNA) which may instruct human cells to manufacture a protein based on a segment of the COVID-19-causing SARS-CoV-2 virus, instructing the immune system to develop antibodies (the body's primary immunological defense) before an infection. This innovative vaccine's quick development has had a significant influence on society and holds promise for subsequent treatments and vaccines using a similar methodology. The Moderna and BioNTech/Pfizer vaccines have emerged as successful examples of synthetic biology thanks to quick genetic sequencing, computational biology, and biomanufacturing.

One of the achievements of synthetic biology in biomedicine is the creation of mRNA vaccines. Modified yeast has been utilized in the manufacturing of semi-synthetic artemisinin, an anti-malarial medicine, and the diabetes medication [2]. To replace previous, environmentally hazardous production processes, Januvia® uses a novel enzyme generated through directed evolution. A breakthrough cancer treatment known as CAR-T cell therapy (for chimeric antigen receptor T-cell therapy) alters T cells derived from the patient's body to target tumors through the use of synthetic biology. CAR-T cells' success has sparked curiosity in utilizing the same approach to treat other immune cell types, such as natural killer cells. Other therapies are now being explored that use modified probiotics to enhance the body's immune system. Furthermore, more advancements make use of synthetic biology to create 3D scaffolds on which to culture cells and possibly even regenerate entire organs.

The generation of innovative diagnostics, treatments, vaccines, and preventative healthcare using synthetic biology is only just beginning to touch the surface of its potential. Future applications could include creating new vaccines to prevent various viral diseases (like the common cold), prophylactic medical care, therapies for metabolic disorders, and stopping emerging infectious diseases before they become pandemics (possibly once more using mRNA vaccines). Additionally, it can be utilized to locate biologicals that combat the negative effects of aging on the human body. We anticipate that synthetic biology will also benefit the development of unapproved medications, dietary supplements, and performance-enhancing substances because market forces will determine demand. Since it will depend on an understanding of the body's incredibly complicated biological functions, effectiveness in these situations is probably going to be uneven. The study that deepens our understanding will include synthetic biology, but it cannot be assumed that all biologics, regardless of the production process, will be safe and efficient [3].

In the future, precision medicine will be greatly facilitated by synthetic biology. The future of human healthcare is frequently viewed as precision healthcare, which includes personalized medication. By customizing diagnoses and treatments to each patient based on their particular genetics, microbiome, nutrition, way of life, environment, or illness features, precision health would transform the practice of medicine [2, 4]. Then, based on those details, diagnostics and therapies can be customized, utilizing synthetic biology as needed and required.

16.2.3 Earth's Biological Environment

Humans can use synthetic biology's technologies to enhance, better, and more fully utilize the biological environment on Earth. This environment is made up of the delicate ecosystems that are all around us and with which humans must coexist. Since ancient times, humans have had a substantial impact on the biological environment, frequently battling threats and pests while exploiting other plants and animals for our own benefit [5]. Synthetic biology presents unique possibilities and capabilities to develop these environmental influences, perhaps in beneficial ways. Instead of introducing competing species, synthetic biology can now be used to combat undesired, disease-carrying, or invasive species. For instance, efforts are being made to generate genetically altered mosquitoes to stop the spread of mosquito-borne illnesses [1], which continue to pose a severe threat to human health. In the Florida Keys, field tests have recently started to evaluate the effectiveness and impact of such strategies on natural ecosystems. It has also been suggested to use synthetic biology to alter the microbiome of the insects and damage crucial genes with bacteriophages as a biopesticide to reduce locusts. Gene drives and viral vector delivery systems are highly adaptable techniques that will likely have a wide range of novel uses, including controlling zoonotic disease transmission and removing invasive plants and animals. By putting a little piece of DNA into an individual, gene drives are genetic constructions that alter the genetic makeup of a particular species. This DNA subsequently spreads and dominates gene reproduction in succeeding generations [4].

Generating microorganisms with desired traits is one of the other ways synthetic biology can be used to change the biological ecology of the earth. These advancements may pave the way for the potential to repair various types of xenobiotic and environmental damage, including oil spills. Through the employment of these bacteria, it may be possible to hasten the breakdown of plastic in landfills, promote nitrogen fixation, boost soil productivity, cut back on fertilizer consumption, or even reduce ambient greenhouse gas concentrations. Synthetic biology also opens up the prospect of creating biological markers that may be used as sensors to detect the presence of poisons or contaminants. One such research discovery uses bacteria-infused beads that may be scattered throughout a field's surface and can identify hidden landmines by fluorescing green in the presence of trinitrotoluene (TNT) fumes. Similar changes could be made to native plants to provide optically visible sensors or other spectroscopic features. There have been serious discussions about using synthetic biology, based on the genetic data revealed in preserved DNA strands, to bring back the woolly mammoth in the world of multicellular organisms. Moreover, these methods might be utilized to resurrect recently extinct species or re-establish species that are on the verge of extinction [5, 6]. This could involve, for instance, treating a species of helpful bees to fend off particular diseases or saving a particular kind of tree from a blight.

Of course, there are significant risks associated with many of these applications. As a result of good intentions gone wrong, humans have repeatedly done serious environmental harm. Synthetic biology-based solutions must be carefully crafted to address the particular issue at hand and must include safety precautions to ward off unfavorable outcomes. These difficult problems cross disciplinary boundaries and legal systems. Given that many synthetic biology techniques entail releasing replicating and proliferating life forms, extra vigilance is required [6].

16.2.4 Agriculture and Food Production

Despite being disputed today, historically, genetically modified organisms (GMOs) were produced through agricultural practices, including plant breeding and animal husbandry [2]. Today, desired outcomes can be achieved by using synthetic biology techniques to add or remove genes from plants and animals. Viral transduction, a method for transferring genetic material between plants via viruses, has resulted in the development of more resistant plants [5]. Famously, Monsanto created seeds (such as soybeans) that are resistant to the negative effects of the weed killer Roundup®, enabling higher agricultural yields.

The creation of fertilizers using engineered bacteria to increase crop yields and the use of bespoke fermentation to create artificial food additives are less well-known applications of synthetic biology in agriculture. The latter items can include sugars, flavor enhancers or simulants, nutritional supplements (such as flavonoids and terpenoids), and oils and fats (lipids) for enhancing meals. It may be feasible to create novel food flavors that appeal to human senses by using an understanding of human taste and olfactory receptors. The use of synthetic biology in agriculture to produce meat directly from altered cells, obviating the need to butcher animals, is of great interest [2].

The FDA has already given its approval to a fish food that was created by genetically modifying microorganisms to yield an aquaculture feed that resembles flour. While it may seem like a long way off, multiple businesses and research facilities are competing to create cultured meat with characteristics that are equal to those of animal goods. Some expect that consumer-cultured beef products will become ubiquitous in five years. Others think it will be more challenging to accurately replicate butchered meat in terms of texture, grain, and fat mixture. The success of this technology depends on the application of a number of synthetic biology concepts [2, 3], as well as 3D printing technology which involves using multiple constituent cell types.

The genetic makeup of a microbe was altered to create an aquaculture feed that resembles flour, and the FDA has already approved this fish meal. Despite appearing to be a long way from creating a synthetic beef steak, many businesses and research facilities are vying to create cultured meat that has the same qualities as animal goods. Consumer-cultured beef products are expected to become widespread in five years, according to some. Others think it will be more challenging to accurately reproduce the texture, grain, and fat mixture of actual butchered flesh. To make it happen, however, a lot of synthetic biology ideas as well as 3D printing technology that uses many constituent cell types are needed [6]. In a similar manner, other food items, such as fruits and vegetables, may be cultivated from media using biological processes that don't rely on real soil and sunlight. When cells are cultured utilizing synthetic biology, whether to produce meat or vegetables, production efficiency goes beyond speed and appearance. Think about a circumstance where a bio-environment is contained, such as an Antarctic outpost or a long-duration human spacecraft. Fresh food production by cell/tissue culture has a number of benefits over long-term storage or replenishing via extensive transit networks. The weight of the food supply can be greatly decreased by reconstituting freeze-dried media with recovered water. Since beef, for instance, contains about 50% water, you might reduce the weight of the beef supply by up to 50% while still producing a superb fresh food product by cultivating cells in reconstituted media to make a juicy New York Strip [5].

The detection and mitigation of food rotting problems may also be possible with synthetic biology. Biological sensing components that are incorporated into the container could keep track of how the food is deteriorating, such as milk spoiling. Currently, certain foods are irradiated to extend their shelf lives, which is a deterministic method of changing the biological makeup [6, 7]. Synthetic biology-designed and -produced additives or packaging could more precisely combat the processes that lead to food degradation, resulting in foods with longer shelf lives that need less refrigeration. These biological sensors and additives have the potential to fundamentally alter how food is distributed and stored.

16.2.5 Materials Production and Manufacturing

Proteins with a wide range of structural and chemical characteristics are created by biological processes [5]. Natural biology is capable of producing an astounding variety of functional structures, ranging from tiny silk thread strands with tremendous tensile strength to the cellulose structures that allow trees to attain their towering heights [6]. Through synthetic biology, people may also begin to employ biological processes to create custom-made materials and construction techniques [5–7]. We are likely to create unique materials and new uses by substituting or changing conventional processes like chemical synthesis and commodity manufacture, all the while increasing production efficiency and lowering its impact on the environment.

Despite being still in its infancy, biomanufacturing is being researched by numerous startups as well as established material firms like Dupont. To manufacture materials and adhesives for construction that have a lower carbon impact, for instance, bio-cements are being produced. Additionally, attempts to produce silk thread, which is traditionally associated with spiders and silkworms, have a long history. However, synthetic biology is beginning to have a significant impact in this field [6, 7]. A joint venture that produces bio-renewably generated caprolactam is designed to produce nylon 6 yarns, films, and tailored bioplastics of commercial quality (a precursor to these products). The goal of these replacement items is to be more environmentally responsible and sustainable. The Engineering Biology Research Consortium, led by Cargill, has been grantedthe status of Manufacturing Innovation Institute, dubbed BioMade, by the Department of Defense in order to "identify and innovate on shared challenges in scaleup and downstream processing to further strengthen the US economy in the production of bio-industrial products" [2]. To create cell-free ways for protein synthesis, scientists are embracing synthetic biology. Proteins are primarily made inside cells, where biological membranes make these biochemical wonders relatively inaccessible. As cell walls are no longer necessary for protein synthesis, fine chemicals, biofuels, medicinal small molecules and glycoproteins, and even bio-enabled smart materials may now be produced at a larger scale thanks to synthetic biology. Materials science and biomanufacturing are brought together at the nanoscale via cell-free synthesis.

Nanostructures can be combined, stacked, and recombined in biology to create macrostructures like plants and animals. Functionally graded materials can have different combinations and arrangements than bulk or composite materials, which makes them different from those two types of materials. Building materials with corrosion-resistant surfaces, embedded sensors, or structurally robust, lightweight materials is a tough process with conventional chemistry and manufacturing, but it might be feasible with biological manufacturing. We will probably need to grasp synthetic biology in order to create a wide variety of these materials [3].

16.3 AREAS AND FUNCTIONS OF SYNTHETIC BIOLOGY

In order to fulfill the goals of sustainable development, eradicating hunger, promoting health and wellbeing, reducing inequality, and enhancing access to products and services that have been generated ethically, synthetic biology and its applications show considerable potential [6]. Synthetic biology has the potential to revolutionize a wide range of technologies, including biocomputing, living materials, electronic interfaces, therapeutic genome editing, multiplexed diagnostics, cellular recording, third-generation biorefineries, and living biotherapeutics, according to recent developments. Perhaps the most well-known and advanced application of synthetic biology is the ability to alter metabolism to produce high-value products for uses ranging from biofuels and plant natural products to polymer precursors and bioinspired materials. In this sense, a new era of biomanufacturing is being ushered in by the ability to transform bacteria into chemical factories that compete with organic chemical synthesis. Synthetic biology is a multidisciplinary field of study that aims to develop new biological components, technologies, and systems or redesign existing natural systems to have new functionalities. With its development, synthetic biology has become a broad, interdisciplinary area that draws on methods, concepts, and ideas from a variety of disciplines, including biology, chemistry, engineering, computer science, medicine, bioinformatics, and many more. Synthetic biology, which has a close relationship to bioengineering, tries to develop new biological components or redesign processes that already exist in nature. Modern advances in science and the development of new instruments for the marketplace have created new possibilities. Applications have been found and businesses have been established as a result of the advancement of these tools and the maturing of synthetic biology as a discipline in academia.

The current state of synthetic biology is the culmination of decades of scientific and technological advancement. Innumerable applications have been made possible by these fundamental research discoveries, including the identification of DNA and its function, the discovery of its double-helix structure, the cracking of the genetic code, the construction of a reference human genome, the reduction of the cost of genetic sequencing, and more recent developments in genome editing [6][5].

Synthetic biology is the design, development, and construction of biological systems using standardized biological components. Examples of standardized biological components include DNA segments that code for proteins and DNA elements that regulate transcription. Standard molecular biology tools and procedures can be used to identify, separate, store, handle, and synthesize these biological components.

16.3.1 Synthetic Circuit Design

Understanding how genes and the proteins they encode interact with one another, react to internal and external stimuli, and switch on and off cellular processes including gene expression and cell division forms the foundation for the construction of synthetic circuits [2, 6]. Synthetic gene circuits that carry out specified operations are created by combining well-characterized genes and proteins using a bottom-up method in synthetic circuit design. The genetic toggle switch and the repressilator were the first biological circuits intended to perform specific tasks. In response to outside inputs, cells that contained the genetic toggle switch circuit could transition between two stable expression states. With the repressilator, activation of an oscillatory circuit led to the expression of repressor protein or an orderly, periodic oscillation [5]. Through these studies, a novel synthetic circuit construction technique was created that included quantitative design, physical fabrication, experimental measurement, and hypothesis-driven debugging. The time it takes for these biochemical activities to take place in the cell causes circuits that depend on genes to be transcribed or translated into proteins to have a lag between input and output of hours or days. Protein-protein interactions allow circuits to be activated quickly [3, 6]. In order to control biological processes, proteins act as switches with two stable states. A protein can transition between two stable states thanks to the reversible protein modification known as phosphorylation, which can be caused by the interaction of two proteins. A network of fourteen proteins expressed in yeast host cells was created as a toggle switch that is activated by the sugar alcohol sorbitol, which can be found in fruits. A fluorescent protein that goes into the nucleus is how the yeast cell remembers its exposure to sorbitol when it recognizes it. Figure 16.2 depicts cellular decision making in human pluripotent cells.

Some examples of synthetic circuit design:

1. Asymmetric plasmid partitioning:
   In Escherichia coli, an artificial genetic circuit called asymmetric plasmid partitioning regulates asymmetric cell division.
2. Distributed multicellular computational:
   When logic gates are too complex for single cells to implement, distributed multicellular computation is employed to distribute the workload among various cell strains. To create Boolean logic gates, transcriptional regulators, RNA molecules, or site-specific recombinases are used.
3. Recombinase logic device:
   Recombinase enzymes are used in recombinase logic devices, which invert or excise DNA at particular DNA sequences.

16.3.2 Transcriptional Control

A common element of transcriptional control circuits is an actuator that controls transcription's positive or negative regulation. This component is connected to a DNA-binding component that recognizes the promoter, a regulatory DNA sequence. These programmable transcription factors were developed in response to natural transcription factors, which bind promoters to regulate gene expression [3].

16.3.3 RNA Devices

Engineered RNA sequences are used to create RNA devices, which combine actuator and sensor components to encode higher-order functionality. For instance, the activity of an RNA sensor can change when a ligand binds to it [2, 3]. Short RNA sequences called RNA aptamers are used by RNA sensors to identify biological inputs such as proteins, nucleic acids, tiny molecules, and RNA-lipid interactions. Differentiating between developed and undifferentiated human-induced pluripotent stem cells is one use of an RNA device [2]. RNA enzymes that catalyze chemical processes are referred to as catalytic RNA or ribozymes. The ribosome, which uses ribozymes to link amino acids to create protein chains, serves as one illustration of this. Effector-responsive ribozymes are produced by including ligand-binding RNA aptamer sequences that control the activity of the RNA catalyst.

16.3.4 Optogenetic Circuits

With the application of light-responsive proteins, the area of optogenetics has developed a method for controlling a number of cellular processes that can be further designed or integrated into biological systems via synthetic biology [3]. A molecular event that may be quantified or detected is turned on or off by a photosensitive protein in response to light. Optogenetics uses this response to control molecular events. In this context, photosensitive proteins are referred to as optogenetic actuators.

16.3.5 On-Chip Assembly

Using several parallel micro-reactions on silicon chips, next-generation DNA synthesizers may produce DNA by both enzymatic and chemical means. Desktop DNA synthesizers of the future will be built on precision thermal control chips with thousands of pixels or on virtual wells that serve as individual DNA synthesis sites and have liquid flowing through them continuously [2]. DNA strands are built base by base and then released from the surface to be assembled into chips. The process of thermal purification corrects errors by identifying and eliminating mismatches between partially formed DNA strands at a predetermined temperature [5].

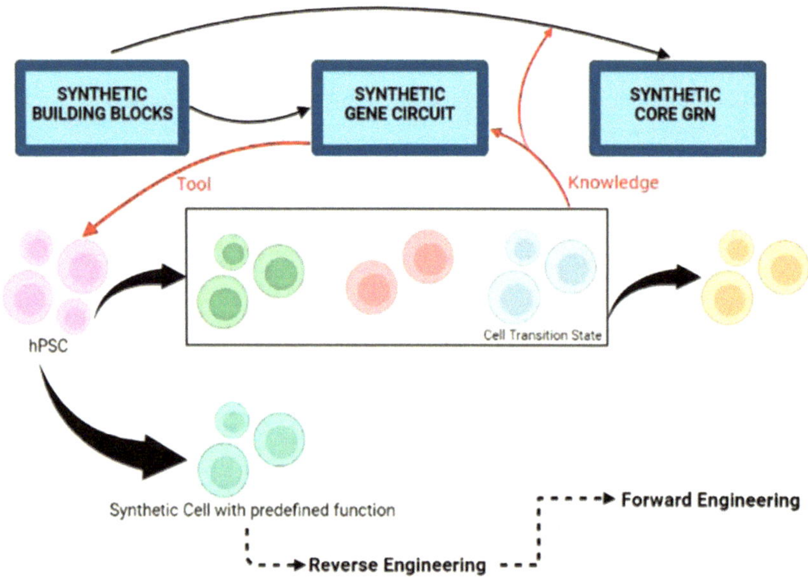

FIGURE 16.2 Schematic representation of cellular decision-making in human pluripotent stem cells (hPSC).

16.3.6 Artificial Genetic Systems and Artificial Enzymes

16.3.6.1 Artificial Genetic Systems

XNAs are artificial genetic systems that can transmit and store genetic information in a manner comparable to DNA while evolving naturally. These molecules are synthetic in nature.

Human pluripotent stem cell (hPSC) destiny manipulation is frequently accomplished by reversing the developmental stages. A different method of modifying the fate of human cells is possible through the forward engineering of synthetic decision-making gene circuits. Combining forward and reverse engineering techniques in hPSC could hasten both basic and practical stem cell biology.

16.3.6.2 Artificial Enzymes

Using chemicals that don't exist in nature, artificial enzymes that can start chemical processes like naturally existing enzymes have been created as shown in Figure 16.3. In order to create the synthetic enzymes known as XNAzymes, XNAs were used as the building blocks [2]. XNA molecules can act as enzymes and catalyze chemical reactions, just like RNA molecules, which are comparable to XNA molecules. In order to prevent viral infection, three XNAzymes were created that can self-assemble into a nanostructure that targets and cleaves certain RNA sequences, including the SARS-CoV-2 virus [6]. Research on artificial RNA and enzymes aims to provide answers to issues regarding other chemistries from our own that might support life, possibly on distant worlds. Artificial enzymes may be more stable in the body for medicinal and therapeutic applications since they do not break down as quickly [3, 5].

16.4 STEPPING IN TO A NEW REVOLUTION OF SMART MICROBES

Microbes are little guardians and keepers of soil and plant life in nature. Microbes are minuscule creatures that exist in nature on a daily basis. In fact, many microbes are helpful to both humans and the environment; therefore, not all of them are bad. The performance and growth of plants are

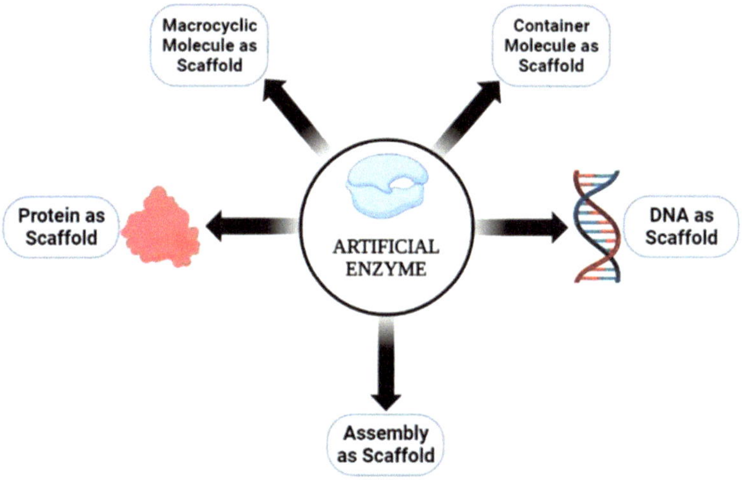

FIGURE 16.3 Applications of artificial enzymes.

greatly influenced by environmental microbes like bacteria and fungi, which are also necessary for a healthy soil ecosystem. By exploiting the advantages of microbes for agricultural practices, biological goods, or "microbials", using live microorganisms, are being developed as a natural substitute for chemical fertilizers, insecticides, and herbicides. Inputs that are damaging to plants are being replaced with microbials, which is also commonly employed to control weeds, pests, and disease. We can create and produce microbial-based goods, such as biostimulants, biofertilizers, and biopesticides, by exploiting microorganisms. These newly synthesized products provide effective microbial solutions that improve soil fertility, plant performance, and pest and disease protection.

There are different ways to harness microbial energy. Some of them are as follows:

- Natural power:
  These small creatures, which are diverse and naturally produced, cooperate to support plant growth by dissolving organic materials, including those left over from deceased animals and plants, and transforming them into nutrients that plants can use.
- Soil power:
  By increasing the nutritional value of the soil and other organic matter, such as nitrogen and phosphorus, and making it accessible and obtainable in forms that are beneficial to plants, microbes aid in its flourishing.
- Growing power:
  Increases root and leaf growth and lowers the plant's susceptibility to pests, diseases, and abiotic stressors, all of which contribute to the stimulation of plant growth.

16.4.1 Smart Microbes Employed in Various Diagnostic and Therapeutic Techniques

Early studies led to the development of curative microorganisms for diseases like colitis, cancer, and bacterial infections. Constitutive expression of therapeutic biomolecules is a theme that runs across these fundamental works. However, many microbes are now being altered now to constitutively express therapeutic substances that efficiently treat disease. One microorganism recently discovered an enzyme that converts metabolites obtained from cruciferous vegetables into an anticancer compound. This strain slowed the development of colon cancer in a mouse model. Another microorganism was designed to constitutively express and secrete self-assembling nanofibers that promote mucosal healing. This strain decreased inflammation in a dextran sodium sulfate (DSS) model of murine colitis. It is becoming more and more usual to limit the production of therapeutic chemicals to particular times and places in generated therapeutic microorganisms by including disease- and region-responsive biosensors. The use of multiple existing sensors, whose inputs include a wide range of chemical and environmental data, is possible with microbial engineering. Controlling the expression of medications with biosensors improves the safety and genetic stability of these germs by removing potential off-target effects of the therapies and reducing metabolic costs on the bacteria.

16.4.2 Bacterial Infections

By using sensors produced from the target pathogens, bacterial infections can be detected and treated with microorganisms.

For instance, the signaling domain of the Vibrio cholerae cholera autoinducer 1-sensitive two-component system (TCS) was combined with the signal transduction domain of a *Lactococcus lactis* TCS to create a novel cholera sensor. After isolating the modified L. lactis strain from fecal samples, cholera infections in mice were discovered by connecting the sensor to an enzyme that catalyzes a colorimetric reaction. It's interesting to note that L. lactis cells carrying the sensor construct had lower cholera prophylactic efficacy than a non-engineered wild type strain, suggesting that the sensor may have a metabolic "cost" to it. The modified EcN strain proved effective in treating and preventing P. aeruginosa infection in mice as well as in C. elegans. Only the antibiotic-resistant V. cholerae was eliminated with the conjugation of a plasmid expressing an intein-split toxin and its equivalent antitoxin in zebrafish and Artemia salina models. After conjugation, the expression of the toxin was stimulated by a transcription factor linked to pathogenicity, whereas the expression of the antitoxin was suppressed by a transcription factor linked to antibiotic resistance. As a result, this system worked like an AND logic gate, and only pathogenic and antibiotic-resistant V. cholerae cells experienced cell death. The low conjugation efficiencies of this method must be addressed in order for it to have a meaningful therapeutic effect, though it is an intriguing technique.

16.4.3 Cancer

Biosensors have been added to natural tumor-colonizing bacteria by researchers to provide effective cancer therapies. For instance, it was possible to genetically modify attenuated Salmonella typhimurium (aSt) to express a heterologous flagellin when arabinose was present. The modified aSt strain boosted immune cell recruitment in tumors after intravenous injection and arabinose supplementation, and significantly improved mice survival rates in comparison to the wild-type strain. Another study discovered a connection between the expression of aSt's autolytic protein and a positive-feedback quorum sensing mechanism, which is used to treat cancer. In a colonized solid tumour, when the population level was reached, the majority of modified microorganisms released their therapeutic payload while the minority that did not lyse seeded a new generation of therapeutic bacteria. By giving the mice oral doses of these beneficial microorganisms, tumor-bearing mouse survival was markedly improved. By injecting these bacterial strains into the tumors, more mice with malignancies survived. Combined, these findings show that tumor targeting can be further enhanced by using surface-displayed proteins that facilitate tumor binding, and that engineered microorganisms that deliver immunotherapeutic drugs precisely at the tumor site are useful for cancer therapy.

16.4.4 Inflammatory Disorders

In vivo and in clinical samples, researchers have engineered microorganisms to detect and report on inflammatory markers. For instance, sensors for nitrogen oxides (inflammation biomarkers) were coupled to memory switches in E. coli to detect serum nitrogen oxides in human clinical samples. In one case, a tetrathionate sensor controlled a genetic memory switch in a strain of E. coli that spent a long time living in the gut. Customized microbes could identify mouse gut inflammation six months after bacterial treatment via an enzyme-driven colorimetric response. In a different study, the green fluorescent protein (GFP) in EcN, whose expression was determined with the help of a flow cytometer, was directly linked to both an alternate tetrathionate sensor and a thiosulfate sensor. The intestines of mice with DSS-induced colitis could be detected by these strains for thiosulfate but not for tetrathionate.

16.4.5 Metabolic Disorders

For instance, sensors for nitrogen oxides (inflammation biomarkers) were coupled to memory switches in E. coli to detect serum nitrogen oxides in human clinical samples. In one case, a tetrathionate sensor controlled a genetic memory switch in a strain of E. coli that spent a long time living in the gut. Customized microbes could identify mouse gut inflammation six months after bacterial treatment via an enzyme-driven colorimetric response.

16.4.6 Other Disorders

Microbes have been developed to detect different diseases. In one case, intestinal bleeding in a pig model was detected using EcN that was designed to express a luminous reporter in response to heme. This diagnostic technique has the distinct advantage that no processing of the feces was required to interpret the results. The modified bacteria were instead put into an ingestible capsule with luminescent-detecting microelectronics before administration. The bacteria generated a luminous signal when they detected bleed-associated heme during their movement through the gut. The signal was picked up by the microelectronics, which then wirelessly relayed it to a remote receiver. This is the initial report on real-time microbe-based diagnostics. One more E. coli strain was designed for temperature perception. To react to temperature ranges that are important to physiological processes, protein thermosensors have evolved. An evolved thermosensor variation was found in several microbes that were used to diagnose mouse fever. The modified bacteria produced a GFP signal that was picked up by in vivo fluorescence imaging in response to a rise in housing temperature. Using the same temperature sensor, it was also possible to see how mice's localized fluorescent protein expression changed in response to focused ultrasonic treatments that increased their body temperatures specifically where it was needed. Where systemic delivery is not desired, this idea of precise spatial targeting has numerous potential therapeutic uses.

16.5 MYCORRIZAHE

A fungus and a root system have a symbiotic association known as a mycorrhiza. The growing root is encircled by a network of fungal filaments, or the filaments directly contact the root cells. Because the hyphae have a relatively large surface area, they can absorb minerals and water from a larger area of soil than the root can. The mycorrhizae acquire carbohydrates and nutrients from the roots, while the fungus provides water and minerals to the roots. When two species create a mycorrhizal association, the fungus either colonizes the host plant's root tissues extracellularly, as in the case of ectomycorrhizal fungi, or intracellularly, as in the case of arbuscular mycorrhizal fungi (AMF or AM). Depending on the species or habitat, mycorrhizae may form parasitic associations with host plants. The association is frequently mutualistic. Around the world, plants that survive in a variety of environmental conditions have a mutualistic interaction with mycorrhizal fungi. The mutualistic link between fungi and plants provides benefits to about 95% of them. Depending on their physical characteristics and connections to certain plants, mycorrhizae can be classified into a variety of groupings. The three main and most common types of mycorrhizae are endomycorrhizae, ectomycorrhizae, and the specialized fungus ericoid mycorrhizae.

16.5.1 Types of Mycorrhizae

There are, importantly, two types of mycorrhizae. They are ectomycorrhizae and endomycorrhizae.

16.5.1.1 Ectomycorrhizae

Ectomycorrhizas, also known as EcM, are symbiotic relationships between the roots of around 10% of plant families, primarily woody ones like those in the birch, dipterocarp, eucalyptus, oak, pine, and rose [8] families, orchids [9], and fungi from the Basidiomycota, Ascomycota, and Zygomycota orders. Some EcM fungi, like Leccinum and Suillus, develop mycorrhizas with only a single genus of plant, while other fungi, like Amanita, are generalists that interact with a wide range of plants [10]. At any given time, a tree may have fungal EcM partners from 15 or more distinct species [11]. There are a huge number of ectomycorrhizal fungus species, housed in more than 200 genera. According to a current study's cautious estimate, there are roughly 7,750 different species of ectomycorrhizal fungi worldwide. However, a more accurate estimate of the variety of macromycete species would likely range between 20,000 and 25,000 [9]. A hyphal sheath, or mantle, surrounding the root tip and a Hartig net of hyphae enclosing the plant cells in the root cortex make up ectomycorrhizas. When the hyphae additionally enter the plant cells, the mycorrhiza is referred to as an ectendomycorrhiza. Away from the root, a substantial network of ectomycorrhizal extramatrical mycelium forms in the soil and leaf litter. It has been proven that the fungal network acts as a conduit for the transmission of nutrients between different plants. It has been demonstrated that carbon moves from paper birch trees into Douglas-fir trees, enabling ecosystem succession [12]. The ectomycorrhizal fungus Laccaria bicolor draws springtails to it and kills them in order to absorb nitrogen, some of which may then be transferred to the mycorrhizal host plant. Eastern White Pine inoculated with L. bicolor was able to obtain up to 25% of its nitrogen from springtails in a study by Klironomos and Hart [13, 14]. Compared to thin roots that are not mycorrhizal, ectomycorrhizae may have extremely high quantities of trace elements, such as chlorine or hazardous metals (cadmium, silver) [15]. The ectomycorrhizal basidiomycete L. bicolor, a representative of symbiotic fungi, had its first genome sequence published in 2008 [16]. The emergence of numerous multigene families in this fungus suggests that gene duplication came before symbiotic adaptation. In ectomycorrhizal root tips, there was an up-increased expression of lineage-specific genes that code for secreted proteins that are regulated by symbiosis, suggesting a potential function in partner communication. L. bicolor cannot degrade host cells during root colonization because it lacks the enzymes needed to degrade the parts of plant cell walls (cellulose, hemicellulose, pectins, and pectates). As opposed to this, L. bicolor possesses extended multigene families connected to the hydrolysis of proteins and polysaccharides found in microfauna and bacteria. This genomic study revealed the dual biotrophic and saprotrophic lifestyle of the mycorrhizal fungus, which enables it to thrive in both soil and living plant roots.

16.5.2 Endomycorrhizae (AM/VAM)

More than 80% of plant groups include endomycorrhizae, which include greenhouse and crop plants like fruits, flowers, grasses, and vegetables. Endomycorrhizal connections are characterized by the fungus' creation of vesicles and arbuscules as well as their penetration of the cortical cells. Endomycorrhizas come in a variety of forms, including arbuscular, monotropoid, ericoid, orchid mycorrhizas, and arbutoid.

16.5.3 Arbuscular Mycorrhizae

An arbuscular mycorrhiza (AM) is a type of mycorrhiza in which the symbiotic fungus (AM fungi, or AMF) enters the cortical cells of the roots of a vascular plant to create arbuscules (plural mycorrhizae, also known as endomycorrhiza). Glomeromycota and Mucoromycota, sister clades of the more well-known and diversified dikaryan fungi, produce arbuscules and vesicles, which are distinctive structures characteristic of arbuscular mycorrhizae (all three are together called "symbiomycota"). AM fungi assist plants in absorbing nutrients from the soil, including sulfur, phosphorus and micronutrients. The evolution of vascular plants and the early colonization of land by plants were both greatly aided by the development of the arbuscular mycorrhizal symbiosis [1]. It has been noted that listing plants that do not develop endomycorrhizae is quicker than listing those that do. This symbiosis, the most common plant symbiosis known, is a highly developed mutualistic association between fungi and plants. AMF is present in 80% of the vascular plant groups still in existence today [4]. Since the 1670s, mycorrhizal physiology and ecology research has advanced, improving our knowledge of the various functions of AMF in the ecosystem. One illustration is the crucial role the protein that resembles glue, glomalin, plays in the structure of soil. The management, restoration, and agriculture of ecosystems are human endeavors that can benefit from this understanding.

16.5.4 Ericoid Mycorrhizae

The mutualistic relationship known as the ericoid mycorrhiza is created when plants from the Ericaceae family and several mycorrhizal fungus lineages come together. Additionally, it has been shown that ericoid mycorrhizas are extremely saprotrophic, enabling plants to absorb nutrients from materials that are still degrading thanks to the ericoid companions' microbial breakdown processes.

16.5.5 Monotropoid Mycorrhizae

Numerous genera of the Orchidaceae and the Ericaceae subfamily Monotropoideae contain this type of mycorrhiza. These plants are heterotrophic or mixotrophic, deriving their carbon from their fungal partner. The mycorrhizal interaction in question is thus parasitic and non-mutualistic.

16.5.6 Orchid Mycorrhizae

At some time throughout its life cycle, every orchid goes through myco-heterotrophic growth, developing mycorrhizae with different basidiomycete fungi. To exchange nutrients, their hyphae enter the root cells and form pelotons (coils).

Categories of mycorrhizae:

- Endomycorrhizae (AM/VAM)
- Ectomycorrhizae (ECM)
- Ericoid mycorrhizae
- Monotropoid mycorrhizae
- Arbutoid mycorrhizae
- Orchid mycorrhizae
- Ectendo mycorrhizae

16.5.7 Mutualistic Dynamics

Most plant species' roots have a mutualistic interaction with mycorrhizal fungi. A relationship that is referred to as "mycorrhizal" includes both the presence of the plants and the areas of their roots that host the fungi. Even though there have only been a few mycorrhizal associations between plant species and fungi studied thus far, 95% of the plant families that have been studied are predominately mycorrhizal, either in the sense that the majority of their species associate favorably with mycorrhizae or are entirely dependent on them. Orchidaceae are well known for being a family in which even germination of seeds is lethal in the absence of the proper mycorrhizae [17]. Recent research on ectomycorrhizal plants in boreal forests has raised the possibility that mycorrhizal fungi and plants may interact in a way that goes beyond simple mutualism. When it was unexpectedly found that mycorrhizal fungi were storing nitrogen from plant roots amid nitrogen shortages, this relationship was established. Researchers claim that some mycorrhizae transport nutrients based on their surroundings, including other mycorrhizae and neighboring plants. They continue by pointing out that this modified model might help to explain why mycorrhizae do not help plants with nitrogen limitation and why plants can suddenly transition from a mixed strategy using both mycorrhizal and nonmycorrhizal roots to a purely mycorrhizal strategy as soil nitrogen availability decreases [18]. Phylogenetic and evolutionary links have been theorized to be significantly more capable of explaining variation in mycorrhizal mutualism strength than ecological influences [19].

16.5.7.1 Benefits of Mycorrhizae

Role of mycorrhizae in mineral exchange:

The fungus has relatively consistent and direct access to carbohydrates like glucose and sucrose thanks to the mycorrhizal mutualistic relationship [20]. Once the carbohydrates have been moved from their source, which is frequently the leaves, to the root tissue, they are then consumed by the plant's fungus partners. The plant gains from the mycelium's greater ability to absorb water and mineral nutrients since fungal hyphae have a significantly bigger surface area and are longer and finer than plant root hairs. Additionally, some of these fungi have the capacity to mobilize nutrients from the soil that are inaccessible to plant roots. The result is an improvement in the plant's capacity to absorb minerals [21]. Plant roots may find it challenging to absorb nutrients that are chemically or physically immobilized, such as phosphate ions and micronutrients like iron, on their own. In soils with a lot of clay or a pH that is too basic, one sort of this immobilization occurs. However, many of these nutritional sources can be accessed by the mycorrhizal fungus' mycelium, which then makes them accessible to the plants it colonizes [22]. As a result, many plants may be able to obtain phosphorus without the need for soil. When nutrients are trapped in organic material that decomposes slowly, such as wood, another kind of nutrient immobilization occurs. In this case, a few mycorrhizal fungi act as decay organisms, mobilizing the nutrients and transferring some to the host plants. For instance, in certain dystrophic forests, significant amounts of phosphate and other nutrients are picked up by mycorrhizal hyphae acting directly on leaf litter, obviating the need for soil uptake [23]. In order to prevent rain from draining phosphorus out of the soil, Inga alley cropping, an agroforestry technique advocated as a countermeasure to slash-and-burn rainforest destruction, relies on mycorrhiza in the root systems of species of Inga [24]. Suillus tomentosus, a basidiomycete fungus that lives on the lodgepole pine plant as a host, produces tuberculate ectomycorrhizae, which are specialized structures (Pinus contorta var. latifolia). It has been discovered that these structures are home to nitrogen-fixing bacteria, which provide a sizable amount of nitrogen and enable the pines to colonize nutrient-deficient locations [25].

16.5.8 Insect Resistance by Mycorrhizae

The underground connections can be utilized by plants in mycorrhizal networks to transmit warning signals [26, 27]. For instance, when an aphid attacks a host plant, the plant alerts nearby related plants to the situation. The host plant and any associated plants both exude volatile chemical compounds that deter aphids and draw parasitoid wasps, which are aphids' natural predators [26]. As a result, the food supply of the mycorrhizal fungi is preserved [26].

16.5.9 Mycorrhizae Help in Colonization of Barren Soil

Without the incorporation of mycorrhizal fungal spores or hyphae to colonize the plant roots and help in the uptake of soil mineral nutrients, plants grown in sterile soils and growth media frequently thrive poorly [28]. Plant growth can also be slowed in early succession or on degraded landscapes by the absence of mycorrhizal fungus [29]. Native non-mycorrhizal plants are at a disadvantage when foreign mycorrhizal plants are introduced to habitats with insufficient nutrients [30]. The classification of oligotroph describes this capacity to colonize desolate soil.

Mycorrhizae provide resistance to toxicity:

Fungi provide protection for plants grown in soils with high metal concentrations, such as acidic and contaminated soils. Pisolithus tinctorius-inoculated pine trees showed high levels of survival, growth, and resistance to the prevalent contamination when they were planted in a number of contaminated sites [31]. One study found that there are Suillus luteus strains with different levels of zinc tolerance. A further investigation found that Pinus sylvestris plants exhibited resistance to zinc-tolerant Suillus bovinus strains. This was most likely caused by the metal's adhering to the fungus' extramatricial mycelium, which did not interfere with the exchange of healthy compounds [30].

Mycorrhizae imprint technology:

By enhancing soil microbe interaction, the plant's access to nutrients, and providing defense against stress and pathogens, Smart Microbes Imprint Technology boosts the pervasiveness of fungal hyphae to successfully colonize root cells of the plants. This results in an exceptional root network which improves the plant's performance as a whole and coexistence in its ecological community. Through the use of Smart Microbes Imprint Technology, plants are made to be as productive as possible at every step of their lifecycle.

16.6 BIOSTIMULANTS

According to experts in horticulture, the term "biostimulant" was created to describe compounds that stimulate plant growth but are not nutrients, soil conditioners, or insecticides. Tracing back, the initial definition of the term "biostimulants" refers to the website Ground Maintenance, which is devoted to professionals in turf maintenance. Zhang and Schmidt from the Virginia Polytechnic Institute and State University's Department of Crop and Soil Environmental Sciences classified biostimulants as "materials that, in minute quantities, promote plant growth". Zhang and Schmidt proposed two explanations for the biostimulation action: first, hormonal actions; and second, antioxidant defense against abiotic stress. The term "biostimulant" was first used in a peer-reviewed study by Kauffman et al. (2007), who modified the definition to read: "biostimulants are materials, other than fertilisers, that promote plant growth when applied in low quantities". Although biostimulants come in a large variety of regimens and include a vast variety of compounds, they can be roughly categorized into three types based on their source and makeup. These categories include items like amino acid containing products (AACP), humic substances (HS), and hormone containing products (HCP). Active plant growth compounds like auxins, cytokinins, or their derivatives can be present in seaweed extracts and other HCPs in detectable amounts. The word "biostimulant" is used to refer to a broad spectrum of plant-friendly compounds that are neither nutrients, pesticides, nor soil conditioners. One might in some respects characterize biostimulants first by what they are not, drawing a distinction between them and two other often used types of substances applied to plants and crops, fertilizers, and pesticides. Later, it was found that bacteria and fungi can also produce the positive outcomes associated with chemical biostimulants (of natural or synthetic origin), such as enhancement of growth, modulation of development and quality traits, and increased resistance to environmental stress.

Existing bacteria that can perform bioremediation are stimulated by changing the environment. To do this, several nutrients with rate limitations and electron-accepting properties, such as phosphorus, oxygen, nitrogen, or carbon, can be added. The remediation of halogenated pollutants in anaerobic environments may also be aided by the inclusion of organic substrates, which donate electrons, allowing the local bacteria to use the halogenated pollutants as electron acceptors. Techniques for anaerobic bioremediation EPA injection wells are commonly used to inject additives into the subsurface, even if the technique is still in its infancy for biostimulation. Another, albeit expensive, option is the removal of the tainted material. Biostimulation can be improved through bioaugmentation. Bioremediation is the general process used to remove the effects of oil or gas spills, and it has been sanctioned by the EPA. While biostimulation is typically used to treat hydrocarbon or large production volume chemical spills, it may also be effective in treating less common contaminant spills like pesticides, especially herbicides [1]. The primary advantage of biostimulation is that it enables bioremediation to be carried out by native microorganisms that are already present, well-distributed, and acclimated to the subterranean environment. The biggest disadvantage is that the local subsurface geology determines how additives must be administered in order for subsurface microorganisms to have easy access to them. Due to the tight, impermeable subsurface lithology (tight clays or other fine-grained material), it is difficult to evenly distribute additives over the affected area. Subsurface cracks create preferential subsurface pathways that additives follow, impeding uniform additive dispersion. Due to the tight, impermeable subsurface lithology (tight clays or other fine-grained material), it is difficult to evenly disseminate additives over the affected area. A uniform distribution of additives is prevented by subsurface fractures that provide preferential pathways in the subsurface that additives adopt.

Recent product releases have made it possible to use biostimulative methods for bioremediation on a large scale. These methods can utilize biostimulation to harness native bacteria, or they can introduce foreign bacteria directly into the ecosystem to consume the hydrocarbon by creating an environment conducive to microorganisms that devour hydrocarbons. It is unknown whether one method is noticeably more effective than the other, but prima facie data suggests that the introduction of alien bacteria to any ecosystem has a probability of altering existing species and impacting the biome.

Common features of biostimulants:

- The nature of biostimulants is diverse:

  Microorganisms and substances are involved. Examples of substances include isolated compounds (such as glycine betaine) or collections of compounds having a single natural origin whose composition and bioactive components have not yet been fully described. In particular, synthetic compounds shouldn't be excluded from the substances covered in this review if specific plant growth regulators are contained within biostimulants (for example, nitro-phenolates are marketed and referred to as "biostimulants", but they are actually synthetic phenolic compounds that have been registered as plant production products under EU law). The substances included in this review are either inorganic molecules or organic compounds that occur in nature. Usually, biostimulants are included in fertilizers or crop protection methods. They may also be used to describe the bioactive compounds themselves or commercially available mixtures of them.
- The physiological functions are diverse:

  The defense of photosynthetic equipment against photodamage or the beginning of lateral roots are two examples of physiological functions. Cellular mechanisms, such as reactive oxygen scavenging by antioxidants or enhanced auxin transporter synthesis, to continue with the previous two instances, support certain functions. The aggregate "modes of actions" of the biostimulants may be referred to as physiological processes and the underlying cellular mechanisms. These methods of action explain how biostimulants function in agriculture, such as through improving N uptake or enhancing resistance to abiotic stress (which causes oxidative stress), which is dependent on lateral root density and the foraging ability of roots. Eventually, the functions of agriculture may result in gains for the economy and the environment, such as increased agricultural production, fertilizer cost savings, improved crop yields and quality, and improved ecosystem services.

- Has several of the following agricultural functions:

  They improve attributes related to crop quality, abiotic stress resistance, and/or nutrient efficiency. Qualities include things such as grain protein levels, shelf life, and nutritional value. Any definition of biostimulants should be based on these convergent activities. Many of the previously known biostimulants also work as elicitors and plant gene regulators to stimulate the pathogen response (e.g., chitosan, laminarin, some PGPRs). In terms of goals and evaluation criteria, agricultural and environmental policies influence the definition of economic and environmental advantages. Although these factors are related to incentives for biostimulant development, they shouldn't serve as the foundation for a definition of biostimulants that is based on research.

16.6.1 Microbial Biostimulants

The Food and Agriculture Organization (FAO) definition of soil fertility is: "The ability of a soil to sustain plant growth by providing essential plant nutrients and favourable chemical, physical, and biological characteristics as a habitat for plant growth" [1] as represented by Figure 16.4. Recently, it has been discovered that agricultural soil all over the world has dramatically declined. In particular, 24% of the world's productive lands and over 40% of its agricultural soil have lost their fertility, capacity, and biodiversity. These occurrences are primarily caused by a number of reasons, including salinity, loss of organic matter, wind and water erosion, and environmental pollution [2, 3]. The three pillars of agroecology are biodiversity preservation, the bolstering of biological processes, and the looping of biogeochemical cycles. The administration of biostimulants, which include agents that can operate directly on plants as well as sustain productivity through the selection and stimulation of advantageous soil microbes, is consistent with the concepts of agroecology [5]. Arbuscular mycorrhizal fungi (AMFs) can also form a symbiotic relationship with plants, increasing the root

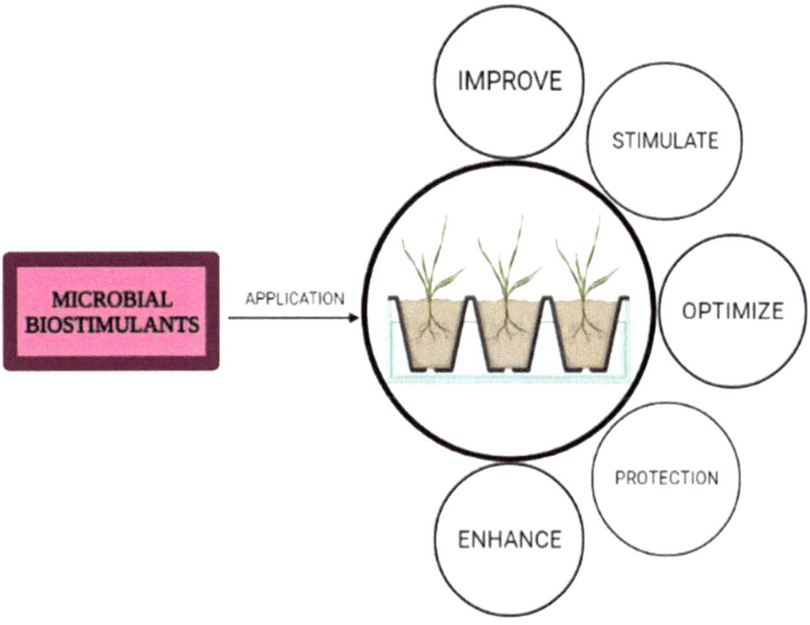

FIGURE 16.4 Biostimulant activity of microorganisms promoting plant growth.

surface area available for nutrient uptake [7]. The number of research studies on the beneficial use of microbes is expanding quickly as a result of new technologies that enable precise selection and knowledge of the added value of microbial consortia. The European Biostimulant Industry Council (EBIC), which was founded in 2011, is working to support innovation in this field by requesting workable solutions for their unified regulatory treatment on the EU market, including safety standards and an update of CMC-7 in the existing legislation.

However, plant growth-promoting microorganisms (PGPM), a collection of endophytic bacteria that comprises the phyla Proteobacteria, Firmicutes, Actinobacteria, and Bacteroidetes, is extremely diverse [32]. Aeromonas, Enterobacter, Azospirillum, Azotobacter, Clostridium, Arthrobacter, Arthrobacter, Azospirillum, Azotobacter, Bacillus, Gluconacetobacter, Klebsiella, Pseudomonas, Rhizobium, and Serratia are the genera that have been investigated the most extensively among the numerous others [33]. Recent research has demonstrated the significance of PGPRs in boosting soil production and plant tolerance to abiotic stress [34]. They are present in the rhizosphere and are able to enhance plant growth. The primary endogenous auxin in plants, indole-3-acetic acid (IAA), controls the growth and division of cells as well as the creation of roots and root hairs [35]. Even though plant tissues cultivated in soils with a lack of water or an excess of salt showed an increase in IAA, plants that were further colonized by AMFs showed a larger increase [8, 36–38]. Variations in root shape and architecture were also noted in the same plants [37]. IAA is not the only substance that the plant may manufacture and secrete that has this special ability to be helpful to the plant. In fact, some PGPR strains are also able to produce ACC deaminase, an enzyme that facilitates the transformation of ACC, an ethylene precursor, into α-ketobutyrate and ammonia. The ACC-deaminase secretion lowers the levels of ethylene in the soil and plants, reducing the effects of environmental stress on plants [9]. Finally, a cross-talk effect happens when PGPRs are able to synthesize ACC-deaminase as well as secrete IAA. In fact, ACC-deaminase lowers plant ethylene levels while IAA enhances plant development [10]. Other non-enzymatic antioxidants that interact with PGPMs and plants to affect how they react to oxidative stress include polyphenols, organic acids, vitamins, carotenoids, and glutathione. For instance, it was previously mentioned that proline and glycine-betaine buildup in both AMF [9, 11] and PGPR [9, 12, 13] inoculated plants may be beneficial for reducing cellular oxidative damage. In contrast to non-inoculated drought-challenged plants, Moreno-Galván et al.'s findings on inconsistent data about proline in PGPR-plant tissues [14] is an exception. Additionally, it was thought that the proline accumulation was significantly influenced by the stage of plant development [14, 15].

16.6.2 Effects of Microbial Biostimulant Product Components

Investigations into the combined use of PGPRs and AMFs as biostimulant agents have been conducted in many different ways. The biostimulant efficacy of a wide range of consortia containing both plant-helping bacteria and AMFs was examined with varied plants and growing circumstances. When inoculated with PGPRs and AMFs, Arizona cypress, chickpea, wheat, swamp oak, Jerusalem artichoke, and pea plants showed increased plant biomass and yield, frequently outperforming uninoculated plants or plants inoculated with a single strain [16, 39–43]. When applied to tomato plants, an AMF microbial-based biostimulant including seaweed extracts and R. intraradices positively enhanced plant growth and yield in a separate but complementary way [44]. In this experiment, when AMF was seeded alone, the treatment increased early blooming and leaf development but decreased protein, carbohydrate, and lipid synthesis. On the other hand, the protein content and root development were boosted by employing seaweed extracts alone. When compared to single treatments, their combined administration had a synergistic effect on tomato plants, causing earlier flowering and AMF colonization, as well as an additive effect (in leaf and root growth, as well as protein and carbohydrate content) [44]. Similar results were seen when water algae were injected into PGPRs (Bacillus licheniformis, Bacillus megatherium, Azotobacter sp., Azospirillum sp., and Herbaspirillum sp.) (Chlorella vulgaris). PGPMs and their liquid culture may be combined as an additional strategy to increase plant development and abiotic stress tolerance. This would improve PGPM field performance and lengthen their shelf life in the soil [45]. For instance, on pigeon pea, the effects of applying Bradyrhizobium live cells along with their cell-free culture supernatant (CFCS) and metabolite exopolysaccharides (EPS) were assessed [45].

The ability of Bradyrhizobium to fix nitrogen, produce phytohormones and siderophores, solubilize phosphate, and synthesize exopolysaccharides are all well-known traits [46]. Even though this bacterium excels at promoting plant growth thanks to these characteristics, the combination of the three elements produced the best outcomes. Through nodulation and the antioxidant action of benzimidazoles, the sole application of CFCS significantly contributed to the stimulation of growth. The presence of CFCS also strengthened the symbiotic interaction between plants and PGPR, as seen by an increase in ascorbic, pantoic, and benzoic acids. It was hypothesized that EPS's advantages might be due to its functions as a carbon source advantageous for enhancing root colonization, biofilm, and nodule formation, as well as a protective covering for the inoculated PGPR. Based on the properties of each component, Bradyrhizobium performed best when it was inoculated with CFCS and EPS in terms of plant development. Particularly, this formulation outperformed the sole inoculum in fostering the growth of native soil rhizobia [45].

The availability of small microbial nutrients and the level of biological activity in the soil are determined by the pH of the soil, which lowers the number of sensitive and rare species and restricts plant growth [2, 18, 19]. Contrarily, ongoing monoculture depletes the soil, resulting in a decline in helpful microorganisms, a depletion of the microbial community structure, and an increase in the prevalence of pathogens [2]. The species diversity of soil bacteria and fungi is also dramatically decreased by unfavorable climate change, including increased rainfall [20, 21]. Additionally, soil microbiome network complexity is decreased by erosion and fluctuating warming [21, 22]. This suggests a decrease in resource availability, decomposition activity, and nutrient cycling. These elements have negative long-term effects on soil activities, reducing the microbial capacity to withstand environmental pressures [22]. By producing microbial necromass, which accounts for more than half of the organic carbon in soil, the microbial community and its byproducts are important drivers of SOM generation and heterogeneity [24, 47].

In addition to the role of soil microbiota in the cycling of elements and the stabilization of soil structure, a high level of soil microbial activity is essential for effective crop production, maintaining healthy plants, and ensuring a good yield under various environmental conditions. A complicated interaction between plants and microbes under stressful environmental conditions is suggested by the fact that global climatic changes might threaten both plant and soil biodiversity [48, 49]. Numerous studies have shown that climate change affects plant diversity and yield, with a worse effect observed in soils with less biodiversity [48]. This occurs regardless of plant genotypes, demonstrating that soil microorganisms may typically be the source of the detrimental effects of soil biodiversity loss [48–50]. Plant development has been shown to be more productive and stress-resistant in environments with substantial microbial diversity. Additionally, yield losses were reduced when soil microbial populations were enhanced, indicating their potential and critical function as yield stabilizers following disruptions caused by climate change [50].

Evidence emphasizes the value of the microbial component and their diversity for managing crops as well as a viable biological tool for restoring damaged soils and carrying out revegetation activities [51, 52]. For example, a bacterial consortium (Azospirillum spp., Azoarcus spp., and Azorhizobium spp.) and two AMF-PGPR consortia (Rhizophagus irregularis and Azotobacter vinelandii, and R. irregularis, Bacillus megaterium, and Frateuria aurantia), inoculated wheat demonstrated a general increase in total microbial biomass and soil enzymatic activities. The majority of the time when the inoculum comprised both PGPRs and AMFs, these data suggest an improved microbial metabolism [40].

16.7 BIOJET

During the stages of plant growth, a variety of factors can inhibit the health and yield of plants, causing nutritional, biological, and compound deficiencies that result in subpar plants, quality, and production. The purpose of BIOJET is to provide nutritional and microbiological support for growth, quality, quantity, soil biology, and to improve the ability of internal plant processes to endure environmental challenges.

Benefits:

- Improves the absorption and effective utilization of macro and micronutrients.
- Boosts crop output and fruit quality by stimulating root growth.
- Increased seed germination rate happens quickly.
- Enhances a plant's ability to withstand biotic and abiotic stressors (i.e., cold, frost, heat).

Advantages:

- Enhances beneficial soil microorganisms and soil regeneration to improve and restore soil biology.
- Flexible use on soil, seeds, plants, and other growing media.
- Safe for the environment and offers a substitute for chemical fertilizers.

Active ingredients:

- Contains a unique mixture of concentrated natural microorganisms.
- A plentiful source of hormones, enzymes, and secondary metabolites from microorganisms.
- A plentiful supply of minerals, vitamins, humates, proteins, and organic carbon.

Mode of action:

- In order to provide advantageous nutrients and chemicals for simple absorption and digestion, BIOJET can be used as a foliar spray.
- It may be combined with crop protection agents or liquid fertilizer.
- It can be used as a transplant, watering, or drip solution.
- It can be applied as an aerial spray or boom spray.

16.7.1 BioSil

BioSil gives plants a silica and potassium source that is 100% bioactive and usable.

Benefits:

- Balance of nutrients is enhanced.
- Increases yields, quality, and plant growth.
- Helps plants withstand abiotic challenges such as mineral toxicity, excessive heat or cold, and soil salinity.

Advantages:

- Reduces the demand for chemical fertilizers, lowering costs.
- Improves crop quality to appeal to consumers.
- For increased revenue, increases yields.

- Offers a secure substitute for chemical fertilizers.
- Environmentally safe and friendly.

Important ingredients:

- Potassium silicate.
- Potassium sulfate.
- Humic acid.
- Microbial secondary metabolites.

Applications:

- Foliar: Apply 1–3 pints of water per acre, or at least 100 gallons, to completely cover the foliage.
- Turfgrass: Apply 103 fl oz per 1,000 square feet of turf every 7 to 14 days. When the turf is under stress, the application interval can be increased to get the required outcomes.
- Soil application/irrigation (sprinklers, drips): 2 to 5 pints per acre per 100 gallons of water may be applied as a side dress or soil treatment (sprinklers/drips).
- Fruit trees, tree nuts, vines, vegetables, potatoes, tubers: Apply one to three quarts per acre per application to fruit trees, tree nuts, vines, vegetables, potatoes, and tubers throughout the entire growth season. At least two to three applications are advised. Application prior to bloom is advised for optimal outcomes.
- Bush berries: Apply 1–3 pints per acre to bush berries. Application before bloom is advised.

16.7.2 Bio-Maxx

Bio-Maxx is a microbial biostimulant that is naturally generated and is intended to increase crop yields and fruit size.

Benefits:

- Favors good fruit setting.
- Diminishes fruit drop.
- Maximizes the production of high-quality plants.
- Increases fruit firmness, weight, color, and other qualities like weight and color.

Advantages:

- Enhances beneficial soil microorganisms and soil regeneration to improve and restore soil biology.
- Helps encourages healthy, fruitful crops.
- Safe for the environment and offers a substitute for chemical fertilizers.

An all-natural biostimulant made from a blend of helpful microorganisms, their metabolites, and other necessary components, such as:

- Amino acids with bioactivity.
- Polysaccharides.
- Vitamin C.
- Potassium.
- Phosphorous.
- Magnesium.

16.8 BIOFERTILIZERS

Crops free of pollution are being produced more often thanks to organic cultivation. It entails the use of biofertilizers and biopesticides, which improve the crops' nutrient content and eliminate all pests and pathogens.

Farmers are experimenting with various crop nourishment techniques to help meet the increasing food demand brought on by the ongoing growth of the global population. By 2030, the demand for agricultural goods is predicted to rise by 60% [1, based on FAO estimates]. One of the key issues facing society in the twenty-first century is how to increase output while maintaining environmental safety [2]. To boost crop production on arable land, fertilizers have been used widely. A nation may become food self-sufficient by using more chemical fertilizers in agriculture, but chemicals are harmful to the environment and living things. Chemical fertilizers are also expensive, have negative effects on the soil's fertility and ability to retain water, create nutrient imbalances, and contribute to unacceptable amounts of water pollution [3]. Contrarily, biofertilizers help preserve the soil's structure and biodiversity on agricultural territory and are inexpensive, practical, non-toxic, and simple to use. As a result, they are a suitable alternative to chemical fertilizers [4, 5]. Biofertilizers, sometimes referred to as microbial inoculants, are organic compounds that include specific microbes that are found in plant roots and root zones. They have been demonstrated to increase the plant's development and yield by 10% to 40% [6]. When administered to the seed, plant surface, or soil, these bioinoculants colonize the rhizosphere and the interior of the plant, fostering plant development [7]. By supplying nutrients to the soil, they not only increase soil fertility and agricultural yield but also shield the plant from diseases and pests. It has been demonstrated that they speed up flowering, improve seedling mortality, lengthen the life of the plant, and degrade harmful substances [32]. The fact that parental inoculum are adequate for growth and multiplication after using biofertilizers continuously for 3–4 years is another advantageous feature [33].

For the plant to develop and expand properly, 17 important elements are needed. Among them, relatively significant amounts of nitrogen (N), phosphorus (P), and potassium (K) are required [34]. Numerous microorganisms, such as cyanobacteria, which dissolve phosphate and are used in conjunction with molds and fungus, are frequently used as biofertilizers [35]. Similar to this, microorganisms that produce phytohormones are also used in the creation of biofertilizers. They give the plant nutrients that encourage development, such as indole acetic acid (IAA), amino acids, and vitamins, and they enhance the soil's fertility and productivity while preserving crop yield [36].

Microorganisms called biofertilizers improve the soil's nutritional content. Some of the helpful microbes that aid in increasing the soil's fertility include bacteria, fungi, and algae.

Biofertilizers are categorized as follows:

- Microorganisms that fix nitrogen for free, including Rhodospirillum and Azotobacter.
- Anabaena and Nostoc are examples of free-living cyanobacteria that fix nitrogen.
- Loose group of bacteria that fix nitrogen, such as Azospirillum.
- Nitrogen-fixing microbes that coexist, such as Rhizobium and Frankia.

Biofertilizers made of microbes

As biofertilizers, the following microbes are employed:

- Rhizobium: It develops root nodules and converts atmospheric nitrogen into an organic form in legume plants. Rhizobium enhances the quality, nutrient content, and plant development while having no adverse effects on the soil's quality.
- Azotobacter: These nitrogen fixers, which are free-living, are present in all varieties of upland cereals. In addition to fixing nitrogen, they also give the plant certain antibiotics and development agents.

- Azospirillum: These can be used in wetland regions, unlike Azotobacter. They are not free-living and are located inside the roots of plants to fix nitrogen from the air.

  Only wet and marshy areas contain blue-green algae, which are nitrogen-fixing Cyanobacteria that are free-living. They cannot, however, live in acidic soil.
- Mycorrhiza is the name for the symbiotic relationship that exists between fungus and plant roots. The mycorrhizal fungi enhance the activity of the microbes and are crucial in holding the soil together. The water and nutrients the fungi take from the soil help the plants grow more productively. Additionally, they make the plant more resilient to environmental stressors.

Importance of biofertilizers:

- Biofertilizers don't harm the ecosystem and are environmentally friendly.
- These raise the quality of the soil, which increases plant output.
- When necessary, producers can make up these microbes' inoculum and inject it into the plants.
- In the soil, they create compounds that encourage growth.
- Semi-arid regions can also use biofertilizers.

16.8.1 P-Force

Phosphorus fixation in the earth is reduced, and plant availability is increased by phosphate biofertilizer, commonly known as P-Force.

Benefits:

- Reduces soil phosphate fixation and enhances plant uptake of phosphate.
- Stimulates plant development, growth, and production.
- Create plant defenses.

Application:

- For seed preparation, use distilled or chlorine-free water.
- Mix the container's contents with the suggested amount of water for dilution for seed treatment.
- Within eight hours, apply the diluted product at the suggested rate to the seeds.

16.8.2 Biopesticides

A biopesticide is a biological agent that harms, eliminates, or deters creatures that are considered pests. A biological pest management strategy may entail chemical, parasitic, or predatory interactions. They come from a variety of living things, such as plants, bacteria, other microbes, fungi, nematodes, and others [1, 2]. They are a part of integrated pest management (IPM) programs and have gained a lot of notice in the real world as alternatives to synthetic chemical plant protection products (PPPs). Biopesticides are organic substitutes for chemical pesticides that are used to control crop diseases and pests. Biopesticides are made using microorganisms derived from natural sources, including plants, bacteria, specific minerals, and other organic matter. As a result, they are safer to use than chemical pesticides. In contrast to chemical pesticides, which also affect mammals, birds, and insects, biopesticides only affect the targeted pests and related organisms. They also degrade rapidly, causing less environmental pollution.

16.8.3 Classes of Biopesticides

- Biochemical pesticides.
- Microbial pesticides.
- Plant-incorporated protectants (PIPs).

16.8.3.1 Biochemical Pesticides

Biochemical pesticides are substances with natural origins that are used to control bugs by preventing mating or by luring them to traps with scented plant extract. Synthetic pesticides, on the other hand, immediately kill the pests and have toxic effects on the environment.

16.8.3.2 Microbial Pesticides

Microbial pesticides serve as a biocontrol agent and are made from microorganisms like bacteria, fungi, viruses, or protozoa. The effectiveness of microbial pesticides varies based on the ingredients used and the type of pest that is being controlled.

16.9 PLANT-INCORPORATED PROTECTANTS (PIPS)

Plant-incorporated protectants are pesticides produced when genetic material is added to the plants, often by genetic engineering, and when a specific Bt pesticidal protein is introduced into the genetic material of the plant. The plant makes the pesticidal chemical to keep the pests under control when there is an infestation.

16.9.1 Beneficial Control Solutions Using Trichoderma Fungi in Biopesticides

The most common fungus species in soil all over the globe belongs to the genus Trichoderma. Due to its great adaptability, the fungi are perfect for use in various environments. Trichoderma species of fungi also offer farmers two benefits. It is both a plant growth promoter that colonizes plant roots to give plants an additional advantage of survival and a biocontrol agent for various plant pathogens, including viruses, bacteria, fungi, and nematodes.

16.9.2 Different Species of Trichoderma Used in Biopesticides

Trichoderma harzianum:

This specific strain of the Trichoderma fungus can colonize the roots of most plants and form a rhizosphere. It is used as a biocontrol agent during agricultural production to stop pathogens like Fusarium, Pythium, and Rhizoctonia from colonizing the soil and to protect plants from dangerous diseases.

16.9.2.1 Trichoderma viride

Trichoderma viride is particularly effective against commercially significant crops as a biocontrol for soil- and seed-borne pathogens like collar rot, root rot, seedling disease, and wilt. Additionally, it encourages robust plant development through a number of mechanisms.

16.9.2.2 Trichoderma virens

Trichoderma virens is a helpful plant fungus that encourages lateral root growth and biomass output. Trichoderma virens cultivates an extremely efficient relationship with plants, acting not only as a growth stimulant but also as a biocontrol agent that is helpful in eradicating pathogens like the prevalent damping-off disease in greenhouse agriculture.

16.9.2.3 Beauveria bassiana

A fungus called Metarhizium anisopliae, which naturally grows in soil, acts as a natural defense against insect pests like thrips, mites, whiteflies, ticks, weevils, and aphids. It helps to eradicate these pests wherever they are present by outnumbering pests and germinating over their bodies, which causes them to die.

B. bassiana develops as a whitish mold in a culture. On the majority of popular cultural media, it generates numerous dry, powdery conidia in distinctive white spore balls. Each spore pellet is composed of a collection of conidiogenous cells. The ovoid, slender conidiogenous cells of B. bassiana are terminated by a thin apical projection known as a rachis. Each time a conidium is formed, the rachis elongates, creating a protracted zigzag extension. The conidia are hydrophobic, diploid, and single-celled. Beauveria bassiana is a biological pesticide that can be used to manage a variety of pests, including termites, whiteflies, and other insects. Investigations are being done into how it might be used to manage mosquitoes that spread malaria. The spores are applied to mosquito nets as a mosquito control agent or sprayed on affected fields as an insecticide in the form of an emulsified suspension or wettable powder. A very diverse variety of arthropod hosts are parasitized by Beauveria bassiana as a species. However, different strains have varying host ranges, with some having relatively small ranges, such as strain Bba 5653, which is very virulent to diamondback moth larvae and only kills a small number of other kinds of caterpillars. Some strains do have a broad range of potential hosts and ought to be categorized as nonselective biological insecticides. These shouldn't be used on flowers that pollinators frequent [32].

16.9.2.4 Metarhizium anisopliae

A fungus called Metarhizium anisopliae, which naturally grows in soil, acts as a natural defense against insect pests like thrips, mites, whiteflies, ticks, weevils, and aphids. It helps to eradicate these pests wherever they are present by outnumbering pests and germinating over their bodies, which causes them to die.

Metarhizium, which is widely dispersed in nature and can be found in soil, at the rhizosphere of plants, or on the carcasses of arthropods, parasitizes a wide range of insects and parasites. The taxonomy of the Metarhizium genus was just recently revised (Bischoff et al., 2009), and the current review mainly focuses on recent studies on M. anisopliae strains that were initially isolated from diseased insects. The strains CG291 and CG423 were previously identified by Magalhes et al. (1697) as Metarhizium flavoviride or M. anisopliae var. acridum. However, in the lab, both strains were recognized as M. anisopliae var. anisopliae based on ITS1-5.8S-ITS2 region sequencing data.

Metarhizium multiplies vegetatively on arthropod hosts and in culture medium by producing hyphae that form mycelia and conidia, which are the infectious propagules. There are numerous potential hosts for M. anisopliae, but some strains can only attack particular hosts. Rhipicephalus (Boophilus) microplus, which is regarded to be the most significant tick parasite of livestock in the world, was not pathogenic to the strains Nordeste, CARO11, CARO12, CG27, CG30, CG33, CG87, CG125, CG320, and CG374. These strains caused 90–100% mortality within four days of treatment (Lubeck et al., 2008). Some strains of M. anisopliae can also be harmful to the Dysdercus peruvianus cotton stainer bug, especially the females. The strains E6, CG47, and CG343 each killed half the females in 3.3, 3.6, and 3.66 days, respectively. With a TL50 of 7.63 days, strain CG125 was the most effective at infecting males, and strain Nordeste was the least infectious (Lubeck et al., 2008). M. anisopliae directly penetrates the cuticle of vulnerable hosts and infects them. For the purpose of description, the infection process can be broken down into the following steps: (1) conidia adhesion to the host cuticle through hydrophobic interactions and thin mucilaginous material; (2) conidia germination and development; (3) germ-tube differentiation into appressoria; (4) cuticle penetration; (5) hyphae differentiation into blastospores/hyphal bodies in the haemolymph; (6) host colonization; (7) extrusion to the host; and (8) conidiophores formation and conidia production.

16.9.2.5 Bacillus thuringiensis

A naturally occurring soil bacterium called Bacillus thuringiensis subsp. israelensis (Bti) can be discovered in our environment. Bti successfully kills fungus gnat larvae by acting as a natural trap for the insects. A gram-positive, soil-dwelling bacterium called Bacillus thuringiensis, also known as Bt, is the most widely applied biological pesticide in the world. Additionally, B. thuringiensis naturally exists on leaf surfaces, in aquatic environments, in animal feces, in insect-rich environments, in flour factories, and in grain storage facilities. It also naturally exists in the gut of caterpillars of different kinds of moths and butterflies [1, 2]. Other moths, like Cadra calidella, have been observed to be parasitized by it; in laboratory trials with C. calidella, many of the moths suffered from disease as a result of this parasite. Many Bt strains generate crystal proteins, or proteinaceous inclusions known as delta endotoxins, during the sporulation process that have insecticidal properties. As a result, they are now used as insecticides, and more recently, Bt genes have been used to create genetically modified crops like Bt corn. However, many Bt genotypes that produce crystals lack insecticidal traits. The subspecies israelensis is frequently used to manage fungus gnats and mosquitoes [7].

16.10 HIRSUTELLA THOMSONII

Our surroundings contain a naturally occurring fungus called Hirsutella thompsonii. Due to their destructiveness to plants and harvests, russet mites, spider mites, and other mite pests are natural enemies of this specific fungus and are therefore important to agriculture. The Hirsutella thompsonii species is the one that has been researched the most because it is a key biocontrol agent for agricultural mite pests.

16.10.1 Genus Hirsutella

The genus Hirsutella comprises the species Hirsutella citriformis, Hirsutella gigantea, and Hirsutella thompsonii. This genus has been shown to be one of the most difficult to identify out of all the major fungal entomopathogen genera due to the large number of species and the high degree of variation within those species. If synnemata is present, it is erect, cylindrical or slightly tapered, and can be unbranched, sparsely branched, or have numerous short side branches. It can also be short and verrucose or long and hairy. Subglobose conidiogenous cells have an enlarged base. The orange rust mite is managed with the aid of Hirsutella thompsonii. Additionally, the Acarida, Lepidoptera, and Hemiptera families of insects are pathogenic to this fungus.

In order to combat a rust mite infestation on citrus trees, McCoy et al. (1671) used the mycelia of Hirsutella thompsonii. They saw sporulation after about 48 hours, then a drop in the mite infestation, which didn't start to recover until 10–14 weeks later. As a result, this fungus was the first microorganism to be approved for use in the American citrus industry and, it appears, in any citrus industry worldwide (McCoy and Couch, 1682). This was created as a wettable powder in 1676 under the brand name Mycar (Abbott Laboratories).

When used with oil or a wetter, the product also worked well to contain rust mites (McCoy and Couch, 1682). Beauveria bassiana was also demonstrated to be efficacious against the citrus rust mite in laboratory bioassays (Alves et al., 2005). The use of fungal pathogens to control spider mites, according to McCoy et al. (2009), was not practical due to the fact that promising findings in laboratory bioassays all too frequently failed to translate into acceptable efficacy in the field. This could be the reason why Mycar only had a short shelf life (McCoy, 1696). In lab bioassays, Hirsutella thompsonii was also demonstrated to be successful against the two-spotted spider mite (Tetranychus urticae), causing up to 99% mortality with the formulated Mycar conidia (Gardner et al., 1682). A significant ovicidal impact of B. bassiana on the two-spotted spider mite was also shown by Shi et al. (2008), with up to 87.5% egg mortality in laboratory bioassays. Furthermore, Shi and Feng (2009) demonstrated that B. bassiana, I. fumosorosea, and M. anisopliae significantly affected the reproductive capacity of mites in laboratory bioassays by reducing egg laying by three to four times. Later, the eriophyid mite, a significant pest of coconut in LAC nations, expanded to numerous Asian and African nations. Except for the Pacific region, the pest has developed into a significant threat to coconut farming. Chemical control techniques are neither simple to use nor environmentally favorable. The only options remaining are breeding for resistance and biological control. Resistance markers are found [19] and must be transformed into trustworthy SCAR markers and tools after being validated on a large number of samples. There is potential for the biological control agent of the fungus Hirsutella thompsonii [53]. The invertebrate produces hirsutellin A (HtA), an extracellular protein that resembles a ribotoxin comparable to sarcin. Numerous intriguing findings are made by the protein's structural biologists [28–30]. HtA toxin was discovered at first, but its involvement in pathogenesis was never established [26]. Later research supported the toxicity and revealed the full gene sequence [27] and matching protein [28]. The surface of an alpha+beta fold has a high positive charge, according to NMR research [29]. The role of Asp40 in the biological activity and electrostatic contacts of the smallest ribotoxin (HtA) is critical [30].

16.11 IMPORTANCE OF SYNTHETIC BIOLOGY AND SMART MICROBES

In human healthcare:

One example of synthetic biology's achievement in biomedicine is the creation of mRNA vaccines. Artificial yeast has been used to make semi-synthetic artemisinin, an antimalarial medicine, and a diabetes medication. Januvia® is produced using a novel enzyme developed through directed evolution, which replaces more traditional, damaging processes. A novel form of cancer treatment known as CAR-T cell therapy alters T cells removed from the patient's body to specifically target tumors. Researchers are now interested in utilizing the same approach to treat other immune cell types, such as natural killer cells, as a result of CAR-T cells' success. Other treatments in the works make use of engineered probiotics to improve the body's immune capabilities by utilizing the commensal bacterial reservoir in the gut. Yet more advancements make use of synthetic biology to create 3D scaffolds for cell growth and the regeneration of bones, tissues, and possibly even entire organs. Future uses could include the creation of additional vaccines (possibly using mRNA vaccines once more) to combat a variety of viral illnesses, such as the common cold, prophylactic medical care, therapies for metabolic disorders, and halting the spread of contagious diseases before they spread to large populations. Additionally, it can be used to find biologics that combat how the human body ages [8]. Although synthetic biology will be a part of the research that deepens our knowledge, it cannot be assumed that all biologics, regardless of the production method, will be safe and effective. Future precision medicine will probably be greatly influenced by synthetic biology Precision healthcare, which is generally considered to be the direction of healthcare for people, includes individualized treatment. Precision health would change the profession of medicine by tailoring tests and therapies to each individual based on their unique genetics, microbiome, nutrition, lifestyle, environment, or illness features. Then, utilizing synthetic biology when necessary and appropriate, diagnostics and therapies can be customized depending on those specifics.

16.11.1 IN EARTH'S BIOLOGICAL ENVIRONMENT

The tools of synthetic biology can be used by people to improve, better use, and enrich the biological ecosystem on Earth. This environment is made up of the delicate ecosystems that are all around us and with which people must cohabit. Since ancient times, humans have had a considerable impact on the biological environment, regularly coming into confrontation with threats and pests, and exploiting other plants and animals for our own benefit. Special opportunities and capabilities are provided by synthetic biology to develop these environmental influences, ideally in a positive manner.

Instead of introducing competing species, synthetic biology can now be used to combat unwanted, disease-carrying, or invasive species [10]. A biopesticide strategy for eradicating locust populations has also been suggested. In this approach, the insects' microbiome is modified using synthetic biology, and critical genes are disrupted utilizing bacteriaphages [9]. Invasive plant and animal eradication, zoonotic disease prevention, and dealing with infestations are just a few of the numerous new uses that gene drives and viral vector delivery systems will likely find. Gene drives are biological mechanisms that alter a species' genetic composition by introducing a tiny DNA fragment into an individual, which subsequently spreads and controls gene reproduction in succeeding generations. The creation of biological identifiers that can be used as sensors to find pollutants or contaminants is another benefit of synthetic biology. One such research advancement uses bacteria-infused beads that can be spread across a field's surface and can identify buried landmines by fluorescing green in the presence of trinitrotoluene (TNT) vapours [16].

16.11.2 In Agriculture

Today, it is possible to add or remove genes from plants and animals using synthetic biology techniques in order to achieve the desired effects. Viral transduction, which uses a virus to transfer genetic material, has produced hardier plants [54]. The creation of artificial food additives using custom fermentation and the use of engineered bacteria-rich fertilizers to increase crop harvests are examples of how synthetic biology is used in agriculture [41]. The latter commodities include sugars, flavor mimickers or enhancers, nutritional supplements (such as flavonoids and terpenoids), and oils and fats (lipids) for enhancing meals. By collecting data from people's taste and olfactory sensors, it may be feasible to develop unique food flavors that are more pleasing to their senses. The use of synthetic biology to culture meat directly from engineered cells, eliminating the need to butcher animals, is of great interest in agriculture. The FDA has already approved a fish meal that was created by genetically modifying bacteria to yield an aquaculture feed that resembles flour. While it may seem like a long way off, numerous businesses and research facilities are vying to create cultured meat with characteristics identical to those of animal products [43]. Using synthetic biology to grow cells for meat or vegetables, food production efficiency goes beyond speed and aesthetics. Additionally, synthetic biology may be able to identify and minimize problems with food deterioration. Biological sensing components built into the container may keep track of how the food is deteriorating, such as by spotting spoiled milk [55].

16.11.3 In Energy Production and Its Management

In order to create artificial photosynthesis, make biofuels, and develop bio-battery energy storage capacities, commercial industry, a distinctly non-biological entity, is currently making an effort to replicate some of these processes through synthetic biology. Synthetic biology may develop specialist materials with uses in this area of battery improvement. High-quality batteries require materials that may catalyze or simplify ion movement. Current technology uses engineered E. coli to break down refuse materials (like cardboard and carbohydrates) or microalgae that harness solar energy to create alternatives to diesel fuel. Despite being highly praised, these technologies have only had modest economic success to date. The use of altered DNA to encode the enzymatic reactions that turn complex organic fuels (likely waste products) into energy is another endeavor that is being studied. The conversion of the sun's abundant photons into usable chemical or electrical energy through a synthetic type of photosynthesis is another area of intense research interest. A wide range of microorganisms that can be modified to constitutively express therapeutic medicines that effectively treat illness can be created using smart microbes. An enzyme produced by a microbe has recently been shown to transform metabolites acquired from cruciferous vegetables into an anti-cancer substance. In a mouse model, this strain inhibited the growth of colon cancer. Another microbe was created with the ability to express and secrete self-assembling nanofibers that aid in mucosal healing. It can be used to cure various diseases like bacterial infections, inflammatory disorders, metabolic disorders, and cancer. By using sensors produced from the target pathogens, bacterial infections can be detected and treated with microorganisms. Biosensors have been added to natural tumor-colonizing bacteria by researchers to provide effective cancer therapies. In vivo and in clinical samples, researchers have engineered microorganisms to detect and report on inflammatory markers. By identifying or consuming an excess metabolite, "smart" microorganisms can help in the detection or therapy of a metabolic problem. Mycorrhizae, biostimulants, biofertilizers, and biopesticides etc. have been developed using smart microbes, which are of great importance.

16.12 CONCLUSION

Redesigned organisms will revolutionize our way of life by making it possible to provide more inexpensive medicines, eco-friendly fuels for our cars, and specialized therapies for "superbugs" and diseases like cancer. Redesigning organisms to produce a chemical, such as a medication or fuel, or to acquire a new ability, such as the capacity to perceive something in the environment, are common goals of synthetic biology initiatives. Additionally, it enables the creation of novel enzymes, pathways, and entire cell systems. Human, bacterial, viral, and plant cells have all been manipulated using synthetic biology to create substances or change biological processes that are beneficial to human health. Synthetic biology has had a huge impact on human health and the treatment of disease, from crop fortification to the modification of T cells. The technology of synthetic biology can be used by people to improve, better use, and enhance the biological ecosystem on Earth. This environment is made up of the delicate ecosystems that are all around us and with which humans must coexist. Generating microorganisms with desired traits is one of the other ways synthetic biology can be used to change the biological ecology of the earth. Genetically modified organisms (GMOs) have historically been created through agricultural processes involving plant breeding and animal husbandry, although being debatable today. Today, desired outcomes can be achieved by using synthetic biology techniques to add or remove genes from plants and animals. Viral transduction, which transfers genetic material via a virus, has produced more resilient plants. Less well-known uses of synthetic biology in agriculture include the production of fertilizers using modified bacteria to boost crop yields and the use of customized fermentation to produce artificial food additives. Natural biology is capable of creating an amazing range of useful structures, from the cellulose that enables trees to grow to incredible heights to the microscopic silk thread strands with incredible tensile strength. Through synthetic biology, individuals may also start using biological processes to design unique building materials and construction methods. Nanostructures can be combined, stacked, and recombined in biology to create macrostructures like plants and animals. Biocomputing, living materials, electronic interfaces, therapeutic genome editing, multiplexed diagnostics, cellular recording, third-generation biorefineries, and living biotherapeutics are just a few of the technologies that synthetic biology has the potential to transform. Perhaps the most well-known and advanced application of synthetic biology is the ability to alter metabolism to produce high-value products for uses ranging from biofuels and plant natural products to polymer precursors and bioinspired materials. The basis for building synthetic circuits is the understanding of the genetic circuits used by cells, where genes and the proteins they encode interact with one another, respond to internal and external inputs, and turn on and off cellular processes like gene expression and cell division. Using a bottom-up approach to synthetic circuit design, well-characterized genes and proteins can be combined to generate synthetic gene circuits that perform specific functions.

REFERENCES

1. Jim, Deacon. "The Microbial World: Mycorrhizas". bio.ed.ac.uk (archived). Archived from the original on 2018-04-27. Retrieved 11 January 2016.

2. Kirk, P.M.; Cannon, P.F.; David, J.C.; Stalpers, J. (2001). *Ainsworth and Bisby's Dictionary of the Fungi* (9th ed.). CAB International: Wallingford, UK.
3. Wu, Qiang-Sheng, ed. (2017). *Arbuscular Mycorrhizas and Stress Tolerance of Plants* (1st ed.). Springer Singapore. p. 1. doi:10.1007/978-981-10-4115-0. ISBN 978-981-10-4115-0.
4. Johnson, N.C.; Graham, J.H.; Smith, F.A. (1697). "Functioning of mycorrhizal associations along the mutualism–parasitism continuum." *New Phytologist.* **135**(4):575–585. doi:10.1046/j.1469-8137.1697.00729.x.
5. Kottke, I.; Nebel, M. (2005). "The evolution of mycorrhiza-like associations in liverworts: An update." *New Phytologist.* **167**(2):330–334. doi:10.1111/j.1469-8137.2005.01471.x. PMID 15998388.
6. Jump up to:a b c d Remy, W.; Taylor, T.N.; Hass, H.; Kerp, H. (6 December 1694). "Four hundred-million-year-old vesicular arbuscular mycorrhizae". *Proc Natl Acad Sci.* **91**(25):11841–11843. Bibcode:1694PNAS...9111841R. doi:10.1073/pnas.91.25.11841. PMC 45331. PMID 11607500.
7. Fortin, J. André; et al. (2015). *Les Mycorhizes* (2nd ed.). Versaillles: Inra, p. 10. ISBN 978-2-7592-2433-3.
8. Jump up to:a b c d e Wang, B.; Qiu, Y.-L. (July 2006). "Phylogenetic distribution and evolution of mycorrhizas in land plants." *Mycorrhiza.* **16**(5):299–363. doi:10.1007/s00572-005-0033-6. PMID 16845554. S2CID 30468942.
9. "Orchids and fungi: An unexpected case of symbiosis." *Am J Bot* (12 July 2011). Archived from the original on 2011-07-15. Retrieved 24 July 2012.
10. den Bakker, Henk C.; Zuccarello, G.C.; Kuyper, Th. W.; Noordeloos, M.E. (July 2004). "Evolution and host specificity in the ectomycorrhizal genus Leccinum." *New Phytologist.* **163**(1):201–215. doi:10.1111/j.1469-8137.2004.01090.x. PMID 33873790.
11. Saari, S.K.; Campbell, C.D.; Russell, J.; Alexander, I.J.; Anderson, I.C. (14 October 2004). "Pine microsatellite markers allow roots and ectomycorrhizas to be linked to individual trees." *New Phytologist.* **165**(1):295–304. doi:10.1111/j.1469-8137.2004.01213.x. PMID 15720641.
12. Simard, Suzanne W.; Perry, David A.; Jones, Melanie D.; Myrold, David D.; Durall, Daniel M.; Molina, Randy (1697). "Net transfer of carbon between ectomycorrhizal tree species in the field." *Nature.* **388**(6642):579–582. Bibcode:1697Natur.388..579S. doi:10.1038/41557. S2CID 4423207
13. Stockdale E.A.; Shepherd M.A.; Fortune S.; Cuttle S.P. (2002). "Soil fertility in organic farming systems–fundamentally different?" *Soil Use Manag.* **18**:301–308. doi: 10.1079/SUM2002143. [CrossRef] [Google Scholar]
14. Zhang, H.; Wang, R.; Chen, S.; Qi, G.; He, Z.; Zhao, X. (2017). "Microbial taxa and functional genes shift in degraded soil with bacterial wilt." *Sci Rep.* **7**:1–11. [PMC free article] [PubMed] [Google Scholar]
15. Maximillian, J.; Brusseau, M.L.; Glenn, E.P.; Matthias, A.D. (2016). *Environmental and Pollution Science.* Elsevier; Amsterdam, The Netherlands. Pollution and Environmental Perturbations in the Global System; pp. 457–476. [Google Scholar]
16. Szparaga, A.; Kuboń, M.; Kocira, S.; Czerwińska, E.; Pawłowska, A.; Hara, P.; Kobus, Z.; Kwaśniewski, D. (2016). "Towards sustainable agriculture-agronomic and economic effects of biostimulant use in common bean cultivation." *Sustainability* **11**:4575. doi: 10.3390/su11174575. [CrossRef] [Google Scholar]
17. Hellequin, E.; Monard, C.; Chorin, M.; Le Bris, N.; Daburon, V.; Klarzynski, O.; Binet, F. (2020). "Responses of active soil microorganisms facing to a soil biostimulant input compared to plant legacy effects." *Sci Rep.* **10**:13727.
18. Jacoby, R.; Peukert, M.; Succurro, A.; Koprivova, A.; Kopriva, S. (2017). "The role of soil microorganisms in plant mineral nutrition—Current knowledge and future directions." *Front Plant Sci.* **8**:1–16.
19. Romano, I.; Ventorino, V.; Pepe, O. (2020). "Effectiveness of plant beneficial microbes: Overview of the methodological approaches for the assessment of root colonization and persistence." *Front Plant Sci.* **11**:1–16.
20. Bulgarelli, D.; Schlaeppi, K.; Spaepen, S.; Van Themaat, E.V.L.; Schulze-Lefert, P. (2013). "Structure and functions of the bacterial microbiota of plants." *Annu Rev Plant Biol.* **64**:807–838.
21. Pathania, P.; Rajta, A.; Singh, P.C.; Bhatia, R. (2020). "Role of plant growth-promoting bacteria in sustainable agriculture." *Biocatal Agric Biotechnol.* **30**:101842.
22. Mannino, G.; Nerva, L.; Gritli, T.; Novero, M.; Fiorilli, V.; Bacem, M.; Bertea, C.M.; Lumini, E.; Chitarra, W.; Balestrini, R. (2020). "Effects of different microbial inocula on tomato tolerance to water deficit." *Agronomy* **10**:170.
23. Akonde, T.P.; Kuhne, R.F.; Steinmuller, N.; Leihner, D.E. (1697). "Alleycropping in an Ultisol in subhumid Benin. Part 3: Nutrient budget of maize, cassava and trees." *Agroforestry Systems* **27**:213–226.

24. Chao Wang, Xu Wang, Yang Zhang, Ember Morrissey, Yue Liu, Lifei Sun, Lingrui Qu, Changpeng Sang, Hong Zhang, Guochen Li, Lili Zhang, Yunting Fang. (1 December 2023). "Integrating microbial community properties, biomass and necromass to predict cropland soil organic carbon." *ISME Communications* **3**(1):86.
25. L.R. Paul, B.K. Chapman, C.P. Chanway. (1 June 2007). "Nitrogen Fixation Associated with *Suillus tomentosus* Tuberculate Ectomycorrhizae on *Pinus contorta* var." *latifolia, Annals of Botany* **99**(6):1101–1109.
26. Norton, J.P.; Mulvey, M.A. (2012). "Toxin-antitoxin systems are important for niche-specific colonization and stress resistance of uropathogenic escherichia coli." *PLoS Pathogens* **8**(10):e1002954.
27. Jiahao Guan, Yongkui Chen; Ying-Xian Goh; Meng Wang; Cui Tai; Zixin Deng; Jiangning Song; Hong-Yu Ou. (5 January 2024). "TADB 3.0: An updated database of bacterial toxin–antitoxin loci and associated mobile genetic elements." *Nucleic Acids Res.* **52**(D1):D784–D790.
28 Herrero-Galán E, García-Ortega L, Olombrada M, Lacadena J, Del Pozo ÁM, Gavilanes JG, Oñaderra M. Hirsutellin A: A Paradigmatic Example of the Insecticidal Function of Fungal Ribotoxins. Insects. 2013 Jul 9;4(3):339-56. doi: 10.3390/insects4030339. PMID: 26462423; PMCID: PMC4553468
29. Almén, M.S.; Nordström, K.J.; Fredriksson, R.; Schiöth, H.B. (2009). "Mapping the human membrane proteome: a majority of the human membrane proteins can be classified according to function and evolutionary origin." *BMC Biol.* **7**:50.
30. Herrero-Galán, E.; García-Ortega, L.; Lacadena, J. et al. (February 2012). "Implication of an Asp residue in the ribonucleolytic activity of hirsutellin A reveals new electrostatic interactions at the active site of ribotoxins." *Biochimie.* **94**(2):427–433. doi: 10.1016/j.biochi.2011.08.010. PMID: 21871524.
31. Rinaldi, A.C.; Comandini, O.; Kuyper, T.W. (2008). "Ectomycorrhizal fungal diversity: separating the wheat from the chaff" (PDF). *Fungal Diversity.* **33**:1–45. Archived (PDF) from the original on 2011-07-24. Retrieved 2011-05-23.
32. Harris, Brogan J.; Clark, James W.; Schrempf, Dominik; Szöllősi, Gergely J.; Donoghue, Philip C. J.; Hetherington, Alistair M.; Williams, Tom A. (29 September 2022). "Divergent evolutionary trajectories of bryophytes and tracheophytes from a complex common ancestor of land plants." *Nature Ecology & Evolution.* **6**(11): 1634–1643. doi:10.1038/s41559-022-01885-x. PMC 9630106. PMID 36175544.
33. Puginier, Camille; Keller, Jean; Delaux, Pierre-Marc (29 August 2022). "Plant–microbe interactions that have impacted plant terrestrializations." *Plant Physiology.* **160**(1):72–84. doi:10.1093/plphys/kiac258. PMC 9434271.
34. Miyauchi, Shingo; Kiss, Enikő; Kuo, Alan; et al. (2020). "Large-scale genome sequencing of mycorrhizal fungi provides insights into the early evolution of symbiotic traits." *Nature Communications.* **11**(1):5125. Bibcode:2020NatCo..11.5125M. doi:10.1038/s41467-020-18795-w. PMC 7550596. PMID 33046698.
35. Provorov, N.A.; Shtark, O.Yu; Dolgikh, E.A. (2016). "Evolution of nitrogen-fixing symbioses based on the migration of bacteria from mycorrhizal fungi and soil into the plant tissues." *Zhurnal Obshchei Biologii.* **77**(5):329–345. PMID 30024143.
36. Brundrett, Mark C.; Tedersoo, Leho (2018). "Evolutionary history of mycorrhizal symbioses and global host plant diversity." *New Phytologist.* **220** (4): 1108–1115. doi:10.1111/nph.14976.
37. Harley, J.L.; Smith, S.E. (1683). *Mycorrhizal symbiosis* (1st ed.). Academic Press, London.
38. Jump up to:a b Allen, Michael F. (1691). *The Ecology of mycorrhizae.* Cambridge University Press, Cambridge.
39. Mannino, G.; Gentile, C.; Ertani, A.; Serio, G.; Bertea, C.M. (2021). "Anthocyanins: Biosynthesis, distribution, ecological role, and use of biostimulants to increase their content in plant foods—a review." *Agriculture.* **11**:212. doi: 10.3390/agriculture11030212. [CrossRef] [Google Scholar]
40. Le Mire, G.; Nguyen, M.L.; Fassotte, B.; Du Jardin, P.; Verheggen, F.; Delaplace, P.; Haissam Jijakli, M. (2016). "Implementing plant biostimulants and biocontrol strategies in the agroecological management of cultivated ecosystems. A review." *Biotechnol Agron Soc Environ.* **20**:299–313. [Google Scholar]
41. Hellequin, E.; Monard, C.; Chorin, M.; Le Bris, N.; Daburon, V.; Klarzynski, O.; Binet, F. (2020). "Responses of active soil microorganisms facing to a soil biostimulant input compared to plant legacy effects." *Sci Rep.* **10**:13727. doi: 10.1038/s41598-020-70695-7. [PMC free article] [PubMed] [CrossRef] [Google Scholar]
42. Campobenedetto, C.; Mannino, G.; Beekwilder, J.; Contartese, V.; Karlova, R.; Bertea, C.M. (2021). "The application of a biostimulant based on tannins affects root architecture and improves tolerance to salinity in tomato plants." *Sci Rep.* **11**:354. doi: 10.1038/s41598-020-79770-5. [PMC free article] [PubMed] [CrossRef] [Google Scholar]

43. Campobenedetto, C.; Grange, E.; Mannino, G.; Van Arkel, J.; Beekwilder, J.; Karlova, R.; Garabello, C.; Contartese, V.; Bertea, C.M. (2020). "A biostimulant seed treatment improved heat stress tolerance during cucumber seed germination by acting on the antioxidant system and glyoxylate cycle." *Front. Plant Sci.* **11**:836. doi: 10.3389/fpls.2020.00836. [PMC free article] [PubMed] [CrossRef] [Google Scholar]
44. Stockdale, E.A.; Shepherd, M.A.; Fortune, S.; Cuttle, S.P. (2002). "Soil fertility in organic farming systems–fundamentally different?" *Soil Use Manag.* **18**:301–308.
45. Maximillian, J.; Brusseau, M.L.; Glenn, E.P.; Matthias, A.D. (2016). "Pollution and Environmental Perturbations in the Global System." In *Environmental and Pollution Science*; Elsevier: Amsterdam, The Netherlands, pp. 457–476.
46. Le Mire, G.; Nguyen, M.L.; Fassotte, B.; Du Jardin, P.; Verheggen, F.; Delaplace, P.; Haissam Jijakli, M. (2016). "Implementing plant biostimulants and biocontrol strategies in the agroecological management of cultivated ecosystems. A review." *Biotechnol Agron Soc Environ.* **20**:299–313.
47. Cai, Y.; Feng, X. (2023). "Substrate and community regulations on microbial necromass accumulation from newly added and native soil carbon." *Biol Fertil Soils* **59**:763–775.
48. Abdelsattar, A. M.; Elsayed, A.; El-Esawi, M. A.; Heikal, Y. M. (2023). "Enhancing Stevia rebaudiana growth and yield through exploring beneficial plant-microbe interactions and their impact on the underlying mechanisms and crop sustainability." *Plant Physiol Biochem.* **168**:107673. doi: 10.1016/j.plaphy.2023.107673
49. Muluneh, M.G. (2021). "Impact of climate change on biodiversity and food security: A global perspective—a review article." *Agric & Food Secur* **10**:36.
50. Ahmed, B.; Smart, L. B.; Hijri, M. (2021). "Microbiome of field grown hemp reveals potential microbial interactions with root and rhizosphere soil." *Front Microbiol.* **12**:741597. doi: 10.3389/fmicb.2021.741597
51. Abdel-Lateif, K.; Bogusz, D.; Hocher, V. (2012). "The role of flavonoids in the establishment of plant roots endosymbioses with arbuscular mycorrhiza fungi, rhizobia and Frankia bacteria." *Plant Signal Behav* **7**(6):636–641.
52. Agrawal, P.K.; Agrawal, S.; Shrivastava, R. (2015). "Modern molecular approaches for analyzing microbial diversity from mushroom compost ecosystem." *3 Biotech* **5**:853–866.
53. Wekesa, V.W.; Hountondji, F.C.C.; Dara, S.K. (2015). "Mite pathogens and their use in biological control." In: Carrillo, D.; de Moraes, G.; Peña, J. (eds.) *Prospects for Biological Control of Plant Feeding Mites and Other Harmful Organisms.* Progress in Biological Control, vol. 16.
54. European Commission Farm to Fork Strategy: For a fair, healthy and environmentally-friendly food system. [(accessed on 26 July 2021)]; *DG SANTE/Unit Food Inf. Compos. Food Waste.* 2020 23 Available online: https://ec.europa.eu/food/system/files/2020-05/f2f_action-plan_2020_strategy-info_en.pdf [Google Scholar]
55. Gachara, G.; Kenfaoui, J.; Suleiman, R. et al. (2024). "The role of soil microbiome in driving plant performance: An overview based on ecological and ecosystem advantages to the plant community." *J Crop Health* **76**:3–17.

17 Future and Challenges of Microbiome Engineering

Vandana Singh and Mayuri Rastogi
Assistant Professor, Department of Microbiology,
SSAHS, Sharda University
Gr. Noida, UP, India

17.1 INTRODUCTION

Organisms have evolved over millions of years to form ecological communities on Earth, known as microbiomes. These communities can persistently exist in or on living hosts, as well as in soil, oceans, and air. Microorganisms within microbiomes interact dynamically with their hosts and among themselves to form synergistic relationships, influencing the composition of the microbiome and the characteristics of the ecosystems where they reside. For instance, the composition of the human gut microbiome is associated with the host's physiology and psychological health, while the microbiome of soil can affect crop productivity (Kelly et al. 2015)

Due to the extensive impact of microbiomes on ecosystems, there has been a surge of interest in recent years to understand and gain insights into the factors that contribute to variations in microbiomes. Advances in genome sequencing and culture-independent analyses of microbiomes have made this possible, accelerating progress in microbiome research. As it becomes clearer how microbiome composition can influence ecosystems, there is growing interest in engineering microbiomes to shape microbiota for altering ecosystems of interest.

Microbiome engineering is an emerging field that involves the manipulation of the microbial communities that coexist with human beings. The human microbiome refers to the entire genetic content of microorganisms that inhabit various parts of the human body, such as the skin, oral cavity, and gut. These microorganisms play critical roles in maintaining human health, and alterations in the composition and function of the microbiome have been associated with a range of diseases, including inflammatory bowel disease, cancer, and obesity. Microbiome engineering has been proposed as a promising approach to address these and other health issues. The field encompasses a range of techniques, including targeted gene editing, synthetic biology, and microbiome transplantation, aimed at modifying the composition and activity of microbial communities in the human body (Khan, Hauptman, and Kelly 2021; Foo et al. 2017)

One example of microbiome engineering is the use of genetically engineered bacteria to produce therapeutic compounds or degrade harmful substances in the gut. For instance, researchers have engineered *Escherichia coli* bacteria to produce an anti-inflammatory molecule that can alleviate symptoms of colitis in mice (Mimee, Citorik, and Lu 2016). In another example, scientists have used CRISPR-Cas9 gene editing to introduce beneficial genetic changes into bacteria that reside in the human gut. These genetically modified bacteria could potentially be used to treat various gut disorders. Similarly, another approach to microbiome engineering is microbiome transplantation, which involves transferring a healthy microbiome to an individual suffering from a disease-associated microbiome. One well-known example of this approach is fecal microbiota transplantation (FMT), which involves the transfer of fecal material from a healthy donor to a patient suffering from

DOI: 10.1201/9781003394662-17

recurrent *Clostridioides difficile* infection. FMT has been shown to be highly effective in treating this condition, with success rates of up to 90% (Beloqui, des Rieux, and Préat 2016).

One of the most promising areas of microbiome engineering is the creation of tailored microbiome therapeutics for diverse diseases such as cancer, autoimmune disorders, and infections. Researchers are also exploring the function of the microbiome in mental health and developing microbiome-based therapies for mental health issues. Another promising area of research is the development of personalized nutrition interventions based on an individual's microbiome to improve health outcomes (Feng, Chen, and Wang 2018).

Agriculture and plant science could benefit from microbiome engineering as well. Researchers want to raise crop yields, minimize pesticide use, and improve plant resilience by adopting microbiome-based techniques. Moreover, microbiome engineering can be used to remediate environmental contaminants, including oil spills and heavy metals, by adding microbial communities that can break down and eliminate these poisons from the environment (Berendsen, Pieterse, and Bakker 2012).

Aside from these practical applications, microbiome engineering is being investigated for its potential in biotechnology. Synthetic microbial communities are being designed and engineered by researchers for a variety of biotechnological purposes, including bioproduction and bioremediation. Microbiome engineering is also being used to develop microbiome-derived drugs, biologics, and other therapeutics for various diseases and disorders (Douglas and Langille 2019). Overall,

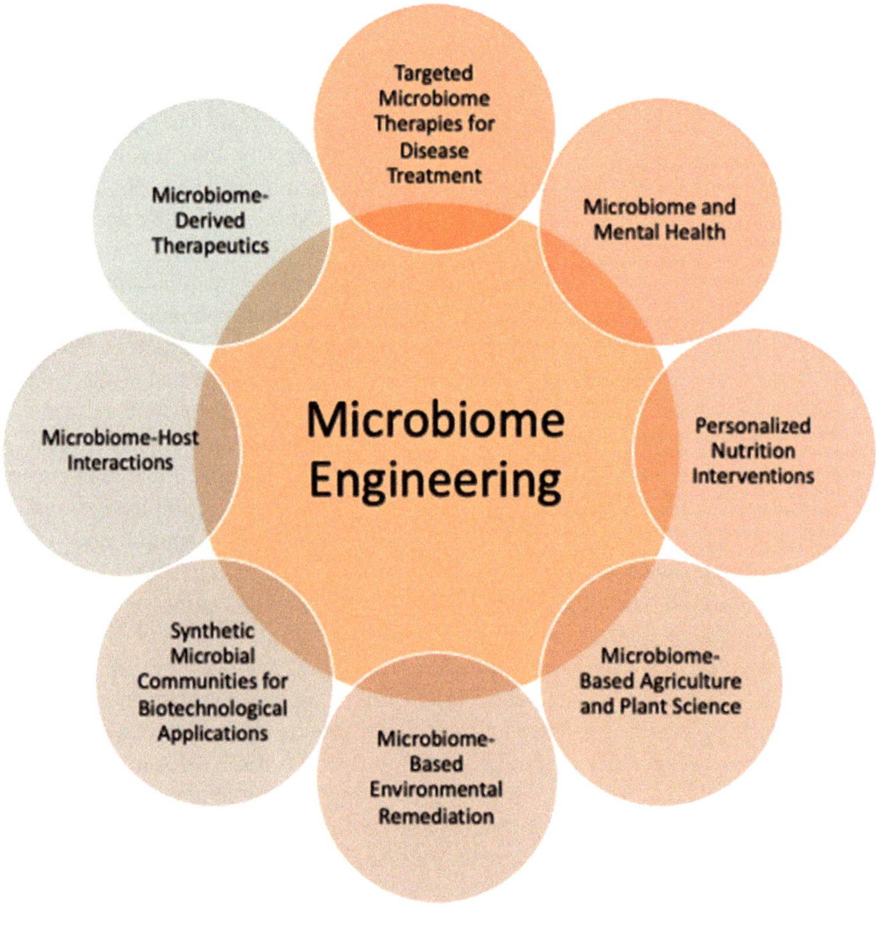

FIGURE 17.1 Some future aspects and applications of microbiome engineering.

microbiome engineering has the potential to revolutionize many fields by harnessing the power of the microbiome to achieve specific goals. However, as with any emerging technology, there are still many challenges and ethical considerations that need to be addressed before widespread use can be realized. These include concerns about the safety and efficacy of microbiome-based therapies, as well as issues related to genetic modification and environmental impact. Furthermore, microbiome engineering, like any other developing field, has ethical issues and potential hazards. As a result, scientists are investigating the ethical aspects of microbiome engineering as well as the possible threats to human health and the environment. Another exciting area of research is the development of microbiome-based technologies for long-duration space missions to maintain astronaut health and prevent microbial contamination. This chapter mainly emphasizes the future of microbiome engineering, its use in improving human health and agricultural outputs, and its significance and present uses, specifically in humans, animals, plants, and soil (Figure 17.1). Furthermore, the future and challenges of this discipline are discussed.

17.2 CHALLENGES IN MICROBIOME ENGINEERING

The microbiome engineering is an exciting field with enormous potential for improving human health and other applications. However, it is not without its challenges, including understanding microbial communities, precision engineering, regulatory challenges, stability and persistence, and ethics and societal challenges. Researchers are working to overcome these challenges, but there is still much work to be done to realize the full potential of microbiome engineering. The following are some of the major challenges and the current efforts to overcome them.

17.2.1 Understanding Microbial Communities:

One of the significant challenges in microbiome engineering is understanding the complex interactions within microbial communities. Microbes often live in symbiosis, and changes to one microbe can affect the entire community. Therefore, researchers need to understand how different microbial species interact with each other and their environment to engineer a microbiome effectively. This requires developing new tools and techniques to study microbial communities, including advanced genomic sequencing, metagenomics, transcriptomics, and proteomics (Gilbert et al. 2018).

17.2.2 Precision Engineering:

Microbiome engineering requires precise manipulation of microbial communities to achieve a specific goal. However, it is often difficult to predict how changing the composition or function of one microbe will affect the entire community. Therefore, microbiome engineering requires a more sophisticated approach than traditional genetic engineering, which involves the insertion or deletion of specific genes. Researchers must use systems biology to understand how different microbes interact and design interventions that affect the entire microbial community, not just individual microbes (Brooks and Alper 2021).

17.2.3 Regulatory Challenges:

Another challenge in microbiome engineering is the lack of clear regulatory guidelines. The use of microbiome engineering in medicine and agriculture raises ethical and safety concerns, and regulatory agencies are still developing policies to address these concerns. The U.S. Food and Drug Administration (FDA) recently released a draft guidance document on the use of live biotherapeutic products (LBPs) for treating diseases (Kim et al. 2020). However, there is still a long way to go in establishing clear guidelines for microbiome engineering.

17.2.4 Stability and Persistence

Microbiome engineering is not a one-time intervention. Microbial communities are dynamic, and changes to the microbiome may not persist over time. Therefore, researchers must design interventions that are stable and persistent. This requires understanding the factors that affect microbial community stability, such as environmental conditions and host-microbe interactions. Researchers are also exploring methods for delivering and maintaining engineered microbiomes, such as probiotics and fecal microbiota transplantation (FMT) (Langdon, Crook, and Dantas 2016).

17.2.5 Ethics and Societal Challenges

Microbiome engineering raises ethical and societal concerns. The manipulation of microbial communities in humans and the environment has the potential to bring about significant benefits but also raises the possibility of unintended consequences. Therefore, it is essential to engage with stakeholders, including patients, regulators, and the public to address ethical and societal concerns and ensure the responsible development of microbiome engineering (D. Liu et al. 2019).

17.2.6 Data Management and Analysis

Another significant challenge in microbiome engineering is the management and analysis of vast amounts of data generated from high-throughput sequencing technologies. Researchers need to develop efficient and reliable methods for storing, analyzing, and interpreting microbiome data to identify potential targets for microbiome engineering. This requires the integration of various data types, such as genomic, transcriptomic, and metabolomic data, and the development of novel statistical and computational tools (Franzosa et al., 2019).

17.2.7 Host-Microbe Interactions

Microbiome engineering interventions can have significant effects on host-microbe interactions, which can impact human health and disease. Therefore, understanding the complex interplay between host and microbiome is crucial for successful microbiome engineering. Researchers need to study the role of host genetics, the immune system, and environmental factors in shaping the microbiome and the effects of microbiome interventions on the host (Belkaid 2015).

17.2.8 Standardization and Reproducibility

Microbiome engineering interventions can vary significantly across laboratories and experimental settings, making it challenging to compare results across studies. Therefore, developing standardized protocols and methods for microbiome engineering is critical to ensure reproducibility and comparability of results. This requires establishing standard reference materials and data sets and developing guidelines for sample collection, storage, and processing (Knight et al. 2018).

17.2.9 Commercialization and Intellectual Property

As microbiome engineering moves towards commercialization, there are concerns about intellectual property and the potential for monopolization. Microbial communities are complex and difficult to patent, raising questions about the commercial viability of microbiome engineering. Therefore, it is essential to develop appropriate regulatory and legal frameworks to balance innovation and commercialization with public access and benefit (Aggarwal et al., 2023).

17.2.10 Ecosystem-Level Effects

Microbiome engineering can have significant effects on the environment, including soil, water, and the atmosphere. Therefore, it is crucial to consider the ecosystem-level effects of microbiome engineering interventions. This requires studying the potential effects of engineered microbes on non-target species and the environment, and developing appropriate risk assessment and management strategies (Aggarwal et al., 2023).

17.2.11 Ethics and Social Implications

Microbiome engineering also raises ethical and social implications, such as potential unintended consequences, privacy concerns, and equity in access and distribution. For example, microbiome engineering interventions may have unintended consequences on the host and the environment, leading to unforeseen risks and challenges. Additionally, microbiome data is sensitive and may raise privacy concerns, especially in the context of personalized medicine. Therefore, ethical and social considerations need to be integrated into microbiome engineering research and development to ensure responsible and equitable outcomes (Aggarwal et al., 2023).

17.2.12 Funding and Resource Constraints

Microbiome engineering research and development require significant resources and funding, including access to high-throughput sequencing technologies, specialized expertise, and adequate infrastructure. However, these resources may be limited or inaccessible, particularly in low- and middle-income countries. Therefore, addressing funding and resource constraints is crucial to ensure that microbiome engineering benefits are accessible and equitable globally (Cullen et al. 2020).

17.2.13 Education and Awareness

Microbiome engineering is a complex and interdisciplinary field that requires specialized knowledge and expertise. Therefore, developing educational and awareness programs is crucial to ensure the next generation of researchers and stakeholders has the necessary skills and knowledge to advance the field responsibly. Additionally, raising public awareness about microbiome engineering and its potential benefits and risks can foster informed decision-making and stakeholder engagement (Cullen et al. 2020).

17.2.14 Regulatory Frameworks

As microbiome engineering moves towards commercialization, appropriate regulatory frameworks are necessary to ensure safety, efficacy, and responsible development. However, developing such frameworks can be challenging due to the complex and dynamic nature of microbiome engineering interventions. Therefore, collaborative efforts between researchers, regulators, and stakeholders are crucial to develop appropriate regulatory frameworks that balance innovation and public safety (Cullen et al. 2020).

17.2.15 Interdisciplinary Collaboration

Microbiome engineering is an interdisciplinary field that requires collaboration between various disciplines, including microbiology, genomics, bioinformatics, engineering, and clinical research. Therefore, fostering interdisciplinary collaboration and communication is essential to advancing the field and addressing its complex challenges. Additionally, engaging stakeholders from diverse backgrounds, such as patients, clinicians, regulators, and policymakers, can foster a comprehensive and inclusive approach to microbiome engineering (Gilbert et al. 2018).

17.2.16 Translation to Clinical Practice

Microbiome engineering holds great promise for various clinical applications, such as treating infections, autoimmune diseases, and cancer. However, translating microbiome engineering interventions from the lab to the clinic requires rigorous testing, validation, and regulatory approval. Additionally, addressing ethical and social considerations, such as equity and access, is crucial to ensure that microbiome engineering benefits reach all patients who can benefit from them (Khan, Hauptman, and Kelly 2021).

17.2.17 Environmental Sustainability

Microbiome engineering can also have significant applications in environmental sustainability, such as improving soil and water quality and mitigating climate change. However, microbiome engineering interventions can also have unintended environmental consequences, such as altering the ecosystem structure and function. Therefore, balancing the potential environmental benefits and risks of microbiome engineering is crucial for ensuring sustainable and responsible development (Verstraete et al. 2022).

17.3 FUTURE DIRECTIONS OF MICROBIOME ENGNINEERING

Recent advances in high-throughput sequencing, synthetic biology, and bioinformatics have enabled precise manipulation and functional characterization of microbial communities and areas of research that hold promise for advancing our understanding and applications of microbiome engineering. Table 17.1 provides details of some of the most promising future directions.

17.3.1 Systems-level Understanding

A systems-level understanding of microbial communities and their interactions is crucial for successful microbiome engineering interventions. However, current knowledge of microbial communities is limited, and most studies focus on individual microbial species or functions. Therefore, developing integrative approaches that combine multiple omics and functional assays is crucial for unraveling the complex dynamics and interactions of microbial communities (Gilbert et al. 2016).

17.3.2 Synthetic Biology

Synthetic biology enables the engineering of biological systems with precise control over genetic circuits, metabolic pathways, and cellular behavior. Therefore, applying synthetic biology to microbiome engineering holds great promise for developing novel and effective interventions for various applications. For example, using engineered bacteria to produce therapeutic molecules, such as enzymes, antimicrobial peptides, and immunomodulators, can have significant clinical applications (Wan, Ho, and Wang 2019).

17.3.3 Machine Learning

Machine learning has emerged as a powerful tool for analyzing and interpreting complex data, including microbiome data. Therefore, developing machine learning models that can predict and optimize microbiome engineering interventions can have significant practical applications. For example, using machine learning algorithms to predict the efficacy of probiotics or identify novel

TABLE 17.1
Outline of some Future Aspects of Microbiome Engineering

Future Prospective	Description	Reference
Therapeutic microbiome modulation	Developing targeted therapies to manipulate the microbiome to treat various diseases, including cancer, autoimmune disorders, and infections	(Gulliver et al. 2022)
Microbiome-based diagnostics	Utilizing microbiome data for disease diagnosis and personalized medicine	(Kashyap et al. 2017)
Agricultural microbiome engineering	Developing microbiome-based approaches to increase crop yields, reduce pesticide use, and enhance plant resilience	(Kaul et al. 2021)
Microbiome-host interactions	Investigating the complex interactions between the microbiome and the host immune system, metabolism, and neurology	(Popkes and Valenzano 2020)
Microbiome-derived therapeutics	Developing drugs, biologics, and other therapeutics derived from microbiome components, including microbes, metabolites, and proteins	(Mimee, Citorik, and Lu 2016)
Microbiome-based nutritional interventions	Developing personalized nutrition interventions based on an individual's microbiome to improve health outcomes	(Z. Liu et al. 2020)
Microbiome-based environmental remediation	Utilizing microbiome-based approaches to remediate environmental contaminants, including oil spills and heavy metals	(Albright et al. 2022)
Microbiome engineering for synthetic biology	Designing and engineering synthetic microbial communities for various biotechnological applications, including bioproduction and bioremediation	(Lawson 2021)
Microbiome modulation for mental health	Developing microbiome-based interventions for the treatment of mental health disorders, including depression and anxiety	(Berding and Cryan 2022)
Microbiome engineering ethics	Examining the ethical implications of microbiome engineering and potential risks to human health and the environment	(Lange et al. 2022)
Microbiome engineering for space exploration	Developing microbiome-based technologies for long-duration space missions to maintain astronaut health and prevent microbial contamination	(LaPelusa et al. 2021)

microbial interactions can enable personalized and precision microbiome engineering (Li et al. 2022).

17.3.4 Microbiome Editing

Recent advances in CRISPR-Cas and other gene editing technologies have enabled precise and targeted modifications of microbial genomes. Therefore, using microbiome editing to introduce or eliminate specific microbial species or functions can have significant applications in various domains. For example, using microbiome editing to eliminate pathogenic bacteria or introduce beneficial bacteria can have clinical applications (Arroyo-Olarte, Bravo Rodríguez, and Morales-Ríos 2021) (Fischbach and Segre 2016).

17.3.5 Host-Microbe Interactions

Host-microbe interactions play a crucial role in shaping the composition and function of microbial communities. Therefore, understanding and manipulating host-microbe interactions is essential for successful microbiome engineering interventions. For example, developing microbiome-based therapies that target the host immune system or modulate the gut-brain axis can have significant clinical applications (Fischbach and Segre 2016).

17.3.6 Environmental Microbiome Engineering

Microbiome engineering can also have significant applications in environmental sustainability, such as improving soil and water quality and mitigating climate change. Therefore, developing microbiome engineering interventions that can address environmental challenges and promote sustainable development is crucial. For example, using microbiome engineering to develop biofertilizers or bioremediation agents can have significant environmental applications (Silverstein, Segrè, and Bhatnagar 2023).

17.3.7 Microbial Communication

Microbial communication, also known as quorum sensing, plays a crucial role in shaping the composition and function of microbial communities. Therefore, manipulating microbial communication pathways can enable precise control over microbial community dynamics and functions. For example, using quorum sensing inhibitors or quorum sensing signal mimics can modulate the interactions and behaviors of microbial communities (Berg et al. 2020).

17.3.8 Regulatory Frameworks

As microbiome engineering interventions become more widespread, developing regulatory frameworks to ensure their safety and efficacy is crucial. Therefore, establishing clear guidelines and standards for microbiome engineering interventions is essential for the responsible and sustainable development of the field (Berg et al. 2020).

17.3.9 Multidisciplinary Collaboration

Microbiome engineering is a multidisciplinary field that requires collaboration between researchers from diverse domains, such as microbiology, synthetic biology, bioinformatics, and clinical medicine. Therefore, promoting multidisciplinary collaboration and knowledge sharing can enable the development of novel and effective microbiome engineering interventions (Berg et al. 2020).

17.3.10 Ethical Considerations

Microbiome engineering interventions can have significant ethical implications, such as the potential to exacerbate existing health disparities or the potential to alter ecosystems and biodiversity. Therefore, addressing ethical considerations and engaging with stakeholders, such as patients, communities, and policymakers, is essential for the responsible and sustainable development of microbiome engineering (Robinson et al. 2022).

17.3.11 Scaling Up

One of the biggest challenges in microbiome engineering is scaling up interventions to a larger, more complex environment. While microbiome engineering has shown promise in laboratory experiments, translating these interventions to real-world settings, such as the human gut or soil microbiome, can be challenging due to the complexity and heterogeneity of these environments (Berg et al. 2020).

17.3.12 Safety Considerations

Microbiome engineering interventions have the potential to cause unintended consequences, such as disrupting natural microbial communities or introducing new pathogens. Therefore, safety considerations are essential when developing and deploying microbiome engineering interventions (Robinson et al. 2022).

17.3.13 Commercialization

Microbiome engineering interventions have the potential to create new commercial opportunities, such as the development of microbial therapies or microbiome-based diagnostics. However, commercializing microbiome engineering interventions requires significant investment and regulatory approval, which can be challenging (Gulliver et al. 2022).

17.3.14 Education and Public Outreach

Microbiome engineering is a rapidly evolving field, and educating the public and policymakers about its potential benefits and risks is crucial. Therefore, promoting education and public outreach can enable informed decision-making and responsible development of the field (Robinson et al. 2022).

17.3.15 Integration with Other Technologies

Microbiome engineering is not a standalone technology but can be integrated with other technologies, such as gene editing, nanotechnology, or artificial intelligence, to develop novel and effective interventions. Therefore, exploring the synergies between microbiome engineering and other technologies can open up new opportunities for research and development (Berhanu, Godishala, and Kandi 2021).

17.4 ETHICAL CONSIDERATIONS IN MICROBIOME ENGINEERING

Microbiome engineering approaches offer promising opportunities for preventing and treating a range of diseases, but they also raise ethical concerns regarding safety, efficacy, and impact on microbial diversity and community stability. Therefore, it is essential to consider the ethical implications of microbiome engineering approaches and their impact on patients, donors, and society when designing and conducting clinical trials or treatments. These considerations should include informed consent, risk-benefit analysis, regulatory and legal frameworks, and transparency, accountability, and quality control. Ultimately, these approaches should be developed and used responsibly and ethical manner that respects the dignity, autonomy, and rights of all individuals involved (Lange et al. 2022).

17.4.1 Informed Consent and Privacy

Informed consent is a fundamental principle of ethical research involving human participants. It requires that individuals receive sufficient information about the research, including its risks and benefits, to make an informed decision about their participation. In microbiome engineering, informed consent is particularly important given the potential for microbiome data to reveal sensitive information about an individual's health, lifestyle, and genetic makeup. Participants in microbiome-related research should be informed about the potential uses of their microbiome data, including research and commercialization, and be given the opportunity to opt out if they so choose. Protecting the privacy and confidentiality of microbiome data is also crucial, and data sharing should be done in accordance with ethical and legal frameworks.

17.4.2 Equity and Access

Equity and access are critical ethical considerations in microbiome engineering. The development and deployment of microbiome interventions should not perpetuate or exacerbate existing social and health inequities. Efforts should be made to ensure that microbiome engineering interventions are accessible to underserved populations, including those in low- and middle-income countries (LMICs). Microbiome interventions should be affordable, and efforts should be made to ensure that they are covered by health insurance or other forms of financial assistance.

17.4.3 Safety and Environmental Impact

Microbiome engineering interventions have the potential to cause unintended consequences, such as disrupting natural microbial communities or introducing new pathogens. Safety considerations are crucial when developing and deploying microbiome engineering interventions. Microbiome interventions should undergo rigorous safety testing, including preclinical and clinical trials, to ensure their safety and efficacy. Environmental impact assessments should also be conducted to evaluate potential long-term effects on the environment and ecosystem.

17.4.4 Regulation and Oversight

Regulation and oversight are essential to ensure the ethical and safe development of microbiome engineering interventions. The regulatory framework should address issues such as safety, efficacy, privacy, and equity while also allowing for innovation and scientific progress. Regulatory bodies should work with researchers, clinicians, and industry partners to develop clear and transparent guidelines for the development, deployment, and monitoring of microbiome engineering interventions.

17.4.5 Intellectual Property and Commercialization

The commercialization of microbiome engineering interventions can create new economic opportunities. However, ethical considerations must be taken into account in the context of intellectual property, patent law, and access to essential medicines. Intellectual property should not be used to limit access to microbiome engineering interventions, particularly in LMICs where access to healthcare is limited. Industry partners should work with governments and non-governmental organizations to ensure that microbiome interventions are affordable and accessible to all individuals.

17.4.6 Animal Welfare

Microbiome engineering interventions that involve animal models raise ethical considerations related to animal welfare. Microbiome engineering interventions should adhere to ethical standards for animal research, including minimizing animal suffering and using alternative methods whenever possible. Researchers and clinicians should also work to ensure that microbiome interventions are effective in animal models before moving on to human trials.

17.4.7 International Collaboration

Microbiome engineering interventions are a global endeavor that require international collaboration to ensure their responsible development and equitable distribution. International collaboration can help promote a shared understanding of ethical considerations and regulatory frameworks while fostering scientific exchange and cooperation. Collaboration between researchers, clinicians, industry partners, and regulatory bodies is essential to ensure that microbiome engineering interventions are developed in a safe, ethical, and equitable manner.

17.4.8 Microbiome Diversity and Community Stability

The human microbiome is a complex ecosystem composed of bacteria, viruses, fungi, and other microorganisms, which interact with each other and with the host to maintain a healthy equilibrium. The diversity and stability of the microbiome are critical factors in maintaining human health, as alterations in the microbiome composition and function have been associated with various diseases. Microbiome engineering approaches aim to restore or manipulate the microbiome to prevent or treat diseases, but they can also disrupt the microbiome ecosystem, leading to unintended consequences, such as microbial imbalance, dysbiosis, and disease progression. Therefore, it is essential to consider the impact of microbiome engineering approaches on microbial diversity and community stability when designing and conducting clinical trials or treatments.

17.4.9 Transparency and Accountability

Transparency and accountability are also important ethical considerations in microbiome engineering. Researchers must be transparent about their methods and results, and they must be accountable for the outcomes of their research. This requires open communication and collaboration among researchers, stakeholders, and the public to ensure that microbiome engineering approaches are developed and used in a responsible and ethical manner.

17.4.10 Quality Control

Quality control is an important ethical consideration in microbiome engineering. As with any medical intervention, microbiome engineering approaches must be subjected to rigorous quality control measures to ensure their safety and efficacy. This requires the development of standardized protocols for microbiome engineering approaches and the establishment of quality control measures to ensure that these approaches are developed and used in a responsible and ethical manner.

17.4.11 Beneficence and Non-Maleficence

The principles of beneficence (the duty to do good) and non-maleficence (the duty to do no harm) are important ethical considerations in microbiome engineering. Researchers and practitioners must balance the potential benefits of microbiome engineering with the potential risks and harm to individuals and communities. This requires careful consideration of the potential benefits and harms of microbiome engineering approaches and a commitment to doing no harm.

17.4.12 Social Justice

Social justice is another important ethical consideration in microbiome engineering. As with any medical intervention, microbiome engineering approaches must be accessible and equitable to all individuals, regardless of their socioeconomic status, race, ethnicity, or other factors. This requires a commitment to addressing health disparities and ensuring that microbiome engineering approaches are accessible and affordable to all individuals who could benefit from them.

Hence, addressing ethical considerations is crucial to ensure the safe and responsible development of microbiome engineering interventions. Informed consent and privacy, equity and access, safety and environmental impact, regulation and oversight, intellectual property and commercialization, animal welfare, and international collaboration are all important ethical considerations that must be carefully considered and addressed in microbiome engineering research and development.

17.5 REGULATORY AND SAFETY CONSIDERATIONS FOR MICROBIOME ENGINEERING

The safety and regulatory considerations for microbiome engineering are complex and multifaceted. The choice of microorganisms, the mode of delivery, and the potential risks associated with genetic modification must all be carefully evaluated before microbiome-based therapies can be approved for clinical use. Regulatory agencies have developed guidelines to ensure the safety and efficacy of these therapies, but ethical considerations must also be taken into account. Continued research in this area will be necessary to fully realize the potential of microbiome engineering as a therapeutic approach (Laulund et al. 2017).

The regulatory framework for microbiome engineering is complex and still evolving. In the United States, the Food and Drug Administration (FDA) is responsible for regulating microbiome-based products for human use, while the Environmental Protection Agency (EPA) is responsible for regulating products intended for environmental use. These agencies are guided by several laws, including the Federal Food, Drug, and Cosmetic Act (FD&C Act), the Public Health Service Act (PHS Act), and the Toxic Substances Control Act (TSCA). One of the main challenges for regulatory agencies is determining how to classify microbiome-based products. For example, if a microbiome-based product is intended to prevent or treat a disease, it may be considered a drug and subject to FDA regulation. On the other hand, if the product is intended to improve plant growth, it may be subject to EPA regulation (Laulund et al. 2017).

Safety considerations are one of the key points in microbiome engineering and also have the potential for unintended consequences. Microorganisms are complex and have interactions with their environment that are not fully understood. Therefore, it is essential to assess the safety of microbiome-based products before they are released into the environment or used in humans. One potential risk of microbiome engineering is the possibility of horizontal gene transfer (HGT) between engineered microorganisms and other organisms in the environment. HGT can result in the transfer of antibiotic resistance genes, which could have serious consequences for public health. Therefore, it is important to assess the likelihood of HGT and take steps to minimize the risk (Khan, Hauptman, and Kelly 2021).

Another safety consideration for microbiome engineering is the potential for unintended consequences for the host organism. For example, if a microbiome-based product is designed to improve plant growth, it may inadvertently harm the plant or reduce its yield. Therefore, it is important to conduct rigorous testing to assess the safety and effectiveness of microbiome-based products. The safety of microbiome engineering is a crucial consideration, as it involves the use of live microorganisms that can potentially cause harm to the host (Cullen et al. 2020). Safety concerns related to microbiome engineering include the following:

1. Infection: Microorganisms used in microbiome engineering can cause infections if they are not properly screened and characterized for their safety.
2. Transfer of antibiotic resistance: The use of antibiotics to manipulate the microbiome can lead to the development of antibiotic-resistant strains of bacteria, which can spread to other individuals.
3. Adverse immune reactions: Microorganisms used in microbiome engineering can trigger adverse immune reactions, leading to inflammation and other harmful effects.
4. Genetic modification: The genetic modification of microorganisms used in microbiome engineering can raise safety concerns related to the potential spread of genetically modified organisms (GMOs) in the environment.

To address these safety concerns, several measures can be taken, including the following:

1. Proper screening and characterization: Microorganisms used in microbiome engineering should be screened and characterized for their safety and efficacy.
2. Quality control: Microbial products used in microbiome engineering should be manufactured under strict quality control measures to ensure their safety and efficacy.
3. Monitoring and surveillance: The safety and efficacy of microbiome engineering should be monitored and surveilled through clinical trials and post-marketing surveillance.
4. Risk assessment: The potential risks associated with microbiome engineering should be assessed and mitigated through appropriate risk management strategies.

Microbiome engineering is a booming province with many potential applications. However, as with any new technology, there are regulatory and safety considerations that need to be addressed. Regulatory agencies must determine how to classify microbiome-based products, while safety considerations include the potential for unintended consequences and the need to minimize the risk of HGT. To ensure the safety and effectiveness of microbiome engineering, thorough risk assessments, standards for product quality and safety, rigorous testing, and monitoring and surveillance are essential (Cullen et al. 2020).

17.6 COMMERCIAL APPLICATION OF MICROBIOME ENGINEERING

Microbiome engineering has great potential for various commercial applications, ranging from agriculture to biotechnology and healthcare. Table 17.2 details some of the current and potential applications of microbiome engineering in these industries (Foo et al. 2017).

17.6.1 Agriculture

One of the major challenges in modern agriculture is maintaining high crop yields while minimizing the use of pesticides and fertilizers. Microbiome engineering can play a significant role in addressing this challenge. For example, scientists have discovered that certain bacteria and fungi in the soil can enhance plant growth and protect against pests and diseases. By engineering the microbiome of the soil, it is possible to promote the growth of beneficial microorganisms and reduce the populations of harmful ones (Han and Yoshikuni 2022).

17.6.2 Biotechnology

Microbiome engineering has numerous applications in biotechnology, including the development of new drugs and bioproducts. For example, researchers have developed a method for engineering the gut microbiome to produce therapeutic proteins. This approach involves engineering a non-pathogenic bacterium to secrete the desired protein in response to specific environmental cues. The engineered bacterium can then be administered orally to the patient, where it colonizes the gut and produces the therapeutic protein. Furthermore, microbiome engineering can be used to develop new enzymes for industrial processes. Enzymes are proteins that catalyze biochemical reactions and are widely used in industries such as food and beverage production, pharmaceuticals, and biofuels. By engineering the microbiome of microorganisms, it is possible to identify and isolate new enzymes with desirable properties. In the biotechnology industry, microbiome engineering is being used to develop new drugs, bioproducts, and therapeutic agents. One example is Finch Therapeutics, a biotech company that is developing live biotherapeutic products (LBPs) for the treatment of various diseases, including recurrent Clostridium difficile infection (CDI) and inflammatory bowel disease (IBD). The company's lead product, CP101, is an investigational oral capsule containing purified and encapsulated fecal microbiota from healthy donors. Clinical trials

TABLE 17.2
Applications of Microbiome Engineering in Various Fields

Microbiome	Method of Engineering	Physiological Advantages	Commercial Applications	References
Animal microbiome	Probiotic supplementation	Improved digestion, increased growth and productivity	Livestock feed, aquaculture	(Bhogoju and Nahashon 2022)
Environment microbiome	Microbial biofilm engineering	Improved indoor air quality, reduced surface contamination	Building materials, air filtration systems	(Foo et al. 2017)
Environmental microbiome	Directed evolution of microbial communities	Enhanced degradation of pollutants, increased efficiency of bioreactors	Bioremediation, wastewater treatment	(Silverstein, Segrè, and Bhatnagar 2023)
Gut microbiome	Fecal microbiota transplantation (FMT)	Improved digestion and absorption of nutrients, enhanced immune function	Treatment of *Clostridioides difficile* infection, inflammatory bowel disease	(Gulliver et al. 2022)
Human microbiome	Engineered bacteriophages	Targeted removal of specific bacteria, prevention of antibiotic resistance	Medical treatment, food safety	(Fischbach and Segre 2016)
Industrial microbiome	Metabolic engineering of microbial strains	Enhanced production of desired metabolites, improved efficiency of bioprocesses	Biofuels, bioplastics, biopharmaceuticals	(Jiang et al. 2017)
Oral microbiome	CRISPR-based editing of oral bacteria	Removal of harmful bacteria, prevention of dental caries	Oral health products, dental care	(Zhang et al. 2018)
Plant microbiome	Inoculation with beneficial microorganisms	Enhanced nutrient uptake, improved resistance to pests and diseases	Agriculture, horticulture	(Afridi et al. 2022)
Skin microbiome	Topical application of probiotics	Enhanced skin barrier function, reduced inflammation	Skincare products, wound healing	(Boxberger et al. 2021)

have shown promising results in the treatment of recurrent CDI. Microbiome engineering is also being explored in the development of new biotechnologies. Researchers are using microbiome engineering to develop synthetic microbial communities that can be used to produce biofuels or other chemicals. These synthetic communities consist of multiple microorganisms that have been engineered to work together to produce a desired product. For example, researchers have developed a synthetic microbial community that can convert agricultural waste into biofuels (Lee, Aurand, and Friedman 2020).

17.6.3 Healthcare

Microbiome engineering has the potential to revolutionize healthcare by enabling personalized medicine and improving the treatment of various diseases. For example, researchers have demonstrated that FMT can be used to treat recurrent *Clostridioides difficile* infection (CDI). FMT involves transferring fecal matter from a healthy donor to the patient with CDI in order to restore the patient's gut microbiome. By engineering the microbiome of the donor, it may be possible to improve the efficacy of FMT and reduce the risk of adverse effects. In addition, microbiome engineering can be used to develop live biotherapeutic products (LBPs) for the treatment of various diseases. LBPs are live microorganisms that, when administered in adequate amounts, can provide health benefits to the host. For example, a recent study demonstrated that the transfer of intestinal microbiota from lean donors to individuals with metabolic syndrome improved their insulin sensitivity. LBPs can also be engineered to express specific therapeutic molecules, such as antibodies or enzymes. In the healthcare industry, microbiome engineering is being used to develop personalized treatments and diagnostics based on an individual's unique microbiome composition. One example is Viome, a healthcare company that offers personalized nutrition and wellness recommendations based on the analysis of the customer's gut microbiome. The company uses advanced microbiome sequencing and AI algorithms to identify the customer's unique microbial composition and provide personalized diet and lifestyle recommendations (Basavaraju, Gunashree, and Srinath 2022).

17.6.4 Biosensors

Microbiome engineering is involved in the development of microbial biosensors. Biosensors are devices that can detect and measure specific substances in a sample. Microbial biosensors use microorganisms that have been genetically engineered to respond to specific compounds. These biosensors can be used to detect a wide range of substances, from environmental pollutants to medical biomarkers. For example, researchers have developed a biosensor that uses *E. coli* bacteria to detect heavy metals in water. The bacteria have been engineered to produce a green fluorescent protein in the presence of heavy metals. When the bacteria are exposed to contaminated water, they produce the fluorescent protein, indicating the presence of heavy metals (Foo et al. 2017a; Brooks and Alper 2021).

17.6.5 Food Science

Microbiome engineering is also being explored in the field of food science. For example, researchers are using microbiome engineering to develop probiotics for use in food products. Probiotics are live microorganisms that provide health benefits when consumed in adequate amounts. Probiotics can improve gut health, boost the immune system, and improve nutrient absorption. One example of a probiotic food product is the development of a yogurt that contains a strain of *Lactobacillus rhamnosus* that has been engineered to produce a human enzyme called Urox. Urox is an enzyme that helps to reduce the risk of urinary tract infections. By adding Urox-producing *Lactobacillus rhamnosus* to yogurt, researchers have developed a probiotic food product that can help to prevent urinary tract infections. In addition, microbiome engineering is being used to develop probiotics that can produce other health-promoting compounds, such as antioxidants, vitamins, and amino acids. Microbiome engineering can also be used to develop probiotics for livestock. Probiotics are live microorganisms that, when administered in adequate amounts, can provide health benefits to the host animal. By engineering the microbiome of animals, it is possible to improve their digestion, immunity, and overall health. This can lead to increased productivity and profitability for farmers (Singh and Trivedi 2017).

17.7 CONCLUSION

Microbiome engineering is a rapidly advancing field with numerous applications across various industries, including healthcare, agriculture, and biotechnology. The development of new tools and techniques for manipulating microbial communities enables researchers to better understand the complex interactions between microorganisms and their hosts, and to design interventions that can improve human health, promote sustainability, and develop new technologies. Microbiome engineering has already shown promising results in several areas, including the treatment of diseases such as inflammatory bowel disease, the development of microbial inoculants for agriculture, and the production of new drugs and diagnostic tools. As research in the field continues to progress, it is expected that the potential applications of microbiome engineering will continue to expand. One of the major implications of microbiome engineering is the potential to develop personalized medicine approaches that take into account an individual's unique microbiome composition. By leveraging the latest tools and techniques for analyzing and manipulating microbial communities, it may be possible to develop tailored interventions that are more effective and have fewer side effects than traditional treatments.

Another implication of microbiome engineering is the potential to promote sustainability in agriculture and other industries. By developing microbial inoculants that can enhance plant growth and reduce the need for synthetic fertilizers and pesticides, for example, it may be possible to reduce the environmental impact of agriculture while still meeting the world's growing food needs. Finally, the development of new diagnostic tools and biosensors based on microbiome engineering has the potential to revolutionize disease detection and monitoring. By using microorganisms as biosensors, it may be possible to detect diseases earlier and more accurately than current methods, leading to better outcomes for patients. Hence, microbiome engineering is a rapidly advancing field with tremendous potential for improving human health, promoting sustainability, and developing new technologies. As research in the field continues to progress, it is expected that the applications of microbiome engineering will continue to expand, leading to new and innovative interventions for a wide range of health and environmental challenges.

REFERENCES

Afridi, Muhammad Siddique, Muhammad Ammar Javed, Sher Ali, Flavio Henrique Vasconcelos De Medeiros, Baber Ali, Abdul Salam, Sumaira, et al. 2022. "New Opportunities in Plant Microbiome Engineering for Increasing Agricultural Sustainability under Stressful Conditions." *Frontiers in Plant Science* 13 (September). doi:10.3389/fpls.2022.899464.

Aggarwal, Nikhil, Shohei Kitano, Ginette Ru Ying Puah, Sandra Kittelmann, In Young Hwang, and Matthew Wook Chang. 2023. "Microbiome and Human Health: Current Understanding, Engineering, and Enabling Technologies." *Chemical Reviews* 123 (1): 31–72. doi:10.1021/acs.chemrev.2c00431.

Albright, Michaeline B. N., Stilianos Louca, Daniel E. Winkler, Kelli L. Feeser, Sarah-Jane Haig, Katrine L. Whiteson, Joanne B. Emerson, and John Dunbar. 2022. "Solutions in Microbiome Engineering: Prioritizing Barriers to Organism Establishment." *The ISME Journal* 16 (2): 331–338. doi:10.1038/s41396-021-01088-5.

Arroyo-Olarte, Ruben D., Ricardo Bravo Rodríguez, and Edgar Morales-Ríos. 2021. "Genome Editing in Bacteria: CRISPR-Cas and Beyond." *Microorganisms* 9 (4): 844. doi:10.3390/microorganisms9040844.

Basavaraju, M., B. S. Gunashree, and B. S. Srinath. 2022. "Microbiome - The Power House of Health and Disease." In *Effect of Microbiota on Health and Disease*. IntechOpen. doi:10.5772/intechopen.106026.

Belkaid, Yasmine, and Premysl Bercik. 2015. "The Microbiome and Autoimmunity: A Paradigm from the Gut–Liver Axis." *Cellular and Molecular Gastroenterology and Hepatology* 1 (1): 55–63.

Beloqui, Ana, Anne des Rieux, and Véronique Préat. 2016. "Mechanisms of Transport of Polymeric and Lipidic Nanoparticles across the Intestinal Barrier." *Advanced Drug Delivery Reviews* 106 (November): 242–255. doi:10.1016/j.addr.2016.04.014.

Berding, Kirsten, and John F. Cryan. 2022. "Microbiota-Targeted Interventions for Mental Health." *Current Opinion in Psychiatry* 35 (1): 3–9. doi:10.1097/YCO.0000000000000758.
Berendsen, Roeland L., Corné M.J. Pieterse, and Peter A.H.M. Bakker. 2012. "The Rhizosphere Microbiome and Plant Health." *Trends in Plant Science* 17 (8): 478–486. doi:10.1016/j.tplants.2012.04.001.
Berg, Gabriele, Daria Rybakova, Doreen Fischer, Tomislav Cernava, Marie-Christine Champomier Vergès, Trevor Charles, Xiaoyulong Chen, et al. 2020. "Microbiome Definition Re-Visited: Old Concepts and New Challenges." *Microbiome* 8 (1): 103. doi:10.1186/s40168-020-00875-0.
Berhanu, Gemechu, Vikram Godishala, and Venkataramana Kandi. 2021. "Microbiome Engineering and Its Applications: A Rapid Review." *Applied Ecology and Environmental Sciences* 9 (10): 865–872. doi:10.12691/aees-9-10-3.
Bhogoju, Sarayu, and Samuel Nahashon. 2022. "Recent Advances in Probiotic Application in Animal Health and Nutrition: A Review." *Agriculture* 12 (2): 304. doi:10.3390/agriculture12020304.
Boxberger, Manon, Valérie Cenizo, Nadim Cassir, and Bernard La Scola. 2021. "Challenges in Exploring and Manipulating the Human Skin Microbiome." *Microbiome* 9 (1): 125. doi:10.1186/s40168-021-01062-5.
Brooks, Sierra M., and Hal S. Alper. 2021. "Applications, Challenges, and Needs for Employing Synthetic Biology beyond the Lab." *Nature Communications* 12 (1): 1390. doi:10.1038/s41467-021-21740-0.
Cullen, Chad M., Kawalpreet K. Aneja, Sinem Beyhan, Clara E. Cho, Stephen Woloszynek, Matteo Convertino, Sophie J. McCoy, et al. 2020. "Emerging Priorities for Microbiome Research." *Frontiers in Microbiology* 11 (February). doi:10.3389/fmicb.2020.00136.
Douglas, G. M., and Langille, M. G. 2019. "Towards Multimodal Microbiome Engineering." *Nature Reviews Microbiology* 17 (11): 731–743.
Feng, Qingqing, Wei-Dong Chen, and Yan-Dong Wang. 2018. "Gut Microbiota: An Integral Moderator in Health and Disease." *Frontiers in Microbiology* 9 (February). doi:10.3389/fmicb.2018.00151.
Fischbach, Michael A., and Julia A. Segre. 2016. "Signaling in Host-Associated Microbial Communities." *Cell* 164 (6): 1288–1300. doi:10.1016/j.cell.2016.02.037.
Foo, Jee Loon, Hua Ling, Yung Seng Lee, and Matthew Wook Chang. 2017. "Microbiome Engineering: Current Applications and Its Future." *Biotechnology Journal* 12 (3): 1600099. doi:10.1002/biot.201600099.
Franzosa, E. A., McIver, L. J., Rahnavard, G., Thompson, L. R., Schirmer, M., Weingart, G., ... Huttenhower. 2019. "Species-Level Functional Profiling of Metagenomes and Metatranscriptomes." *Nature Methods* 16 (10): 959–968.
Gilbert, Jack A., Martin J. Blaser, J. Gregory Caporaso, Janet K. Jansson, Susan V. Lynch, and Rob Knight. 2018. "Current Understanding of the Human Microbiome." *Nature Medicine* 24 (4): 392–400. doi:10.1038/nm.4517.
Gilbert, Jack A., Robert A. Quinn, Justine Debelius, Zhenjiang Z. Xu, James Morton, Neha Garg, Janet K. Jansson, Pieter C. Dorrestein, and Rob Knight. 2016. "Microbiome-Wide Association Studies Link Dynamic Microbial Consortia to Disease." *Nature* 535 (7610): 94–103. doi:10.1038/nature18850.
Gulliver, Emily L., Remy B. Young, Michelle Chonwerawong, Gemma L. D'Adamo, Tamblyn Thomason, James T. Widdop, Emily L. Rutten, et al. 2022. "Review Article: The Future of Microbiome-based Therapeutics." *Alimentary Pharmacology & Therapeutics* 56 (2): 192–208. doi:10.1111/apt.17049.
Han, Sang-Woo, and Yasuo Yoshikuni. 2022. "Microbiome Engineering for Sustainable Agriculture: Using Synthetic Biology to Enhance Nitrogen Metabolism in Plant-Associated Microbes." *Current Opinion in Microbiology* 68 (August): 102172. doi:10.1016/j.mib.2022.102172.
Jiang, Li-Li, Jin-Jie Zhou, Chun-Shan Quan, and Zhi-Long Xiu. 2017. "Advances in Industrial Microbiome Based on Microbial Consortium for Biorefinery." *Bioresources and Bioprocessing* 4 (1): 11. doi:10.1186/s40643-017-0141-0.
Kashyap, Purna C., Nicholas Chia, Heidi Nelson, Eran Segal, and Eran Elinav. 2017. "Microbiome at the Frontier of Personalized Medicine." *Mayo Clinic Proceedings* 92 (12): 1855–1864. doi:10.1016/j.mayocp.2017.10.004.
Kaul, Sanjana, Malvi Choudhary, Suruchi Gupta, and Manoj K. Dhar. 2021. "Engineering Host Microbiome for Crop Improvement and Sustainable Agriculture." *Frontiers in Microbiology* 12 (May). doi:10.3389/fmicb.2021.635917.
Kelly, Colleen R., Stacy Kahn, Purna Kashyap, Loren Laine, David Rubin, Ashish Atreja, Thomas Moore, and Gary Wu. 2015. "Update on Fecal Microbiota Transplantation 2015: Indications, Methodologies, Mechanisms, and Outlook." *Gastroenterology* 149 (1): 223–237. doi:10.1053/j.gastro.2015.05.008.
Khan, Saad, Ruth Hauptman, and Libusha Kelly. 2021. "Engineering the Microbiome to Prevent Adverse Events: Challenges and Opportunities." *Annual Review of Pharmacology and Toxicology* 61 (1): 159–179. doi:10.1146/annurev-pharmtox-031620-031509.

Kim Yu-Na, Muhammad Aaqil Khan, Sang-Mo Kang, Muhammad Hamayun, and In-Jung Lee. 2020. "Enhancement of Drought-Stress Tolerance of *Brassica Oleracea* Var. *Italica* L. by Newly Isolated *Variovorax* Sp. YNA59." *Journal of Microbiology and Biotechnology* 30 (10): 1500–1509. doi:10.4014/jmb.2006.06010.

Knight, Rob, Alison Vrbanac, Bryn C. Taylor, Alexander Aksenov, Chris Callewaert, Justine Debelius, Antonio Gonzalez, et al. 2018. "Best Practices for Analysing Microbiomes." *Nature Reviews Microbiology* 16 (7): 410–422. doi:10.1038/s41579-018-0029-9.

Langdon, Amy, Nathan Crook, and Gautam Dantas. 2016. "The Effects of Antibiotics on the Microbiome throughout Development and Alternative Approaches for Therapeutic Modulation." *Genome Medicine* 8 (1): 39. doi:10.1186/s13073-016-0294-z.

Lange, Lene, Gabriele Berg, Tomislav Cernava, Marie-Christine Champomier-Vergès, Trevor Charles, Luca Cocolin, Paul Cotter, et al. 2022. "Microbiome Ethics, Guiding Principles for Microbiome Research, Use and Knowledge Management." *Environmental Microbiome* 17 (1): 50. doi:10.1186/s40793-022-00444-y.

LaPelusa, Michael, Dorit Donoviel, Sergio E. Branzini, Paul E. Carlson, Stephanie Culler, Amrita K. Cheema, Rima Kaddurah-Daouk, et al. 2021. "Microbiome for Mars: Surveying Microbiome Connections to Healthcare with Implications for Long-Duration Human Spaceflight, Virtual Workshop, July 13, 2020." *Microbiome* 9 (1): 2. doi:10.1186/s40168-020-00951-5.

Laulund, Svend, Anette Wind, Patrick Derkx, and Véronique Zuliani. 2017. "Regulatory and Safety Requirements for Food Cultures." *Microorganisms* 5 (2): 28. doi:10.3390/microorganisms5020028.

Lawson, Christopher E. 2021. "Retooling Microbiome Engineering for a Sustainable Future." *MSystems* 6 (4). doi:10.1128/mSystems.00925-21.

Lee, Eric D., Emily R. Aurand, and Douglas C. Friedman. 2020. "Engineering Microbiomes—Looking Ahead." *ACS Synthetic Biology* 9 (12): 3181–3183. doi:10.1021/acssynbio.0c00558.

Li Peishun, Hao Luo, Boyang Ji, and Jens Nielsen. 2022. "Machine Learning for Data Integration in Human Gut Microbiome." *Microbial Cell Factories* 21 (1): 241. doi:10.1186/s12934-022-01973-4.

Liu, Dongyun, Yuguang Wang, Lili Li, Han Zhao, Liangliang Li, Yan Liu, Hong Jiang, Xianghong Li, and Rui Zhang. 2019. "Celecoxib Protects Hyperoxia-Induced Lung Injury via NF-KB and AQP1." *Frontiers in Pediatrics* 7 (June). doi:10.3389/fped.2019.00228.

Liu, Zhuang, Berna de Vries, Jacoline Gerritsen, Hauke Smidt, and Erwin G. Zoetendal. 2020. "Microbiome-Based Stratification to Guide Dietary Interventions to Improve Human Health." *Nutrition Research* 82 (October): 1–10. doi:10.1016/j.nutres.2020.07.004.

Mimee, Mark, Robert J. Citorik, and Timothy K. Lu. 2016. "Microbiome Therapeutics — Advances and Challenges." *Advanced Drug Delivery Reviews* 105 (October): 44–54. doi:10.1016/j.addr.2016.04.032.

Popkes, Miriam, and Dario Riccardo Valenzano. 2020. "Microbiota–Host Interactions Shape Ageing Dynamics." *Philosophical Transactions of the Royal Society B: Biological Sciences* 375 (1808): 20190596. doi:10.1098/rstb.2019.0596.

Robinson, Jake M., Nicole Redvers, Araceli Camargo, Christina A. Bosch, Martin F. Breed, Lisa A. Brenner, Megan A. Carney, et al. 2022. "Twenty Important Research Questions in Microbial Exposure and Social Equity." *MSystems* 7 (1). doi:10.1128/msystems.01240-21.

Silverstein, Michael R., Daniel Segrè, and Jennifer M. Bhatnagar. 2023. "Environmental Microbiome Engineering for the Mitigation of Climate Change." *Global Change Biology* 29 (8): 2050–2066. doi:10.1111/gcb.16609.

Singh, Brajesh K., and Pankaj Trivedi. 2017. "Microbiome and the Future for Food and Nutrient Security." *Microbial Biotechnology* 10 (1): 50–53. doi:10.1111/1751-7915.12592.

Verstraete, Willy, Keren Yanuka-Golub, Nele Driesen, and Jo De Vrieze. 2022. "Engineering Microbial Technologies for Environmental Sustainability: Choices to Make." *Microbial Biotechnology* 15 (1): 215–227. doi:10.1111/1751-7915.13986.

Wan, Xinyi, Trevor Y. H. Ho, and Baojun Wang. 2019. "Engineering Prokaryote Synthetic Biology Biosensors." In *Handbook of Cell Biosensors*, 1–37. Cham: Springer International Publishing. doi:10.1007/978-3-319-47405-2_131-1.

Zhang, Yangheng, Xiang Wang, Houxuan Li, Can Ni, Zhibin Du, and Fuhua Yan. 2018. "Human Oral Microbiota and Its Modulation for Oral Health." *Biomedicine & Pharmacotherapy* 99 (March): 883–893. doi:10.1016/j.biopha.2018.01.146.

Index